Introductory Organic Chemistry

Introductory Organic Chemistry

J. T. GERIG

UNIVERSITY OF CALIFORNIA
SANTA BARBARA, CALIFORNIA

ACADEMIC PRESS New York and London

A SUBSIDIARY OF HARCOURT BRACE JOVANOVICH, PUBLISHERS

ACADEMIC PRESS, INC.
111 Fifth Avenue, New York, New York 10003

United Kingdom Edition published by
ACADEMIC PRESS, INC. (LONDON) LTD.
24/28 Oval Road, London NW1

Library of Congress Cataloging in Publication Data

Gerig, J T
Introductory organic chemistry.

Includes bibliographical references.
1. Chemistry, Organic. I. Title.
QD251.2.G47 547 73-9424
ISBN 0–12–280750–2

PRINTED IN THE UNITED STATES OF AMERICA

to my parents and teachers
and to b. and t.

CONTENTS

PREFACE

This text is intended for nonmajors' use in an introductory organic chemistry course of two quarters or one semester in length. I have had the opportunity to teach such a course several times over the past few years. Though my students were drawn mostly from the biological sciences, I have also had majors from the departments of home economics, electrical engineering, ergonomics, physics, history, psychology, and even religious studies. Coming with diverse backgrounds, preparations, and interests, the members of these classes shared several common characteristics: they were in the class primarily because their major department required it; they seemed to be convinced by their introductory chemistry courses that chemistry is an abstract, possibly irrelevant subject area; and they were rather anxious about a topic as "complicated" as organic chemistry. At the same time I found these students just as bright as chemistry majors and perhaps more open-minded.

The text is intended to provide a descriptive overview of the subject of organic chemistry and to indicate how modern organic chemistry is practiced. It is hoped that it will offer a broad enough foundation so that each instructor by his lectures can fit his course more precisely to the needs of his particular class without the necessity of sticking closely to the discussion in the text.

I am indebted to a number of my colleagues for their suggestions and counsel and to my undergraduate students who provided (often unwittingly) comments on the effectiveness of the presentations used in this book. I am grateful to those people who tolerated unexplained absences while the text was in preparation and to Mrs. Rosemary Toby and Mrs. Joyce Sato for their skillful typing. And finally thanks are due to my wife, Elizabeth, without whose support and assistance it would not have been written.

I will be most happy to receive comments on matters of fact and pedagogy in this book.

J. T. GERIG

1

A Review of Some Fundamental Ideas

To study profitably modern organic chemistry, it is essential that you, the student, have a firm grasp of the far-reaching ideas that were presented in your beginning chemistry course. It is our intention to review in this chapter the most prominent of these ideas. This review must necessarily be brief; if you encounter difficulties with a particular section, restudy of your beginning chemistry text or one of the references is in order.

Atoms

If one should accept only the gross evidence that is provided by his sensory mechanisms, he is likely to conclude that the structure of the matter in our environment is continuous. A desk top or this page of paper appears to consist of solid, connected material and we might surmise that even when examined with a powerful microscope, this material would appear as a continuous mass. Many of the early Greeks accepted a contrary idea, however. They held that at the most microscopic level, matter was essentially discontinuous—that matter consists of small particles which cannot be subdivided. Virtually all modern scientific knowledge supports, in form at least, this latter idea, which we recognize to be a forerunner of modern atomic theory. The existence of small, discrete entities called atoms and the existence of different kinds of atoms, called elements, is a fundamental postulate of this theory. The atomic theory, like all scientific theories, coordinates a large body of information into a consistent explanation of certain phenomena, and we will accept as correct the notions of the atomic theory in all that follows.

Contemporary high-energy physics has provided evidence for the existence of over 100 fundamental particles. These particles are the "stuff" of which atoms are constructed. Fortunately for the chemist, there are only three particles that need to be considered in most chemical discussions. These are the proton, the neutron, and the electron. The characteristic masses and electrical properties of these particles are listed in Table 1.1. Atoms consist of a nucleus, composed primarily of protons and neutrons, and electrons which circulate in some manner around the nucleus. It is the number of protons in the nucleus that determines the identity of the atom. An atom with 2 protons in the nucleus is a helium atom; an atom which contains 26 protons in the nucleus is an iron atom; and so forth. The number of protons in a nucleus is called the atomic number of the nucleus; each atomic number corresponds to a particular chemical element.

Table 1.1
Properties of Some Fundamental Particles

Particle	*Mass* (g)	*Charge*
Proton	1.7×10^{-24}	+
Neutron	1.7×10^{-24}	0
Electron	9.1×10^{-28}	−

An atom can contain a variable number of neutrons in its nucleus. The number of neutrons and protons essentially determines the mass of a nucleus, and because of the small mass of the electrons, the mass of an atom is nearly equal to the total mass of its nucleus. Most elements are found to exist as atoms with a variable number of neutrons in the nucleus; these mass-different species are called isotopes. For example, atoms of chlorine (atomic number = 17) can be found in two stable isotopic forms, chlorine-35 and chlorine-37. (The number after the element name indicates the total mass of the atom.) Each of these atoms has 17 protons in its nucleus (thus making the atom a chlorine atom), but a variable number of neutrons, as is indicated by the numbers in the names above.

It is possible to obtain isotopes of chlorine and other elements that are unstable. These materials decay to stable isotopes by various processes known collectively as radioactivity. Radioactive isotopes are useful to the chemist because the physical effects of their decay processes provide a way to follow the atoms through a series of chemical reactions. Table 1.2 lists several isotopes of some elements of interest to the organic chemist.

Table 1.2
Some Isotopes of Common Elements

Element	*Stable*[a]	*Unstable*[b]
Hydrogen	^{1}H, ^{2}H[c]	^{3}H[d]
Boron	^{11}B, ^{10}B	
Carbon	^{12}C, ^{13}C	^{14}C
Nitrogen	^{14}N, ^{15}N	
Oxygen	^{16}O, ^{17}O, ^{18}O	
Fluorine	^{19}F	
Phosphorus	^{31}P	^{32}P
Sulfur	^{32}S, ^{33}S, ^{34}S	^{35}S
Chlorine	^{35}Cl, ^{37}Cl	^{38}Cl
Bromine	^{79}Br, ^{81}Br	^{82}Br
Iodine	^{127}I	^{129}I

[a] The isotope of greatest natural abundance is listed first.
[b] Isotopes with decay rates slow enough that they can be used as radioactive tracers.
[c] ^{2}H, often symbolized D, is called deuterium.
[d] ^{3}H is also called tritium (T).

The Electronic Structure of Atoms

It is the number of circulating electrons that to a large extent determines the chemical properties of an atom. To describe the chemical behavior of an atom adequately, it will be necessary to describe the wanderings of the electrons about the atomic nucleus. This is not a simple problem and, to make any progress, a description of the moving electron must employ the methods of a possibly unfamiliar branch of science known as quantum mechanics. Excruciating mathematical difficulties are encountered when attempts are made to apply these methods to relatively complicated chemical problems, but a description of electronic behavior in the hydrogen atom is accessible. This simple atom consists of a single proton at the nucleus and one planetary electron. An important result of the quantum-mechanical treatment of the motion of this electron is that the energy of the electron is shown to have only certain permissible values. These allowed values of energy are called energy levels, or eigenstates. Figure 1.1 shows schematically some of the permitted energy levels of the electron in a hydrogen atom. Each state has a traditional name that is also given in labels on the figure.

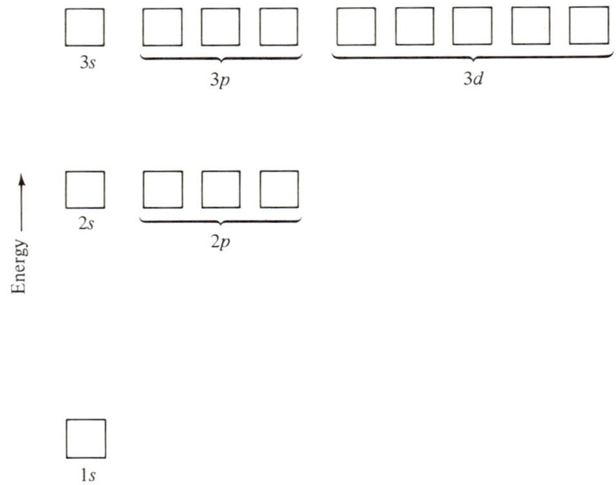

Fig. 1.1. The lowest allowed energy levels of the hydrogen atom (schematic).

Another result of the quantum-mechanical description of the hydrogen atom is an indication of how the electronic charge is distributed about the nucleus. For each of the allowed energy levels of the electron in this atom, there is a particular and characteristic electron distribution. The computed distribution of electron charge for several of the hydrogenic energy levels is shown in Fig. 1.2.

The mathematical function which can be used to describe electron distributions is commonly called an atomic orbital or wavefunction. The shape of an electronic distribution defined by an atomic orbital can most easily be visualized by surrounding the nucleus with a three-dimensional surface within which the probability of finding the electron is high. The 1*s* orbital and all other *s* orbitals are spherical in shape, while the 2*p* orbitals have a dumbbell shape (Fig. 1.3). The three 2*p* states differ only in the orientation of the "dumbbell" in space.

The basic ideas we have just developed for the simplest atom, hydrogen, can be utilized to provide insight into the electronic structures of more complicated atoms. The electronic energy of these atoms is also quantized; that is, it can have only certain discrete values. However, the energy level diagrams for atoms more complicated than hydrogen are only qualitatively similar to Fig. 1.1. For neutral atoms in the first two rows of the periodic chart, the lower electronic energy levels are spaced approximately as shown in Fig. 1.4. The general shape of the *s* and *p* orbitals in larger atoms is similar to those shown in Figs. 1.2 and 1.3.

A quantum-mechanical rule known as the Pauli principle states that no more than two electrons can occupy a given orbital simultaneously. With the aid of

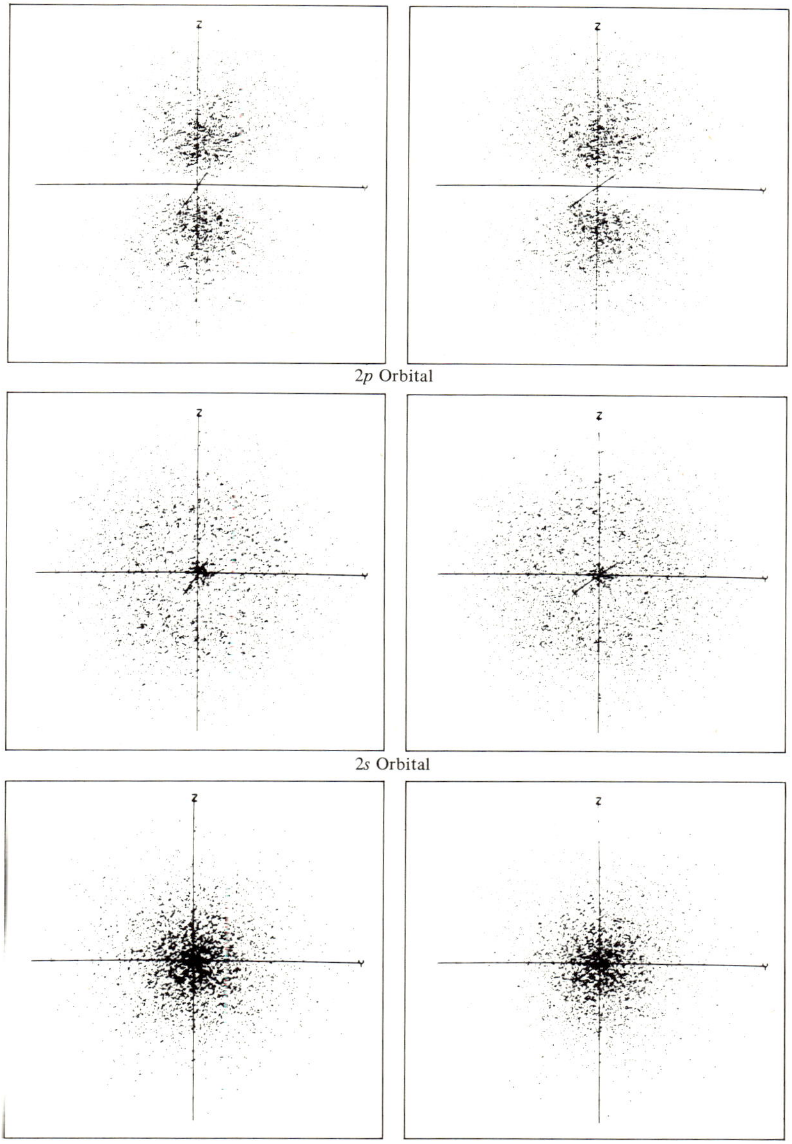

Fig. 1.2. Three-dimensional representation of electron density in some hydrogenic atomic orbitals. Each set represents a stereoscopic pair. The three-dimensional effect is most easily achieved with a viewer; if not available, hold diagram at arm's length, look past the drawing and allow the eyes to focus at infinity. Glance at figure; each pair should merge into a single image. Concentrating on the central image, move book slowly to bring this image into focus. Practice helps obtain the necessary relaxation of the eye muscles. (Illustrations courtesy of D. T. Cromer.)

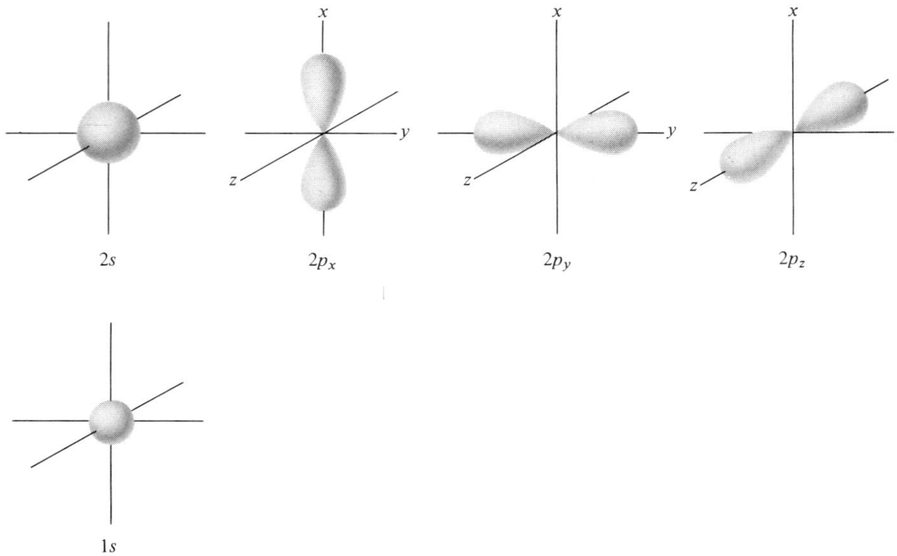

Fig. 1.3. Two-dimensional representation of *s* and *p* orbitals.

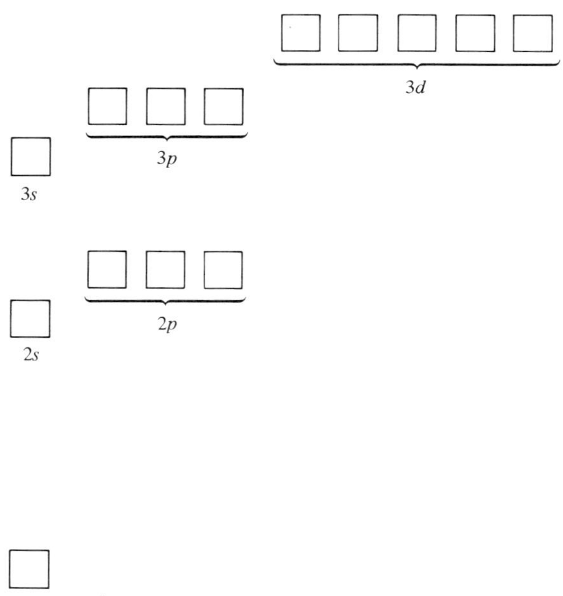

Fig. 1.4. Schematic representation of atomic energy levels in atoms larger than hydrogen.

the diagram in Fig. 1.4 and the Pauli principle, we can deduce the qualitative nature of the electronic states of larger atoms. The electronic structure, in terms of energy level diagrams, for several atoms of particular interest to the organic chemist is shown in Fig. 1.5. It should be noted that in each atom the lowest-lying electronic energy levels are filled first.

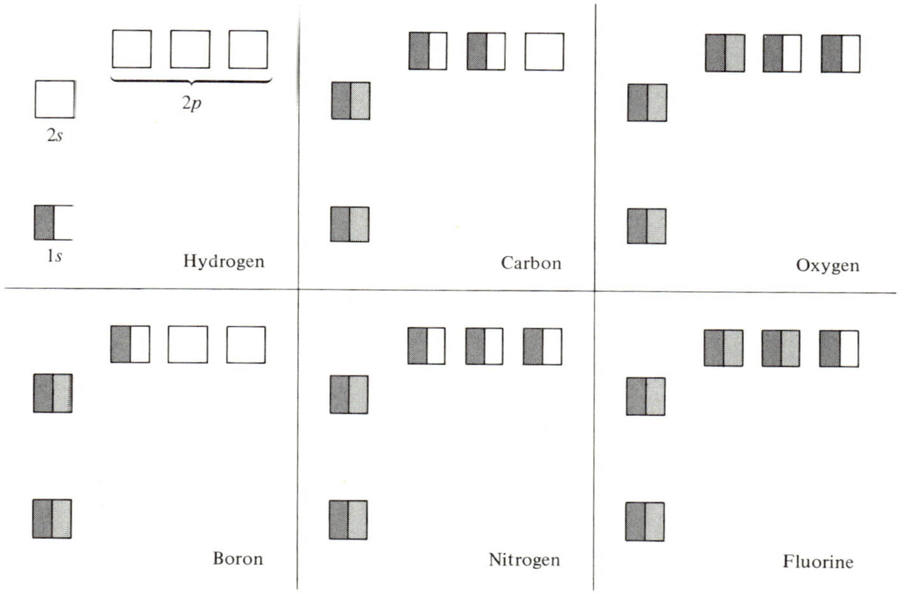

Fig. 1.5. Qualitative energy level diagrams for several atoms.

As atoms become progressively more massive their electronic charge clouds become increasingly more diffuse. The interactions between electron and nucleus are relatively weaker and the result is that an atom with a large number of electrons tends to fill more space than an atom with fewer electrons. The relative sizes of atoms can be estimated in a number of ways. A common expression of the size of an atom is the *atomic radius*, usually considered to be the distance from the nucleus to the outer edge of appreciable electron density. Table 1.3 records the approximate atomic radii of many of the atoms that are important to organic chemistry. The essential point here is that atoms behave as if they occupy space and that the effective "size" of an atom is a function of the number of electrons and protons that make up the atom.

Table 1.3
Approximate Atomic Radii

Element	*Radius* (Å)[a]
Hydrogen	0.52
Boron	0.89
Carbon	0.77
Nitrogen	0.70
Oxygen	0.66
Fluorine	0.64
Phosphorus	1.10
Sulfur	1.04
Chlorine	0.99
Bromine	1.11
Iodine	1.28

[a] An angstrom unit, abbreviated Å, is 10^{-8} cm.

Molecular Structure

Let us first consider the formation of a chemical bond from an intuitive viewpoint, using as an example the chemical bond in the simple molecule, H_2. If the distance between the two nuclei is very large, the mutual interactions between the two atoms of this molecule will be weak and the total energy of the molecule will be essentially twice the energy of one hydrogen atom. We know that as the internuclear distance becomes smaller, the energy of the molecule must first decrease to some minimum value and then increase as the internuclear distance is further diminished. This behavior is sketched in Fig. 1.6. The energy must initially decrease because chemical bond formation is an energy-lowering process. (If it were not, molecules would not exist!) The energy reaches a minimum value at an internuclear distance corresponding roughly to twice the atomic radius of hydrogen; this distance is called the equilibrium bond length. Decreasing the internuclear distance to values much smaller than the bond length is firmly resisted by the strong repulsive forces between the like-charged nuclei of the two atoms.

We must persist in the use of a quantum-mechanical description of electronic behavior when we consider molecules since electrons are the "glue" that holds the atoms of a molecule together. In order to form a chemical bond between two hydrogen atoms, the atomic orbitals of the electrons on each atom must interact or overlap one another. Intuition suggests a reason why this overlap of electronic charge eventually leads to a reduction in energy. As a result of orbital overlap a region of relatively high electron density is produced between the two nuclei

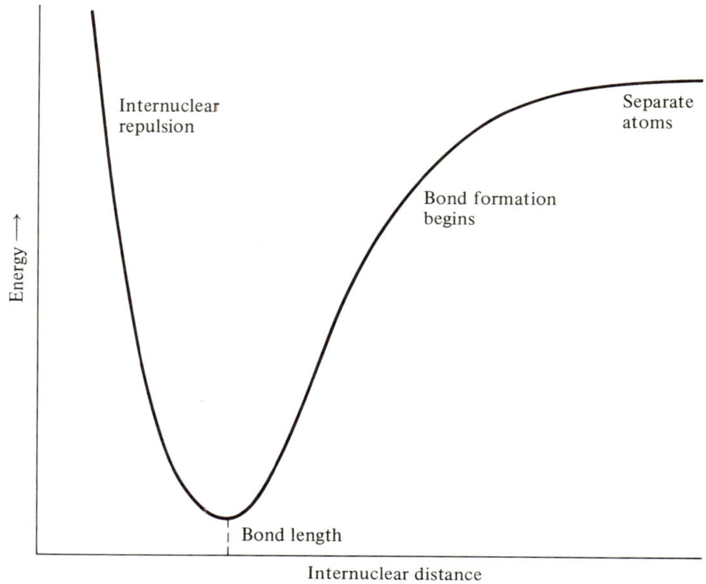

Fig. 1.6. The change in energy of a diatomic molecule as a function of internuclear distance.

and it is likely that it is the mutual and simultaneous attraction of the positively charged nuclei for this region of high electron density which holds the molecule together.

This picture of chemical bonding suggests a generalization known as the *principle of maximum overlap* which states that chemical bonds are strongest when the atomic orbitals on each of the constituent atoms overlap each other to the maximum possible extent.

The approach used above to describe bond formation in the hydrogen molecule can also be used with more complex molecules. For example, if we wished to describe the chemical bonds in the water molecule (H_2O), it would be necessary to choose appropriate atomic orbitals of the oxygen atom and the hydrogen atoms to be overlapped in forming the two O—H bonds. Recall the electronic configuration of the oxygen atom shown in Fig. 1.5. One of the singularly occupied $2p$ orbitals of the oxygen atom and a $1s$ orbital from a hydrogen can be utilized to form each O—H bond.

By the principle of maximum overlap, an O—H bond will be strongest (lowest in energy) when the hydrogen and oxygen nuclei lie along the line that forms the axis of the $2p$ orbital. That is, the arrangement A, shown on p. 10, is more stable than arrangement B. We can use this fact to predict the geometrical properties of simple molecules like water. The entire system of $2p$ electronic

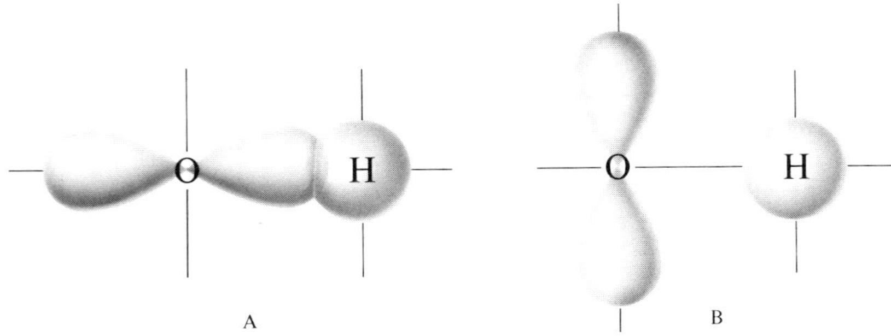

orbitals of oxygen is represented in Fig. 1.7. The $2p_x$ orbital already contains two electrons and is thus unavailable for the formation of additional bonds.* The $2p_y$ and $2p_z$ orbitals of oxygen can each form a bond to a hydrogen atom by the process described above. If the principle of maximum overlap is followed then these two O—H bonds should be oriented at right angles to each other. From this argument, the bond angle, α, defined as the angle between these two covalent bonds, is predicted to be 90°. The experimental value is 104°, in moderate agreement with this prediction. This line of reasoning can often be used to

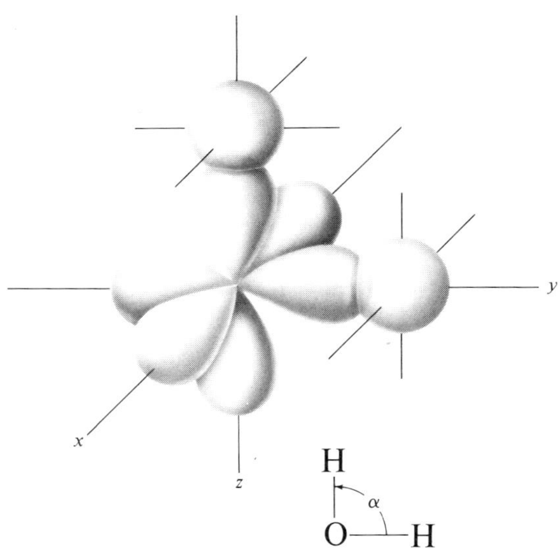

Fig. 1.7. The expected arrangement of the atomic orbitals used to form the molecular orbitals of the water molecule.

* The choice of the $2p_x$ for this lot is arbitrary; we reach the same conclusions using the $2p_y$ or $2p_z$ orbitals since these orbitals differ only in their orientation in the coordinate system.

deduce the approximate geometrical properties of an organic molecule. The important message here is that molecules have three-dimensional attributes that are a direct reflection of the properties of the chemical bonds that hold the molecules together.

Molecular Orbitals

One method of seeking a more detailed and quantum-mechanically acceptable description of a chemical bond is called the molecular orbital theory. This theory supposes that there are molecular analogs for the atomic energy levels and atomic orbitals which were discussed previously. The mathematical difficulties involved are even worse than those encountered with descriptions of atoms and approximations must be introduced much earlier into the computations. A common approximation made in molecular orbital calculations is to assume that the molecular orbitals retain the characteristics of atomic orbitals in regions of space near the atoms of the molecule. The molecular orbitals are then approximated as sums of atomic orbitals.

Hybridization of Atomic Orbitals in Carbon

The element carbon, upon which much of the attention of this text will be focused, provides embarrassing exceptions to the generalizations presented above. The atomic energy level diagram for carbon (Fig. 1.5) predicts that this element should form two chemical bonds via two available $2p$ atomic orbitals and that, as in the case of water, the angle between these bonds should be about 90°. In fact, stable carbon compounds almost always have four covalent bonds to each carbon atom and, although the bond angles in the compounds of carbon are variable, they usually fall into three classes, characterized by bond angles of about 109°, 120°, and 180°, respectively. L. Pauling and J. C. Slater have suggested a way of getting around these obvious discrepancies between the prediction from theory and the experimental facts. They showed that the $2s$ and $2p$ atomic states of carbon can be mixed in various ways to give *hybrid atomic orbitals*. The geometrical properties of these hybrid orbitals depends upon the number of $2p$ orbitals in the mixture and are in better accord with the experimental bond angles found in the compounds of carbon.

One type of hybrid orbital arises when the carbon $2s$ state and all three $2p$ states are mixed to give four identical hybridized orbitals. This hybridization is termed sp^3 and is consistent with the energy level diagram shown in Fig. 1.8.

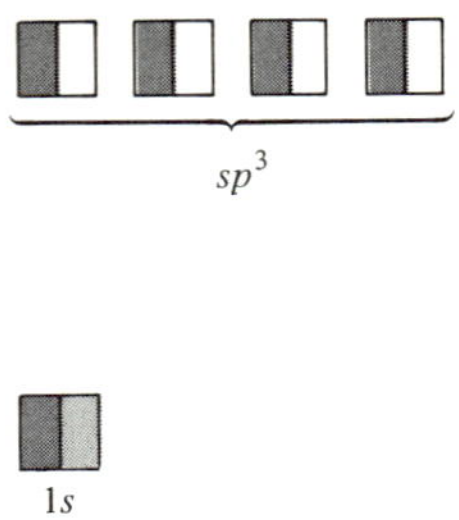

Fig. 1.8. Schematic energy level diagram for an sp^3 hybridized carbon atom. The unhybridized $2s$ energy level is slightly below the level of the hybrids, while the energy of the $2p$ states is slightly above the energy of the hybrids.

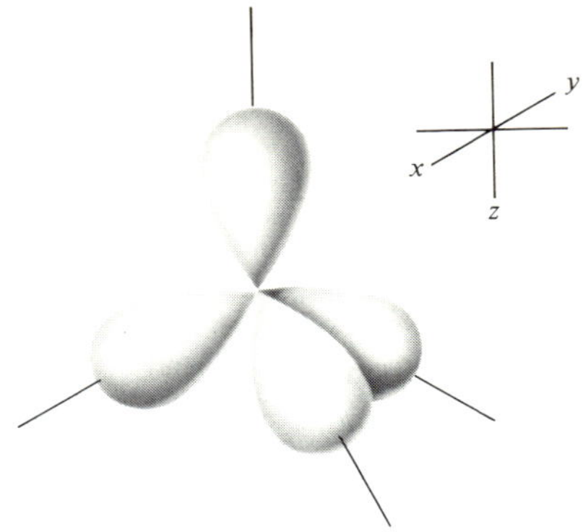

Fig. 1.9. The arrangement of the sp^3 hybrid orbitals in space.

The four sp^3 hybrid orbitals are energetically equivalent and differ only in their orientation in space (Fig. 1.9). These hybrid atomic orbitals point along the axes of a regular tetrahedron. The interior angle between each sp^3 orbital is 109°, so that carbon orbitals hybridized in this manner should be used in our discussion of compounds of carbon which have bond angles of this magnitude. These discussions begin in Chapter 2.

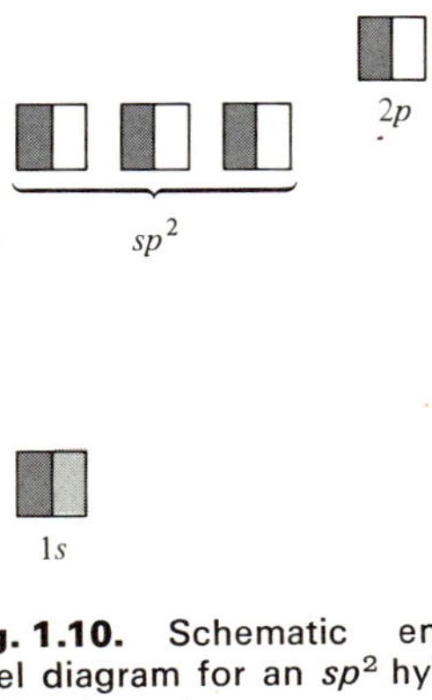

Fig. 1.10. Schematic energy level diagram for an sp^2 hybridized carbon atom.

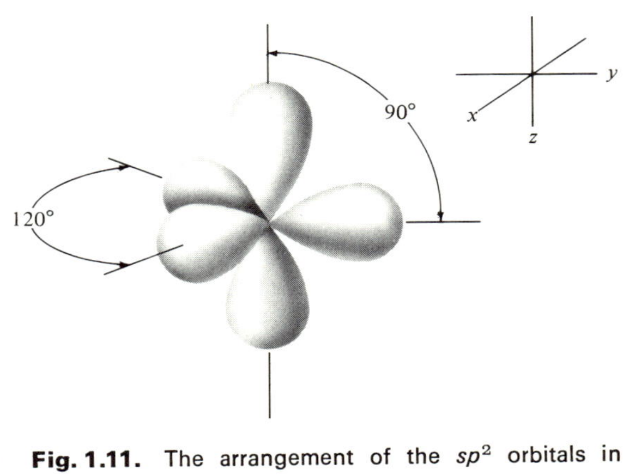

Fig. 1.11. The arrangement of the sp^2 orbitals in space.

A second type of hybridization is termed sp^2 and results from mixing a $2s$ state with two $2p$ states of carbon to give three identical sp^2 hybridized orbitals. An energy level diagram for this situation is shown in Fig. 1.10. In the sp^2 hybridized carbon atom there are three hybrid orbitals of equivalent energy with one $2p$ state essentially untouched by the orbital mixing process. The sp^2 orbitals point toward the corners of an equilateral triangle (Fig. 1.11), while the unhybridized $2p$ orbital appears perpendicular to the plane defined by these three orbitals. Carbon orbitals hybridized in the sp^2 fashion should be used to describe molecules in which the bond angles between the groups bonded to the carbon atom are about 120°.

For those carbon atoms involved in bonding situations that result in linear molecules (bond angle = 180°), sp hybridization is most useful. These hybrids are formed by mixing the carbon $2s$ orbital with a single carbon $2p$ orbital. The characteristic energy level diagram (Fig. 1.12) shows the two $2p$ orbitals that are unaffected by the hybridization process. The geometrical properties of the sp hybrids sketched in Fig. 1.13 are similar to those of a $2p$ state, except that now each lobe or half can be used in forming chemical bonds.

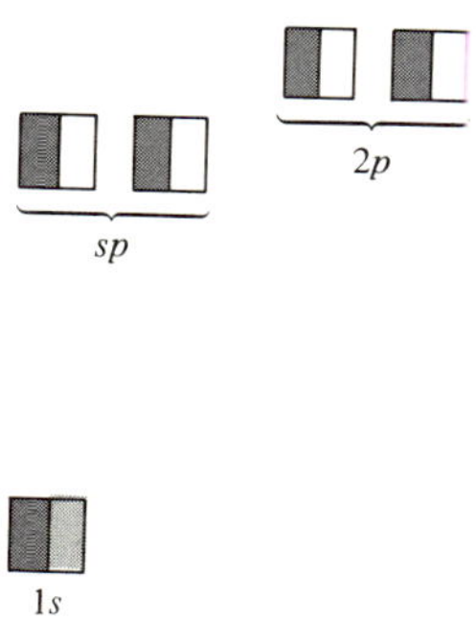

Fig. 1.12. Schematic energy level diagram for an *sp* hybridized carbon atom.

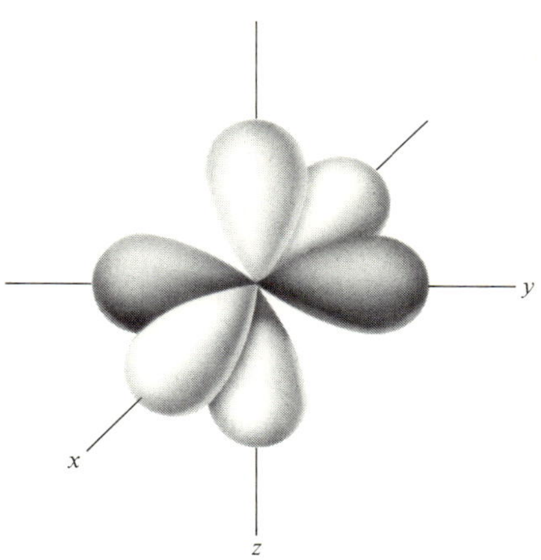

Fig. 1.13. The arrangement of *sp* orbitals in space.

The use of these various hybrid orbitals will become clearer as we progress with the study of carbon compounds. It should be recognized that, for our purposes, the idea of hybridization is merely a very convenient device for systematizing our knowledge of the structure of carbon-containing compounds and that it is not necessarily rooted in reality.

Chemical Equilibrium

Equilibrium is defined as a state of balance between opposing forces or actions and is a concept that is an accepted part of everyday life. Riding a bicycle or surfboard is successfully accomplished only after a proper state of balance between opposing forces is obtained. In chemical equilibria, the opposing actions are chemical changes. Among the first experiments that suggested the idea of chemical equilibrium were the investigations of M. Bodenstein (1894). He found that a mixture of hydrogen gas and iodine gas, at high enough temperatures, soon contained an appreciable quantity of hydrogen iodide. Moreover, when a pure sample of hydrogen iodide was exposed to the same conditions, it was shortly contaminated with hydrogen and iodine gases. Thus, two reactions can take place between these three molecules, one reaction having an effect exactly the opposite of the other. This is a chemical equilibrium and can be expressed by Eq. (1.1). The double arrows indicate that both a forward ($H_2 + I_2 \rightarrow 2\ HI$) and a backward ($2\ HI \rightarrow H_2 + I_2$) reaction take place.

$$H_2 + I_2 \rightleftharpoons 2\ HI \tag{1.1}$$

There are two features of chemical equilibria that are of interest. One of these is the relative amounts of the materials present at equilibrium and the other is the rate at which the equilibrium condition is achieved. The relative quantities of materials present at equilibrium in a chemical system define the "position" of the equilibrium and can be expressed in a quantitative manner by using an equilibrium quotient. Consider the general system at equilibrium shown in Eq. (1.2), where the capital letters represent molecular species and the lower case

$$a\mathrm{A} + b\mathrm{B} \rightleftharpoons c\mathrm{C} + d\mathrm{D} \tag{1.2}$$

letters stand for stoichiometric coefficients. The equilibrium quotient, K, for this reaction is defined by Eq. (1.3),

$$K = \frac{[\mathrm{C}]^c[\mathrm{D}]^d}{[\mathrm{A}]^a[\mathrm{B}]^b} \tag{1.3}$$

where the square brackets mean "the concentration of" the species inside the brackets.* At constant temperature and pressure, it is an experimental fact that the value of K is a constant when the system has achieved equilibrium.†

* More rigorously, one should use the activities of the various molecules rather than the concentrations. We shall presume that activities and concentrations can be used interchangeably for our discussion.

† We should note that the numerical value of an equilibrium quotient can depend upon the units used to express the concentrations. There is no inconsistency here, however. The width of this book is a constant, but the numerical value of the constant will depend on the unit of measurement used to express this dimension.

The equilibrium constant, K, for the hydrogen iodide reaction (1.1) would be written

$$K = \frac{[HI]^2}{[H_2][I_2]} \tag{1.4}$$

according to the prescription given above. If the value of K is large it means that the concentration of HI is large relative to the concentration of H_2 and I_2 at equilibrium. The equilibrium in this case would be said to lie to the right. If the reactants H_2 and I_2 were the predominant species at equilibrium, K would be correspondingly smaller. When the reactants are highly favored in an equilibrium, the reaction is said to lie to the left.

Whether described quantitatively by an equilibrium constant or by the qualitative terms mentioned above, the position of an equilibrium depends on the standard free energy difference, $\Delta G°$, between the components written on the left side of the equilibrium and those written on the right ($\Delta G° = G°_{\text{products}} - G°_{\text{reactants}}$). The smaller $\Delta G°$ (algebraic sign considered) the more likely it is that, at equilibrium, the products will predominate. We use the term "free energy" because not all of the energy of a system of chemical compounds is available to affect the position of a chemical equilibrium. The tendency of all natural processes to move toward a state of lower energy is one factor, but not the only factor, that can influence chemical events. These events, like human events often do, tend to proceed in the direction that leads to the most chaotic condition and a highly disordered or disorganized chemical system is more probable than one that is highly ordered. It is these two competing factors—the tendency of a system of molecules to minimize its energy and to maximize its disorder—that determine the position of equilibrium. It is the effect of molecular disorder that makes it impossible for only the total energy of the system to be considered in determining positions of equilibrium and forces us to use the "free energy" idea. The above discussion can be summarized by Eq. (1.5),

$$\begin{aligned} \Delta G° &= \Delta H° - T\,\Delta S° \\ \Delta H° &= H°_{\text{products}} - H°_{\text{reactants}} \\ \Delta S° &= S°_{\text{products}} - S°_{\text{reactants}} \end{aligned} \tag{1.5}$$

where $\Delta H°$ represents the difference in total energy (often called enthalpy) between reactants and products if both are considered at standard conditions, and $\Delta S°$ is a quantity, called the entropy, which depends upon the difference in the disorder of the products and reactants. A classic example of the effect of entropy in determining the position of an equilibrium is the burning of a log in a fireplace. It might be possible to carry out this combustion in such a way that all of the products and the energy given off during the reaction could be conserved. Nevertheless, it is very highly unlikely that the ashes from a burned log would reassemble themselves to give back the original log; the ashes represent a much more random state than the log and, therefore, a much more probable state, irrespective of the energetics of the combustion reaction.

Any variation in the molecular structure of the reactant or product molecules that tends to change ΔH° or ΔS° so that ΔG° is more negative will make the equilibrium constant for a system larger and, thus, tend to favor the presence of larger amounts of products at equilibrium. This could be a change that lowers the enthalpy (energy) of the product molecules or raises the enthalpy of the reactant species. A structural variation that increases the entropy of the products or decreases the entropy of the reactants would have an effect in the same direction.

Estimating Enthalpy Differences

Table 1.4
Some Covalent Bond Energies[a]

Bond	*Bond energy (kcal/mole at 25°C)*
H—H	−104.2
O=O	−119.1
N≡N	−225.8
Cl—Cl	−58.0
Br—Br	−46.1
I—I	−36.1
H—Cl	−103.2
H—Br	−87.5
H—I	−71.4
H—C	−98.7
H—N	−93.4
H—O	−110.6
C—C	−82.6
C=C	−145.8
C≡C	−199.6
C—N	−72.8
C=N	−147.0
C—O	−85.5
C=O	ca. −175.0
C—F	−116.0
C—Cl	−81.0
C—Br	−68.0
C—I	−51.0

[a] Taken from J. D. Roberts and M. C. Caserio, "Basic Principles of Organic Chemistry." Benjamin, New York, 1964.

A helpful scheme for estimating the enthalpy difference for a chemical system is based upon the supposition that the energetics of chemical bond formation depend primarily on the nature of the two atoms involved and are independent of the nature of the molecule in which these atoms are located. If this condition is satisfied, then the total electronic energy of a molecule can be taken to be approximately the sum of the energies for each covalent bond in the molecule. Thus, the enthalpy of water would be twice the energy associated with one oxygen–hydrogen bond. The bond energies for many covalent bonds of importance to the organic chemist have been estimated by several techniques and are listed in Table 1.4. It is important to note that the assignment of a specific energy to a particular bond in a molecule is an approximation that is useful only for qualitative predictions.

The use of bond energy data to estimate $\Delta H°$ for an equilibrium can be illustrated with the H_2–I_2 system discussed previously. The enthalpy of two hydrogen iodide molecules should be $2 \times -71.4 = -142.8$ kcal, while the enthalpy of 1 mole of H_2 and 1 mole of I_2 would be $-104.2 + (-36.1) = -140.3$ kcal. The enthalpy difference for this equilibrium should be about -2.5 kcal. The entropy content of the products and the reactants would be expected to be about equal since there are two molecules involved on both sides of the equation ($\Delta S° \approx 0$) and one would suggest that the equilibrium $H_2 + I_2 \rightleftharpoons 2\ HI$ should favor products (HI) or be toward the right side.

Sample bond energy calculation:	**H_2 +**	**$I_2 \rightleftharpoons$**	**$2 \cdot$H—I**
Covalent bond energies:	**-104.2**	**-36.1**	**$2 \times (-71.4)$**

Bond energies of products: -142.8 kcal/mole
Bond energies of reactants: -140.3 kcal/mole
Predicated $\Delta H°$ for reactions: -2.5 kcal/mole

Chemical Dynamics

The position of a chemical equilibrium need not have any relevance to the rate at which the equilibrium is achieved. The fact that the production of HI from hydrogen and iodine is favored by about 2.5 kcal/mole does not tell us how long it will take to form the equilibrium concentration of HI if only H_2 and I_2 are present initially. Our everyday experience tells us that the rates of chemical reactions can vary considerably. The rusting of iron is a slow process taking many years, while the explosive reaction of gasoline vapors can be over in a matter of milliseconds. None of the remarks that have been made about chemical equilibria offer any insight into why these two processes should have such different rates and the only statement that can be made about the rates of the reactions involved in an equilibrium system is the obvious one: at equilibrium, the rate of the forward reaction must be equal to the rate of the reverse reaction. The study of the factors which determine the velocity of a chemical reaction form an area of endeavor now called chemical dynamics. It embraces both the experimental and theoretical study of reaction rates.

Intuitively, the rate of a chemical reaction should depend on the number of collisions or encounters that occur between the reactant molecules. In other words, the rate should depend upon the concentrations of the reactants. The mathematical expression that relates the experimentally determined rate of reaction to the concentration of reactants is called the rate law for the reaction. For the general reaction $A + B \rightarrow C$, the rate law would have the form shown in Eq. (1.6),

$$\text{Rate} = k[A]^n[B]^m \tag{1.6}$$

where k is a proportionality constant. The exponents n and m must be determined experimentally and are usually small integers but can be fractions. These exponents need not be related in any obvious way to the stoichiometric coefficients of A and B. The sum of n and m is called the order of the reaction.

The proportionality constant k is usually termed the rate constant. We should always carefully distinguish between the rate of a reaction and the rate constant for the reaction. The latter quantity is the more fundamental of the two, since the former depends on the concentration of reactants *and* the value of the rate constant. Chemical reaction rates normally increase with increasing temperature, so that k also increases with increasing temperature.

For many reactions, the rate law is related in a straightforward way to the stoichiometry of the reaction. The decomposition of ozone ($O_3 \rightarrow O_2 + O$) is a reaction which involves only a single reactant (O_3) and it has been determined by experiment that the rate law for the reaction is rate $= k[O_3]$. This process is an example of a unimolecular decomposition reaction.

A reaction step involving two molecules is a bimolecular reaction. An example is the neutralization of hydronium ions (H_3O^+) by hydroxide ions (OH^-).

$$H_3O^+ + OH^- \longrightarrow 2\,H_2O$$

The experimental rate law for this reaction is rate $= k[H_3O^+][OH^-]$.

Theoretical investigations in chemical dynamics revolve around questions regarding the factors that make a particular reaction proceed at the rate it does. A chemical reaction involves the progression from one free energy state (the reactants) to another free energy state (the products). If the progression is "downhill" as far as the free energy change for the overall process is concerned, then the reaction is favored by the thermodynamic considerations discussed above. But if we examine the process more closely we are forced to conclude that there must be states *higher* in energy than the reactants or products that lie between the reactants and products. This idea is shown diagrammatically in Fig. 1.14. If this were not the case, then all possible reactions in our universe would have occurred long ago. This intuitive conclusion that there must be an energy barrier to the smooth transversal from chemical reactants to reaction products is incorporated into current theories of chemical dynamics.

The state of maximum free energy that separates the reactants and products is called the transition state. Since the transition state is relatively high in energy, it is not possible to isolate these structures; molecules in the transition state configuration have only a fleeting existence and, within the time of one or two molecular vibrations ($\sim 10^{-13}$ seconds), they must either revert to reactant molecules or proceed to product molecules. The overall rate of reaction depends on the number of molecules leaving the transition state in the direction of products.

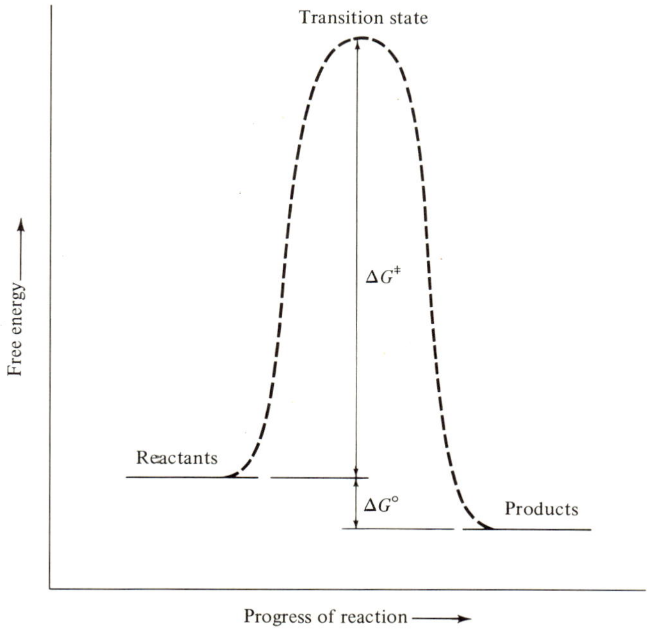

Fig. 1.14. The change in free energy as a chemical system proceeds from reactants to products.

The transition state can be thought of as a chemical entity that is in equilibrium with the reactants (and products). With this assumption, the concentration of molecules in the transition state can be related to the ground state reactant concentration by an equilibrium expression, which can be characterized by an equilibrium constant, $K^\ddagger$. (The symbol ‡ is traditionally affixed to signify quantities which depend on transition state properties.) The value of $K^\ddagger$ depends on the free energy difference between the transition state and ground state, $\Delta G^\ddagger$. In other words, the rate of a reaction depends upon the height of the energy barrier, $\Delta G^\ddagger$. Further development of the theory shows that this free energy change can be related to an enthalpy change and an entropy change by the equation

$$\Delta G^\ddagger = \Delta H^\ddagger - T\,\Delta S^\ddagger \tag{1.7}$$

To a good approximation, the enthalpy change for a reaction is a reflection of the differences in bonding energy in reactants and transition state. As this difference becomes larger, the reaction proceeds more slowly. However, the entropy difference between the reactant state and the transition state can serve to modify the influence of this enthalpy difference. If the transition state is less ordered than the reactant state, then $\Delta S^\ddagger$ will be positive ($S^\ddagger > S_{\text{reactants}}$, $\Delta S^\ddagger = S^\ddagger - S_{\text{reactants}}$) and the term $-T\,\Delta S^\ddagger$ will tend to decrease $\Delta G^\ddagger$, with a concomitant increase in the reaction rate.

It should be noted from Fig. 1.14 that the reverse reaction which converts products back to reactants passes through the *same* transition state that is involved in the forward reaction. If we can deduce certain features of the transition state from a study of the forward reaction, then we have also learned something about the reverse reaction.

Not many reactions of organic molecules involve a simple progression from reactants to products. Species may appear along the reaction pathway that are neither products, reactants, or transition states. These species are called intermediates. An organic reaction scheme may have many intermediates between reactants and products. We should note, however, that the rate of the reaction depends predominantly on the height of the largest free-energy barrier in the scheme. The particular event which corresponds to crossing this barrier is called the *rate-determining* or *rate-limiting step*. An analogy might make the importance of the rate-limiting step clearer. During the noon rush at the cafeteria in the student union, a customer can select a sandwich in 15 seconds, a beverage in 10 seconds, and a salad in 12 seconds. When he gets to the cashier it takes 45 seconds to pay for his choices. The rate at which students get through the serving line depends only on the last (slowest) step in the process; the rate at which food can be selected has no important bearing on the rate at which the line moves.

Similarly, in a multistep chemical reaction, the overall progress of the reaction depends primarily on the rate of the slowest step.

Acids and Bases

A common application of the ideas of chemical equilibria occurs in the study of the dissociation behavior of acids or bases. Since acids and bases often function as catalysts for organic reactions we should have some appreciation for aspects of these equilibria.

A convenient definition of an acid is due to J. N. Brønsted. According to this definition, an acid is a proton source or donor. A base is defined as a proton acceptor. By the very act of proton donation, an acidic species is transformed into a base. In the forward reaction of the equilibrium

$$HCl + H_2O \rightleftharpoons H_3O^+ + Cl^-$$

the molecule hydrogen chloride behaves as an acid but, in the course of this behavior, is transformed into a base, Cl^-. Pairs of chemical species that differ only by the presence of a proton are called conjugates; Cl^- is the conjugate base of HCl, while HCl is the conjugate acid of Cl^-.

Acid–base dissociation equilibria are most often studied in a solvent and the solvent usually participates actively in the equilibrium. Upon dissolution of monohydrogen phosphate ions in water the following equilibrium is established.

$$\underset{\text{Monohydrogen phosphate}}{HPO_4^{2-}} + \underset{\text{(Solvent)}}{H_2O} \rightleftharpoons \underset{\text{Hydronium ion}}{H_3O^+} + \underset{\text{Phosphate}}{PO_4^{3-}}$$

The monohydrogen phosphate and phosphate ions are an acid and its conjugate base, respectively. Similarly, the hydronium ion is the conjugate acid of the water molecule, which functions here as a base. The equilibrium constant for this system would be written as shown in Eq. (1.8):

$$K = \frac{[H_3O^+][PO_4^{3-}]}{[HPO_4^{2-}][H_2O]} \tag{1.8}$$

Most practical problems involving salts dissolved in water are done at low concentrations of the salt ($< 1\ M$) and therefore the concentration of water (55 M at 25°C) is much greater than the concentration of the solute ions. It is customary to consider the concentration of water, $[H_2O]$, as essentially invariant under these conditions and to incorporate this constant concentration into the constant, K.

Thus, the equilibrium constant for the above system would more typically be written

$$K_a = \frac{[H_3O^+][PO_4^{2-}]}{[HPO_4^{2-}]} \tag{1.9}$$

The magnitude of the equilibrium constant for dissociation of an acid depends upon the "strength" of the acid and the "strength" of the base which accepts the proton from the acid. Strong acids are those that ionize easily and are highly dissociated. The equilibrium quotient, K_a, is large for these acids. Table 1.5 gives the dissociation constants for a number of common acids.*

Table 1.5
Some Acid Dissociation Constants in Water at 25°C

Acid	K_a	pK_a
HCl	1.0×10^{7}	−7.0
H_3O^+	5.5×10	−1.7
H_3PO_4[a]	7.1×10^{-3}	2.2
CH_3COOH	1.8×10^{-5}	4.8
H_2S[a]	1.0×10^{-7}	7.0
$H_2PO_4^-$[a]	6.3×10^{-8}	7.2
OH^-	1.0×10^{-25}	25.0

[a] The constants given are for the ionization of a single proton.

Determination of the Structure of Organic Molecules

One of the first questions that discerning beginning students of organic chemistry ask relates to how one "knows" the structure of an organic molecule. Even relatively simple organic compounds can consist of exotic arrangements of covalent chemical bonds so that the assignment of one structure to one particular arrangement rather than an alternative may seem to be a perplexing bit of magic. The structure of an organic molecule is known only insofar as there is experimental evidence that can be cohesively understood in terms of the proposed structure. The proof of the structure of an organic molecule is an intellectual

* Since the numerical value of acid or base dissociation equilibrium constants can vary over a large range (Table 1.5) it is usually represented in exponential or scientific notation. In order to compress this notation further, it has become traditional to define a new quantity, the pK, as the negative logarithm of the equilibrium constant K. That is,

$$pK = -\log K$$

Because of the negative sign in the defining equation above, the strongest acids have the smallest pK_a (Table 1.5).

adventure that resembles in many ways a challenging crossword puzzle or a good detective story.

Before any attempt is made to determine the structure of a compound, the chemist must be certain that the compound is pure. For a very long time virtually the only techniques available to the organic chemist for the separation and purification of compounds were crystallization and distillation. These methods have been supplemented or supplanted by many so-called chromatographic procedures.* The criterion for purity remains the same, however. A sample is considered pure if exhaustive efforts to purify it by a variety of methods fail to change its physical and chemical properties. The physical properties of a compound include crystalline form, color, melting point and/or boiling point, and odor.

One of the first chemical features of a molecule that should be considered in an effort to define its structure is the identity of the atoms that make up the molecule and the number of atoms of each kind that are present. Quantitative analyses for the elements in organic compounds can be done on a few milligrams of material and are now provided by consulting laboratories especially established for this purpose. From the information obtained by a quantitative analysis, it is possible to compute an empirical formula for the compound. An empirical formula is defined as the simplest whole number ratio that expresses the kind and number of atoms that are present in a compound. But the empirical formula does not tell the whole story. Suppose an analysis showed that an organic compound consisted of 85.7% carbon and 14.3% hydrogen, by weight. A calculation taking into account the fact that a carbon atom is 12 times heavier than a hydrogen atom shows that there must be two hydrogen atoms for every carbon atom present in this compound; the *empirical formula* is CH_2. However, the analytical data are equally consistent with the molecule being constructed from four atoms of hydrogen and two atoms of carbon, C_2H_4. The formula which expresses the number of atoms in a molecule, rather than their ratio, is called a *molecular formula*. It differs from the empirical formula by a multiplicative constant, although this constant can be 1. To determine the molecular formula for a compound one needs to know the molecular weight of the compound in addition to the empirical formula.

To obtain confirmatory evidence for a proposed organic structure, the chemist often performs chemical reactions with the material whose outcome is known. If the structures of the products of these reactions are known or can be determined, then the structure of the original compound is largely established.

---

* A discussion of these techniques is beyond the scope of this book; consult one of the many fine laboratory manuals for organic chemistry that are available for details.

Advances in instrumentation and theory made in the past two decades have provided new and informative spectroscopic tools for the study of additional properties of organic structures. These techniques are now important for the biologist as well as the chemist; we shall consider some of them more fully in Chapters 5 and 12.

PROBLEMS

1. Sketch approximate energy level diagrams for the boron, nitrogen, and fluorine atoms and indicate the electronic structure of these atoms.

2. Sketch an appropriate energy level diagram for the electronic configuration of the phosphorus atom. How does a similar diagram for the argon atom differ from it? Which of the two atoms is more likely to form chemical bonds?

3. Suggest what the three-dimensional shape of the ammonia molecule, NH_3, should be. (The experimental H—N—H bond angle is 107°.)

4. All the N—H bond lengths in the ammonium ion, NH_4^+, are identical. Likewise, all H—N—H bond angles are the same. What is the structure of this ion? What type of bonding orbitals might the nitrogen atom use that is consistent with this structure?

5. Calculate an approximate $\Delta H°$ for each of these reactions.

(a) 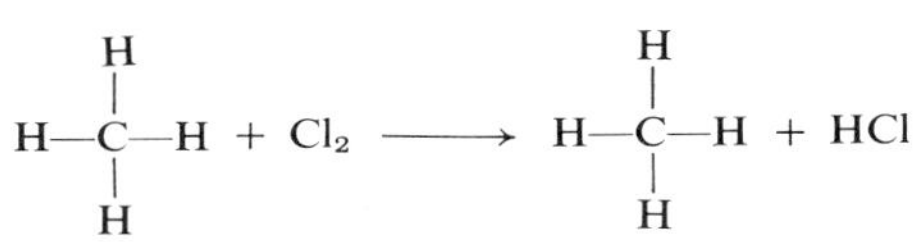

(b) 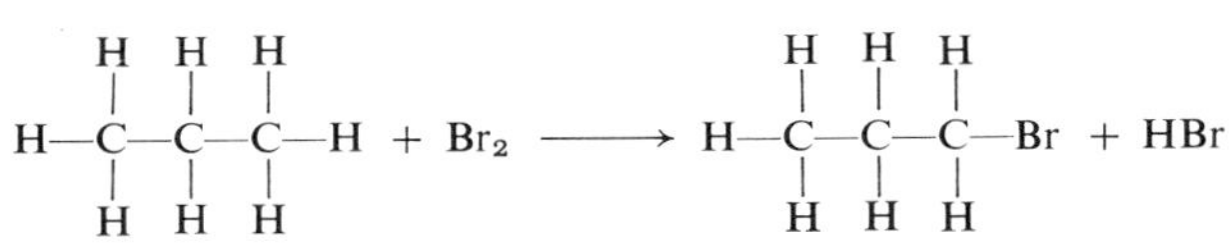

(c) 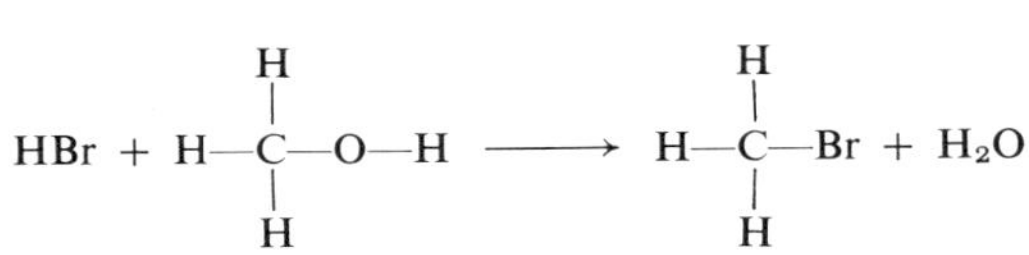

6. Is the following reaction a feasible one?

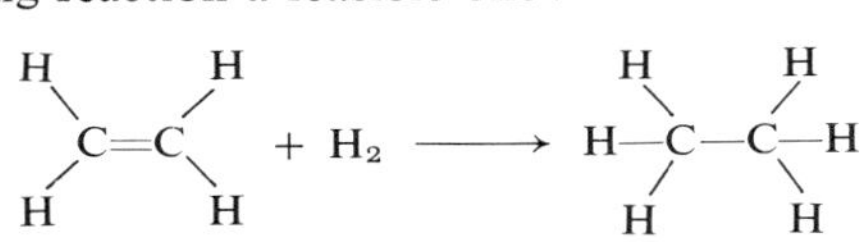

7. Calculate the hydronium ion concentration in a 0.1 M aqueous solution of sodium acetate.

8. Predict the effect of doubling the reactant concentration(s) on the rate of a unimolecular reaction and a bimolecular reaction.

9. What is the hydronium ion concentration in a 1.0 *M* aqueous solution of H_2PO^{2-}?

10. Common table sugar (sucrose) has the molecular formula $C_{12}H_{22}O_{11}$. What percent of the mass of this molecule is carbon?

11. What is the empirical formula for the compound methane if analysis shows that it contains 75% carbon and 25% hydrogen?

12. The percent composition of acetic acid is 40% carbon, 6.7% hydrogen, and 53.3% oxygen. A determination of the molecular weight gives the value 60. What is the molecular formula of acetic acid?

13. Combustion of 0.15 g of an organic compound containing only carbon, hydrogen, and oxygen in an atmosphere of pure oxygen produced 0.33 g of carbon dioxide (CO_2) and 0.18 g of water. What is the empirical formula for this compound?

SUGGESTED READINGS

F. Brescia, J. Arendts, H. Meislich, and A. Turk, "Fundamentals of Chemistry." Academic Press, New York, 1966.

B. H. Mahan, "University Chemistry." Addison-Wesley, Reading, Massachusetts, 1965.

W. H. Slabaugh and T. D. Parsons, "General Chemistry." Wiley, New York, 1966.

C. W. Keenan and J. H. Wood, "General College Chemistry." Harper, New York, 1966.

L. Pauling, "The Nature of the Chemical Bond." Cornell Univ. Press, Ithaca, New York, 1960.

C. A. Coulson, "Valence." Oxford Univ. Press, London and New York, 1961.

J. R. Partington, "A Short History of Chemistry." Harper, New York, 1960.

2

Alkanes and Cycloalkanes

Organic chemistry, by traditional definition, is the chemistry of the compounds formed by carbon. Nurtured in the laboratories of highly imaginative and creative German chemists during the 1800's, this branch of science has grown to one of tremendous factual content. It is estimated that there are over two million organic compounds now known and the current rate of increase of this number is very large.

There are more chemists who call themselves "organic chemists" than any other species. In industrial laboratories, organic chemists focus on the properties of molecules which may be important as pharmaceutical agents, petroleum fuels, lubricants, dyes, flavors and scents, plastic structural materials and fibers, and pesticides and other agricultural chemicals. Moreover, all living things are composed of organic molecules, some of them exquisitely complex. Organic chemistry thus is an important component of what are commonly regarded as the "life sciences" and many areas of biological, behavioral, and medical science draw upon the insights and skills of the organic chemist. The efforts and successes of these workers continually add to the store of knowledge about the compounds of carbon and, in so doing, also increase the burdens that confront the beginning student in this field.

An important organization operated by the American Chemical Society is Chemical Abstracts Service. The Service publishes short abstracts of virtually all articles having to do with chemistry as they appear. Over five million of these abtracts have been published so far. Although it took almost 65 years for this number of abstracts to accumulate, it is estimated that the next five million abstracts will be generated within 10 years. A significant fraction of the papers and patents reported describe organic compounds.

Computers have been enlisted to help keep track of the burgeoning number of organic compounds. Chemical Abstracts Service recently recorded the one-millionth unique chemical structure in its computer-based Chemical Registry System, which is bringing together in machine-searchable form information on the structures, names, and sources of published information for virtually all known chemical substances.

Courtesy of Chemical Abstracts Service.

ALKANES

Both in terms of structural features and reactivity, the simplest organic compounds are those that contain only carbon and hydrogen atoms, the hydrocarbons. The simplest hydrocarbon is methane, CH_4. The three-dimensional structure of this molecule is shown in Fig. 2.1. Each hydrogen atom is located at a corner of a regular tetrahedron so that the atomic orbitals used by the central carbon atom to bind these hydrogen atoms must be sp^3-hybridized.

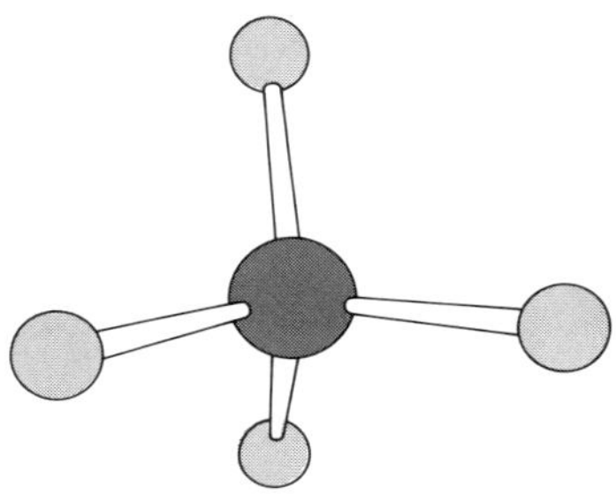

Fig. 2.1. A three-dimensional view of a model of the methane molecule.

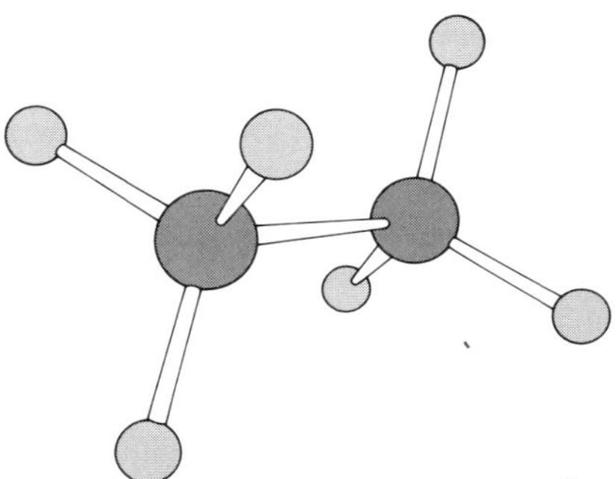

Fig. 2.2. A three-dimensional drawing of a model of ethane.

The feature which most distinguishes carbon from other elements is its ability to form strong covalent bonds to itself as well as to other elements. It is this ability which partially accounts for the very large number of known organic compounds. If two carbon atoms are bound to each other and the remaining bonds are joined to hydrogen, the compound ethane, C_2H_6, is formed. This molecule is illustrated in Fig. 2.2. As this drawing suggests, ethane can be assembled from two partial methane molecules. Three carbon atoms chained together in a similar manner are found in the hydrocarbon propane, C_3H_8 (Fig. 2.3). These three hydrocarbons, and all hydrocarbons of the molecular formula C_nH_{2n+2}, are called saturated hydrocarbons or alkanes.

Representation of the structure of organic compounds on the printed page can not always be done satisfactorily. Conventionally, a dash is used to represent the electron-pair bond formed by the overlap of atomic orbitals between two atoms. If stereoscopic techniques are not available but the three-dimensional aspects of an organic structure are important, heavy lines are used to represent bonds which point out of the page (toward the viewer), and dashed lines those

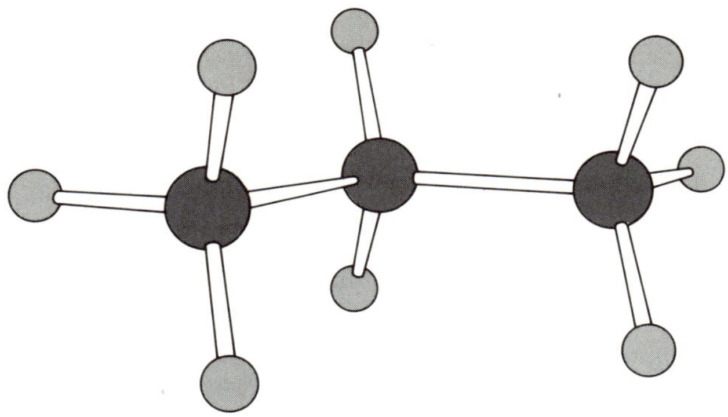

Fig. 2.3. A three-dimensional representation of propane, C_3H_8.

which point into the page (away from the viewer). Also, a shorthand notation called compressed formulas is often used to conserve space. These representations of the three hydrocarbons described so far are illustrated in Fig. 2.4.

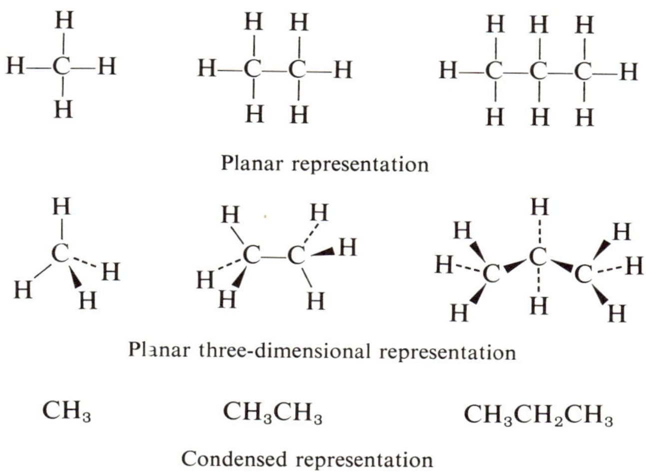

Fig. 2.4. Various two-dimensional representations of methane, ethane, and propane.

Structural Isomers

A hydrocarbon with the formula C_4H_{10} is called a butane. There are two possible structures for butane as shown below. When two or more materials have the same molecular formula, but are not the same compound, they are said to be isomers of one another. The corresponding adjective is *isomeric*. When isomers have different numbers or different types of chemical bonds, they are said to be structural isomers. Thus, the two butanes shown below are structural isomers.

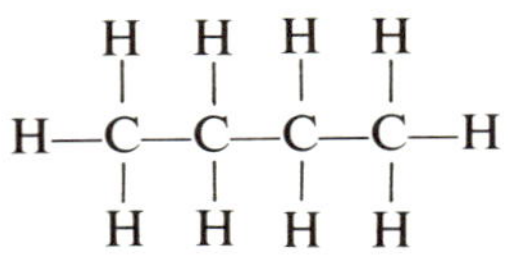

C_4H_{10}
n-Butane

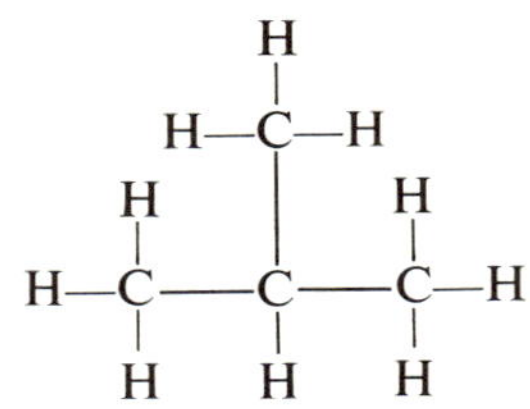

C_4H_{10}
Isobutane

The number of possible structural isomers increases rapidly as the number of carbon atoms in the hydrocarbon increases. Three isomeric pentanes (C_5H_{12}) are known. Their structures are shown below. Alkanes which contain six carbon atoms are called hexanes (C_6H_{14}) and can be found in five structurally isomeric forms. Heptane (C_7H_{16}) can exist as nine structural isomers. There are about 1800 isomers of tetradecane ($C_{14}H_{30}$) and over 4 million tricontanes ($C_{30}H_{62}$).*

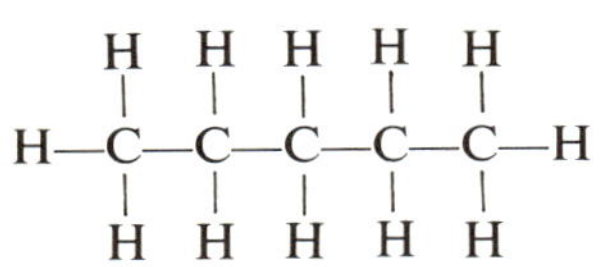

n-Pentane

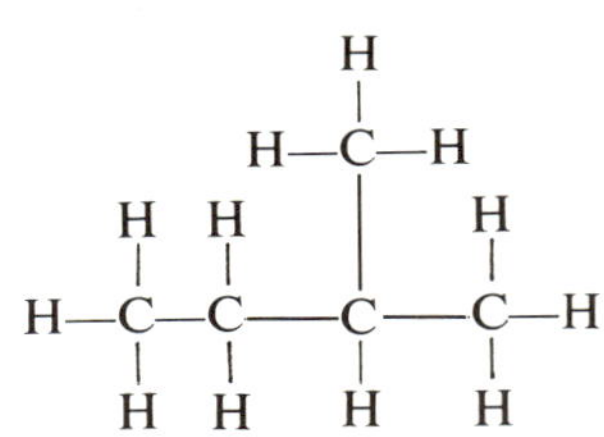

Isopentane

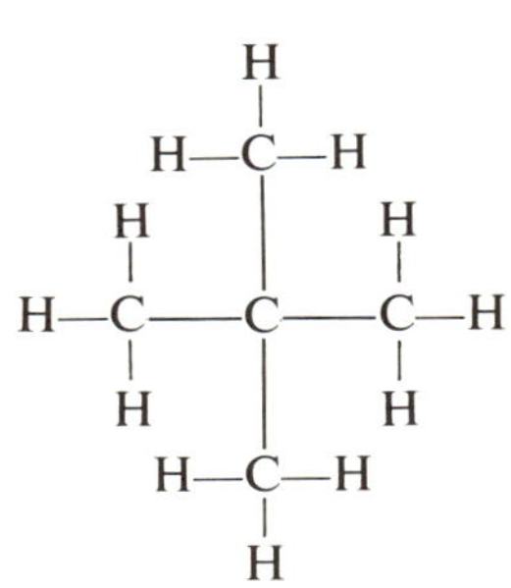

Neopentane

* A discussion of the mathematical methods used to determine the number of isomers of an alkane is given by H. R. Henze and C. M. Blair, *J. Amer. Chem. Soc.* **53**, 3077 (1931).

Nomenclature of Alkanes

The most serious problem that arises as a result of the phenomenon of structural isomerism in alkanes is that of nomenclature. This problem was circumvented for smaller molecules like the butanes and pentanes by adding a prefix such as *iso-* or *neo-* to the common name. But it soon became clear that this procedure could not be continued for larger molecules. Besides challenging the imagination, the continual invention of special prefixes tends to create nondescriptive and unwieldy systems of nomenclature. Early organic chemists recognized this problem and met in 1892 to devise a systematic method for naming organic compounds. Under the auspices of the International Union of Pure and Applied Chemistry (IUPAC) this system has been modified and updated several times since.

Before discussing the IUPAC nomenclature system we should pause to note some of the structural features of the hydrocarbon molecules discussed up to this point. The simplest arrangement of carbon atoms in these molecules is a continuous (nonbranched) chain; the carbon atoms are connected one to another in what appears to be a linear array. The structural isomer of a hydrocarbon that has this continuous, unbranched arrangement of carbon atoms is called the *normal* isomer. The normal isomer is indicated by a lower-case *n* prefixed to the compound name. Examples are the *n*-butane and *n*-pentane molecules discussed above. For hydrocarbons with five or more carbon atoms, the formula name is obtained by adding the suffix *-ane* to the appropriate Greek stem. Thus C_5H_{12} is pentane, C_6H_{14} is hexane, and C_7H_{16} is heptane.

If one hydrogen atom is removed from an alkane, an alkyl *radical* is the result. These radicals have importance in the IUPAC nomenclature system and are designated by replacing the *-ane* syllable in the name of the corresponding hydrocarbon with *-yl*. If one hydrogen atom is removed from methane, for example, the methyl radical, CH_3, results. This is also called a methyl *group*. Similarly, removal of a terminal hydrogen atom from heptane produces $CH_3CH_2CH_2$-$CH_2CH_2CH_2CH_2$, the *n*-heptyl radical or group.

To apply the systematic nomenclature system, one must first find the longest continuous chain of carbon atoms in the molecule. The compound is then named as a derivative of the corresponding alkane. Although the compound below contains eight carbon atoms, it is systematically named as a derivative of heptane because the longest continuous chain contains seven carbon atoms. The next step is to number the carbon atoms in the longest chain; the position of each group that is not a hydrogen atom along this backbone chain of the molecule can be specified by the number of the carbon atom to which it is attached. The name of the compound is formed by using these numbers and the appropriate radical name as prefixes to the name of the backbone hydrocarbon. The compound pictured below is thus called 3-methylheptane. If a particular radical appears more than once along the backbone chain, an additional prefix such as *di-*, *tri-*, or *tetra-* to indicate the number of times the radical appears is added to the radical name. The associated position-specifying numbers are combined

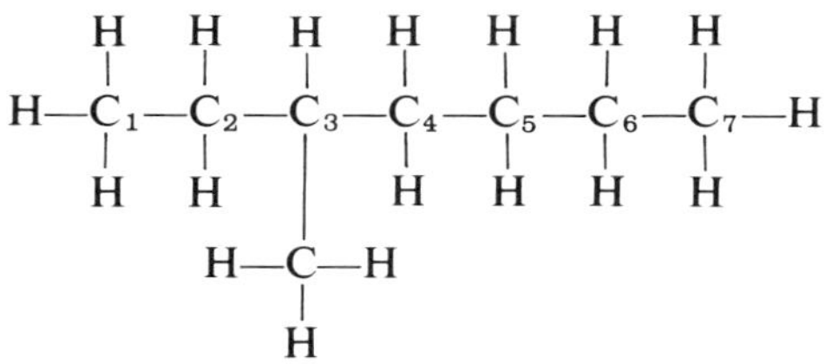

3-Methylheptane

and separated by commas. The branched octane (C_8H_{18}) shown below is named 3,3-dimethylhexane by this convention.

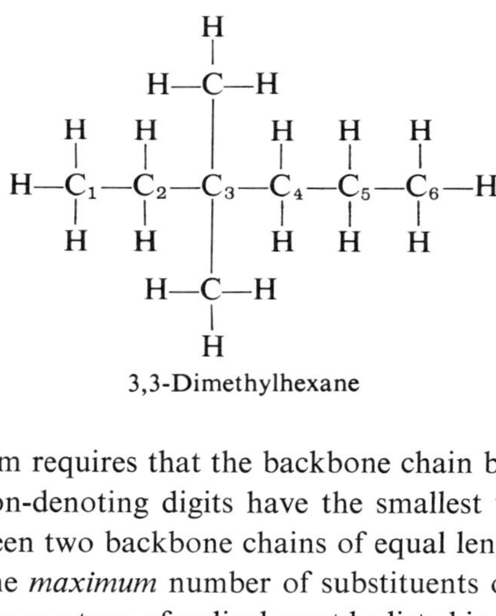

3,3-Dimethylhexane

The IUPAC system requires that the backbone chain be numbered in such a way that the position-denoting digits have the smallest values possible. Also, when a choice between two backbone chains of equal length can be made, that chain which gives the *maximum* number of substituents or side chains is to be selected. If more than one type of radical must be listed in the prefix to the backbone name, they are to appear in alphabetical order or in order of increasing complexity.* Satisfy yourself that the rules mentioned above are followed in the names given to the isomeric octanes shown on p. 33.

* For these purposes the radical containing the greatest number of carbon atoms is regarded as more complex.

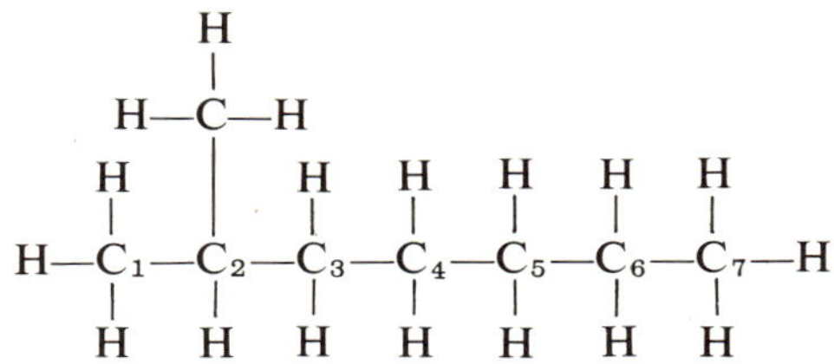

2-Methylheptane

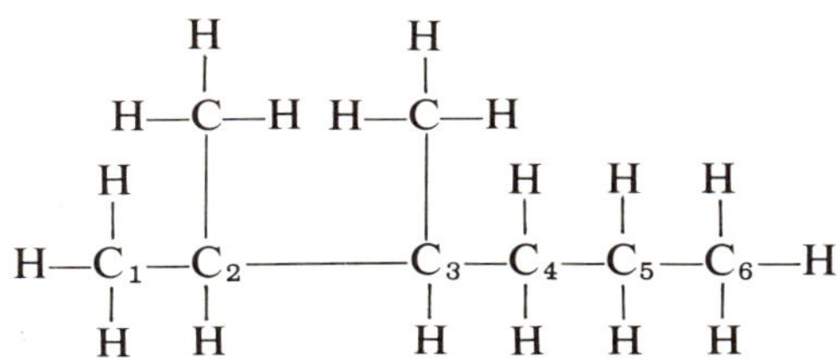

2,3-Dimethylhexane

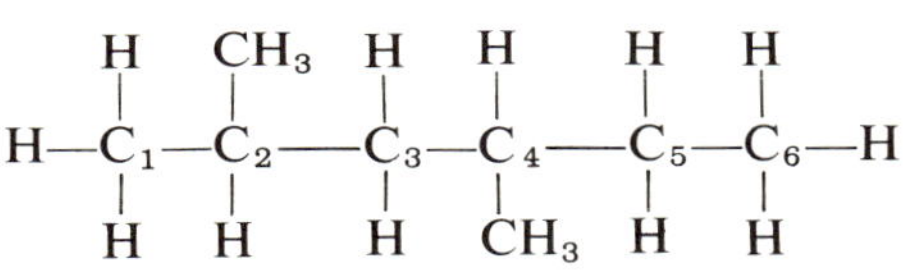

2,4-Dimethylhexane

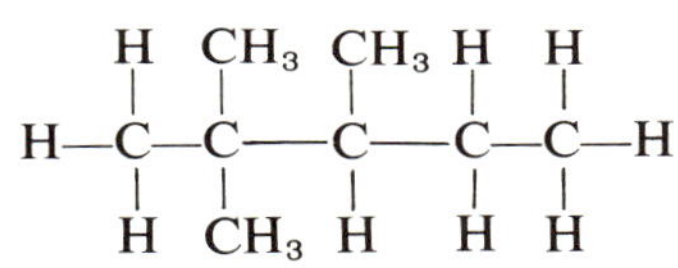

2,2,3-Trimethylpentane

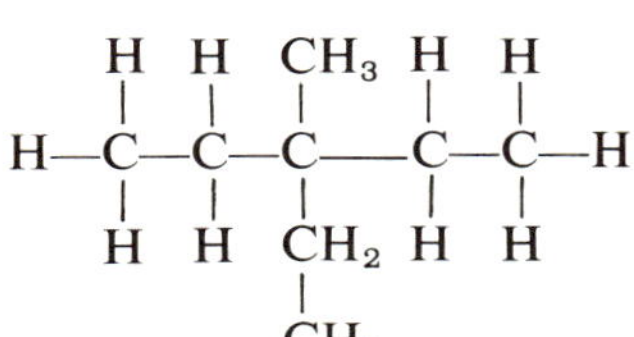

3-Ethyl-3-methylpentane

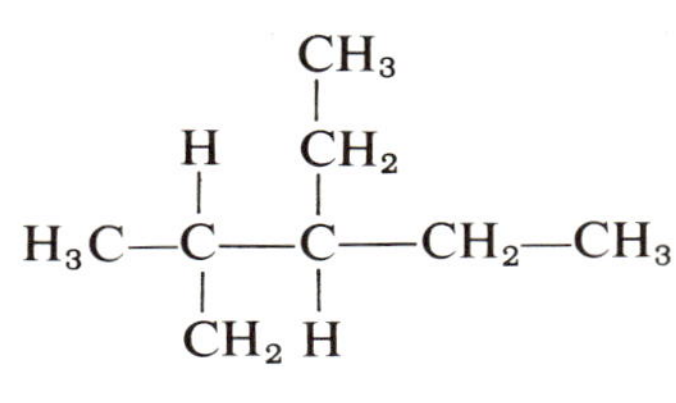

3-Ethyl-2-methylpentane

Molecular Models

A useful tool in the study of the architecture of an organic molecule is a model of the molecule. These models help us to visualize structural features of a molecule that are not apparent from a two-dimensional, planar representation. Molecules are, of course, three-dimensional, space-filling entities and these three-dimensional characteristics are made more evident in a model. Organic chemists use several different types of models. One is of ball-and-stick construction and gives a model similar to the drawings shown above. Various colored balls are used to represent atoms and these "atoms" are connected by some device which plays the role of chemical bonds. Another type of molecular model is constructed of metal or plastic tubes. Each tube represents a chemical bond; the atoms are understood to be present at the intersections and termini of these tubes. These models can be constructed so that bond lengths and bond angles are more accurately represented than is true in the case of the ball-and-stick models. Models of methane, ethane, and propane constructed from commercially available sets of these types of molecular models are shown in Fig. 2.5.

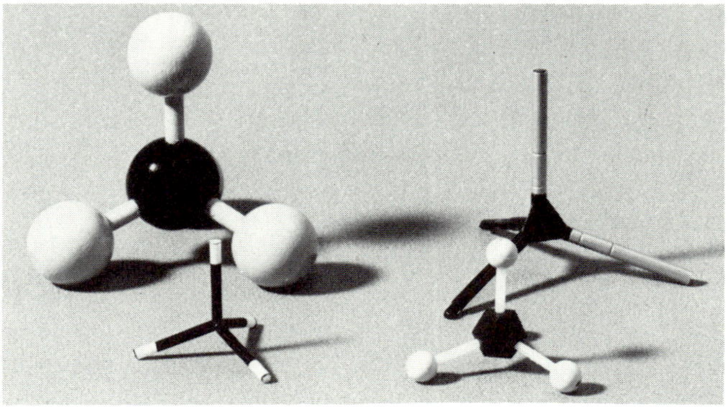

(a)

(b)

Fig. 2.5. Various models of the methane (a), ethane (b), and propane (c) molecules.

(c)

Fig. 2.5.

All of these models have one basic flaw—they underemphasize the sizes of the constituent atoms. Electrons occupy the space around atomic nuclei, and because nonbonding electron–electron interactions are usually repulsive, two atoms in a molecule cannot get much closer than approximately their combined atomic radii. Models which represent the space-filling features of an atom are available. Space-filling models of methane, ethane, and propane are shown in Fig. 2.6 and give one quite a different impression of the "solidity" of molecules. Models of both the skeletal or framework type and space-filling variety are needed in the study of organic chemistry, although neither gives a completely accurate representation of what a molecule is really like. Ball-and-stick kinds of models are useful in the study of the carbon skeleton of a molecule, whereas a space-filling model emphasizes the steric requirements of various parts of a molecule.

Fig. 2.6. Space-filling models of methane, ethane, and propane.

Conformational Isomers

Upon examination of a model of ethane, one is soon convinced that there is a whole family of three-dimensional arrangements for the atoms of this molecule. These can be generated simply by rotating one end of the ethane molecule relative to the other (Fig. 2.7). Each of these arrangements has the same number and kind of chemical bonds and differs only in the dihedral (H—C—C—H) bond angle at the carbon–carbon bond. Such arrangements are called alternatively conformations, conformational isomers, or conformers. All the structures shown in Fig. 2.7 are ethane molecules. They differ only in the way the atoms are arranged relative to one another in space. The energy difference between these conformations is small and they are interconverted at a rapid rate.

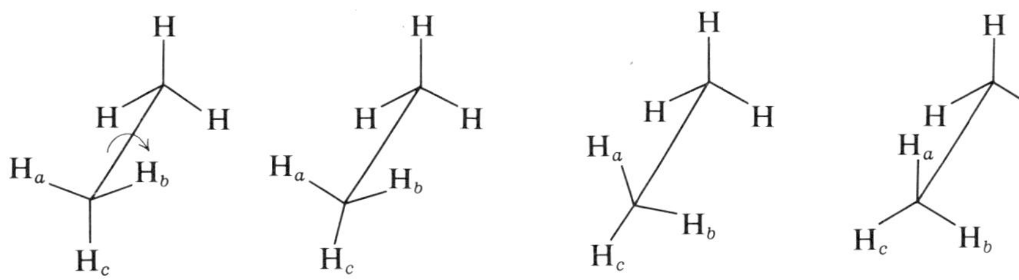

Fig. 2.7. Rotation about the carbon–carbon bond in ethane. The front part of the molecule is rotated while the back portion is held fixed. A three-dimensional view of the staggered form of ethane appears in Fig. 2.2.

Of the many conformational isomers of ethane that can be generated by rotation about the central carbon–carbon bond, one is more probable than the others. This is the staggered conformation. The free energy of this conformer is lower than any other because unfavorable nonbonded interactions between the electronic charge clouds of the hydrogen atoms are minimal in this arrangement. The energy difference between the staggered conformation of ethane and the isomer of maximum energy (the eclipsed form) is about 3 kcal/mole. All conformations of ethane are in equilibrium with each other, but because the free energy of the staggered form is lowest, most ethane molecules in the equilibrium mixture are present in this conformation.

A problem of notation arises when one attempts to represent the conformational isomers of a molecule. The "sawhorse" drawings used in Fig. 2.7 suffice for discussion of simple molecules like ethane. More complicated structures are most profitably discussed with the aid of Newman projection drawings, developed by Professor M. Newman of Ohio State University. Projection representations of ethane are shown in Fig. 2.8. It is presumed that the intersection of lines in the "steering wheel" portion of the drawing corresponds to the carbon atom of a carbon–carbon single bond that is closest to the viewer. Each line represents a bond to hydrogen or another group. The dihedral angle, ϕ, is more clearly evident in this representation.

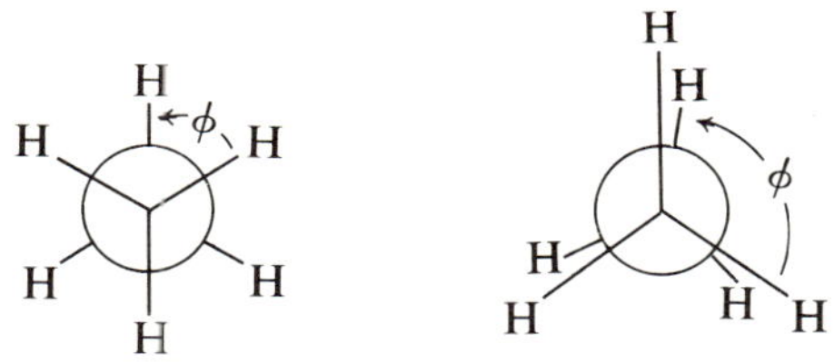

Fig. 2.8. Newman projection drawings of ethane.

Sawhorse and Newman projection drawings of the staggered forms of *n*-butane are depicted in Fig. 2.9. In staggered conformational isomer *a*, called the *trans* isomer, the terminal methyl groups are as far from each other as possible. In form *b*, the *gauche* isomer, these methyl groups interact rather strongly with each other because of their mutual steric requirements. The *trans* isomer is thus lower in free-energy content than the *gauche* isomer and is the one that predominates at moderate temperatures.

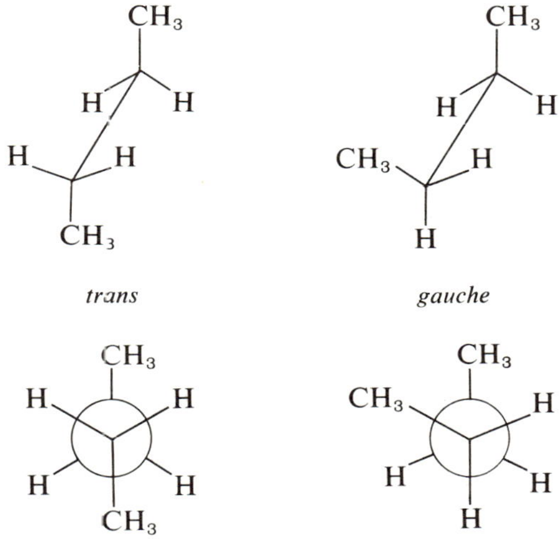

Fig. 2.9. Representations of the two staggered conformers of *n*-butane.

Perturbations of the free energy of a molecule that have their origins in the space-filling properties of atoms or groups of atoms are called steric effects. Steric effects are as simple as the proposition that two people cannot occupy comfortably the same chair. In the *gauche* conformer of butane, two methyl groups must come close to one another and, since repulsive interactions are involved, the energy of these staggered conformations is higher than when the methyl groups are more distant from each other. The photograph in Fig. 2.10 shows space-filling models of several groups that commonly occur in organic compounds. The relative steric requirements of these groups should be noted.

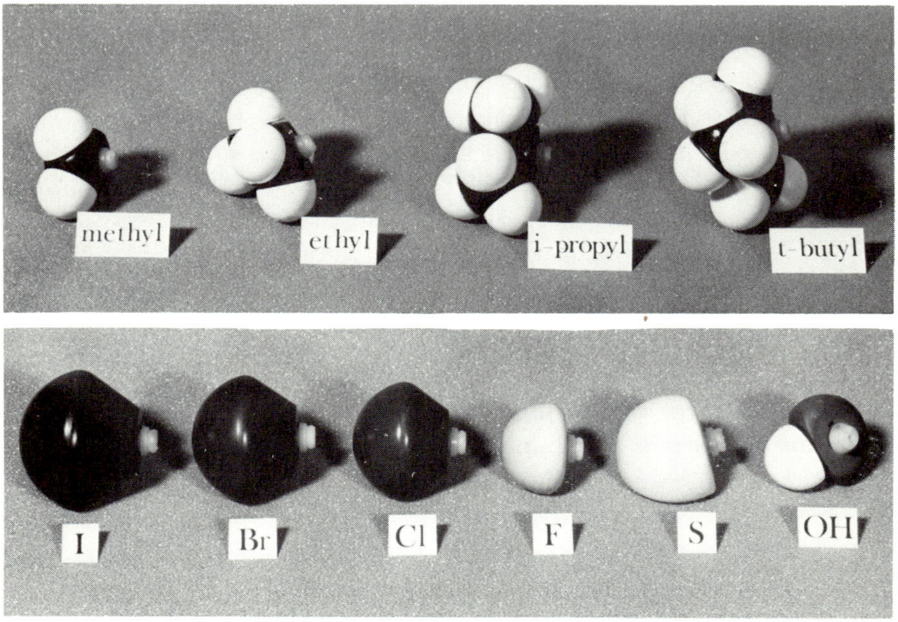

Fig. 2.10. Space-filling models of several common organic groups. (The common name isopropyl for the $(CH_3)_2CH$— group has not been encountered yet.)

CYCLOALKANES

Let us imagine a simple "thought" experiment. Two hydrogen atoms, one from each end, are removed from the hydrocarbon propane and a chemical bond is formed between the two terminal carbon atoms. A new hydrocarbon with the formula C_3H_6 is thus formed. (See top of page 39 for reaction.)

A similar experiment could be done with each of the normal (straight-chain) hydrocarbons to give corresponding new hydrocarbons with the general empirical formula C_nH_{2n}. Such molecules exist and are said to be cyclic hydrocarbons or cycloalkanes because the carbon backbone of the molecules defines a ring

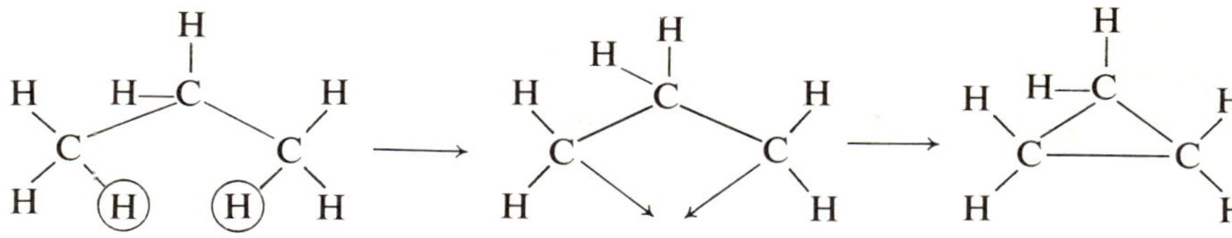

Propane Cyclopropane

structure, rather than a chain. These new materials are named by adding the prefix *cyclo-* to the usual name for the alkane corresponding to the number of carbon atoms in the ring. The three-membered ring "constructed" above is thus called cyclopropane. Other cyclic alkanes are shown in Fig. 2.11. A common abbreviation for the structure of a cycloalkane is a regular polygon with the same number of vertices as there are atoms in the ring.

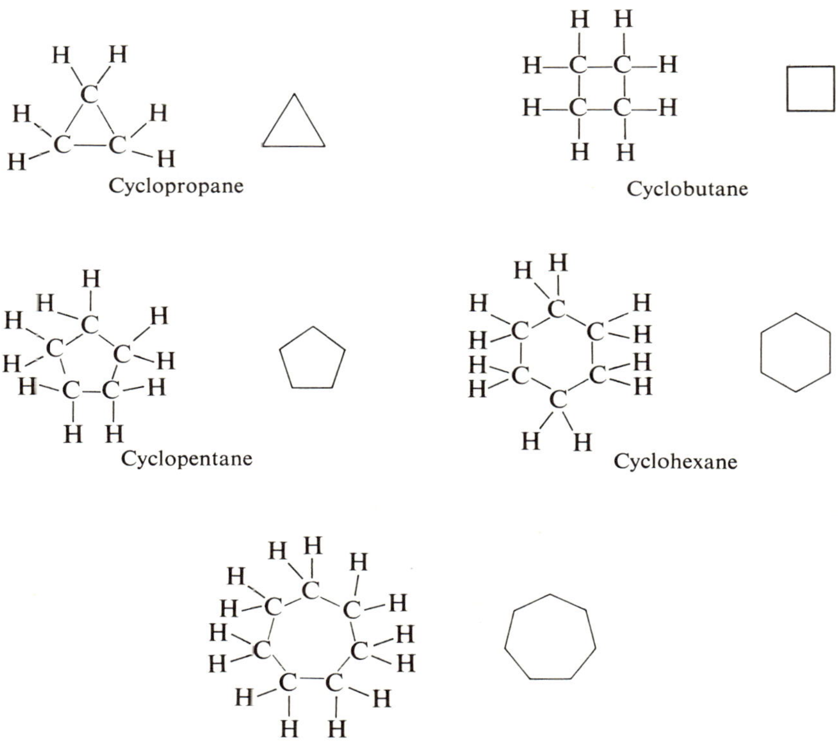

Fig. 2.11. Structures, names, and abbreviations for several cycloalkanes.

The formation of cyclic molecules is another manifestation of the tendency of carbon to form chemical bonds with itself. It should be noted that cyclopropane and cyclobutane have abnormally small C—C—C bond angles. The associated bond energies are also somewhat different so that these small-ring compounds have unusual reactivities relative to those of the corresponding open-chain structure. Rings of carbon atoms larger than these show essentially the same chemical properties as those of the *n*-alkanes; the bond angles in these compounds are close to those expected for materials constructed of sp^3-hybridized carbon atoms.

Nomenclature for the cycloalkanes follows directly from the nomenclature for the open-chain compounds. Substituents are located on the ring skeleton by assigning a number to each position in such a way that the sum of these numbers is a minimum. If the cycloalkane ring is attached to a larger or more complex molecule, it may be named as a substituent of this molecule. The following examples illustrate these ideas.

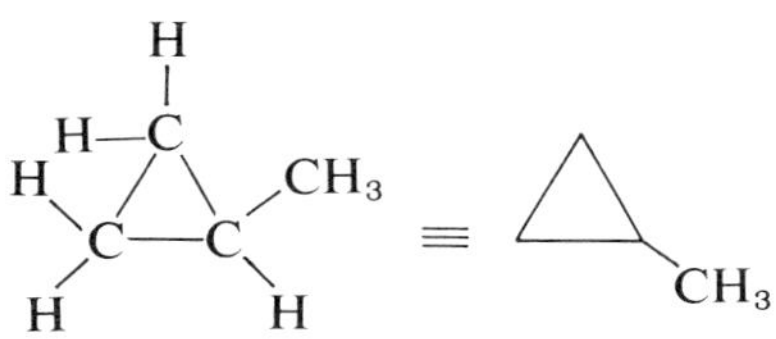

Methylcyclopropane
(The locating number is not needed since there is no ambiguity.)

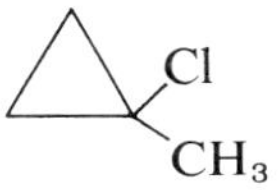

1-Chloro-1-methylcyclopropane

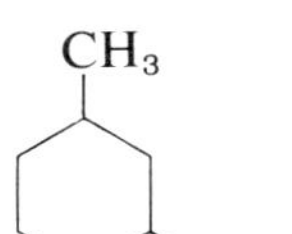

1,3-Dimethylcyclohexane

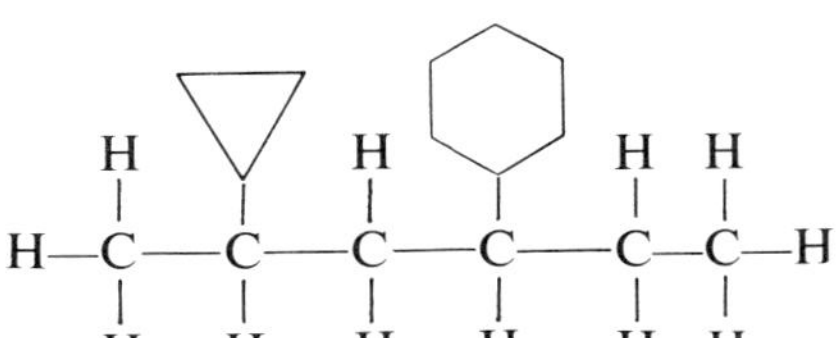

2-Cyclopropyl-4-cyclohexylhexane

Conformations of Cyclohexane

The conformational isomers of the cycloalkanes are particularly interesting and the cycloalkane most studied with regard to its conformational properties is cyclohexane. Inspection of a molecular model of this compound shows one that there are a number of conformations possible for its carbon atom skeleton; each of these has normal bond angles and lengths. Two of them are the so-

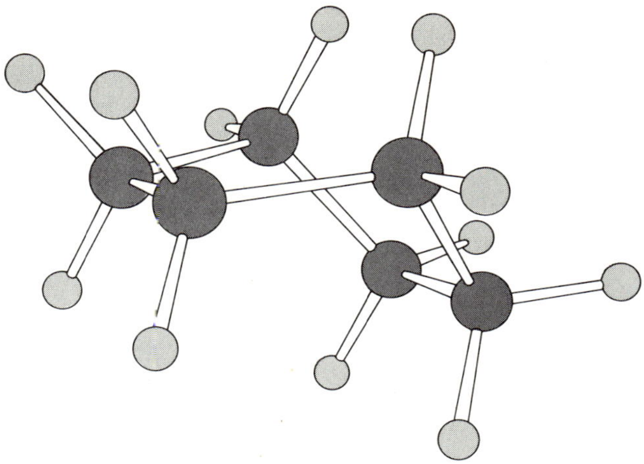

"Boat"

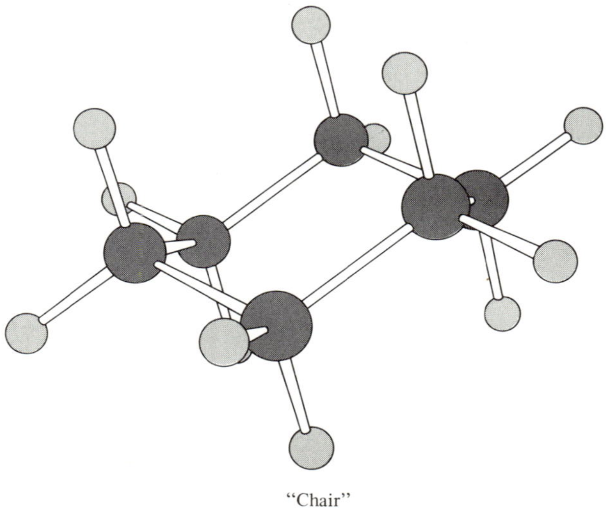

"Chair"

Fig. 2.12. Three-dimensional drawings of two conformations of cyclohexane.

called chair and boat forms, sketched in Fig. 2.12. Of these, the chair conformer is more stable. The hydrogen atoms of chair-form cyclohexane can be in two distinct environments. These are called axial and equatorial. Figure 2.13 attempts to illustrate the two types of hydrogen atoms in the cyclohexane. The figure also shows that when the ring is flexed in the manner indicated, an equivalent chair conformation results, but with the axial and equatorial hydrogens interchanged.

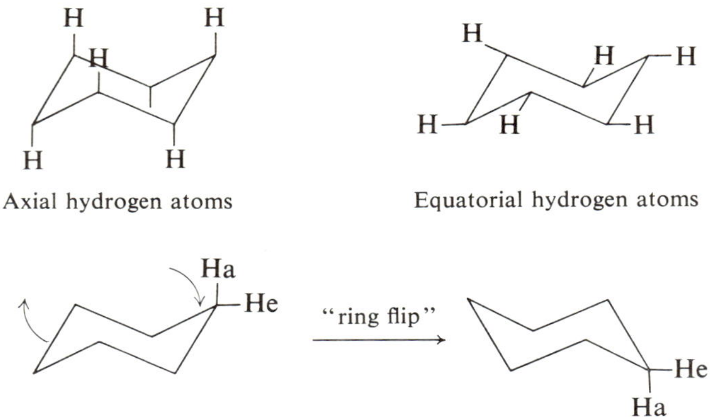

Fig. 2.13. Axial and equatorial hydrogen atoms in cyclohexane. A ring-flipping motion of the type indicated converts hydrogen atoms in one environment to the other.

When a substituent such as a chlorine atom replaces a hydrogen atom in cyclohexane, it can occupy either an equatorial or an axial position (Fig. 2.14). An axial chlorine atom can interact with axial hydrogens that are present on the 3- and 5-carbon atoms. The magnitude of these interactions will depend upon the steric requirements of the substituent; if the substituent is very large these interactions will appreciably increase the energv of this conformation. Steric overcrowding in these cases can be relieved by a "ring-flip" of the compound which changes the axial substituent to an equatorial position.

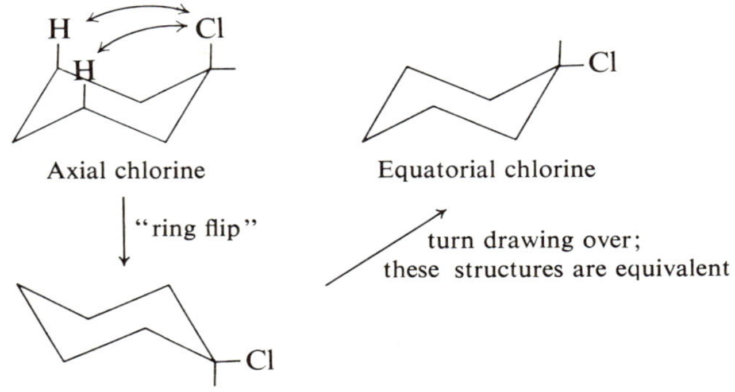

Fig. 2.14. Two conformations of chlorocyclohexane. Note the nonbonded hydrogen–chlorine interactions in the axial compound. Hydrogen–chlorine interactions in the equatorial conformation are much less severe.

Cis-Trans Geometrical Isomers in Cycloalkanes

When there are two or more substituents attached to a cyclohexane ring, a new kind of isomerism is possible. Consider the drawings of 1-fluoro-4-*t*-butylcyclohexane below.

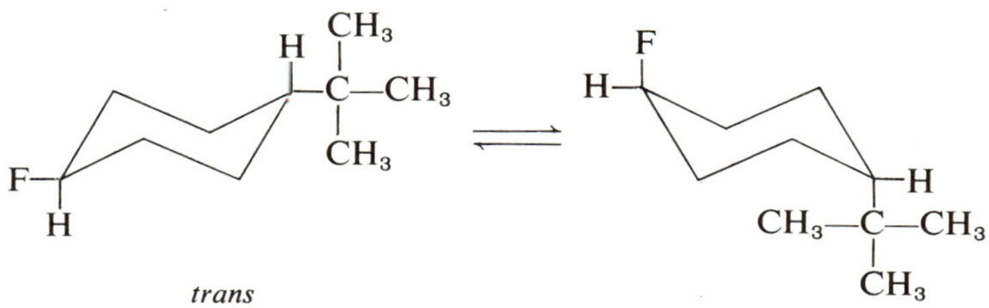

trans

A B

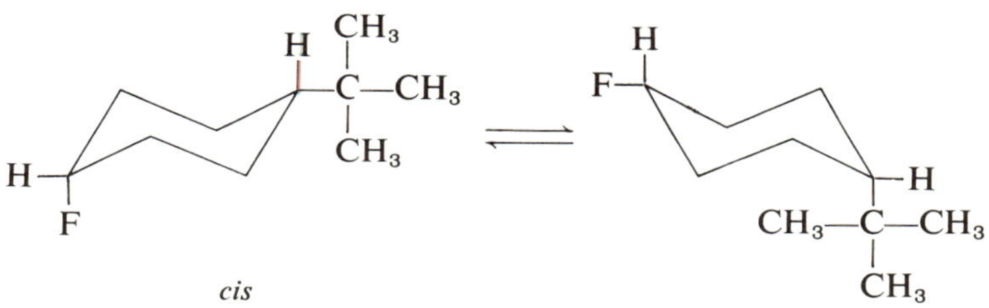

cis

A B

This first compound (*trans*, A) has both the fluorine and the *t*-butyl substituents in equatorial positions. This molecule can undergo the "ring-flipping" conformational motion described above to generate a conformational isomer (*trans*, B) in which both substituents are in axial positions. However, the fluoro- and *t*-butyl substituents could be attached to the ring as shown in the structure *cis*, A. Here, the *t*-butyl group is in the equatorial position and the fluorine is axial.* A ring-flip would lead to *cis*, B, a conformational isomer in which the two substituents have taken up axial and equatorial positions, respectively. The pair of molecules (*trans*, A) and (*cis*, A) have the same molecular formula and the same number and types of chemical bonds, but differ in the way the atoms are arranged in space. Moreover, this difference is a permanent one, being enforced by the nature of the cyclohexane ring system. Such molecules are called geometrical isomers. They are fundamentally different compounds with different

---

* The word *trans* to designate a structural isomer here is used in the same sense that it was used with the staggered rotational isomer of butane; it implies "on opposite sides."

chemical and physical properties; the rate of interconversion between geometrical isomers is normally a very slow process. In the case of cyclohexane derivatives discussed here, conformational isomerism is also possible. Conformational isomers are usually not separable because they interconvert at a rate much faster than the rate at which a chemical separation can be carried out. Thus, a sample of *trans*-1-fluoro-4-*t*-butylcyclohexane will contain the pair of rapidly interconverting conformational isomers *trans*, A and *trans*, B shown above. This sample will be separate and distinguishable from *cis*-1-fluoro-4-*t*-butylcyclohexane, although the latter will also contain a pair of interconverting conformers, *cis*, A and *cis*, B.

Geometrical *cis–trans* isomerism is not limited to 1,4-substituted cyclohexanes as the examples below indicate.

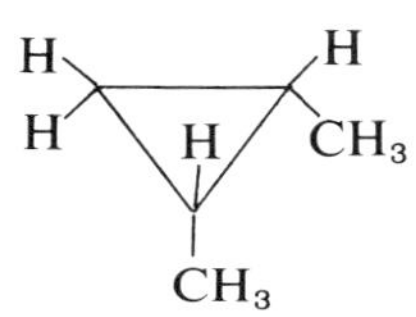

cis-1,2-Dimethylcyclopropane

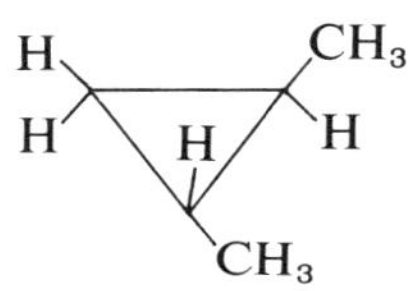

trans-1,2-Dimethylcyclopropane

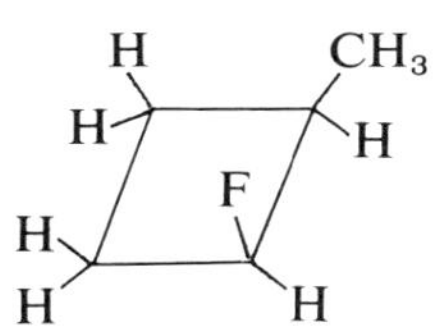

cis-1-Fluoro-2-methylcyclobutane

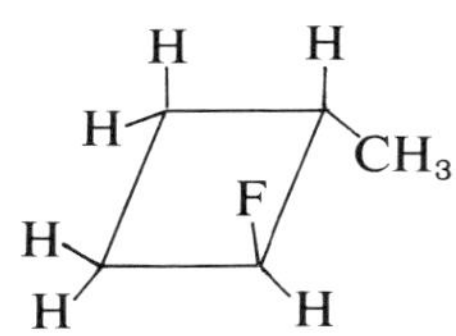

trans-1-Fluoro-2-methylcyclobutane

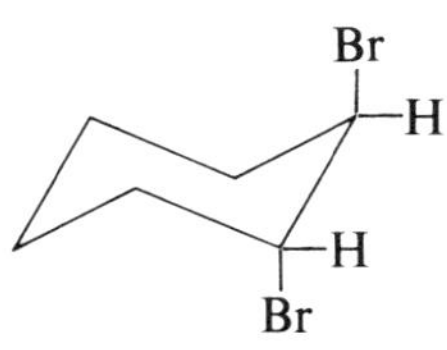

cis-1,2-Dibromocyclohexane

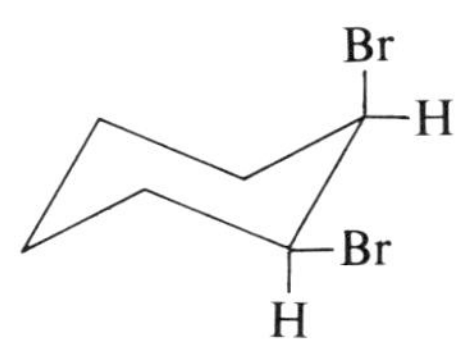

trans-1,2-Dibromocyclohexane

We see that, in addition to the ability of carbon to form strong bonds to itself, another feature of organic chemistry which contributes to the large number of compounds is the ability of organic molecules to often exist in different structurally and conformationally isomeric forms.

POLYCYCLIC ALKANES

The story does not end at this point, however. There is no reason to limit cyclic compounds to those which contain only one ring of carbon atoms. A large number of organic compounds with two or more rings of carbon atoms are known. However, it is beyond the scope of this text to discuss them exhaustively and we shall limit ourselves to several polycyclic alkanes which are important or interesting.

If one starts with *trans*-1,2-dimethylcyclohexane and imagines connecting the methyl groups (less a hydrogen atom on each) by a —CH_2—CH_2— fragment, a bicyclic hydrocarbon is produced (Fig. 2.15). The formula for this species is $C_{10}H_{18}$; it is a bicyclodecane but is commonly known by the trivial name *trans*-decalin. The molecule appears to be constructed of fused, chair-cyclohexane-like fragments. The carbon–carbon bond in common to both rings is called the ring juncture.

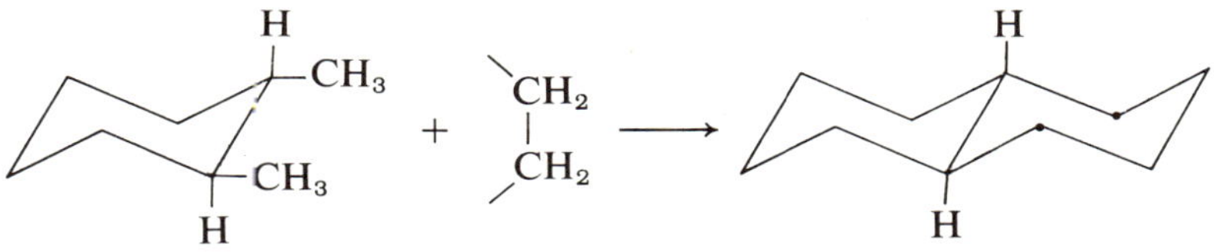

trans-Decalin

Fig. 2.15. The relation of *trans*-decalin to *trans*-1,2-dimethylcyclohexane.

The fusion of several cycloalkane rings to one another can lead to a variety of complicated molecules. One important example, a tetracyclic molecule, is shown in several representations in Fig. 2.16. This molecule, which appears to consist of three cyclohexane rings and a cyclopentane ring, comprises the fundamental carbon skeleton for a variety of biological materials including the sex hormones, constituents of the brain and spinal fluids, and pharmaceuticals such as cortisone, digitalis, and birth control medications. This long, relatively planar and conformationally fixed system has three ring junctures.

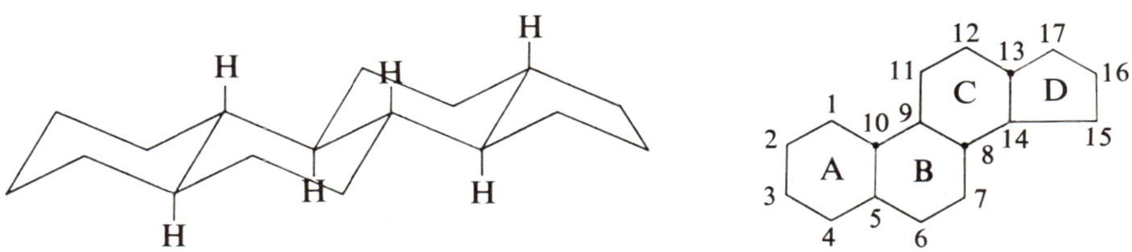

Fig. 2.16. The steroid ring system. Numbers of locating substituents are shown; the rings are usually referred to as A, B, C, and D.

FUNCTIONAL GROUPS

We have previously mentioned how an organic radical can be formed from a hydrocarbon by the removal of a hydrogen atom. The unsatisfied valence at the carbon atom where such a removal occurs can be used to bind other atoms or groups of atoms that may or may not be composed strictly of carbon and hydrogen. If groups of this nature are eventually the site of some chemical reaction, they are often termed *functional groups*. The hydrocarbon part of an organic molecule tends to be inert and it is usually transformations of the functional group that determine the chemistry of the molecule. The dividing line between the part of a molecule which is regarded as the functional group and the remainder of the molecule is a very arbitrary one and is dependent upon the context of the surrounding discussion. Consider the ethane molecule. If one of the hydrogen atoms is abstracted then one has an ethyl radical, CH_3CH_2. Replacement of the departed hydrogen atom by a bromine atom forms the compound CH_3CH_2Br, called ethyl bromide. Most of the interesting chemical reactions that ethyl bromide undergoes are reactions that involve the bromine atom. The bromine atom can thus be regarded as a functional group and, for many purposes, the atoms of the remainder of the molecule (the ethyl group) can be thought of as innocent bystanders to these chemical processes. The importance of the functional group concept is the following: to a crude first approximation, the reactions of organic molecules depend only upon the properties of the functional group and are independent of the hydrocarbon structure which holds the functional group. We expect, on this basis, that the reactions of propyl bromide $CH_3CH_2CH_2Br$ will be very similar to those of ethyl bromide, CH_3CH_2Br. The functional group idea also is a boon to the student of organic chemistry, for if we learn about the reaction of one alkyl bromide, to a large extent we have learned about the reactions of all alkyl bromides. In our discussions throughout the rest of this book there will be space to mention only a few of the many known reactions of each functional group.

PHYSICAL PROPERTIES OF ALKANES

The boiling points and melting points of a number of alkanes and cycloalkanes are given in Table 2.1. It can be seen that the low molecular weight materials (C_1–C_4) are gases at room temperature, while the higher molecular weight hydrocarbons ($>C_{12}$) are solids. These solids have a waxy texture; paraffin waxes consist of alkanes with 22 to 32 carbon atoms in their structure.

Table 2.1
Physical Properties of Alkanes and Cycloalkanes

Name	*Number of carbon atoms*	*Boiling point* (°C)	*Melting point* (°C)
Methane	1	−162	−183
Ethane	2	−89	−183
Propane	3	−42	−188
Cyclopropane	3	−33	−90
n-Butane	4	−1	−135
Cyclobutane	4	13	−90
Isobutane	4	−12	−160
n-Pentane	5	36	−130
Cyclopentane	5	49	−94
Isopentane	5	28	−160
Neopentane	5	10	−17
n-Hexane	6	69	−95
Cyclohexane	6	81	7
n-Heptane	7	98	−91
n-Octane	8	126	−57
n-Nonane	9	151	−54
n-Decane	10	174	−30
n-Undecane	11	196	26
n-Dodecane	12	216	−10
n-Tridecane	13	234	−6.2
n-Tetradecane	14	252	6
n-Eicosane	20	334	36
n-Tricontane	30	446	66

REACTIONS OF ALKANES

The open chain and cyclic alkanes are quite unreactive hydrocarbons. Possibly their most familiar reaction is combustive oxidation, a reaction with oxygen that takes place at high temperatures. The reaction of methane with oxygen is an example. The enthalpy change, $\Delta H°$, for this combustion is −212 kcal/mole.

$$CH_4 + 2\,O_2 \longrightarrow CO_2 + 2\,H_2O$$

Thus, the oxidation of methane and other lower molecular weight hydrocarbons is a convenient source of energy, and huge quantities of hydrocarbons are used as fuels for this reason.

The major commercial source of hydrocarbons is petroleum. Crude petroleum consists of hundreds of different hydrocarbons from methane to those constructed from 40 or more carbon atoms. This complex mixture of substances is separated by fractional distillation as part of a process known as refining.

Commercial natural gas consists primarily of methane and ethane, while automotive gasoline is composed of various hexanes, heptanes, octanes, nonanes, and decanes.

"An abundant supply of low-cost energy is the key ingredient in continuing to improve the quality of our total environment."

"In other words, those zealots who propose going on without . . . increasing energy are not speaking for you or for most of the people of the country."

"They are not speaking for themselves, either, because they go home and turn on the air conditioner."

This exchange took place between a Congressman and a science advisor to the President as concern about the effects of a technological society on the human experience began to mount in the early 1970's. It is apparent that two powerful forces within industrialized societies are headed for a collision that could do damage to all. The first is an insatiable appetite for energy to run the commerce, transportation systems, and gadgets that have come to characterize such societies. The second force sees that these activities have profound, often harmful effects on the fragile environment for human existence provided by the Earth and hopes to reverse these effects. The overall consumption of energy has grown five percent a year since 1965 and demand for electricity, coal, and petroleum products is seriously threatening to outrun supply. Attempts to correct this situation have encountered increasing resistance because of the environmental impact of the proposed solutions. What the outcome of the struggle these two forces will be is not yet clear.

A major share of petroleum products goes for the operation of automotive vehicles. The combustion properties of the complex mixture of hydrocarbons known as gasoline are usually represented by an octane number. Octane ratings are measured by comparing the combustion of a hydrocarbon mixture in a test engine to the combustion of a blend of controlled amounts of 2,2,4-trimethylpentane and heptane. Heptane has been assigned an octane number of 0, while 2,2,4-trimethylpentane, a smooth-burning fuel, has an octane number of 100. If 90 parts of 2,2,4-trimethylpentane and 10 parts of heptane combust in the same way as the test mixture, the mixture is said to have an octane rating of 90. The octane requirements for knock-free operation of an automobile engine depend upon the compression ratio of the engine. At a compression ratio of 4:1 an octane rating of 60 is adequate, but an octane rating of 98 is required for smooth operation at a compression ratio of 10:1.

Until recently, gasoline manufacturers increased the octane rating of their products by adding rather substantial amounts of tetraethyl lead and other additives to their products. Up to three grams of tetraethyl lead per gallon of gasoline have been

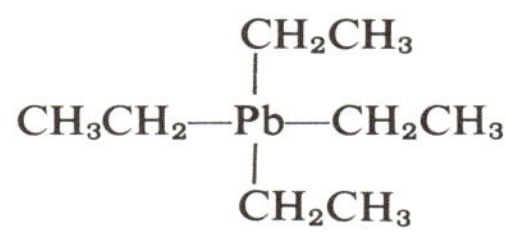

Tetraethyl lead

added to give mixtures with octane numbers up to 100. These added compounds must go somewhere as the gasoline containing them is combusted; a large fraction end up as aerosols in the atmosphere. Increasing concern about lead poisoning and other effects of these aerosols have forced reconsideration of their inclusion in gasoline.

Alkanes and cycloalkanes are attacked by species with unpaired electrons. Common reactions of this type involve the halogens, chlorine and bromine. The reaction of chlorine with ethane proceeds in this manner to yield ethyl chloride. This is a substitution reaction; an atom of chlorine has replaced or

$$CH_3CH_3 + Cl_2 \xrightarrow{h\nu} CH_3CH_2Cl + HCl$$

substituted for one of the hydrogen atoms of the ethane molecule. The symbol $h\nu$ means this reaction proceeds in the presence of light. The actual progress of the reaction is not so simple as this overall equation would indicate, for it appears to involve several steps. The first is the formation of chlorine atoms by the interaction of a chlorine molecule (Cl_2) with light.

$$Cl_2 \xrightarrow{h\nu} 2\,Cl\cdot \tag{2.1}$$

Chlorine atoms have unpaired electrons (an unsatisfied valence) and can be considered radicals in the same sense as the term has been used previously. Species with this electronic configuration are expected to be very reactive. The hydrocarbon ethane reacts with one of these chlorine atoms in the next step of the reaction to give

$$CH_3CH_3 + Cl\cdot \longrightarrow HCl + CH_3CH_2\cdot \tag{2.2}$$

an ethyl radical and hydrogen chloride. The ethyl radical reacts with a second molecule of chlorine to form another chlorine atom and ethyl chloride.

$$CH_3CH_2\cdot + Cl_2 \longrightarrow CH_3CH_2Cl + Cl\cdot \tag{2.3}$$

The concentration of Cl· is generally low, so it is more likely that the ethyl radical will encounter a chlorine molecule rather than a chlorine atom. The chlorine atom generated in this third step can abstract another hydrogen atom from a second molecule of ethane [reaction (2.2)] and the sequence of reactions can start again. This sequence defines a *chain reaction.* Reaction (2.2) generates a reactive species ($CH_3CH_2\cdot$) for use in reaction (2.3), but reaction (2.3) produces a necessary reagent (Cl·) for reaction (2.2). As soon as one chlorine atom is formed by reaction (2.1), a process generally called initiation, the overall reaction of chlorine with ethane is self-sustaining (chain-propagating steps) because of the interconnection of the next two reactions. The overall reaction will continue until all of the ethane is used up or until two of the radicals involved in the reaction encounter one another; two radicals will react very rapidly with one another to give a molecule. These latter events remove radicals from the cyclic part of the reaction and are called *termination steps.* Collecting the above steps, we see that the chain reaction can be summarized in the following way.

$$Cl_2 \xrightarrow{h\nu} 2\,Cl\cdot \qquad \text{(Initiation)}$$

$$\begin{aligned} CH_3CH_3 + Cl\cdot &\longrightarrow CH_3CH_2\cdot + HCl \\ CH_3CH_2\cdot + Cl_2 &\longrightarrow CH_3CH_2Cl + Cl\cdot \end{aligned} \qquad \text{(Chain-propagating steps)}$$

$$\begin{aligned} 2\,Cl\cdot &\longrightarrow Cl_2 \\ CH_3CH_2\cdot + Cl\cdot &\longrightarrow CH_3CH_2Cl \\ 2\,CH_3CH_2\cdot &\longrightarrow CH_3CH_2CH_2CH_3 \end{aligned} \qquad \text{(Possible termination reactions)}$$

A series of fundamental reaction steps that accounts for an overall chemical transformation is called a *mechanism* for the transformation. The reaction of ethane, and other alkanes, with chlorine is said to proceed by a radical-chain mechanism. As suggested above, the fundamental steps in a radical-chain mechanism are initiation, propagation, and termination.

The reaction of an alkane with bromine or chlorine by this mechanism does not necessarily have to halt after the introduction of one halogen atom. In practice, compounds with additional halogen atoms present are formed; the exact nature and amount of these products depends upon the conditions under which the reaction is run.

Other radicals such as $\cdot NO_2$, $\cdot SO_3$, $\cdot OH$, and $\cdot R$ (where R is a general hydrocarbon radical) can abstract hydrogen atoms from alkanes and thus initiate a chain reaction. Radical reactions thus provide a way for the organic chemist to introduce reactive functional groups into a hydrocarbon molecule.

PROBLEMS

1. Give a name for the following compounds.

(a) 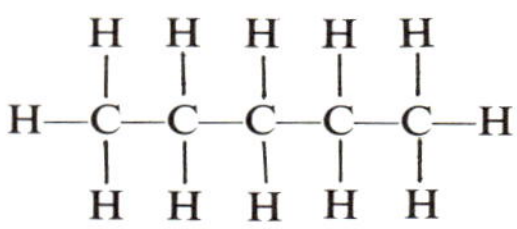

(b) 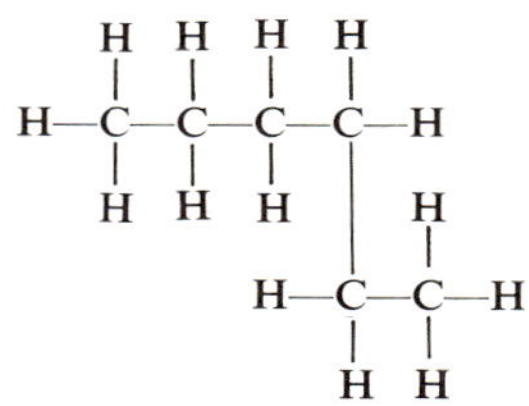

(c) 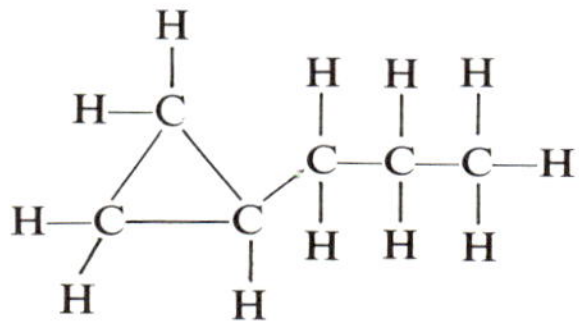

(d) 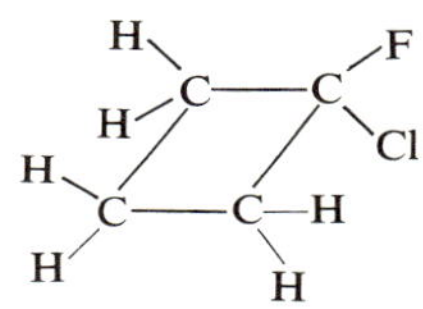

(e) CH_3Cl

(f) CH_2I_2

(g) CH_3CCl_3

(h) CH_2BrCH_2Cl

(i) 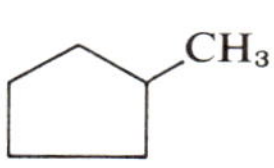

(j) 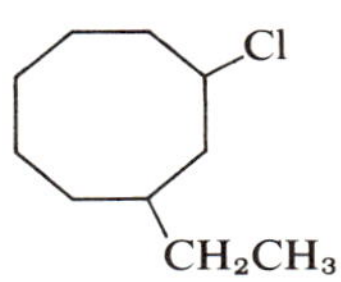

(k) 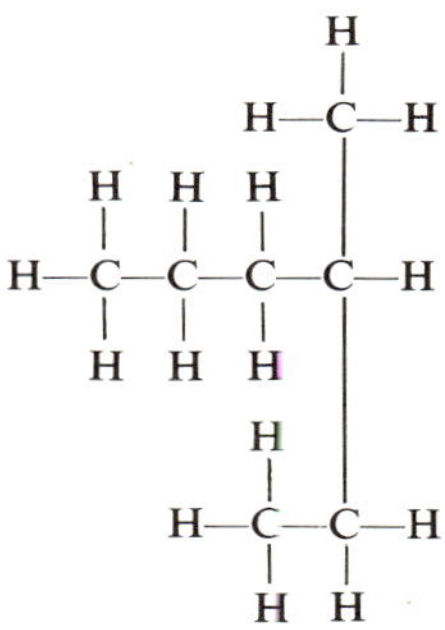

(l) $CH_3CH_2CCl_2CH_3$

(m) 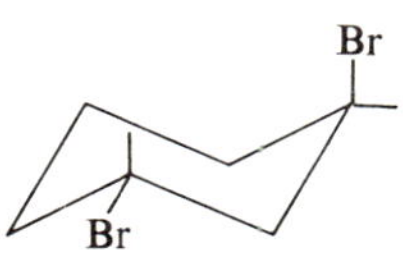

(n) 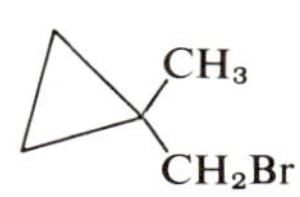

(o) $CH_3(CH_2)_8CH_2Cl$

(p) 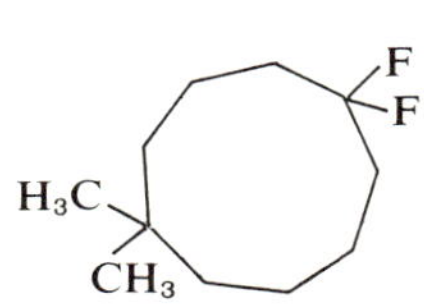

2. Sketch the organic structural formula for the following compounds.
   (a) Cycloheptylchloride
   (b) 3-Ethyl-2,5-dimethylhexane
   (c) 1-Bromo-3-methylpentane
   (d) 1,1-Dibromocyclobutane
   (e) 1-Ethyl-4-methylcyclooctane
   (f) *cis*-1-Ethyl-2-methylcyclohexane

3. Write the structure of all possible bromopentane ($C_5H_{11}Br$) molecules and name them.

4. A compound which contains only carbon, hydrogen, and chlorine atoms gave the following elemental analysis: C, 38.4% and H, 4.8%. What are five possible structures for this compound?

5. Acceptable "ball-and-stick" molecular models can be constructed from gumdrops and toothpicks. Obtain a supply of these materials (with proper allowances for nibbling) and construct models of ethane, cyclopropane, and cyclohexane.

6. Sketch "sawhorse" and Newman projection drawings of ethane, 1,1-diiodoethane, and 1,2-diiodoethane. What is likely to be the preferred conformation of the last molecule?

7. Make a "sawhorse" drawing of 2,2,5,5-tetramethylhexane. What is the conformation of lowest energy?

8. Estimate ΔH° for each step of the radical-chain mechanism for the chlorination of ethane using the table of bond energies in Chapter 1.

9. Estimate how much heat will be evolved upon combustion of one gallon of octane? (This roughly is equivalent to the heat given off when one gallon of gasoline is combusted.)

10. One liter of a gaseous hydrocarbon at STP weighs 1.96 g. What is the molecular weight of the hydrocarbon?

11. Considering the three-dimensional structure, suggest a reason or reasons for the fact that neopentane melts at $-17°C$, while its structural isomer isopentane melts at $-160°C$, over 140°C lower.

12. Draw a free-energy reaction coordinate diagram for the ring-flip "reaction" of cyclohexane. The ground state of the molecule is the chair form and the boat form is an intermediate in the interconversion process.

SUGGESTED READINGS

ANONYMOUS, Pollution. *Chem. Eng. News* **47**, 33 (1969).

O. T. BENFEY, "The Names and Structures of Organic Compounds." Wiley, New York, 1966. (A programmed text.)

E. L. ELIEL, N. L. ALLINGER, S. J. ANGYAL, AND G. A. MORRISON, "Conformational Analysis." Wiley (Interscience), New York, 1965.

B. F. GREEK, Gasoline. *Chem. Eng. News* **48**, 52 (1970).

C. D. HURD, *J. Chem. Educ.* **34**, 43 (1961). (A discussion of organic nomenclature.)
R. T. MORRISON AND R. N. BOYD, "Organic Chemistry," 2nd ed, pp. 3–142. Allyn & Bacon, Boston, Massachusetts, 1966.
W. A. PRYOR, "Free Radicals." McGraw-Hill, New York, 1957.
"Resources and Man," A Report of the NAC-NRC Committee on Resources and Man. Freeman, San Francisco, California, 1969.
J. D. ROBERTS AND M. C. CASERIO, "Basic Principles of Organic Chemistry," pp. 3–22, 63–120. Benjamin, New York, 1964.
E. H. TINLEY, "Naming Organic Compounds." Noyes Press, Pearl River, New York, 1962.
W. S. TRAHANOVSKY, "Functional Groups in Organic Compounds." Prentice-Hall, Englewood Cliffs, N.J., 1971.

3

Alkenes, Cycloalkenes, and Alkynes

In the previous chapter we briefly discussed the properties of a general class of compounds known as alkanes. The single most important distinguishing feature of the alkanes is the fact that all of the carbon atoms in an alkane are sp^3-hybridized, with the possible exception of those in the very small ring compounds. A second important class of hydrocarbons, called alkenes or olefins, is considered in this chapter. These substances are characterized by the occurrence of pairs of sp^2-hybridized carbon atoms in the skeleton of the molecule.

ALKENES

Two directly bonded sp^2 carbon atoms define a chemical bond system known as a double bond. In order to see why this terminology is appropriate, let us consider the manner in which the covalent bonds in the simplest alkene, ethylene (C_2H_4), are formed. The sp^2 orbitals on each carbon atom can be used to bind the hydrogen atoms to the molecule and to form one bond between the carbon atoms, as suggested in Fig. 3.1. This system of single bonds defines a framework for the ethylene molecule. The generalization that carbon always forms four bonds to other atoms can be satisfied in ethylene if a second chemical bond is formed by overlap of the two unhybridized $2p$ orbitals at each carbon atom.

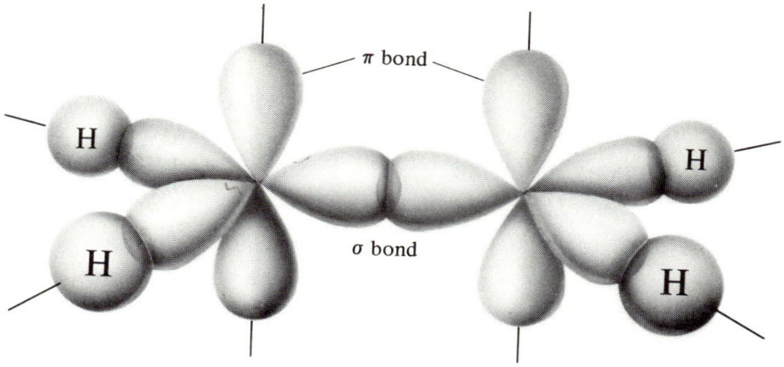

Fig. 3.1. Bonding in the ethylene molecule.

This second carbon–carbon bond is fundamentally different from the first bond which joined the two carbon atoms. The first bond was fashioned by the end-on overlap of two atomic orbitals; bonds created in this way are called σ (sigma) bonds. The second bond, called a π (pi) bond, was constructed by overlapping *p* orbitals in a sideways manner so that there were two points of contact between the atomic orbitals. The electrons in this second carbon–carbon bond tend to be distributed over a larger volume than those of the σ bond. As we shall see shortly, this fact has important implications for the chemical reactivity of the alkenes.

By tradition, compounds in which all of the carbon atoms present are bound to four other groups are called saturated compounds. Materials, like the alkenes, that have one or more multiple bonds are *unsaturated*, the implication being that these substances can react further in some way to give a saturated molecule.

The π-bond component of the double bond is weaker than the σ bond, but the two bonds acting together comprise a system which is more stable than a typical carbon–carbon single bond found in an alkane. It can be noted from Table 1.4 that the bond energy for the carbon–carbon double bond is about 146 kcal/mole. The π bond apparently contributes about 63 kcal/mole of stabilization to a double bond. One consequence of the greater strength of the carbon–carbon double bond is an appreciable shortening of the carbon–carbon internuclear distance. In ethane, this distance is 1.54 Å, but, in ethylene, the carbon–carbon distance is 1.34 Å.

Cis-Trans Isomerization in Alkenes

The ease of conformational interchanges in ethane, brought about by rotation around the carbon–carbon bond, has been previously discussed. A very different situation obtains in the case of ethylene. Recall that the π component of the double bond in ethylene is formed by the overlap of two *p* orbitals. By the principle of maximum overlap, this π bond will be strongest when the orbitals are oriented relative to one another as shown in Fig. 3.1. Any attempt to rotate one end of the ethylene molecule relative to the other end will tend to seriously diminish the extent to which the two electron clouds of the *p* orbitals can overlap (Fig. 3.2). This motion must lead to an increase in the energy of the molecule; the energy barrier to the process should be roughly 60 kcal/mole, the approximate energy of the π bond that is destroyed during rotation. This energy barrier is so large that, at moderate temperatures, rotation about the carbon–carbon bond in ethylene does not occur to any detectable extent, and it is only at very high temperatures or in the presence of certain catalysts that this rotation can take place rapidly.

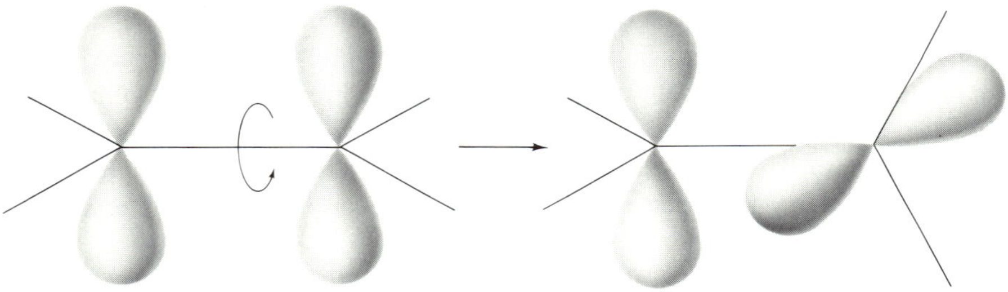

Fig. 3.2. The effect of carbon–carbon rotation on the ethylene π bond.

An immediate consequence of strongly restricted rotation in the ethylene molecule is the occurrence of geometrical isomerism. A simple example is provided by 2-butene. In one form of 2-butene, called the *cis* isomer, both of the terminal methyl groups are on the *same side* of the double bond. In the *trans* isomer, they are on opposite sides of the double bond. The *cis* and *trans* isomers of 2-butene are shown in Fig. 3.3. These two isomers of 2-butene could be interconverted by rotation but, since this process is not significant at ordinary temperatures, these two isomers can be isolated from one another and treated as separate, stable entities.

Steric effects play a part in determining the relative stabilities of geometrical isomers. As the illustration in Fig. 3.4 shows, there are nonbonded interactions between the methyl groups in *cis*-2-butene that are absent in the *trans* compound. As a result the *trans* isomer is more stable than the *cis* isomer. In general, the *trans* isomer of an appropriately substituted olefin is the more stable one.

In order to isomerize or interconvert the *cis* and *trans* forms of a set of olefinic

Fig. 3.3. *Cis–trans* isomerism in 2-butene.

geometrical isomers, it is necessary to break up the π component of the double bond. As mentioned above, this can be done by supplying roughly 60 kcal/mole of energy to the molecule. A way that molecules can obtain extra energy is by the absorption of light. There are high energy or "excited" electronic quantum states for the π electrons in a double bond that are characterized by a distribution of electronic charge that is not conducive to bonding between the constituent atoms. These states are called antibonding states and, when the molecule is in one of these excited electronic states, the π bond is greatly weakened or entirely disappears. When an alkene absorbs light of the correct wavelength it can obtain enough energy to undergo a transition from the π-bonding to the π-antibonding electronic state and isomerization can take place. This is an example of a photochemical reaction; photochemical reactions are reactions promoted by the absorption of light.

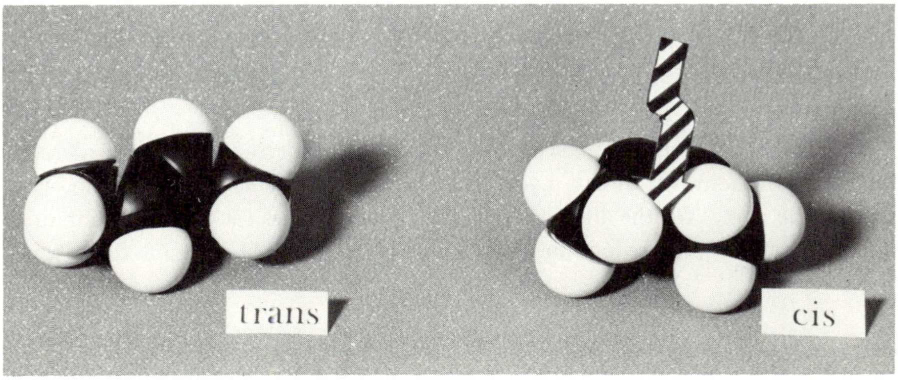

Fig. 3.4. Steric effects in *cis* and *trans* isomers of 2-butene. The arrow indicates the near-contact of the methyl groups in the *cis* isomer.

An interesting application of photochemical isomerization is found in the reactions that are a part of the chemistry of vision. A red pigment called rhodopsin is involved in the process by which light is perceived by the eye. Rhodopsin consists of the olefinic compound shown below in association with a colorless

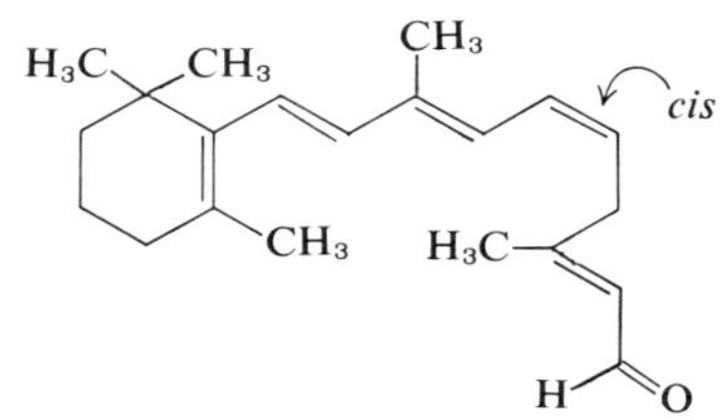

11-*cis*-Retinal (red)

protein called opsin. When light strikes rhodopsin, this olefinic material is photochemically isomerized to the corresponding all-*trans* isomer. The shape of the *trans* molecule is drastically different from that of the crimped *cis* isomer so that the nature of its interaction with the protein, opsin, is considerably altered. Additional chemical changes leading to nerve excitation

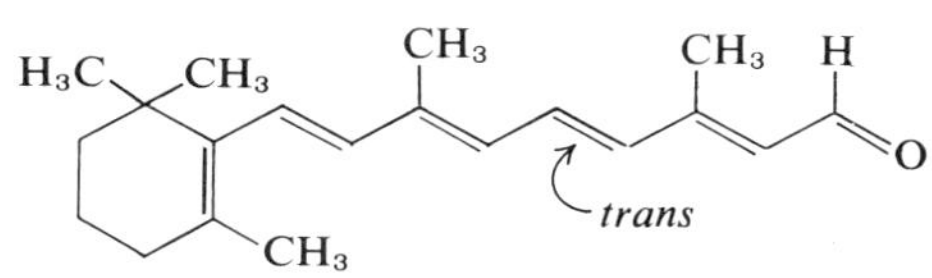

all-*trans*-Retinal

and behavioral response are consequences of this simple photochemical reaction.

Nomenclature for the Alkenes

The alkene, ethylene, can be thought of as being formed from the alkane, ethane, by the removal of two hydrogen atoms from the ethane molecule. The fact that the names for the alkenes are related to the names for the corresponding alkanes should therefore not be surprising. In the IUPAC system for alkene

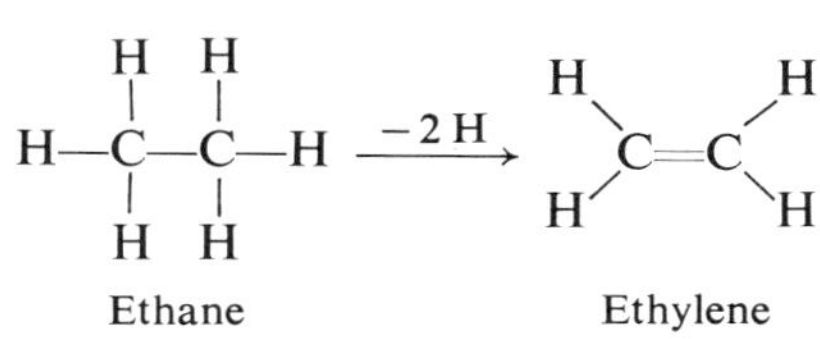

nomenclature, the basic, backbone carbon chain is chosen as the longest continuous chain of carbon atoms which contains the double bond. If there are several double bonds in the molecule, the chain with the maximum number of double bonds is chosen. The carbon atoms in the basic chain are then numbered so that the smallest numbers possible are assigned to the carbon atoms of the double bond(s). The chain is named by changing the suffix *-ane* in the name of the corresponding alkane to *-ene*. If there are two double bonds, the suffix added is *-adiene*, and for three double bonds in the chain, *-atriene* is appended. Finally, a number corresponding to the first carbon atom of the double bond is prefixed to the name so as to unambiguously locate the position of the double bond in the chain. Side chains are designated by the same procedures as used with the alkanes. Some examples of the application of these conventions for the naming of alkenes are shown below.

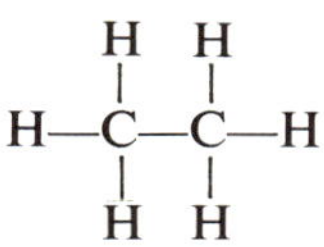
Ethane

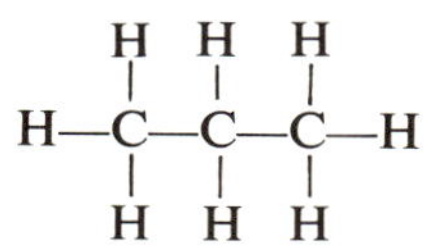
Propane

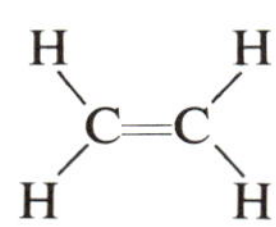
Ethene
(ethylene)*

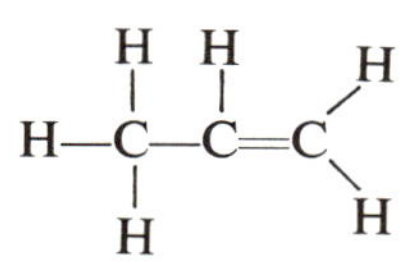
Propene
(propylene)*

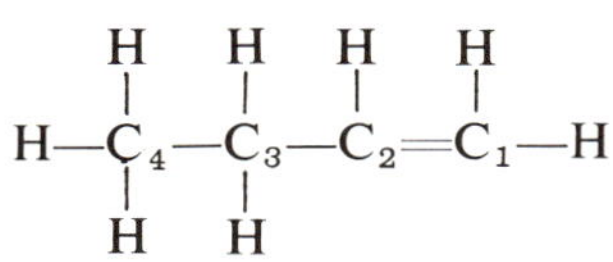
1-Butene

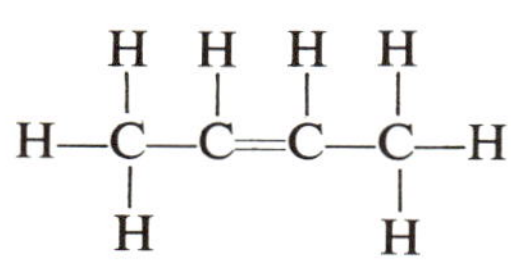
2-Butene

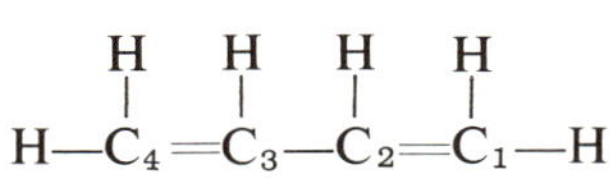
1,3-Butadiene

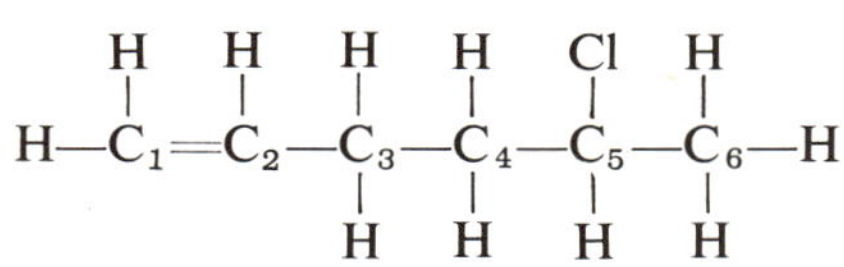
5-Chloro-1-hexene

* These common or trivial names are widely used.

As with the alkanes, one can visualize the abstraction of a hydrogen atom from an alkene to form a hydrocarbon radical. Radicals generated from alkenes are named by changing the final -e in the name of the alkene to *-yl*. The radicals derived from several alkenes have trivial names that are also considered correct in the IUPAC system.

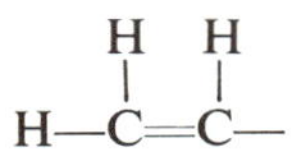

Vinyl

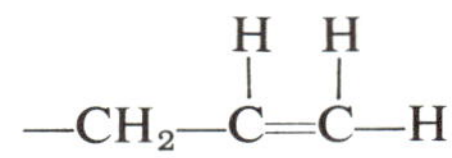

Allyl

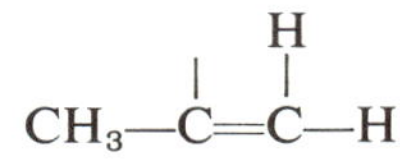

Isopropenyl

Thus, 1-bromoethene is more usually called vinyl bromide while 3-fluoropropene would be referred to as allyl fluoride.

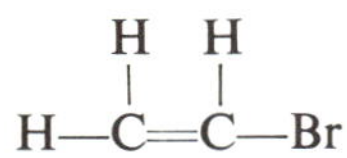

Vinyl bromide
(1-bromoethene)

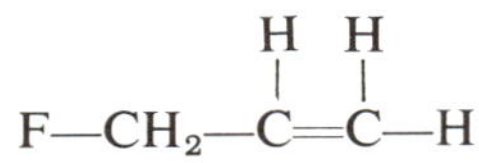

Allyl fluoride
(3-fluoropropene)

The *cis–trans* nomenclature which essentially indicates the positions of groups relative to some reference plane through a molecule, and has been used up to this point to denote the relative orientations of groups attached to cycloalkane rings or to carbon–carbon double bonds, becomes ambiguous if there are more than two substituents other than hydrogen attached to a double bond. For example, *cis*-1,2-dichloro-1-fluoroethylene could mean either of the two molecules shown below, depending upon one's frame of reference.

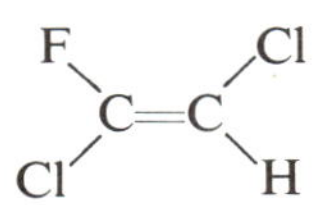

E

or

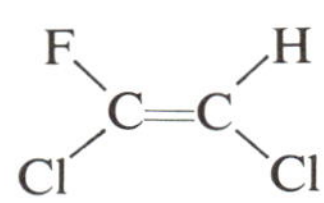

Z

In these cases one examines the substituents attached to the double bond and assigns to each a priority number. Priorities are based on the atomic number of the atom of a group which is bound directly to the double bond; the atom with the largest atomic number has the highest priority. In case two groups are attached to the double bond through the same kind of atom, then the next atom removed from the double bond is used in determining the priority ranking of the substituent. If the two groups of highest priority at each end of the double bond are on the same side, the isomer is designated as *Z* (from the German word *zusammen*, together). If these two groups are on opposite sides of the double bond, the isomer is called *E* (from the German word *entgegen*, opposite). The first molecule shown above is thus (*E*)-1,2-dichloro-1-fluoroethylene, while the other is the *Z*-isomer.

CYCLOALKENES

The carbon–carbon double bond can also be incorporated into cyclic structures. When the number of carbon atoms in the rings is less than eight carbon atoms, these compounds must necessarily have the *cis* configuration. The reason for this can be seen from molecular models and is illustrated schematically below. The loop of carbon atoms which is attached to the double bond must

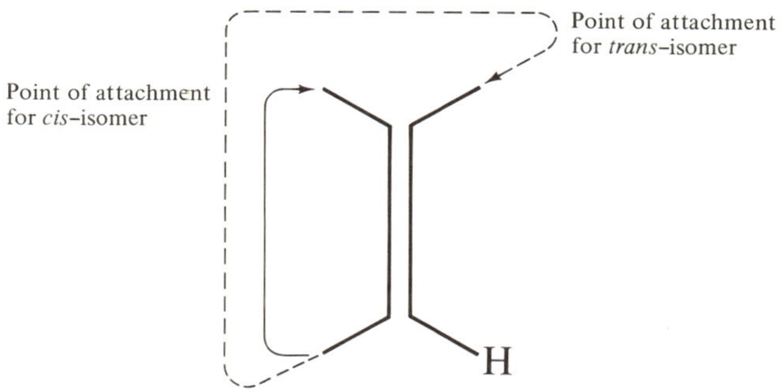

exceed a certain length so that its connection to the double bond can be made in a *trans* manner. It is only when this loop is longer than six carbon atoms that a *trans* hookup can be made.

The smaller cycloalkenes, cyclopropene, cyclobutene, and cyclopentene, are known compounds. The valence angles about the double bond in some of these compounds deviate substantially from the ideal 120° arrangement shown in Fig. 3.1. These materials are relatively high in energy and somewhat more

Cyclopropene Cyclobutene Cyclopentene Cyclohexene

reactive than noncyclic olefins with normal valence angles. In contrast, cyclohexene, though found only as the *cis* isomer, has essentially normal valence angles.

The ring conformation of this latter cycloalkene bears little resemblance to the conformation of the corresponding cycloalkane. The preferred arrangement of the atoms of the cyclohexene ring is shown in Fig. 3.5. This arrangement is often called the envelope conformation. A "ring-flip" motion takes place in the cyclohexene ring, but this motion is more subtle than the ring flip in chair-cyclohexane and occurs at a much higher rate.

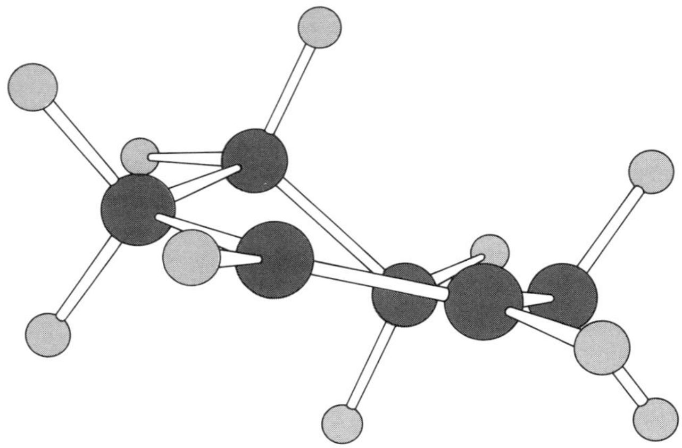

Fig. 3.5. A three-dimensional drawing of the envelope conformation of cyclohexene.

Cycloalkenes can also be formed by incorporating only one carbon atom of a double bond into a ring to give structures with so-called exocyclic double bonds. An example is methylenecyclohexane.

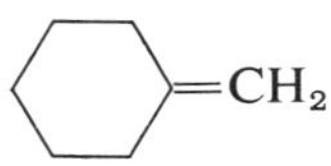

Methylenecyclohexane

Nomenclature for the Cycloalkenes

Systematic nomenclature for the cyclic olefins is a simple extension of the approach used with the open-chain alkenes. The basic name is obtained from the name of the corresponding open-chain alkene by adding the prefix *cyclo*-. If the ring contains only one double bond, then one of the carbon atoms at the double bond is called the 1-carbon. With this convention there is no need to specify the location of the double bond in the compound name. However, if two or more double bonds are present, they must each be specified with a number as in the case of the open-chain compounds. Some illustrative examples of these nomenclature rules follow.

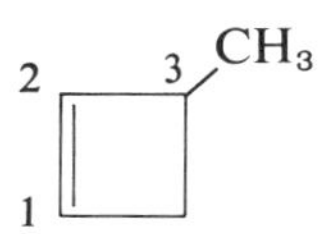

3-Methylcyclobutene

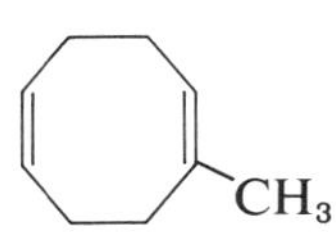

1-Methyl-1,5-cyclooctadiene

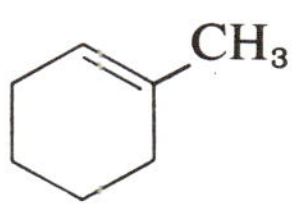

1-Methylcyclohexene

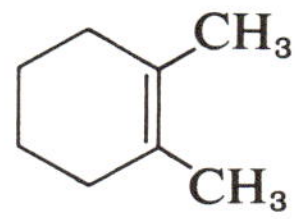

1,2-Dimethylcyclohexene

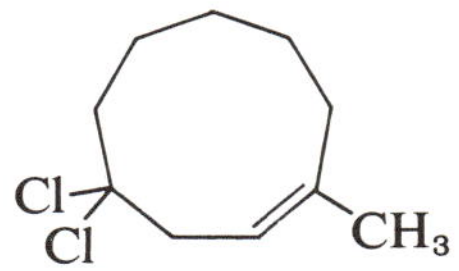

4,4-Dichloro-1-methylcyclononene

REACTIONS

Addition Reactions

The π bond of an alkene is readily attacked by chemical reagents that are attracted to localities of high electron density. Reagents that are particularly electron-hungry are called electrophiles, a name derived from the Greek words meaning "electron lovers." All substances that are positively charged are electrophiles and the chemical reactions of alkenes are more numerous and varied than are the reactions of alkanes. Reagents with unpaired electrons (radicals) also react readily with alkenes.

A common reaction of alkenes involves the destruction of the π component of the double bond followed by the attachment of new groups to the two carbon atoms that were originally connected by the double bond. Concomitant with this addition is rehybridization of the carbon orbitals to the sp^3 configuration. Since atoms are added to the basic skeletal structure of the molecule that initially contained the double bond, these reactions are termed addition reactions.

Addition of Halogens The reaction of bromine with ethylene is a common addition reaction. We note that the two bromine atoms of the starting bromine molecule are added to each carbon atom of ethylene to yield a substituted ethane.

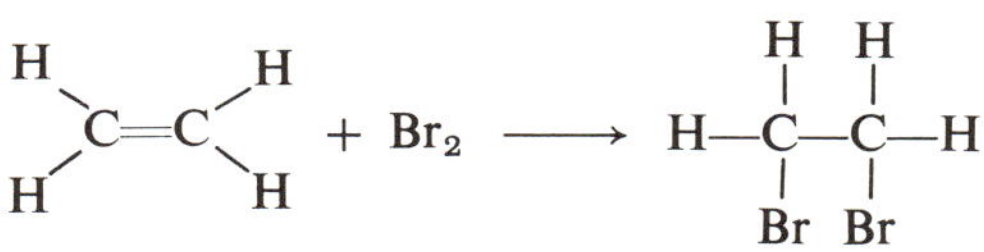

1,2-Dibromoethane

The mechanism of this addition reaction in solution probably involves the attack of an electrophilic bromine molecule on the π bond to give a positively charged intermediate ion with the structure shown. In a second step this positively

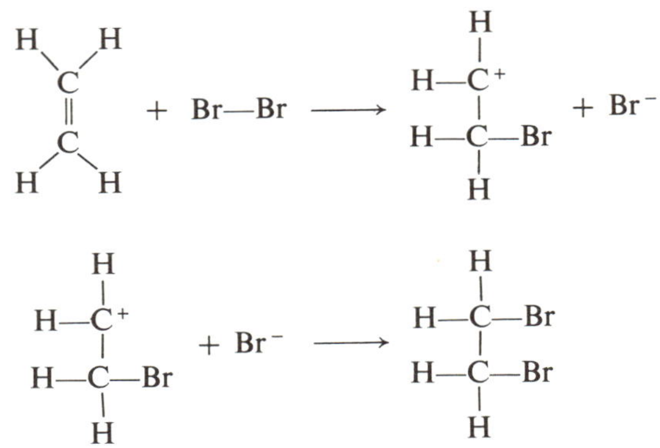

charged species (traditionally called a carbonium ion or, more recently, a carbocation*), is attacked by bromide ion to form the observed product. This process is typical of most organic reactions in at least two respects: (1) it involves several sequential reactions in the formation of the products, and (2) these reaction steps involve chemical species (intermediates) of low stability (high energy) relative to either the starting compounds or the products.

The addition of bromine to carbon–carbon double bonds proceeds smoothly with most olefins. A useful test for unsaturation utilizes a solution of bromine in carbon tetrachloride. In this solvent molecular bromine has a characteristic red appearance which disappears as the addition reaction with an unsaturated molecule takes place, since the dibromoalkane is colorless. The rates at which addition occurs vary with the exact structure of the olefin.

Chlorine and iodine can also undergo addition reactions. Chlorine reacts more rapidly than bromine, while iodine reacts considerably more sluggishly than bromine. Molecular fluorine (F_2) is very reactive and does not undergo addition to alkenes in a controllable way.

Addition of Polar Reagents. Polar reagents add to unsaturated carbon linkages by mechanisms that are similar to the addition reaction with bromine. The addition of hydrogen bromide (HBr) involves an intermediate that is formed by protonation of the olefin. The olefin thus acts like a Brønsted base. Subsequent attack on the protonated olefin completes the addition to give an

* Professor G. A. Olah, a pioneer in the chemistry of positively charged carbon, has suggested a more refined nomenclature for these species: see *J. Amer. Chem. Soc.* **94**, 808 (1972).

alkyl halide. Hydrogen chloride, hydrogen iodide, sulfuric acid, hypohalous acids (HOCl, HOBr), and water (under acidic conditions) will all add to olefins in a manner similar to that used by hydrogen bromide, but not necessarily by

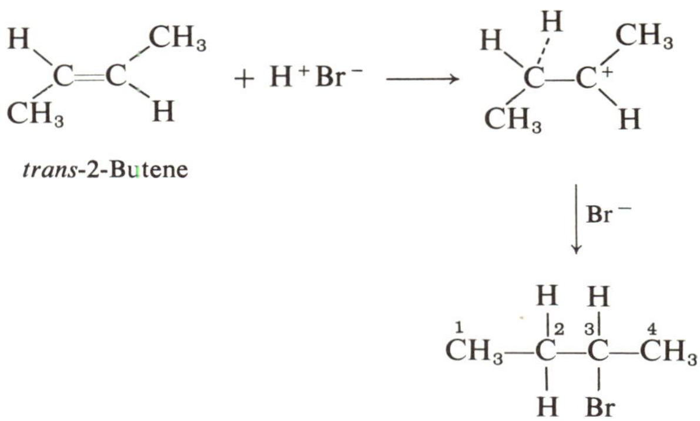

trans-2-Butene

2-Bromobutane

the same mechanism. All of these reagents are unsymmetrical; they are composed of a part which is (formally) positively charged and is thus electrophilic and a portion which is nucleophilic ("nucleus-seeking") by virtue of negative charge. The electrophilic portion of the adding reagent attacks the diffuse π-electronic system of the double bond, followed by the addition of the nucleophilic portion to complete the reaction. Examples of these reactions are shown below.

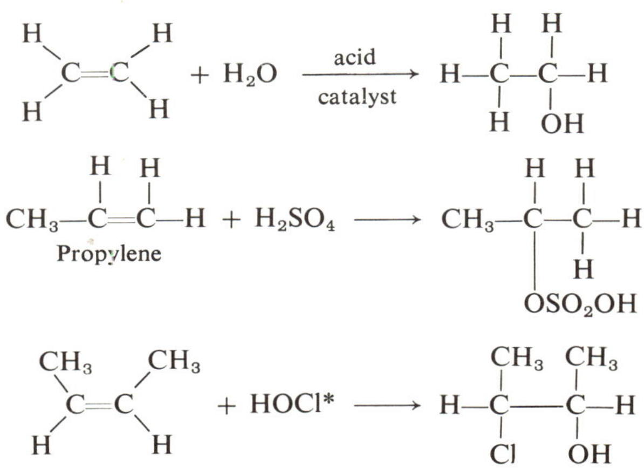

Propylene

cis-2-Butene

---

* The hypohalous acids can be regarded as dissociated according to the equilibrium

$$HOCl \rightleftharpoons HO^- + Cl^+$$

The positively charged halogen functions as an electrophile.

The compounds formed by the addition of sulfuric acid to olefins are called sulfate esters. They are often unstable and decompose to compounds in which the sulfate group (—OSO_2OH) has been replaced by a hydroxyl group (—OH).

A quandary arises when an unsymmetrical reagent like hydrochloric acid (HCl) adds to an unsymmetrical olefin (one that is not equivalently substituted at both carbons of the double bond). There are two possible products for this type of addition; examples are shown below. A generalization known as Markownikoff's rule summarizes the results of addition reactions of this type in which two structurally distinct products can be formed. This rule suggests that the positive portion of an unsymmetrical reagent will add to that carbon atom of a double bond which has the most hydrogen atoms already attached to it. By this rule, 1-chloro-1-methylcyclohexane and 1-chloro-2-hydroxypropane would be expected to be the major products of the two reactions depicted here.

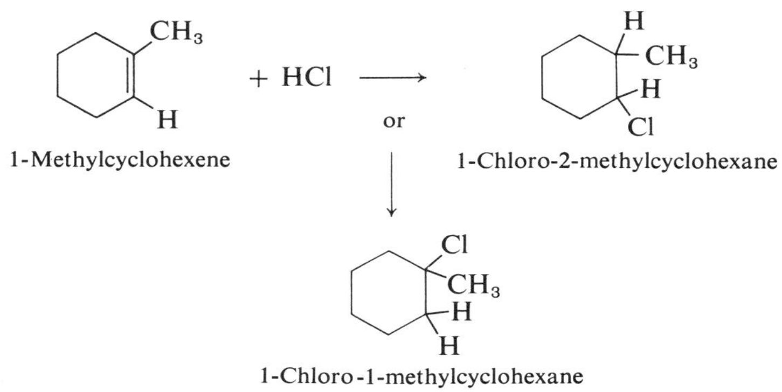

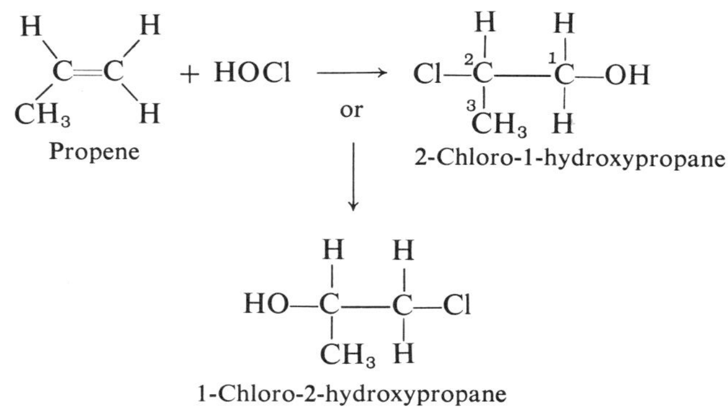

Historically, Markownikoff's rule was based on experimental observations, but it is now also defensible in terms of the expected properties of the intermediates involved in these addition reactions. The relative stabilities of the possible carbonium ions formed in the addition of unsymmetrical reagents to alkenes is the key to determining which of the two possible products will be formed in the greatest amount. There are three general classes of carbocations: primary, secondary, and tertiary ions. Primary carbonium ions have two hydrogen atoms attached to the positively charged carbon atom, secondary carbonium ions have one hydrogen attached, and tertiary carbonium ions, none. (The methyl carbonium ion, CH_3^+, is considered a primary ion.) It has been found

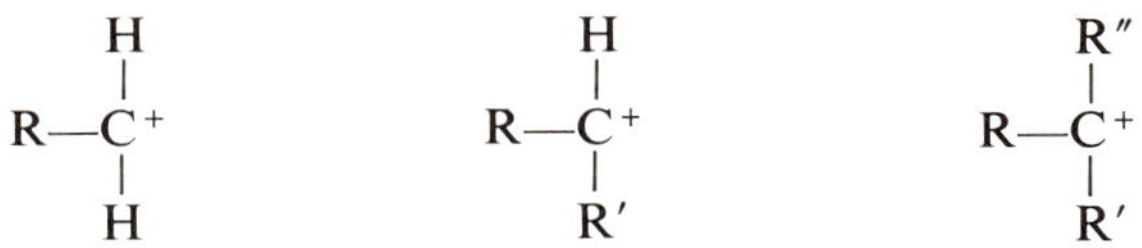

Primary carbonium ion — Secondary carbonium ion — Tertiary carbonium ion

(The symbols R, R′, R″, and the like are often used by organic chemists to represent general substituents)

experimentally that tertiary carbon cations are more stable than are secondary ions which, in turn, are more stable than primary cations. In the addition of HBr to propene (Fig. 3.6), a secondary carbonium ion intermediate would be formed by attachment of the electrophile (H^+) to the carbon at the double bond bearing only hydrogen atoms. If the electrophile becomes attached to the other carbon atom at the double bond, a primary cationic intermediate is formed.

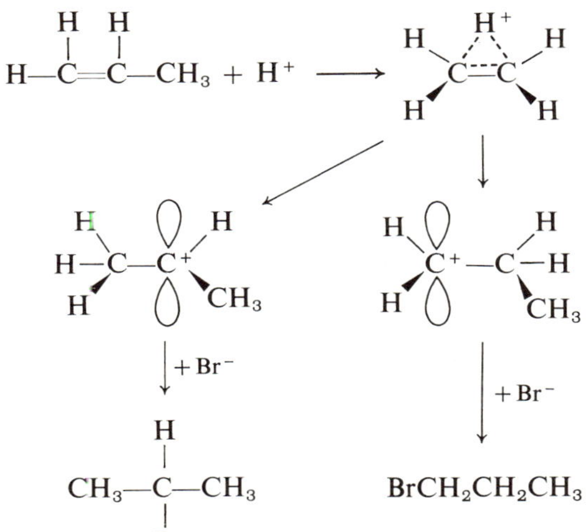

2-Bromopropane — 1-Bromopropane

Fig. 3.6. Possible intermediates and reaction pathways in the reaction of HBr with propene.

Since secondary carbonium ions are lower in energy than are primary carbonium ions, the formation of the first intermediate is greatly favored over the formation of the second and, as a result, the observed product of this electrophilic addition reaction is 2-bromopropane, *not* 1-bromopropane. This is the same product that would be predicted using Markownikoff's rule. This rule may be restated: the addition of an unsymmetrical reagent to an unsymmetrical double bond proceeds by the addition of the electrophile so as to give the product derived from the most stable carbonium ion intermediate.

Addition of Boron Hydrides Boron hydrides are compounds composed of boron and hydrogen. They add readily to alkenes to produce alkyl boron compounds. While the most common reagent of this nature is diborane with the formula B_2H_6, it is convenient to think of the addition reactions of diborane as involving BH_3. With propylene the reaction proceeds in three stages to give, ultimately, tri-*n*-propylborane.

$$CH_3{-}\overset{\overset{\displaystyle H}{|}}{C}{=}\overset{\overset{\displaystyle H}{|}}{C}{-}H + BH_3 \longrightarrow CH_3{-}CH_2{-}CH_2{-}BH_2$$

$$CH_3CH_2CH_2BH_2 + CH_3{-}\overset{\overset{\displaystyle H}{|}}{C}{=}\overset{\overset{\displaystyle H}{|}}{C}{-}H \longrightarrow (CH_3CH_2CH_2)_2BH$$

$$(CH_3CH_2CH_2)_2BH + CH_3{-}\overset{\overset{\displaystyle H}{|}}{C}{=}\overset{\overset{\displaystyle H}{|}}{C}H \longrightarrow \underset{\text{Tri-}n\text{-propylborane}}{(CH_3CH_2CH_2)_3B}$$

Note that boron is *less* electronegative than hydrogen so that it is the "positive" portion of the boron–hydrogen bond. The addition of diborane to alkenes thus follows Markownikoff's rule.

The trialkylboranes, in themselves, are not of great intrinsic interest to the organic chemist. However, they undergo a multitude of reactions and are of importance as synthetic intermediates. For example, they may be oxidized with mild reagents such as hydrogen peroxide (H_2O_2) in alkaline solution to give the corresponding hydroxyl-substituted compounds. In this manner, H_2O may be added to double bonds in a manner contrary to Markownikoff's rule. Treatment of the alkyl boranes with acid at high temperature leads to a reductive cleavage of the carbon–boron bond and the formation of an alkane.

$$(CH_3CH_2CH_2)_3B + 3\ H_2O_2 \xrightarrow{\text{base}} 3\ CH_3CH_2CH_2OH + \text{boric acid}$$

$$(CH_3CH_2CH_2)_3B + H^+ \xrightarrow{\text{heat}} \underset{\text{Propane}}{CH_3CH_2CH_3}$$

Radical Addition Reactions. Radical species (organic or inorganic substances with unpaired electrons) can also add to the carbon–carbon double bond. The products of these reactions are usually those *not* expected to be formed according to Markownikoff's rule because a different reaction mechanism and set of reaction intermediates are involved. An example is the *light-catalyzed* addition of HBr to propylene shown below. Radical addition reactions are usually more rapid than the multistep, polar additions discussed above.

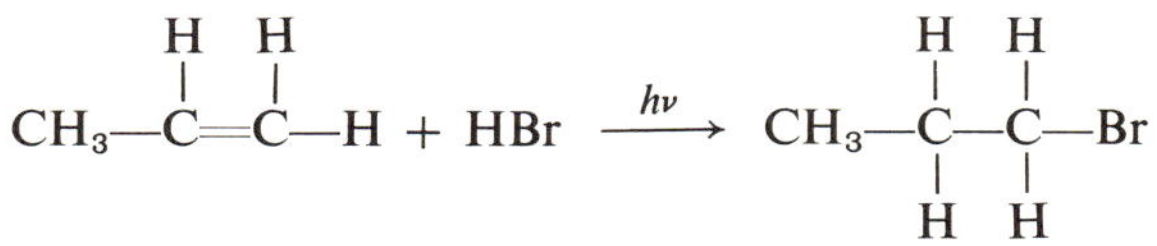

Propylene — *n*-Propyl bromide or 1-Bromopropane

Reduction

The addition of molecular hydrogen to an unsaturated hydrocarbon leads to the corresponding saturated hydrocarbon. From bond energy considerations, one might expect the products to be more stable than the reactants by about 30 kcal/mole. This reaction does not proceed at a noticeable rate in a homogeneous solution of alkene and hydrogen. However, it occurs readily if certain

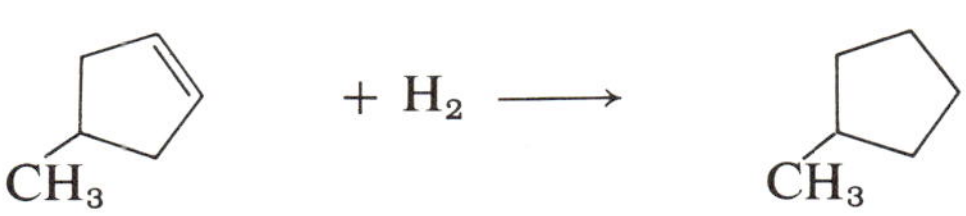

4-Methylcyclopentene — Methylcyclopentane

finely divided metals or alloys are included in the system as catalysts. The catalytic hydrogenation reaction is a heterogeneous one which apparently takes place on the surface of the metal catalyst; the molecular hydrogen–hydrogen bond is probably weakened by the catalyst so that absorbed alkenes (or other unsaturated molecules) can react more readily with it. The addition of both hydrogen atoms to the carbon–carbon double bond takes place from the same side of the double bond; this is called *cis* addition. The alkane molecule that is thus formed is then desorbed from the surface of the metal.

Catalytic hydrogenation is of considerable importance to the synthetic organic chemist. Commonly used catalysts include the metals platinum (Pt), palladium (Pd), nickel (Ni), and rhodium (Rh), and oxides of copper or chromium. These metal catalysts, like all catalysts, do not change the basic thermodynamic relationship between products and reactants, as far as relative stability is concerned, but merely change the rate at which the products are formed from the reactants. Pressures of hydrogen ranging from one to several thousand atmospheres are used.

Oxidation

Alkenes can be oxidized by a number of reagents. When ozone, O_3 (an allotrope of oxygen), is bubbled into a solution of an alkene, a complicated reaction occurs which results in the formation of a relatively unstable intermediate called an ozonide. Ozonides are not usually isolated but reduced (often with zinc and

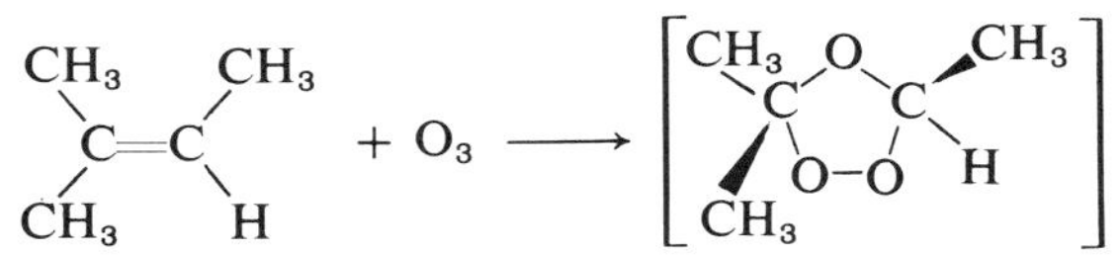

2-Methyl-2-butene An ozonide

acetic acid) to give compounds which contain oxygen atoms double-bonded to the carbon atoms that were initially part of the double bond. The $\rangle C{=}O$ system is called a carbonyl group; carbonyl-containing substances will be examined later in more detail. It is sufficient to note at this point that if the carbonyl compounds can be identified, then the structure of the alkene that was used in the ozonolysis reaction can be deduced.

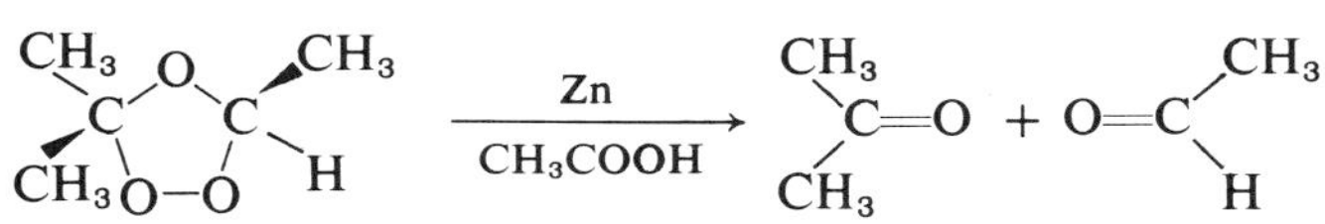

An ozonide

Other oxidizing agents tend to be less vigorous in their reactions with alkenes. Two such reagents are permanganate ion (MnO_4^-) and osmium tetraoxide (OsO_4), both oxides of heavy metals. The oxidation by permanganate can be controlled in such a manner that two separate stages of oxidation can be identified. As illustrated in the reaction shown below, permanganate oxidation in cold, dilute solutions at high pH can be visualized as the addition of "HO—OH" to the double bond. The product of this reaction, a dihydroxy substituted alkane, is usually called a *diol* or sometimes a *glycol*. The diol can be further oxidized under more vigorous conditions, namely, higher temperatures and a more neutral pH. The exact nature of the products of additional oxidation depend on

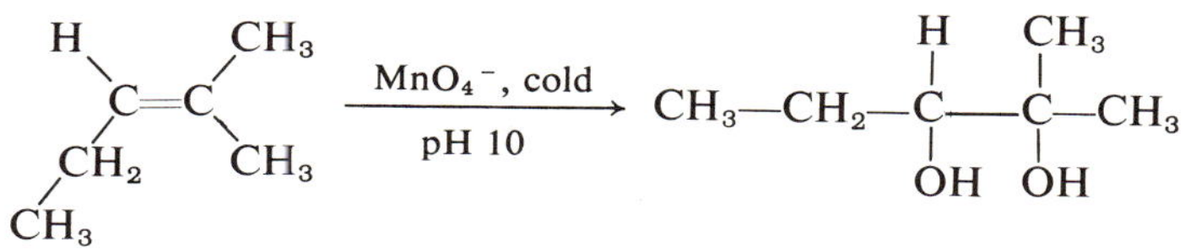

2-Methyl-2-pentene A diol

the structure of the starting alkene. Portions of the double bond which initially had the general structure

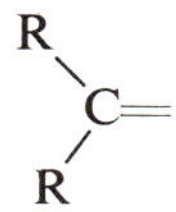

give carbonyl compounds

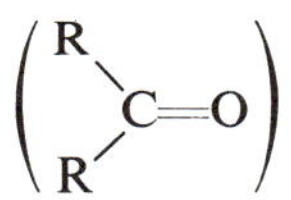

When the double-bond fragment is

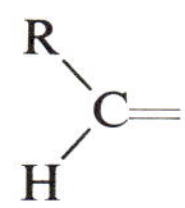

the oxidation product has the form

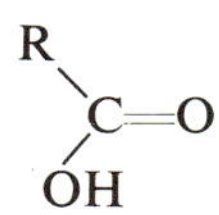

The functional group

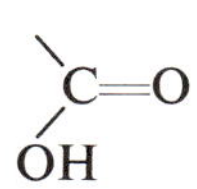

is called the organic acid or carboxyl group. Finally, if the double-bond fragment is

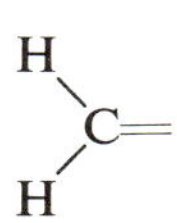

the products of oxidation by permanganate are carbon dioxide and water. These various modes of reaction are illustrated below.

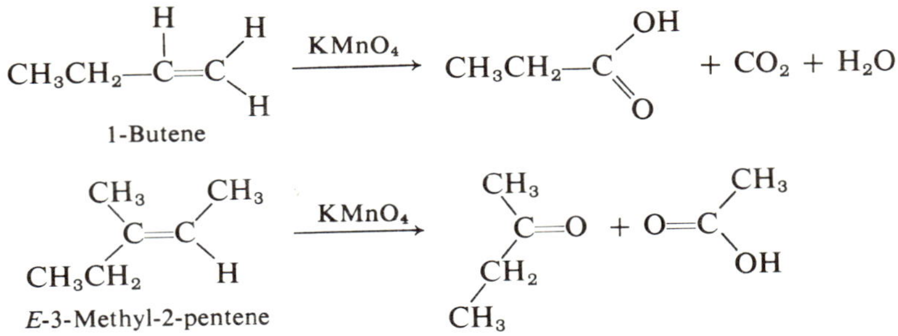

The oxidation of permanganate provides a convenient test for the presence of unsaturation in an organic molecule. As the oxidation reaction proceeds, the familiar purple color of the MnO_4^- ion is replaced by a turbid brown precipitate of MnO_2 which is formed during the reaction.

Osmium tetraoxide, OsO_4, is a colorless, low-melting solid that is soluble in organic solvents. Addition of an alkene, such as cyclohexene, to a solution of this oxidizing agent produces a black precipitate which likely has the structure shown below.

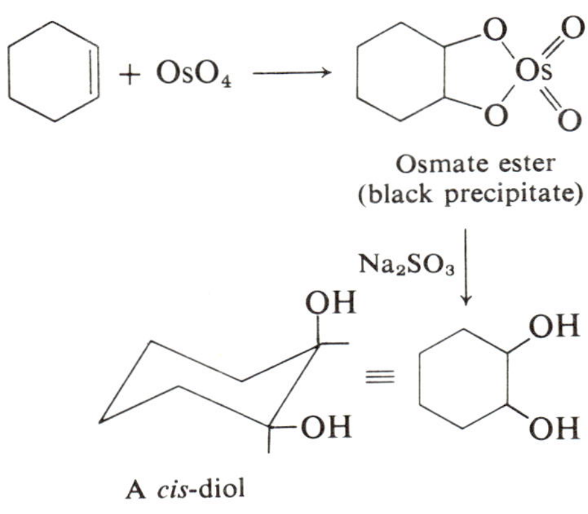

This cyclic intermediate, called an osmate ester, can be isolated. Upon treatment with a mild reducing agent (Na_2SO_3), the osmate ester is cleaved to produce a diol. It is likely that a similar manganate ester is involved in the permanganate oxidations.

Polymerization

A reaction of alkenes that is of very great commercial importance is their conversion to high molecular weight compounds called polymers. A polymer is operationally defined as a long-chain molecule with recurring structural subunits. These subunits are called *monomers*. Two subunits bonded to one another form a *dimer*, three attached subunits is a trimer, and a substance with many units attached together is a *polymer*. The polymerization of ethylene, shown schematically below, affords examples of the use of these terms.

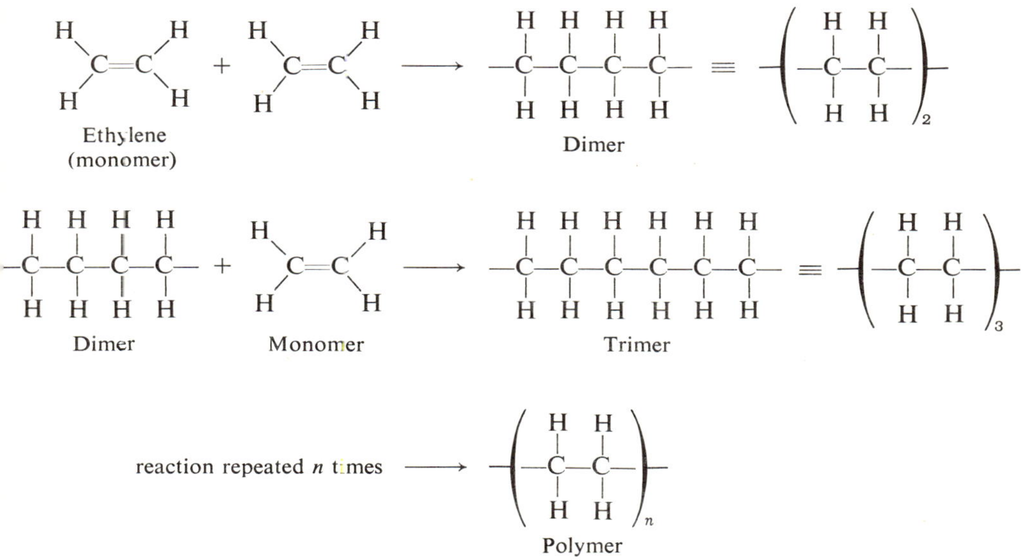

Because of their large molecular weight, polymers tend to be solids or viscous fluids. Polyethylene can be a liquid, a waxy semisolid, or a translucent solid, depending upon the size of the polymer molecule (the value of n) and the chemical reactions used to bring about polymerization. The solid material has a molecular weight of about 50,000.

The use of polymers as plastic structural material is so pervasive in our life that it is difficult to realize that polymers have attained widespread importance only since 1950. Table 3.1 lists the typical uses of polymers based on substituted ethylenes. This list barely begins to describe the number of different types of polymers that are commercially important; synthetic rubbers and fibers (rayon, nylon, orlon) are all polymeric materials and we shall see in subsequent chapters that the basic structural materials which support living systems are also polymeric.

Polymerizations of alkenes occur by chain mechanisms. It will be recalled from Chapter 2 that these mechanisms involve an initiation step, a propagation step, and a termination step. For alkene polymerization the initiator may be a radical, cationic, or anionic species. An example of the use of cationic initiators

Table 3.1
Some Polymers Based on Substituted Ethylenes

Monomer	*Polymer*	*Uses*
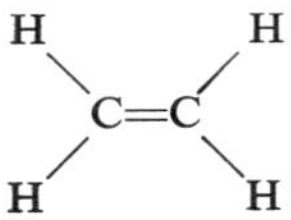 Ethylene	Polyethylene	Sheeting, plastic, bottles, toys, pipes
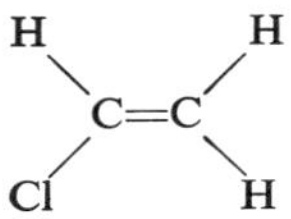 1-Chlorethene vinyl chloride	Polyvinyl chloride	Phonograph records, flooring tile
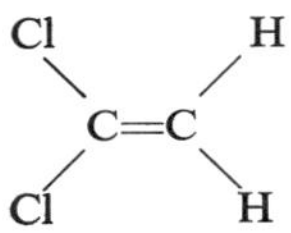 1,1-Dichlorethene	Polyvinylidene chloride	Food packaging (Saran)
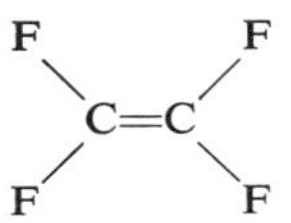 Tetrafluoroethylene	Polytetrafluoroethylene	Teflon
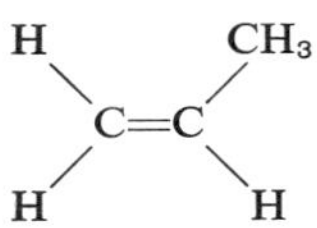 Propylene	Polypropylene	Outdoor–indoor carpet, fiber, films

would be the acid-catalyzed polymerization of ethylene as outlined in Fig. 3.7.

The gross physical properties of large molecules such as the polymers are to a considerable extent determined by the way in which carbon backbones of the molecules are folded and coiled. In synthetic polymers derived from simple alkenes, the conformation of the polymer can be more or less controlled by the manner in which the olefin is polymerized. The conditions include temperature, pressure, the concentration and type of catalyst, and the presence of more than one kind of monomer. We shall not digress at this point to consider the relationship between chain conformation and physical properties. However, we can distinguish three limiting conformational classes for large, chainlike molecules. These are illustrated in Fig. 3.8. In the first conformation, the long polymer chain is extended to its maximum length to give the appearance of a rigid rod. A second conformation would be the opposite of the first, namely, a completely random orientation in space of various segments of the chain. In the third limiting conformation, the chain can be coiled in some regular, often helical, manner.

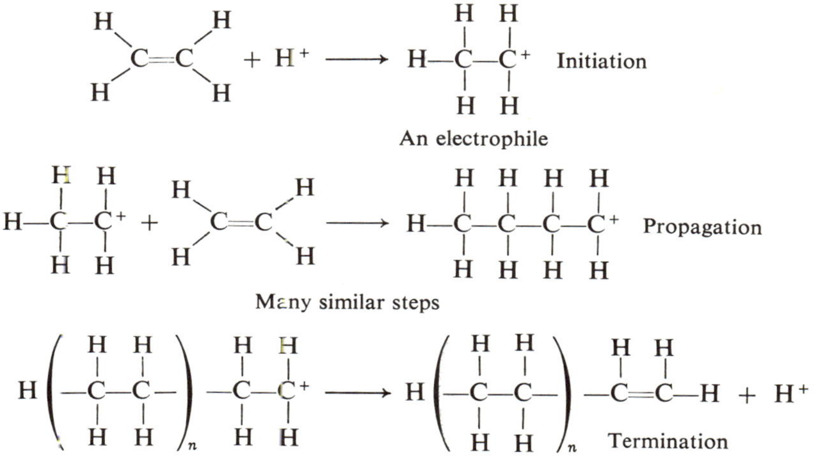

Fig. 3.7. The acid-catalyzed polymerization of ethylene.

It is possible for a polymer chain to have sections that are of one conformational type, e.g., random, and other sections that have different conformational attributes.

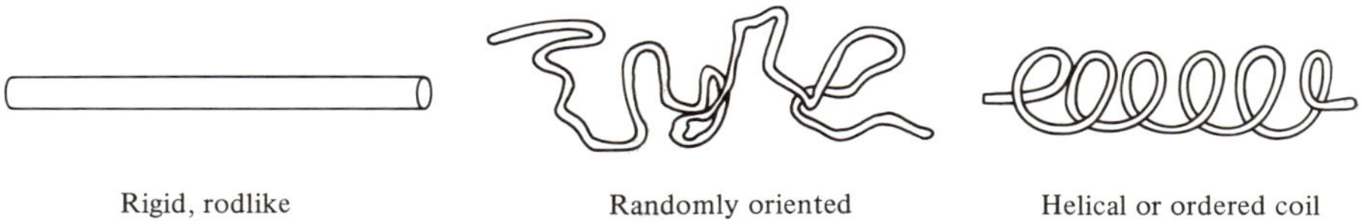

Fig. 3.8. Some conformations for long, flexible molecules.

ALKYNES

Another kind of unsaturated carbon–carbon bond can be formed by the interconnection of two *sp*-hybridized carbon atoms. The result is a molecule with a triple carbon–carbon bond. The simplest molecule with this type of bond is acetylene, C_2H_2, which is a linear molecule. As can be seen from Fig. 3.9, the linear structure of acetylene is a direct result of the fact that *sp*-hybridized carbon orbitals are used in its bonding system. The *sp* hybrid orbitals on each carbon

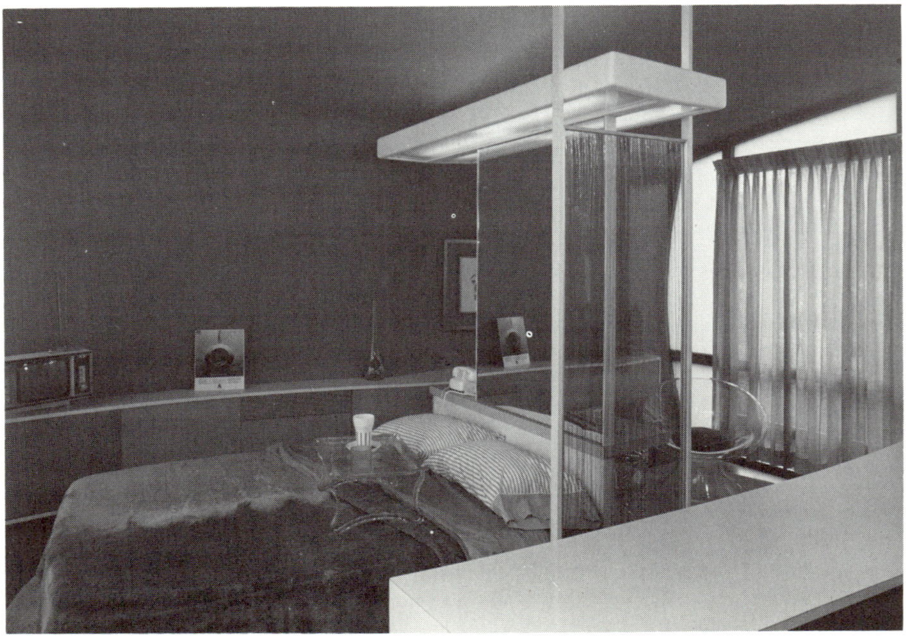

In 1957, the Monsanto Chemical Company erected a "Home of the Future" in Disneyland, Anaheim, California. Constructed almost entirely of plastics, this experimental house was intended to show the potential of plastic building materials. In 1957 the total use of plastics in the construction industry was 511 million pounds; within the next ten years this figure doubled. The "future" caught up with this graceful house and in 1967, after more than 20 million people had walked through the structure, it was torn down. Many of the innovations it had pioneered had become commonplace in the meantime.

bond a hydrogen atom and mutually interact to form a σ bond between the carbon atoms. Two π bonds are formed by overlap of the unhybridized $2p$ atomic orbitals. The triple carbon–carbon bond is considerably shorter (1.20 Å) than the carbon–carbon bond in ethylene or ethane. The triple bond is a stronger bond than either the single or double bond, with a bond energy of nearly 200 kcal/mole.

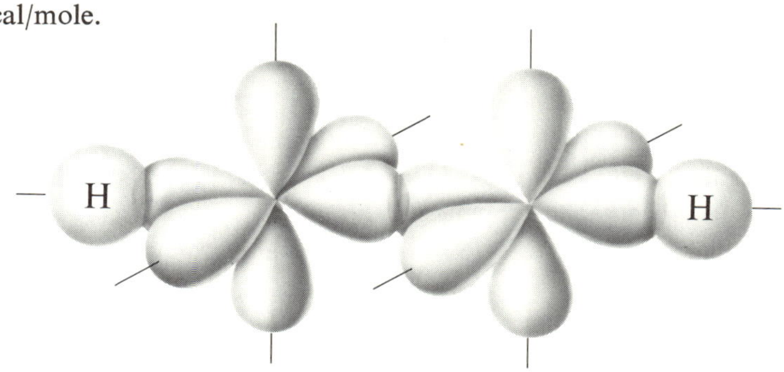

$$\mathrm{H{-}C{\equiv}C{-}H}$$

Fig. 3.9. The electronic structure of acetylene.

Nomenclature for the Alkynes

Compounds with triple carbon–carbon bonds are generically called alkynes or acetylenes. The IUPAC systematic nomenclature for an alkyne uses the syllable *-yne* in place of *-ane* in the name for the corresponding saturated hydrocarbon. If named according to this system, C_2H_2 is called ethyne; the common name, acetylene, is more prevalent, however. The position of the triple bond and the substituents in an alkyne are specified with numbers in the same manner as with the alkenes. When double and triple bonds are present in the structure, the basic chain used to name the molecule should contain both multiple bonds. The ending for the name in this case would be *-enyne*. If more than one double or triple bond is present, this is reflected by a position-specifying number before the name and by an ending syllable such as *-diyne*, *-triyne*, *-dienyne*, according to the number of double and triple bonds present. A double bond has precedence over a triple bond when determining position numbers. The systematic nomenclature of alkynes is illustrated by the examples below.

Propyne
(methylacetylene)

2-Butyne
(dimethylacetylene)

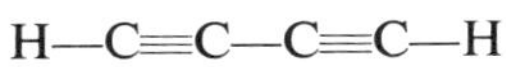

1,3-Butadiyne

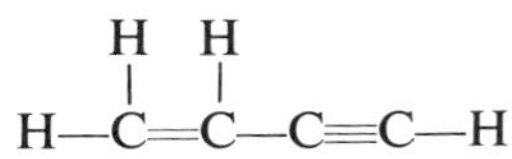

Butenyne

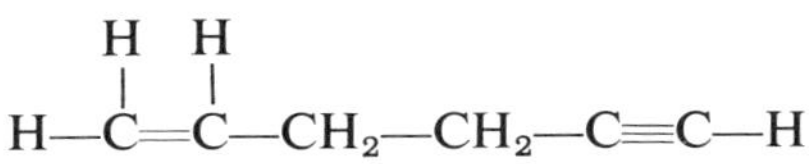

1-Hexen-5-yne

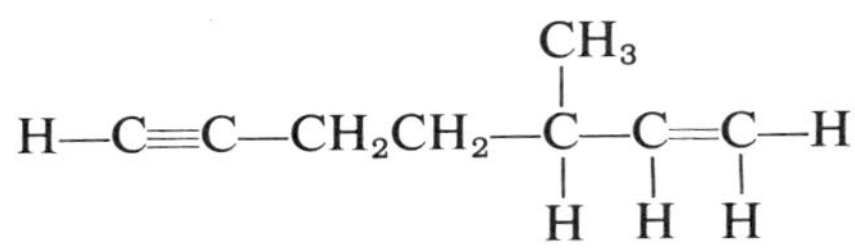

3-Methyl-1-hepten-6-yne

Although this nomenclature system for simple alkynes is straightforward enough, many organic chemists prefer to name the simpler compounds as derivatives of acetylene. By this method, 2-butyne would be called dimethylacetylene and 1-pentyne, *n*-propylacetylene.

A number of simple radicals derived from the alkynes by removal of a hydrogen atom are shown below.

Ethynyl

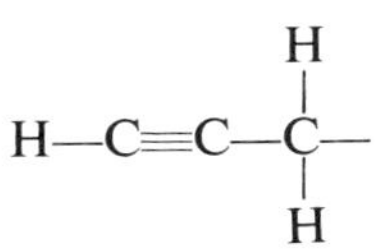

Propargyl
(common name)

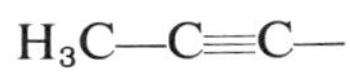

Propynyl

Because of the linear arrangement about the carbon–carbon triple bond, there is no *cis–trans* isomerism in alkynes as was found with the alkenes.

Reactions of Alkynes

Chemically, the reactions of the alkynes resemble those of the alkenes; the alkynes often function as an unsaturated compound with two attackable double bonds. Thus, 2 moles of bromine or chlorine add to 1 mole of dimethylacetylene by a two-step process. The first addition gives *trans*-1,2-dibromo-2-butene, which then adds another molecule of bromine to afford 1,1,2,2-tetrabromobutane. The addition of sulfuric acid followed by hydrolysis of the sulfate ester or the acid-catalyzed addition of water initially gives an *enol*, a compound having a double

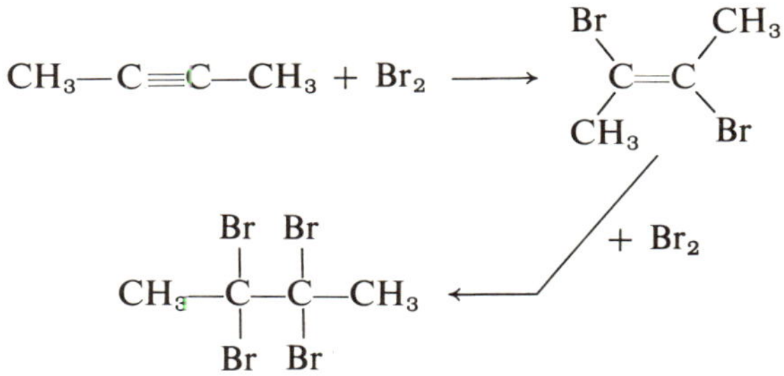

bond and an adjacent hydroxyl group. The enol is in equilibrium with a carbonyl compound, and it is the carbonyl compound that is isolated from the reaction mixture. This type of equilibrium is often called a tautomeric equilibrium. The rate of establishment of the equilibrium condition is usually very rapid.

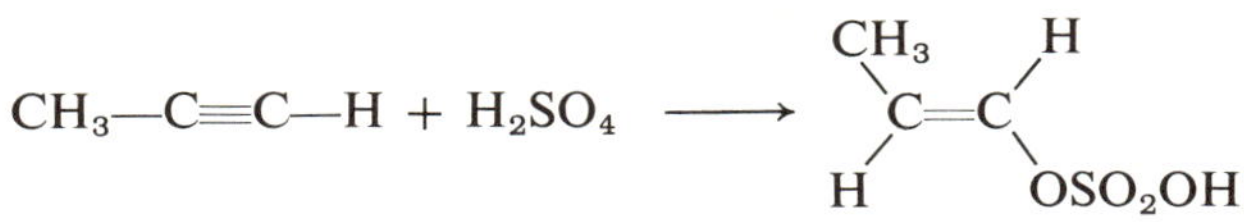

Methylacetylene Sulfate ester

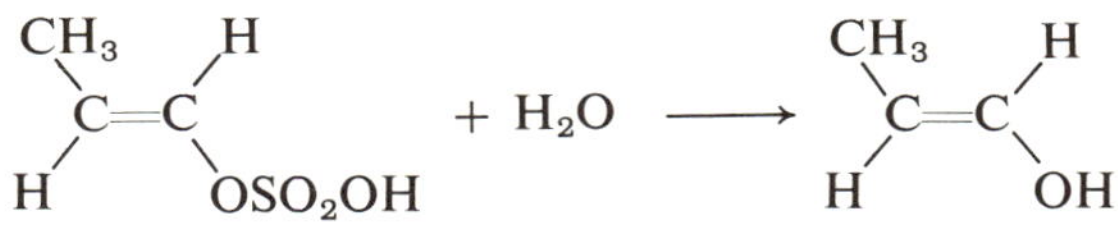

An enol

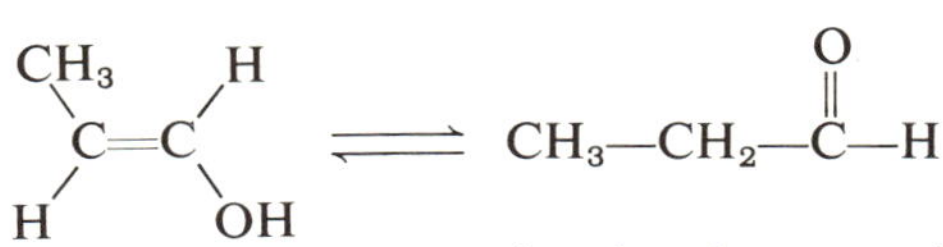

A carbonyl compound

The addition of halogen acids to alkynes is generally quite facile and the direction of such additions normally follows Markownikoff's rule. With care the catalytic hydrogenation of an alkyne can be stopped at the intermediate alkene stage. It appears that the use of palladium catalysts provides the most controllable conditions for this reaction. The alkene obtained is largely the *cis* isomer. This stereochemical result is consistent with the mechanism for catalytic hydrogenation discussed earlier; addition of two hydrogen atoms from the same side of the triple bond would be expected if both hydrogen atoms are held by the surface of the catalyst.

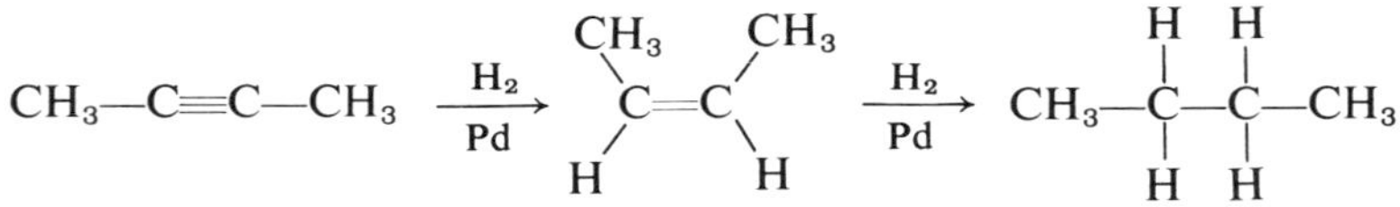

There is an unusual feature of the chemistry of acetylenes that is a reflection of the properties of the *sp* carbon orbitals used to bind the hydrogen atoms to this linkage. Because of the high amount of 2*s* character in this hybrid, electrons in this orbital tend to spend proportionately more of their time near the carbon atom. The result is that the C—H bond in acetylenes tends to be weakly acidic and, given a strong enough base, the hydrogen atom can be removed as a proton, H^+.

It is possible to prepare sodium or potassium salts of the acetylenes by the use of sodium or potassium amides ($NaNH_2$ or KNH_2) with liquid ammonia as the solvent. These salts are useful intermediates in the synthesis of organic compounds.

$$\underset{\text{1-Hexyne}}{CH_3CH_2CH_2CH_2{-}C{\equiv}C{-}H} + NaNH_2 \xrightarrow{NH_3} \underset{\text{A sodium acetylate}}{CH_3CH_2CH_2CH_2C{\equiv}C^{\ominus}Na^{\oplus}} + NH_3$$

In contrast to the alkenes, the alkynes are rarely found in nature. Small amounts of compounds which contain the carbon–carbon triple bond have been found in certain fungi and in higher plants. It is likely that continued investigation will turn up other examples. Because it is relatively less important in this context, we will not have a great deal more to say about the organic chemistry of the triple bond. However, this neglect should not be construed as a minimization of the commercial and synthetic importance of these compounds.

PROBLEMS

1. Using bond energy considerations, calculate $\Delta H°$ for the reaction of ethylene with hydrogen, with chlorine, and with bromine.
2. Provide an acceptable name for the following organic structures.

(a) 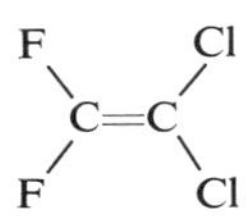

(b) $CH_3{-}CH_2{-}C{\equiv}C{-}CH_3$

(c) 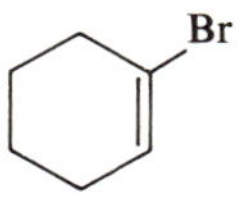

(d) $CH_3-\overset{H_3C}{\overset{|}{C}}=\overset{H}{\overset{|}{C}}-H$

(e) 

(f) 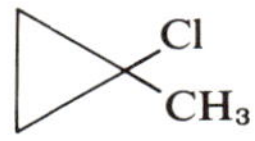

(g) $\underset{CH_3}{\overset{H}{}}C=C\underset{Br}{\overset{CH_2Br}{}}$

(h) $CH_3-\overset{H_3C}{\overset{|}{C}}=\overset{H}{\overset{|}{C}}-CH_3$

3. Write the structure and names for all possible products of hydrogenation of dimethylacetylene.

4. Write a possible method for the synthesis of the following molecules. The starting material should be a hydrocarbon.

(a) CH_3CH_2OH

(b) CH_3CH_2Br

(c) $CH_3CH_2CH_2OH$

(d) $CH_3CH_2\overset{O}{\overset{\|}{C}}H$

(e) CH_2BrCH_2Br

(f) $CH_3-\underset{OH}{\underset{|}{\overset{H}{\overset{|}{C}}}}-CH_3$

(g) $CH_3-\underset{HO}{\underset{|}{\overset{H_3C}{\overset{|}{C}}}}-\underset{Cl}{\underset{|}{\overset{H}{\overset{|}{C}}}}-CH_3$

(h) $CH_3CH_2\underset{OH}{\underset{|}{C}}H-\underset{OH}{\underset{|}{C}}HCH_3$

(i) $H-\overset{O}{\overset{\|}{C}}-CH_2CH_2CH_2CH_2-\overset{O}{\overset{\|}{C}}-H$

(j) 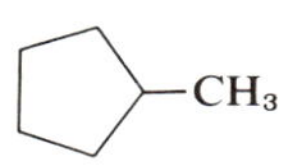

(k) *cis* $CH_3-\overset{H}{\overset{|}{C}}=\overset{H}{\overset{|}{C}}-CH_3$

5. Name the starting materials and the products of the reactions used in your solutions to problem 4.

6. What products would you expect from the treatment of ethylene with an aqueous solution of bromine (saturated, about 0.25 *M*) and sodium chloride (6 *M*)?

7. Draw structures that correspond to the following names.
(a) Allyl bromide
(b) Vinyl chloride
(c) Propylacetylene

(d) *cis*-1,2-Dibromocyclohexane
(e) *cis*-2,5-Dibromocyclohexene
(f) *trans*-Difluoroethylene
(g) Diiodoacetylene
(h) 3-Hydroxypropyne
(i) 2-Methyl-1-butene
(j) Cycloheptatriene
(k) 2-Octene
(l) 1-Nonyne
(m) (*Z*)-3-Chloro-4-methyl-3-hexene

8. Predict the products of the following reactions.

(a) $CH_2{=}CH{-}CH{=}CH_2 \xrightarrow[\text{2. Zn, HOAc}]{\text{1. ozone}}$

(b) 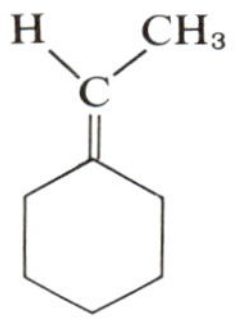 $+ \ HBr \longrightarrow$

(c) $CH_3CH_2CH_2\overset{H}{\overset{|}{C}}{=}\overset{H}{\overset{|}{C}}H + H_2O \xrightarrow{H_2SO_4}$

(d) $CF_3{-}C{\equiv}C{-}CH_3 + H_2 \xrightarrow[\text{catalyst}]{\text{Pd}}$

(e)  + diborane $\longrightarrow$

(f) $CH_3C{\equiv}CH + H_2O \xrightarrow{H_2SO_4}$

(g) $CH_3CH_2\overset{H_3C}{\overset{|}{C}}{=}\underset{\underset{H}{|}}{\overset{CH_3}{\overset{|}{C}}}{-}CH_2 + KMnO_4 \longrightarrow$

(h) 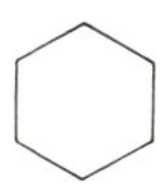 $+ \ Cl_2 \xrightarrow{h\nu}$

(i) $CH_2{=}C(CH_3)(CH_3) + HOCl \longrightarrow$

SUGGESTED READINGS

E. L. ELIEL, *J. Chem. Educ.*, **48**. 163 (1971). (Nomenclature.)

T. A. GEISSMAN, "Principles of Organic Chemistry." Freeman, San Francisco, California, 1968.

W. H. SAUNDERS, JR., "Ionic Aliphatic Reactions." Prentice-Hall, Englewood Cliffs, N.J., 1965.

H. F. MARK, Giant molecules. *In* "Bio-organic Chemistry" (M. Calvin and M. J. Jorgenson, eds.). Freeman, San Francisco, California, 1969.

R. HUBBARD AND A. KROPF, Molecular isomers in vision. *In* "Bio-organic Chemistry" (M. Calvin and M. J. Jorgenson, eds.). Freeman, San Francisco, California, 1969.

A. G. H. DIETZ *et al.*, "Engineering the Plastic 'House of the Future.'" *Modern Plastics*, June and July (1957).

4

Aromatic Hydrocarbons

The hydrocarbons that have been discussed up to this point, whether saturated or not, belong to a broad class of molecules called *aliphatic* hydrocarbons. Another, almost equally broad class of organic molecules is called aromatic hydrocarbons. The adjective aromatic used in this context refers to particular structural features of the molecules rather than to any fragrance that they might have. At one time it was thought that all hydrocarbons with pungent aromas had these structural features, but it is clear now that this need not be true since many aromatic hydrocarbons have no special olfactory properties.

BENZENE

The simplest aromatic hydrocarbon is benzene, C_6H_6, which can be obtained by distillation of coal or petroleum. Benzene is a liquid (bp 80°C) and, although it burns easily, it gives off black smoke upon combustion. This is a reflection of the high carbon content of the molecule. From the molecular formula, and the fact that all of the hydrogen atoms in the molecule behave equivalently in chemical reactions, one might surmise that benzene is an unsaturated molecule and that the structure shown below could be written for it. This hexagonal array of six carbon atoms is used so often in discussions of the chemistry of aromatic

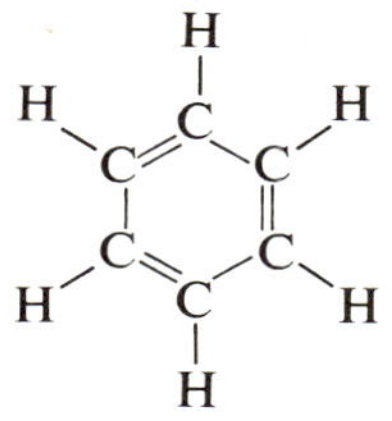

Benzene

Common symbols for benzene

compounds that the shorthand symbols shown are usually used to represent the benzene molecule.

On the basis of the structure drawn one might expect benzene to behave chemically as an olefin, or more precisely, as a triene (a molecule with three double bonds). It is instructive to compare the reactivity of benzene with the isomer of 1,3,5-hexatriene shown below. In the conformation shown, this triene

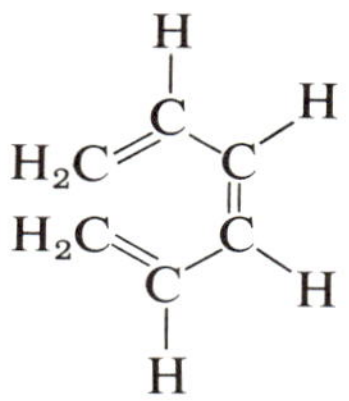

1,3,5-Hexatriene

superficially resembles the benzene structure. Several chemical reactions of benzene and 1,3,5-hexatriene are compared in Table 4.1. It is clear from the observations reported there that benzene does not undergo reactions in the same way that the triene does. The triene reactions are essentially those characteristic of an olefin and we must conclude that benzene does not have much olefinic character but, instead, possesses unusual and unexpected stability. The reasons for this extra dimension of stability become apparent if we examine the electronic system of benzene. The benzene molecule is constructed from a hexagonal arrangement of sp^2 carbon atoms. By overlap of the hybrid carbon orbitals and $1s$ atomic orbitals from the hydrogen atoms, a planar electronic framework for this molecule can be formed (Fig. 4.1). As with ethylene, the unhybridized $2p$ orbital at each carbon atom is perpendicular to this plane. These $2p$ orbitals overlap with one another to form π bonds. Something of a quandary now develops. Consider the $2p$ orbital on carbon-1. Does it overlap with the $2p$ orbital at carbon-2 to form the π component of a double bond between atoms 1 and 2 or does it overlap with the $2p$ orbital on carbon atom 6 for the same purpose?

Table 4.1
A Comparison of Several Reactions of Benzene and 1,3,5-Hexatriene

Benzene	1,3,5,-*Hexatriene*
 $+ 3H_2$ $\xrightarrow[\text{25 hr at 25°C}]{\text{Pt}}$  cyclohexane	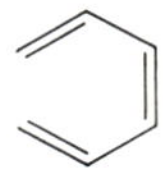 $+ 3H_2$ $\xrightarrow[\text{1 hr at 25°C}]{\text{Pt}}$  hexane
 $+ Br_2$ $\longrightarrow$ no reaction in the absence of catalysts	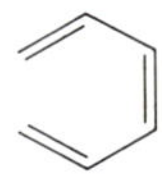 $+ Br_2$ $\xrightarrow[\text{rapid}]{\text{13°C}}$ 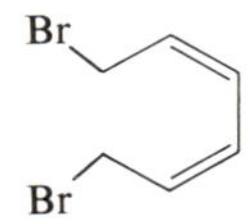
 $+ KMnO_4$ $\longrightarrow$ no reaction at moderate temperature	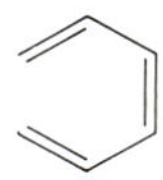 $+ KMnO_4$ $\longrightarrow$ rapid oxidation to many products
 $+ B_2H_6$ $\longrightarrow$ no reaction	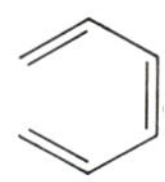 $+ B_2H_6$ $\longrightarrow$ formation of several alkyl boranes
 $+ O_3$ $\longrightarrow$ triozonide	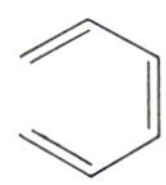 $+ O_3$ $\longrightarrow$ triozonide

Knowledge of the carbon–carbon bond distances in the benzene molecule can resolve this question, since it has been noted previously that the carbon–carbon single bond distance (1.54 Å) is greater than the carbon–carbon double bond distance (1.33 Å). Careful determinations of the structure of benzene show that, in fact, *all* carbon–carbon bond lengths in this molecule are the same, 1.397 Å. A given carbon 2*p* orbital in benzene must, therefore, overlap equally well with *both* of its neighbors and the resulting bond is neither a pure single, nor a pure double bond, but a bond of intermediate character. A more precise description of the π-electronic system in benzene shows that the system of six 2*p* orbitals defines a continuous path for the six *p* electrons. A single *p* electron can be associated with a particular 2*p* orbital only in the most formal sense; these electrons move freely in circular, π-type, molecular orbitals about the molecule.

The π electrons in benzene are therefore said to be delocalized. At first glance delocalized chemical bonds cause some discomfort to those who have been accustomed to the more restrictive two-electron, two-atomic orbital visualization of a chemical bond, but delocalized chemical bonds, formed when molecular orbitals significantly encompass more than two nuclei, are common. (In the most rigorous sense, all molecular orbitals are delocalized over all the atoms of

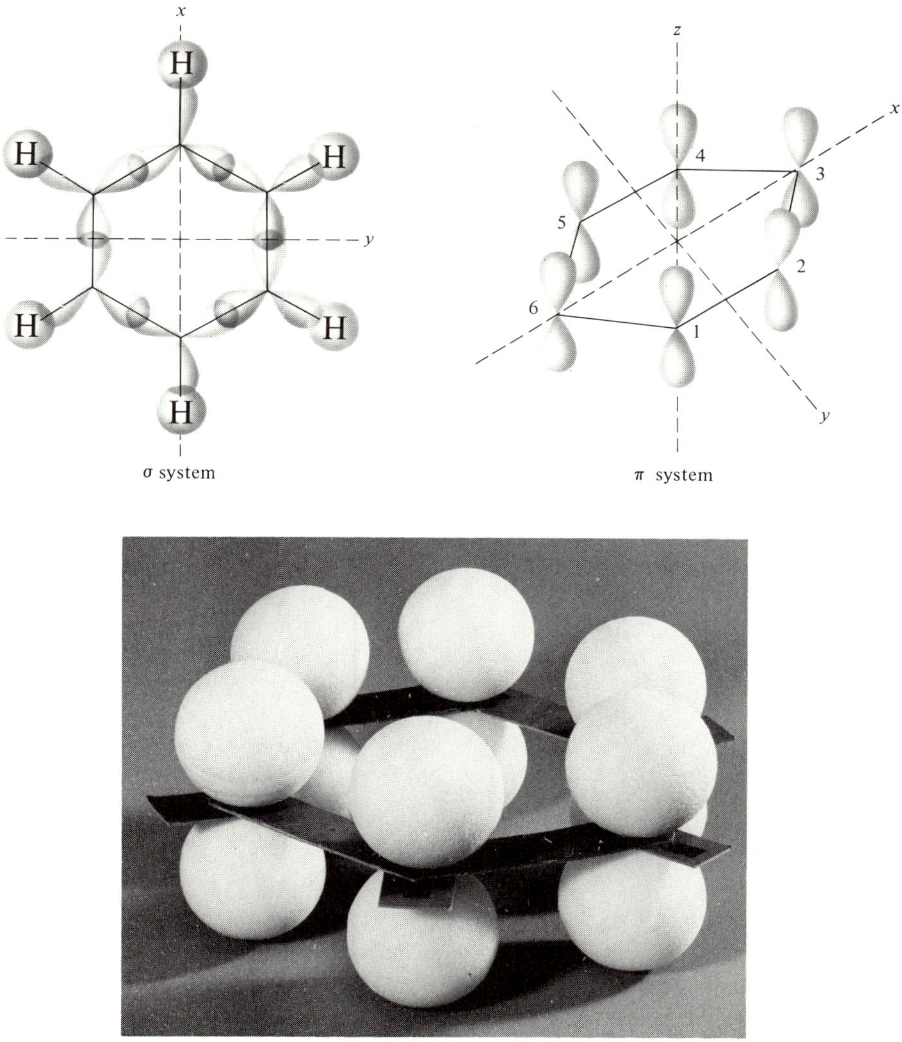

Fig. 4.1. Representations of the electronic system of benzene. The photograph shows a model of the σ electrons of the benzene molecule, represented by the flat framework and the π-electronic system, appearing above and below the plane of this framework.

a molecule but in many cases it is a good approximation to think of the chemical bonds in the molecule as being formed from pairs of atomic orbitals.) The delocalized π-molecular orbitals of benzene provide greater dimensional freedom for the $2p$ electrons and a significant lowering of electronic energy occurs. As a result, benzene and other aromatic hydrocarbons not only are stable enough that they do not have the chemical properties typical of unsaturated molecules, but are also so stable that their basic carbon skeletons usually remain intact throughout all but the most vigorous chemical reactions.

An aromatic molecule is one which is constructed of a planar array of sp^2-hybridized atoms and which is significantly stabilized by electron delocalization. The extra stability that accrues from delocalization of electrons in aromatic molecules is called the *resonance* energy of the molecule. Benzene is about 40 kcal/mole more stable than 1,3,5-cyclohexatriene and this quantity of energy may be equated to its resonance energy.

Nomenclature

The nomenclature of molecules derived from benzene follows to a large extent the systematic IUPAC system. However, a number of substituted benzenes have acquired common names that are used more than their systematic names.

The radical derived by abstracting one hydrogen atom from the benzene ring is called a phenyl radical. Union of a phenyl radical with a methyl radical produces methylbenzene. Although this name is correct, the molecule is much more

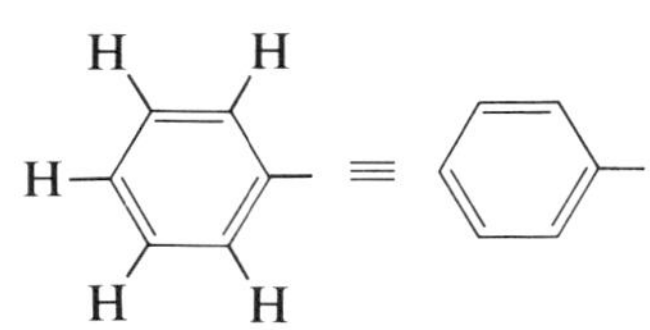

Phenyl radical

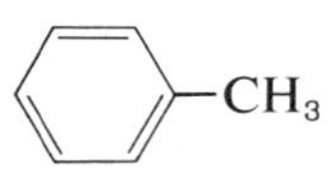

Toluene (methylbenzene)

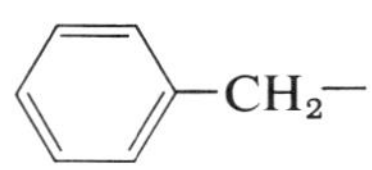

Benzyl radical

commonly known as toluene. The radical formed by removal of a hydrogen atom from the methyl group of toluene is called a benzyl radical. One should be especially careful not to confuse the phenyl and benzyl groups because the common nomenclature for them is not particularly logical. Other important alkyl and alkenyl substituted benzenes are shown below.

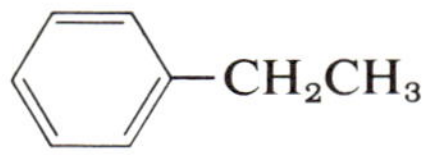

Ethylbenzene

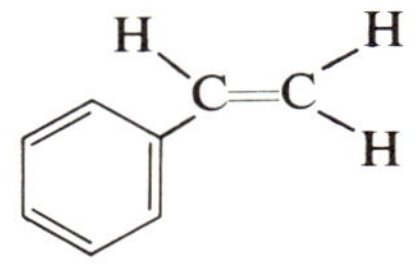

Styrene (vinylbenzene)

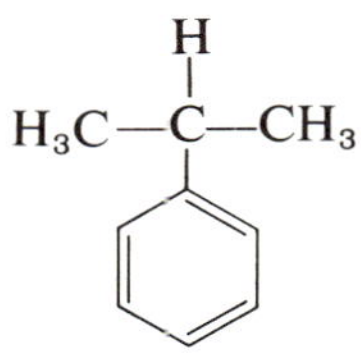

Cumene
(isopropylbenzene)

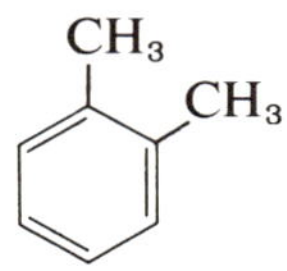

o-Xylene

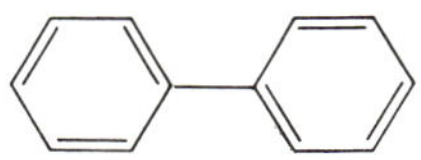

Biphenyl

Structural isomers are possible when benzene is multiply substituted. The three dichlorobenzenes shown below illustrate this. When naming compounds of this nature, the relative positions of substituents in substituted benzenes can be indicated by numbers. Thus, the first compound shown below is 1,2-dichlorobenzene; the second, 1,3-dichlorobenzene; and the third, 1,4-dichlorobenzene.

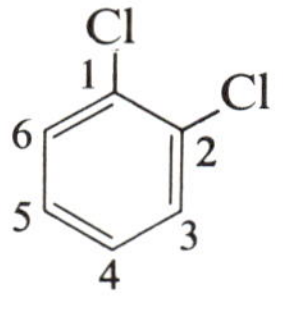

1,2,-Dichlorobenzene

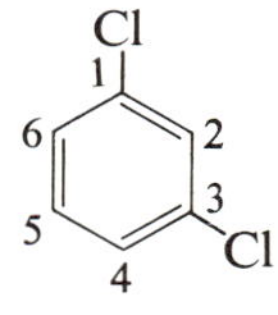

1,3-Dichlorobenzene

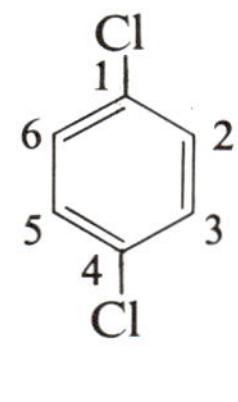

1,4-Dichlorobenzene

The numbers are chosen in such a way that they are as small as possible. The choice of which carbon atom to call carbon 1 is often arbitrary; it seems best to arrange the substituents alphabetically and to select as the 1-carbon that carbon atom which bears the group whose name is closest to the front of the alphabet. Usage of this nomenclature system is further illustrated below.

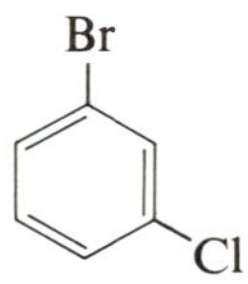

1-Bromo-3-chlorobenzene

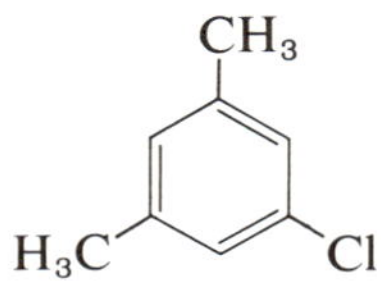

1-Chloro-3,5-dimethylbenzene

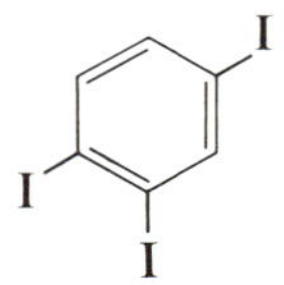

1,2,4-Triidobenzene

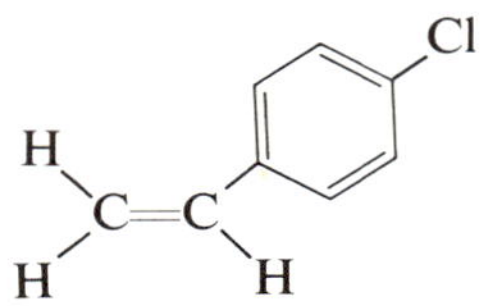

1-Chloro-4-vinylbenzene
(4-chlorostyrene)

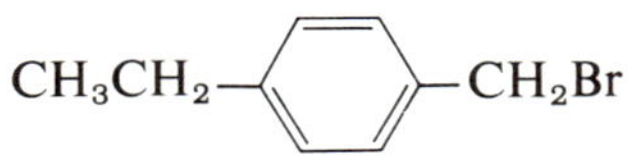

4-Ethylbenzylbromide

1-Chloro-2-fluoro-6-hydroxybenzene

Another convention for naming disubstituted benzenes involves the attachment of a prefix to the name which indicates the relative orientation of the substituents. The prefix *ortho*, usually abbreviated *o*, is used for the 1,2 arrangement, the prefix *meta* (*m*) for the 1,3, and the prefix *para* (*p*) for the 1,4-arrangement. The three dichlorobenzenes above could equally well be called *o*-dichlorobenzene, *m*-dichlorobenzene, and *p*-dichlorobenzene, respectively.

Resonance Structures

Irritating problems arise when we attempt to write down a structural formula for a molecule with delocalized chemical bonds like benzene, for it is not clear how the delocalized nature of the π-electronic system of benzene should be depicted. Some organic chemists prefer the representation shown in Fig. 4.2a where the circle is intended to portray the six π-electrons. Another method, of historical importance and theoretical significance, attempts to describe the molecule in terms of a number of idealized structures. For benzene, some of

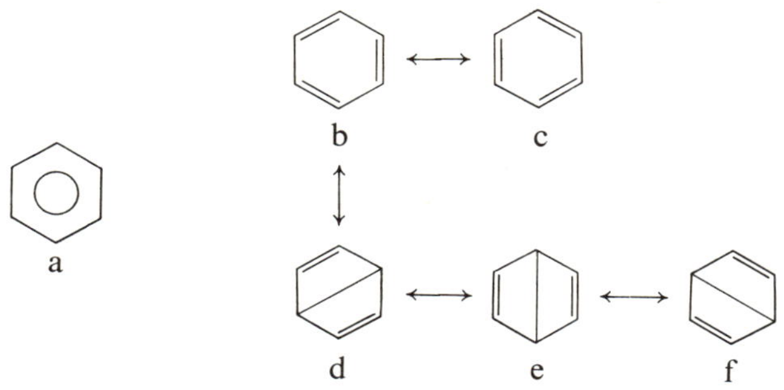

"Resonance structures"

Fig. 4.2. Various pictorial representations of the π electrons of benzene.

these formulations are shown in Fig. 4.2b–f. We know from the structural studies alluded to above that benzene cannot in reality be any of these molecules. What is implied in this series of representations is that benzene is a hybrid of these various extreme forms, much the way that a mule is a hybrid of a donkey and a horse. These extreme representations of the benzene molecule are called resonance structures and the real molecule, benzene, is said to be a resonance hybrid of these various extreme depictions. When more than one resonance structure for a molecule is written, it is customary to connect various forms by double-headed arrows. (These arrows do *not* imply an equilibrium.)

It is important to note that *none* of the resonance structures drawn for an aromatic molecule correspond to a "real" molecule. Rather each structure is intended to portray some feature of the real molecule, so that when all structures are considered together, the properties of the aromatic molecule emerge. The contribution of each of the various resonance structures to the properties of an aromatic molecule can be described in quantum-mechanically correct terms by an alternative formulation of chemical bonding theory called the valence bond approach. The valence bond theory predicts that benzene should be unusually stable and, as a rough guide, it appears that the more resonance structures that can be written for a molecule, the more stable it will be. However, valence bond theory shows that not all resonance structures contribute equally to the hybrid structure; those structures of lowest energy contribute proportionately more toward the properties of the hybrid molecule. Often one can intuitively tell the relative energies of a set of resonance structures and thus infer something about the properties of the molecule under consideration. In Fig. 4.2, the resonance structures with the long and therefore weak central carbon–carbon bond will be relatively high in energy and thus contribute little to the chemical properties of benzene. Resonance structures involving long bonds or considerable separations of electronic charges are not as stable as less extreme structures.

Substituents that have 2*p* orbitals on the atom adjacent to the benzene ring can interact with the delocalized π-electronic system of the aromatic molecule. An example is fluorobenzene. As shown in Fig. 4.3, there is a 2*p* orbital on the fluorine atom so positioned that its electronic cloud can overlap efficiently with the 2*p* orbital on the adjacent carbon. This interaction has the effect of further extending the region in space available to the delocalized electrons and can result

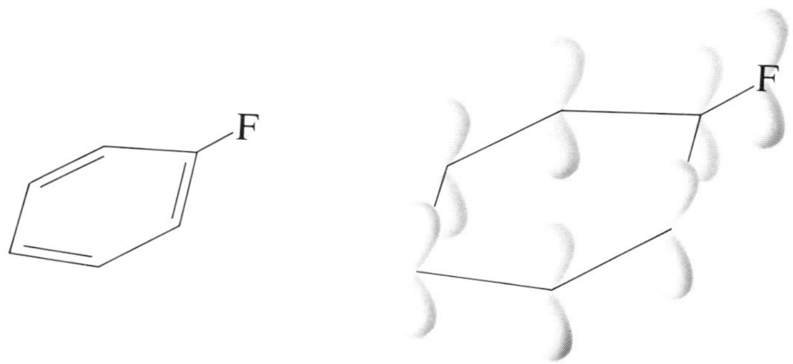

Fig. 4.3. Interaction of a fluorine substituent with the benzene π-electronic system.

in a further lowering of the π-electronic energy. The fluorine 2*p* orbital is occupied by two electrons and in terms of resonance structures, we can represent the pairing of the eight π electrons in fluorobenzene with structures like those shown below.* These various resonance structures indicate that the interaction

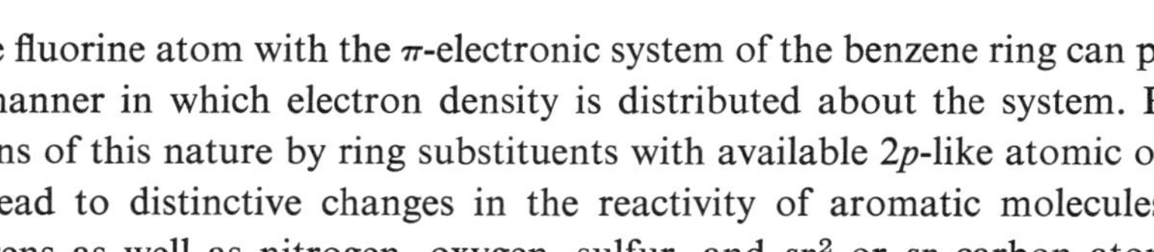

Resonance structures for fluorobenzene

of the fluorine atom with the π-electronic system of the benzene ring can perturb the manner in which electron density is distributed about the system. Perturbations of this nature by ring substituents with available 2*p*-like atomic orbitals can lead to distinctive changes in the reactivity of aromatic molecules. The halogens as well as nitrogen, oxygen, sulfur, and sp^2 or *sp* carbon atoms are potentially capable of resonance interactions of this nature.

* Notice that when writing the various resonance structures of a compound we do not permit any of the atoms to change their relative positions. Resonance structures are used to describe the π-electron system of a particular molecule and a change in atomic position generates a new molecule.

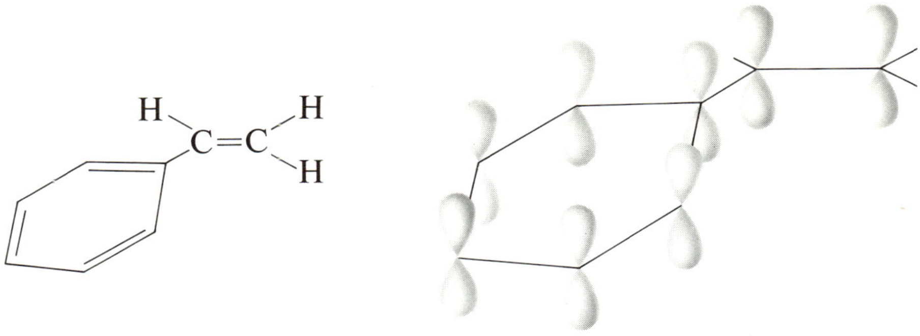

Fig. 4.4. The π-electronic system of styrene.

A resonance interaction would be expected in styrene if the vinyl group could be restrained to coplanarity with the benzene ring (Fig. 4.4). When this condition cannot be met, the overlap of the π-electronic clouds of the first $2p$ orbital of the vinyl group and that of the orbital on the adjoining carbon atom of the

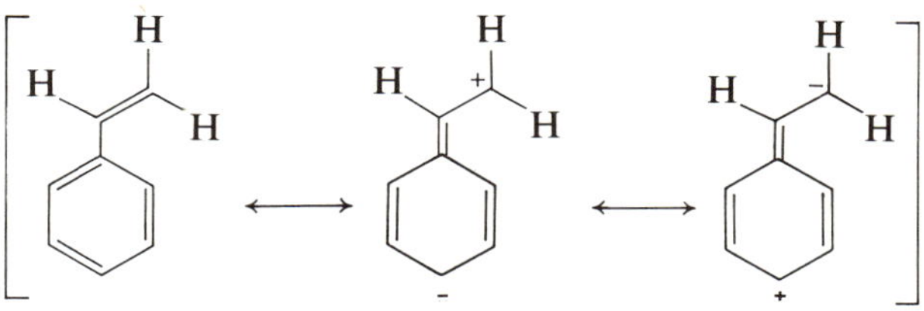

Resonance structures for styrene

benzene ring will be less than ideal; the magnitude of the resonance interaction will be diminished and any effect of this interaction on the chemical or physical properties of the molecule should be correspondingly attenuated. Large substituents in the positions *ortho* to the vinyl group of this molecule will tend to force the vinyl group away from the optimum planar arrangement. Thus resonance effects can be modified by steric effects.

LARGER AROMATIC MOLECULES

Larger and more complicated aromatic molecules can be constructed by merging additional rings of sp^2 carbon atoms to benzene. These materials are often called polynuclear aromatic hydrocarbons and are usually solids. Naphthalene, $C_{10}H_8$, can be thought of as two benzene molecules fused together along

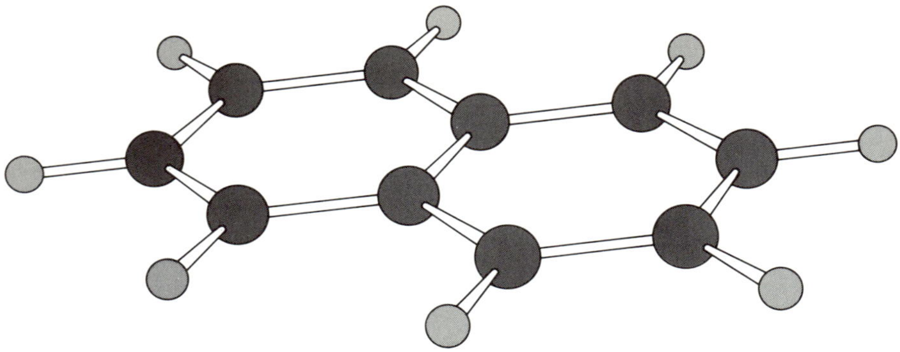

Fig. 4.5. A three-dimensional drawing of naphthalene.

a common side (Fig. 4.5). The π-electronic system of naphthalene is delocalized, as the resonance structures drawn attempt to suggest.

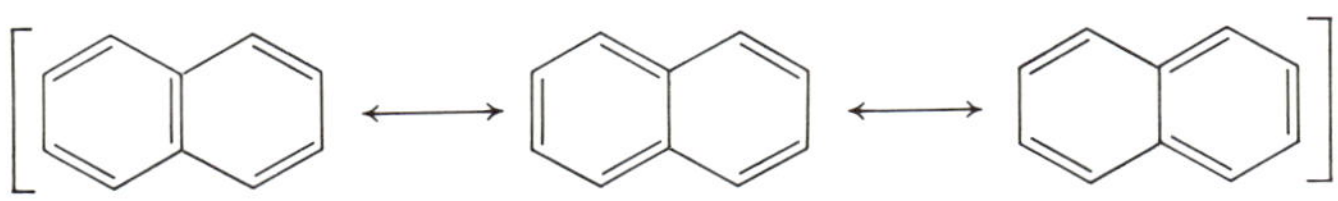

Resonance structures for naphthalene

This delocalization lowers the electronic energy of the molecule as it does in benzene, so that, like benzene, the naphthalene ring system usually remains intact during chemical reactions.

In naming derivatives of naphthalene, the numbering system shown below is used. When there can be no ambiguity, the numbers selected to denote the position of a substituent on the naphthalene rings should be as small as possible. An alternative nomenclature system is occasionally used when there is only one substituent on the naphthalene skeleton. This system employs Greek letters, as

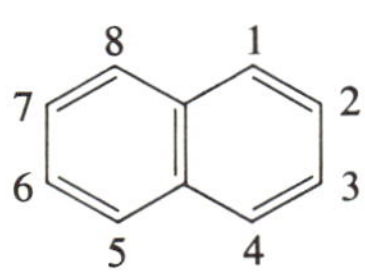

Naphthalene

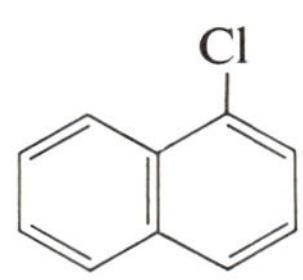

1-Chloronaphthalene
(not 8-chloronaphthalene)
(α-chloronaphthalene)

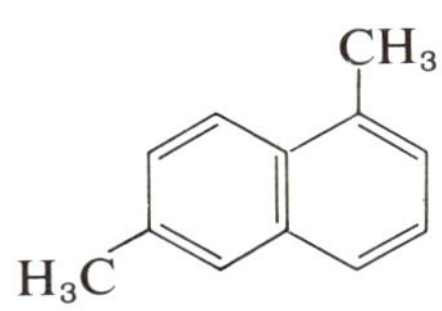

1,6-Dimethylnaphthalene
(not 4,7-dimethylnaphthalene)

illustrated below, to designate the two possible orientations of the single substituent.

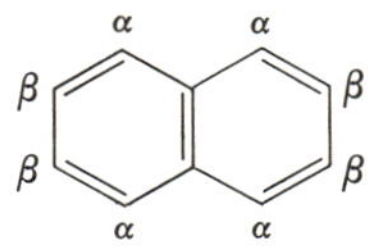

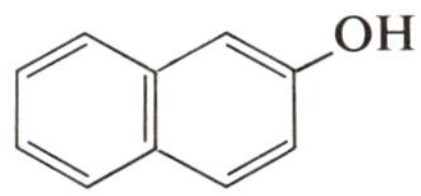

β-Hydroxynaphthalene
(2-hydroxynaphthalene)

REACTIONS

Reduction and Oxidation

Aromatic hydrocarbons can be catalytically hydrogenated to the corresponding cycloalkanes, but vigorous conditions of high temperature and high pressure may be required to effect this reaction. It was noted earlier that the reduction of benzene produces cyclohexane.

There are more than 1200 known organic compounds in tobacco smoke including alkanes, cycloalkanes, alkenes, cycloalkenes, acetylenes and over 100 aromatic hydrocarbons. Many of the larger aromatic hydrocarbons and compounds derived from them have powerful carcinogenic properties. Dibenz[*a,h*]-anthracene, chrysene, benzo[*j*]fluoranthene, and 3,4-dibenzpyrene are among the aromatic hydrocarbons identified in cigarette smoke; very small quantities of 3,4-benzpyrene are sufficient to induce skin tumors in mice and statistical studies have overwhelmingly linked cigarette smoking with the occurrence of lung cancer.

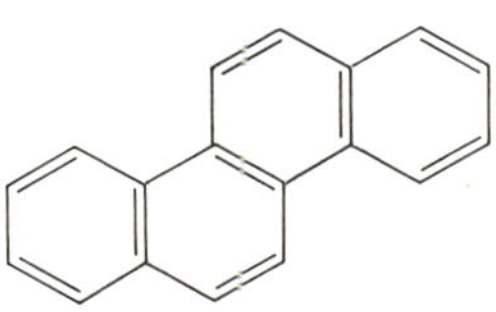

Chrysene

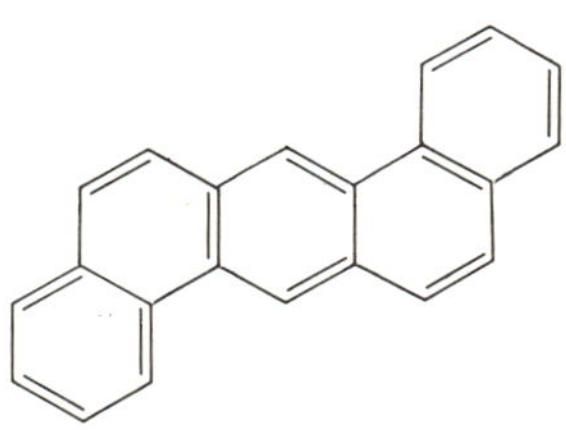

Dibenz[*a,h*]anthracene

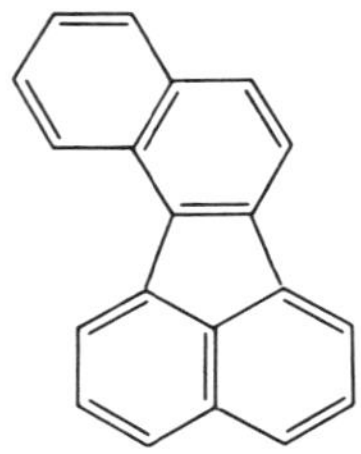

Benzo[*j*]fluoranthene

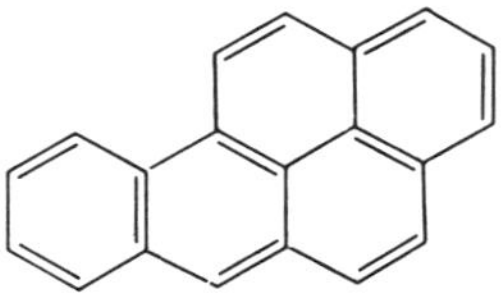

3,4-Benzpyrene

The reverse reaction, dehydrogenation, often occurs when a cycloalkane or cycloalkene is treated with elemental sulfur or selenium. Hydrogenation catalysts that do not contain any absorbed hydrogen can also function as dehydrogenating agents. Thus, treatment of *cis*-1,2-dimethylcyclohexane with molten sulfur at 300°C forms 1,2-dimethylbenzene (*o*-xylene).

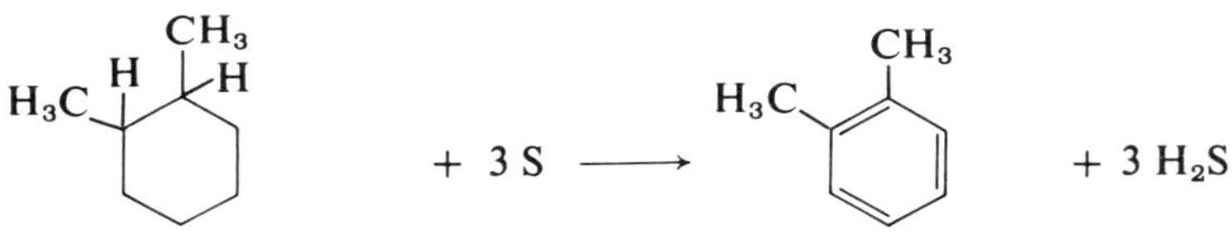

cis-1,2-Dimethylcyclohexane

1,2-Dimethylbenzene (*o*-xylene)

Most aromatic compounds are resistant to oxidation under moderate conditions. If an aliphatic side chain is attached to the aromatic ring system, however, it is oxidized to a carboxylic acid function when the molecule is treated with permanganate or dichromate. The length of the side chain is not important; the same product, an aromatic acid, is observed no matter what the chain

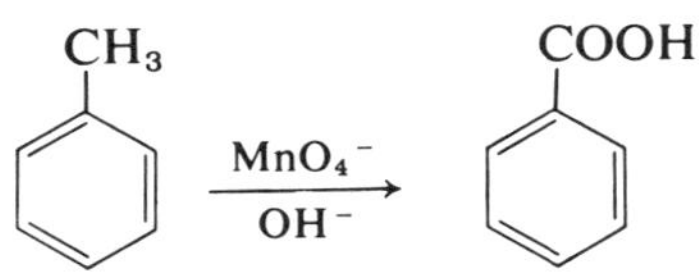

Toluene

Benzoic acid

length. The first carbon atom of the side chain is retained in the carboxyl group of the oxidation product and the remainder of the side chain is digested to ill-characterized byproducts by the reaction solution. This side-chain oxidation reaction is important because it can aid in determining the point of attachment of aliphatic groups to an aromatic compound.

Electrophilic Substitution

The π-electronic system of an aromatic molecule is an attractive site for attack by electrophilic (electron-seeking) species. Generally the product of the reaction of an electrophile with an aromatic ring is a substitution product; the electrophile, in some form, replaces a hydrogen atom or other group attached to the ring. The displaced group is expelled, often as a positively charged ion. Electrophilic substitution reactions occur between molecular chlorine or bromine and aromatic substances, but a catalyst is normally needed for this reaction. The function of the catalyst, usually a Lewis acid (such as ferric chloride, $FeCl_3$, or aluminum chloride, $AlCl_3$, if a chlorination reaction is to be run), is to weaken or even break the halogen–halogen bond. The halogen catalyst complex that results is highly polarized and the halogen is made more electrophilic. It attacks

$$Cl{-}Cl + FeCl_3 \rightleftharpoons \underset{\text{Catalyst–chlorine complex}}{Cl^+ \cdots FeCl_4^-}$$

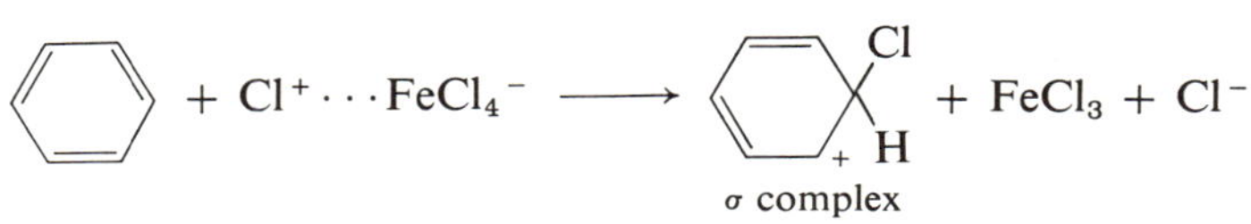

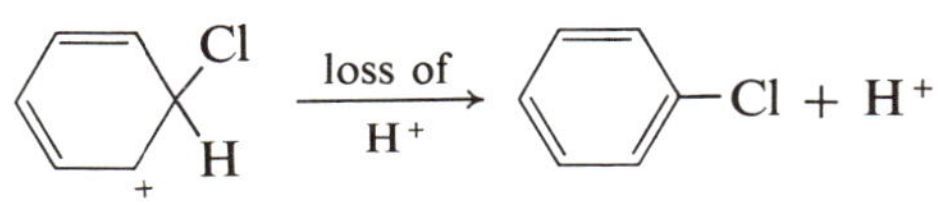

Overall reaction:

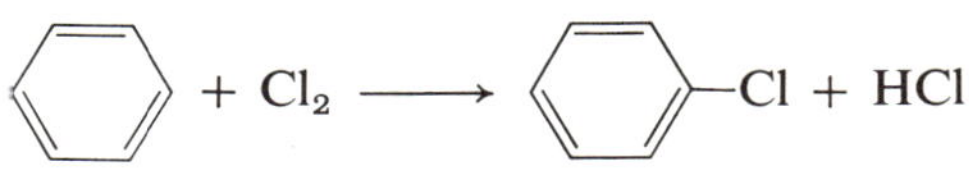

Mechanism for electrophilic substitution on aromatic hydrocarbons

the aromatic π-electronic system, finally disrupting it to form a bond to a carbon atom. The intermediate formed is called a σ complex. It has delocalized π electrons as indicated by the resonance structures below. The resonance energy of the intermediate is not as great as that of benzene so that loss of a hydrogen ion from the complex and reformation of a benzene-like aromatic system is energetically favorable. The substitution product, chlorobenzene, is the result.

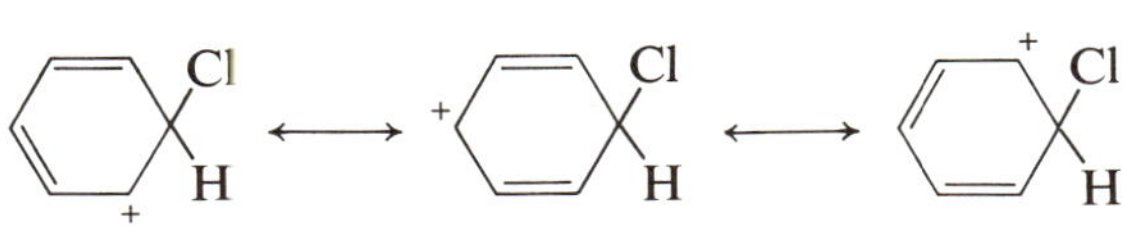

Resonance structures for σ-complex

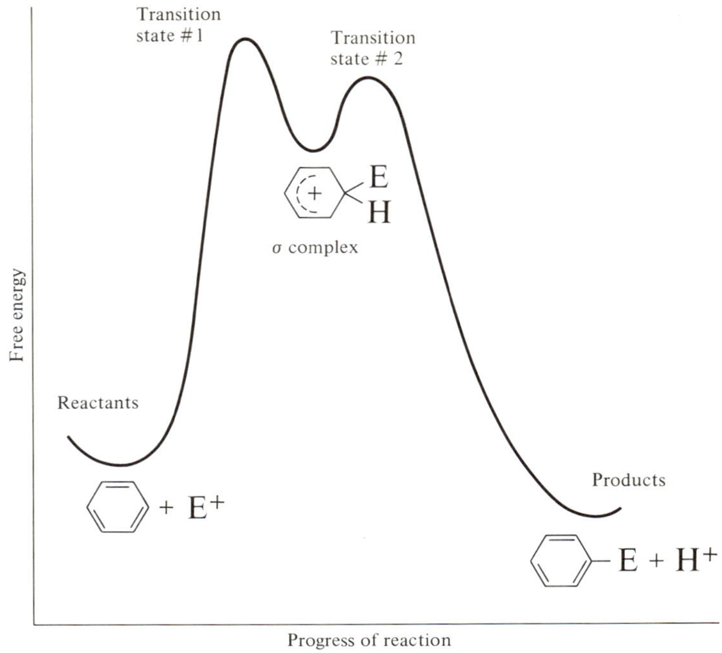

Fig. 4.6. A possible free energy–reaction coordinate diagram for electrophilic substitution on an aromatic ring system.

Table 4.2
Electrophilic Substitution Reactions

Reagent	*Electrophile*	*Product formed with benzene*
Br_2 with $FeBr_3$	Br^{+}[a]	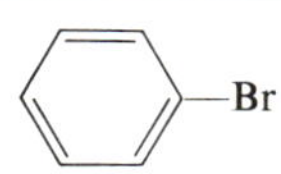 Bromobenzene
HNO_3 in H_2SO_4	NO_2^{+}	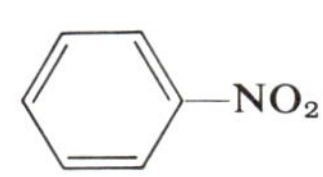 Nitrobenzene
H_2SO_4	SO_3^{+}	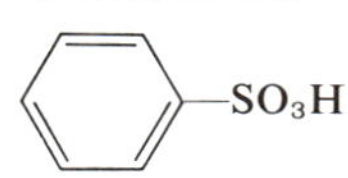 Benzenesulfonic acid
CH_3Cl with $AlCl_3$	CH_3^{+}[a]	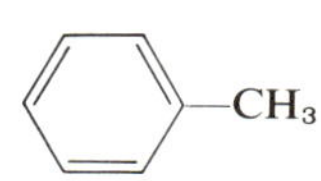 Toluene

[a] Electrophile may be a complex of this species and the catalyst.

The formation of the σ complex is usually the slow step in an electrophilic aromatic substitution reaction. The free energy–reaction coordinate diagram in Fig. 4.6 is plausible for this mechanism. The first energy barrier shown (transition state 1) is the barrier to formation of the σ complex; it is usually the highest energy peak along the reaction coordinate since the stable, delocalized π-electronic structure of the starting material must be destroyed during the formation of the σ complex. With minor embellishments, this mechanism for aromatic electrophilic substitution is followed by a wide range of electrophilic species. Several of these reactions are summarized in Table 4.2.

Substituent Effects on Electrophilic Substitution

What is the course of an electrophilic substitution reaction if there is a substituent already present on the aromatic ring? For example, toluene can potentially be brominated to afford three products: the *ortho*, *meta*, or *para* bromotoluenes. All three isomers are formed in this reaction, although not in equivalent amounts.

One factor that helps determine how much of a particular isomer is formed is the size of the substituent already present on the aromatic ring. Electrophilic substitution reactions with *t*-butylbenzene give essentially only the *para* isomer

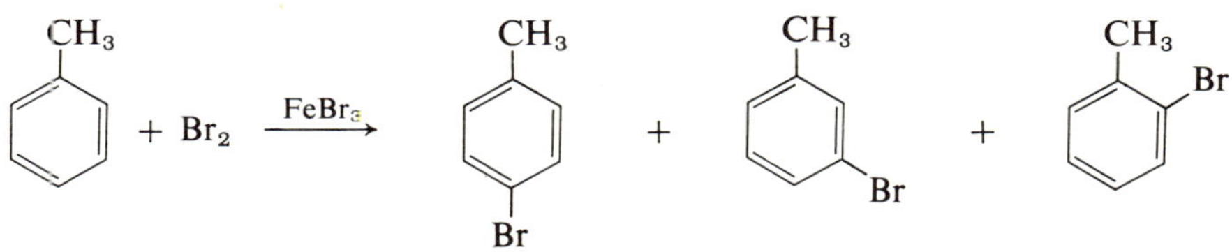

p-Bromotoluene 53%   *m*-Bromotoluene 4%   *o*-Bromotoluene 43%

since the very bulky *t*-butyl group tends to protect the *ortho* position from the approach of the electrophile. But steric effects cannot be the only factor which

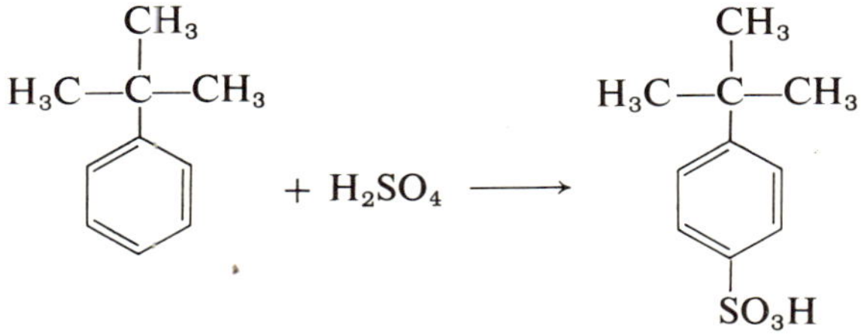

t-Butylbenzene   100%

determines the positioning of the new substituent in these reactions since the *meta* and *para* positions of the *t*-butylbenzene are essentially equally available for approach of an electrophilic reagent.

Other factors which we must consider are related to the way the substituent perturbs the distribution of electrons in the reactant molecule and in the transition state that separates the reactants from the σ complex. Substituents on an aromatic ring have two effects on the electrophilic substitution process. They affect the overall rate of the reaction and they determine the orientation of the incoming electrophile relative to themselves. The two effects are not separate: substituents which direct the electrophile to positions *ortho* and *para* to themselves also accelerate the rate of the reaction relative to the rate with benzene. Other types of substituents slow down the rate of the reaction and may cause the electrophile to attack at a *meta* position or *ortho* and *para*. Table 4.3 lists a number of substituents that fall into these three categories. The preferential formation of *o*-nitrotoluene and *p*-nitrotoluene when toluene is nitrated with nitric acid in sulfuric acid is primarily due to the fact that these two positions react the most rapidly with the electrophile. Similarly the deactivating effect of the nitro group (NO_2) leads to the formation of *meta* isomers because, although all positions are deactivated by the substituent, the *meta* position is deactivated the least and the rate of the electrophilic substitution reaction is most rapid at this position.

Table 4.3
Substituent Effects on Aromatic Electrophilic Substitution Reactions

I.	Substituents which activate the aromatic ring and direct to *ortho* and *para* positions
	$—NH_2$ $—OH$ $—O^-$ $—NR_2$ $—OR$ alkyl aryl
II.	Substituents which deactivate the aromatic ring and direct to *ortho* and *para* positions
	—F —Cl —Br —I $—CH_2Cl$
III.	Substituents which deactivate the aromatic ring and direct to the *meta* position
	$—NO_2$ $—C{\equiv}N$ $—CF_3$ $—CCl_3$ $—\overset{O}{\overset{\Vert}{C}}—OH$

The products of an aromatic electrophilic substitution reaction that are observed are primarily those formed at the fastest rate rather than those which are intrinsically or thermodynamically the most stable. Any factor which lowers the transition state energy will accelerate the reaction as will any factor that raises the energy of the reactant molecules. Transition states have only a fleeting existence and by definition cannot be studied experimentally in isolation from the reactants. How then can we predict what might be the effect of a substituent upon the electronic structure or other properties of a transition state? As a molecular system moves from reactants to transition state to intermediate (or product), there must be a smooth interconnection of the molecular energy levels at each step along the way. The electronic energy levels of the reactants must flow uninterruptedly into the electronic energy levels for the transition state and these energy levels must merge with those of the products. It seems reasonable, then, to expect the levels of the transition state to resemble most the levels of that state (reactants or products) which is closest to it in energy. For the present reaction type, the σ complex is higher in free energy than the reactants and much closer in energy to the transition state than are the reactants. Therefore, we may expect the transition state to resemble most the σ complex; any substituent effects on the stability of the σ complex should be strongly reflected in the energy or stability of the transition state. It is sufficient, then, for qualitative predictions of relative reaction rate to consider the effect of substituent groups on the energy difference between the reactants and the σ complex.

In the nitration reaction of toluene, three reactions leading to the *ortho*, *meta*, and *para* isomers of nitrotoluene are possible; it is the fastest of these three reactions which leads to the observed products.

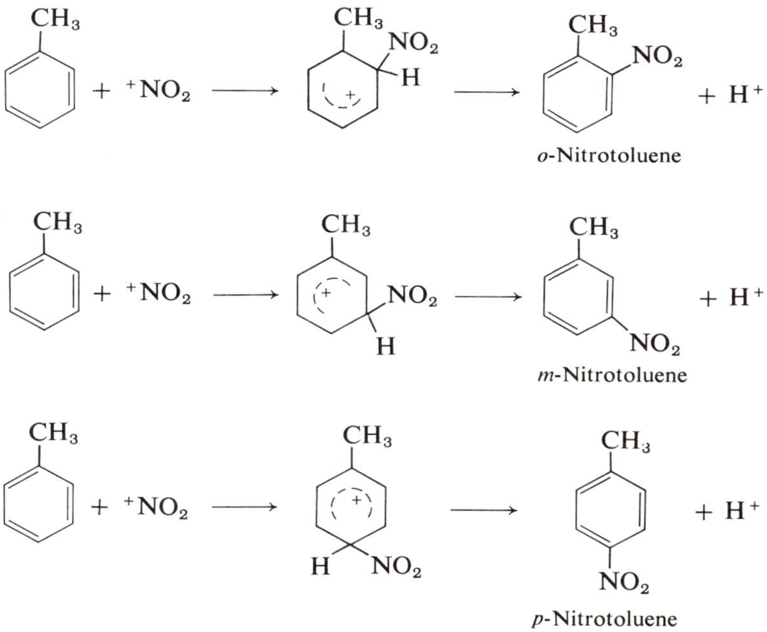

The three possible nitration reactions of toluene

The free energy of the reactants is the same for all three reactions, but the free energy of the σ-complex intermediates will be different in each case. Those energy differences should be reflected in the relative free energies of the transition states for the reactions. Resonance structures for each of the three σ complexes are shown below. Consideration of these resonance forms suggests that the σ complexes which are formed by attack at the *ortho* and *para* positions of the aromatic molecule should be more stable than the complex obtained by attack

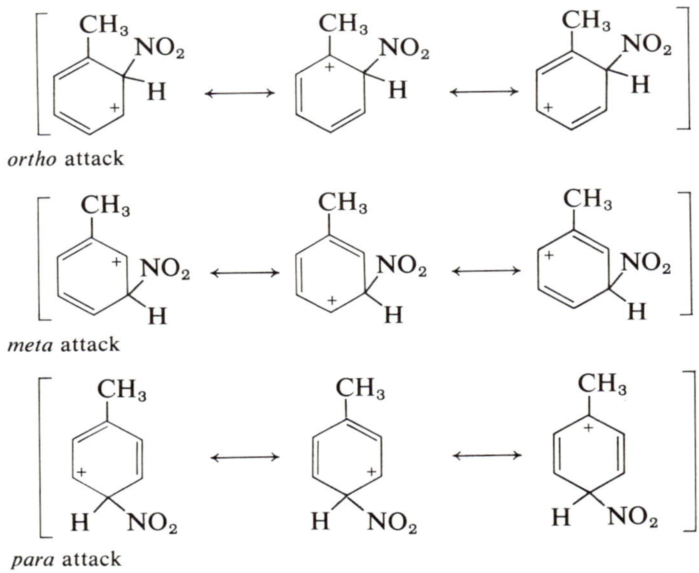

Resonance structures for the σ complexes formed during nitration of toluene

at the *meta* position. In some of these cases a resonance form having the equivalent of a tertiary carbonium ion structure is formed. These resonance structures should contribute strongly to the properties of the σ complexes formed by *ortho* or *para* attack. The energies of the corresponding transition states should be correspondingly lower so that $\Delta G^{\ddagger}$ for the formation of *ortho*- or *para*-substituted products will be less than $\Delta G^{\ddagger}$ for the formation of the *meta* product (Fig. 4.7) and the *ortho* and *para* products will, therefore, be generated most rapidly and become the observed products of the nitration reaction.

The nitro group and the other *meta*-directing groups listed in Table 4.3 are constructed largely of strongly electron-withdrawing atoms. These groups are therefore also strongly electron-withdrawing. Consider the three possible σ complexes formed by the attack of an electrophile on nitrobenzene as a model for substitution reactions in the presence of one of the *meta*-directing groups.

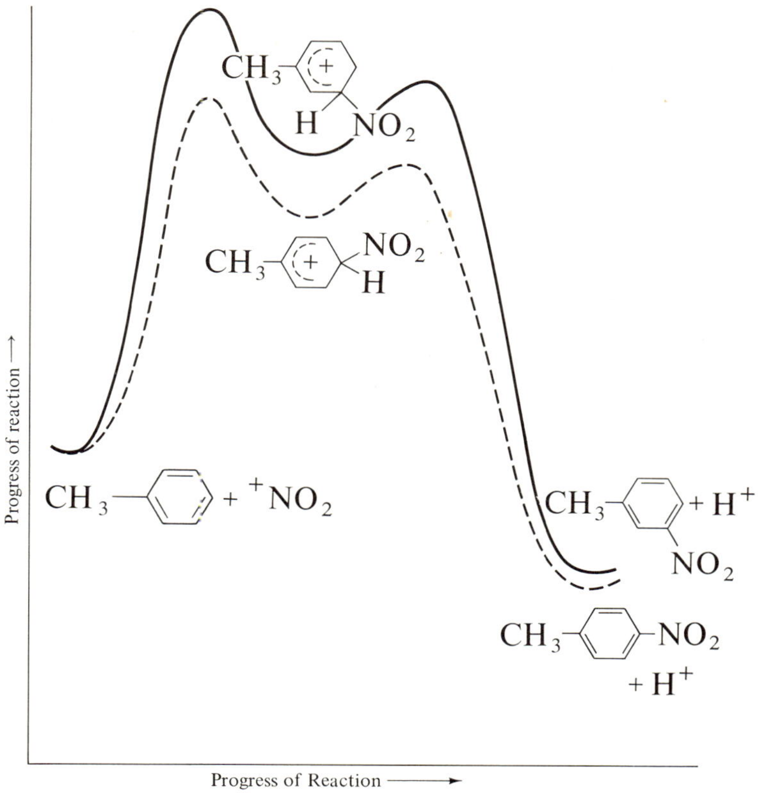

Fig. 4.7. Free energy–reaction coordinate diagrams for the formation of *meta* and *para* nitrotoluenes from toluene and nitric acid.

For concreteness, we can take the bromine–ferric tribromide complex as the electrophile. In all three cases, the σ- and π-electron density on the aromatic ring will be reduced relative to what it is in the case of benzene because of the electron-removing effect of the nitro group. The attractive interaction between the electrophile and the ring will thereby be reduced and the rate of the reaction at any of the three possible positions will be smaller than the rate of reaction with benzene. However, the intermediates for attack of the electrophile at the carbon atoms *ortho* and *para* to the nitro group will be less stable (higher in free energy) than the one which obtains during the formation of the *meta*-substituted product. The generation of a positive charge directly adjacent to an electron-withdrawing group such as nitro is not energetically favorable so that the resonance contributors for the *o* and *p* complexes which have this feature will not have much importance for the total electronic structure of the respective

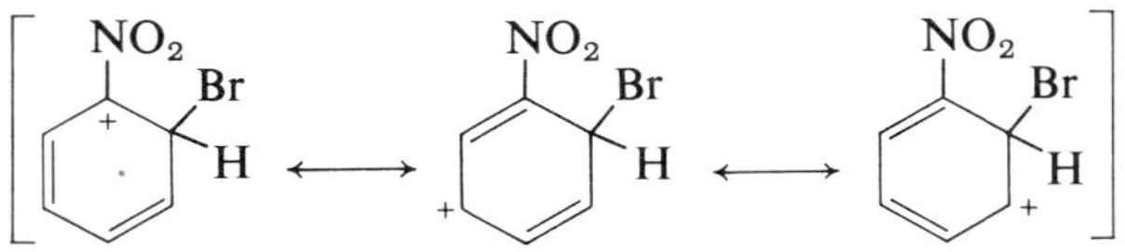

ortho attack

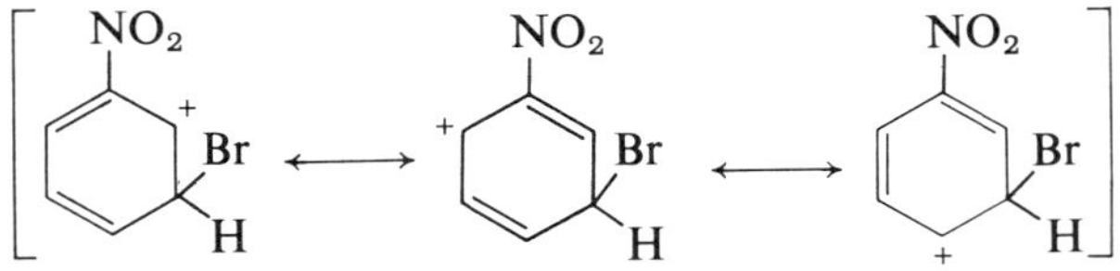

meta attack

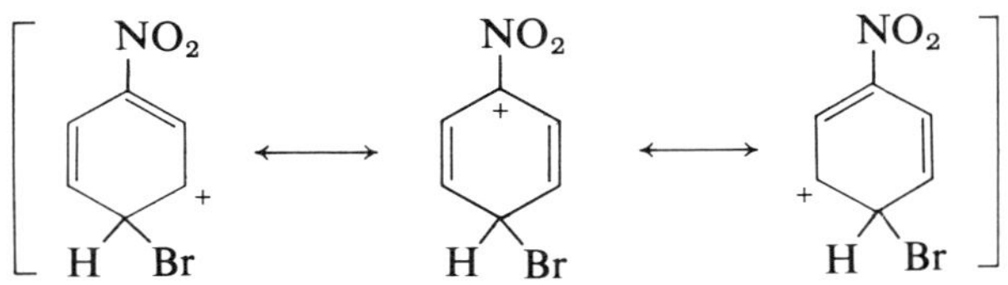

para attack

Resonance structures for the σ complexes formed during the halogenation of nitrobenzene

intermediates. The resonance energy for the complexes leading to *ortho*- and *para*-substituted products is less than the resonance energy of the *meta*-substituted complex. Thus, although the rate of electrophilic substitution is not enhanced by the presence of the nitro or other *meta*-directing group, its influence is less strongly felt at the *meta* position and *meta* substitution occurs more rapidly than substitution at other positions.

The halogen substituents exert two conflicting effects on the electronic system of an aromatic ring. These atoms are electronegative and can remove electron density from the ring. However, the resonance interaction of the *p* electrons on the halogen with the aromatic π-electronic system has the effect of stabilizing the intermediates formed during electrophilic substitution at the *ortho* and *para* positions. Take as an example the reaction of fluorobenzene with the methyl cation (CH_3^+) generated from methyl chloride by the action of aluminum trichloride. (This reaction is an example of the Friedel-Crafts alkylation reaction.) Resonance structures for the three possible intermediates are shown on p. 105. It can be seen that the electronegative halogen substituent interacts favorably with adjacent positive charges so that the *ortho* and *para* intermediates will be expected to be more stable than the intermediate leading to formation of the *meta* product. The net effect of the halogen substituent is to slow the rate of the substitution reaction relative to the rate with benzene at all positions; the effect is felt most strongly at the *meta* position because the favorable counteracting

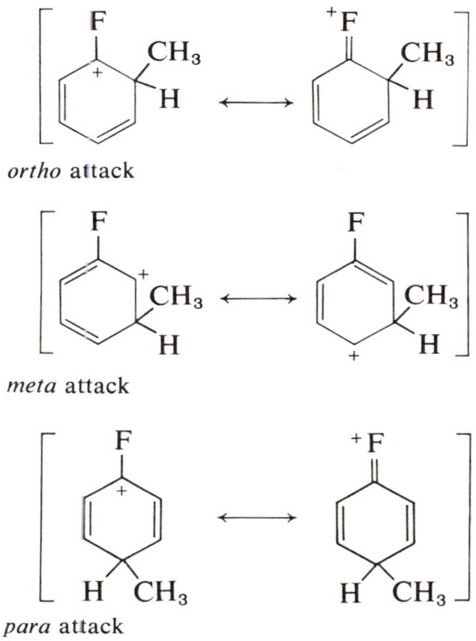

Some resonance structures for the σ-complexes formed during the alkylation of fluorobenzene (additional resonance structures analogous to those shown earlier for fluorobenzene can be drawn)

influence of resonance interaction with the halogen cannot take place within the corresponding σ complex.

The effect a given substituent has on both the rate of formation of product and the orientation of the substituting group can be used to predict the products of aromatic substitution reactions when more than one aromatic-ring substituent is present. Electrophilic substitution on *p*-nitrotoluene gives predominantly the product derived by attack at the position *ortho* to the methyl. Methyl directs *ortho–para*, while the nitro group directs *meta*; both

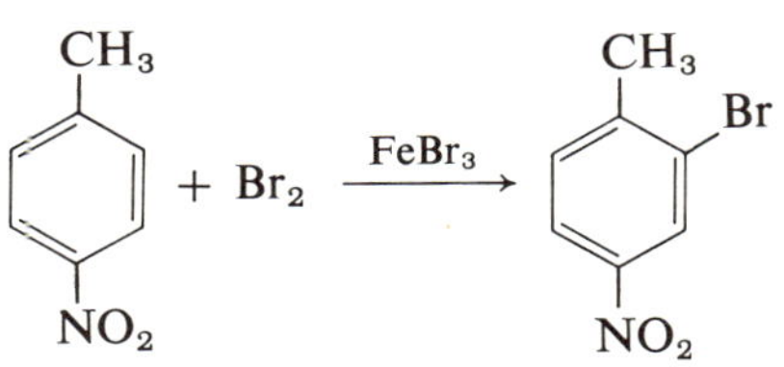

substituents direct to the position taken by the incoming electrophile in this case. With *m*-nitrotoluene, the substituents direct the incoming electrophile to different positions but the methyl group, in addition to directing *ortho–para*, also is an activating substituent. The rates of formation of the products dictated by the positioning effect of the methyl group are greater than the rate leading to the isomer obtained by attacking *meta* to the nitro group. The observed products are the ones shown. (Attack between the two substituents is not likely for steric

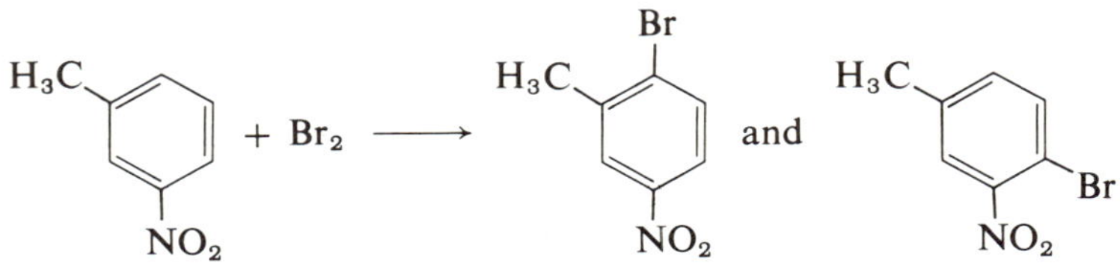

reasons.) The positioning effect of activating groups usually wins out over the effect of deactivating groups because of the rate-enhancing effect of these substituents in aromatic electrophilic substitutions.

PROBLEMS

1. Write the structure that corresponds to each of the following names.
   (a) Ethylbenzene
   (b) *p*-Iodotoluene
   (c) α-*t*-Butylnaphthalene
   (d) 1,3,7-Trichloronaphthalene
   (e) Cyclopentylbenzene
   (f) 1,3-Cycloheptadiene
   (g) 1,2-Diphenylethene
   (h) 1-Benzyl-1-chlorocyclopropane
   (i) 1-Fluoro-2,4-dinitrobenzene
2. Provide a name for each of the following compounds.

(a) 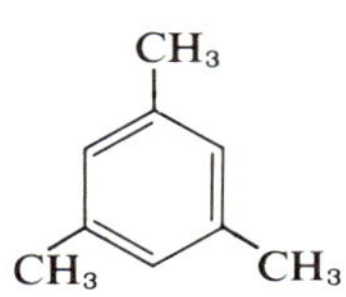

(b) 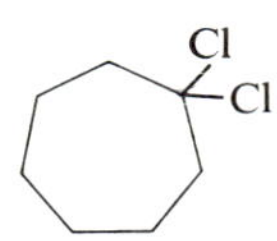

(c) 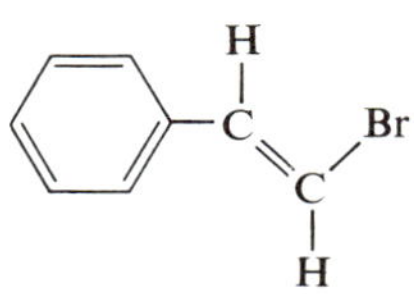

(d) 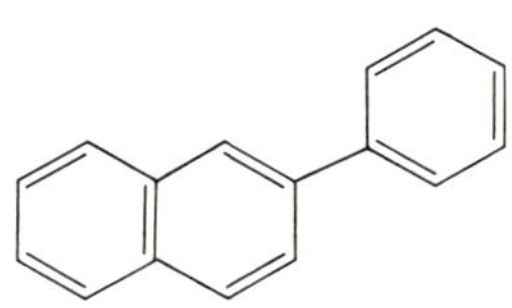

(e) 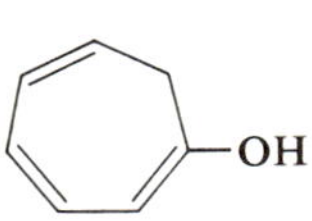

(f) 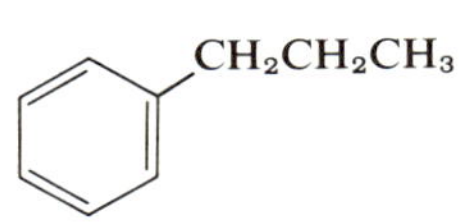

(g) 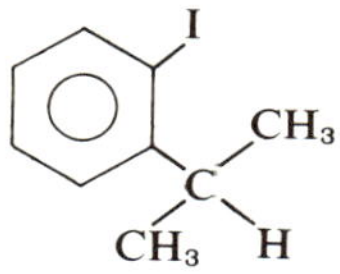

(h) 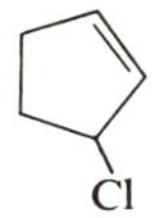

(i) 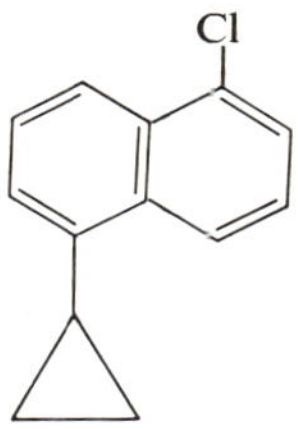

(j) 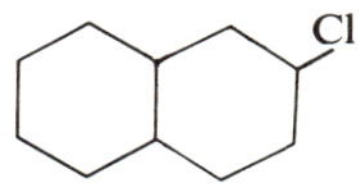

(k) 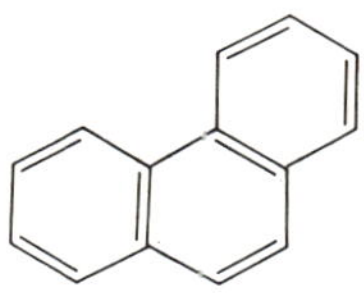

3. Write and name the products of the following reactions.

(a) 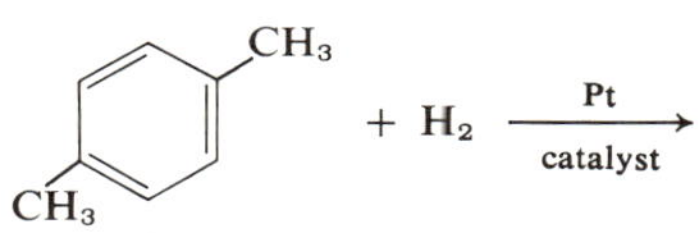

(b) 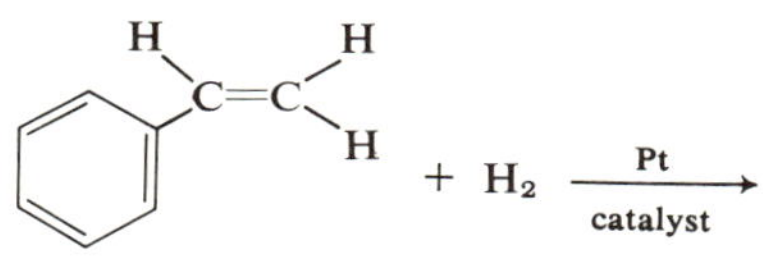

(c) $CH_3—C{\equiv}C—CH_3 + Br_2 \xrightarrow{\text{dark}}$

(d) 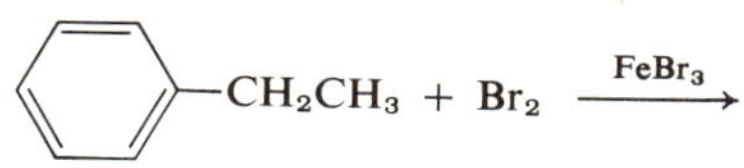

(e) 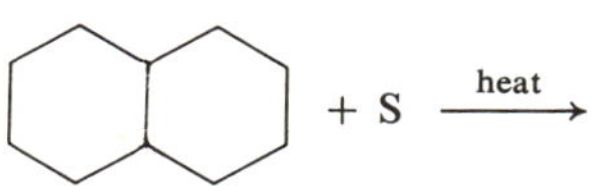

(f) 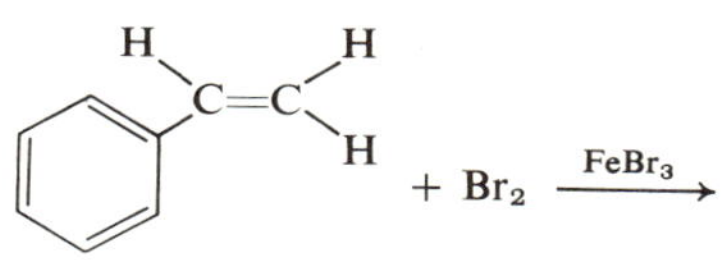

(g) 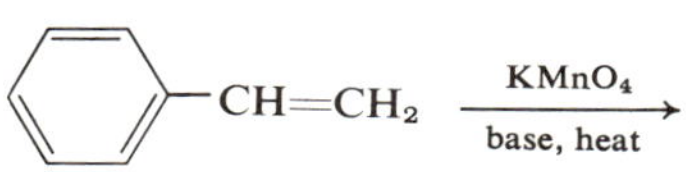

(h) 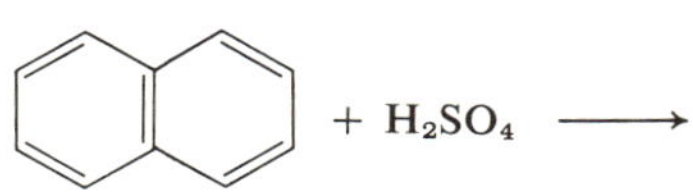

(i) 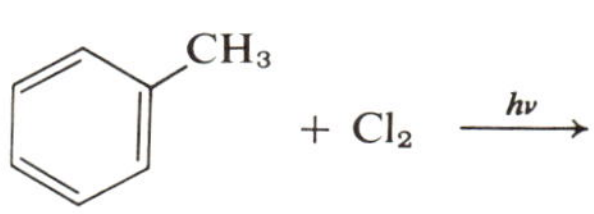

(j) 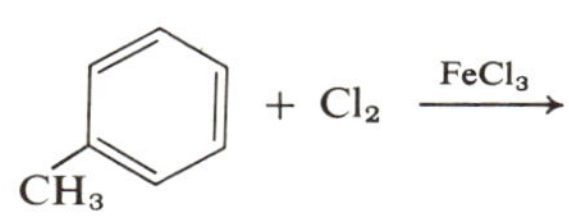

(k) 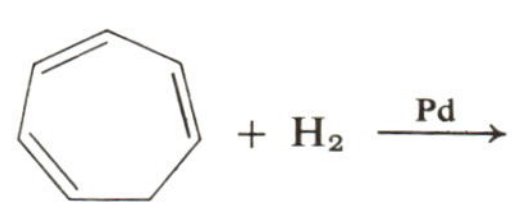

4. Devise a synthesis of the following compounds that uses no more than two reaction steps.

(a) 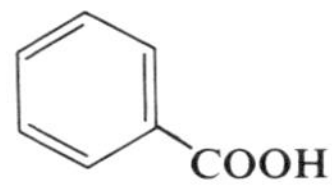

(b) 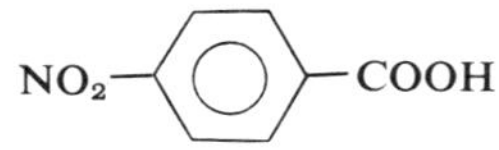

(c) 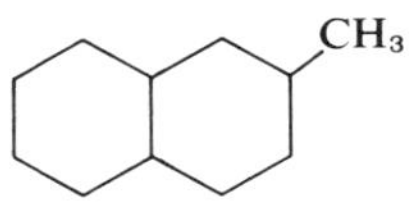

(d) 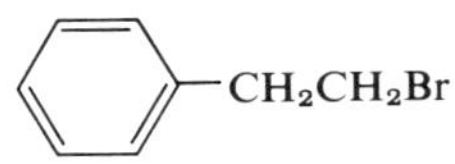

(e) 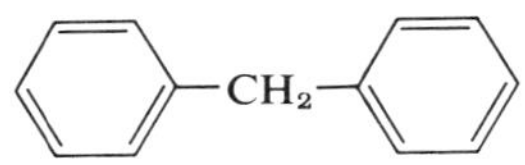

(f) 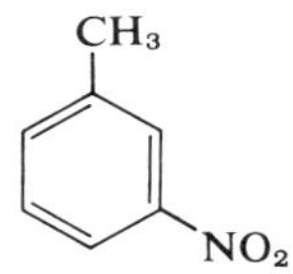

(g) 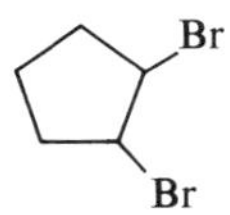

(h) 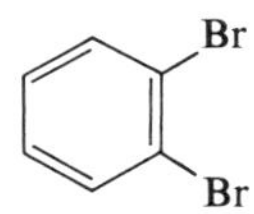

5. Sketch a free energy–reaction coordinate diagram for the addition of bromine to ethylene. Label all intermediates and transition states. Construct a similar diagram for the reaction of bromine with benzene.

6. A compound of formula $C_9H_{10}O_2$ is an acid. Upon vigorous treatment with $KMnO_4$, this material is transformed into another acid of molecular weight 122. What are the two acids?

7. Which compound will be a stronger acid: hydroxybenzene or hydroxycyclohexane?

8. When a mixture of propene and benzene is treated with a strong acid, isopropylbenzene is isolated. Suggest a mechanism by which this reaction may occur.

9. Write as many resonance structures as you can for chrysene and benzo[*j*]-fluoranthene. Which of these aromatic hydrocarbons would you expect to have the largest resonance energy?

10. (a) Using bond energy considerations, estimate the enthalpy change for hydrogenation (reduction) of cyclohexene. The experimental value is -29 kcal/mole.

(b) Estimate the enthalpy change for 1,3,5-cyclohexatriene.

(c) The experimental enthalpy change for reduction of benzene is -50 kcal/mole. Calculate the resonance energy of benzene using these data.

5

Molecular Spectroscopy and Chemical Structure I

Early in this text we mentioned some of the ways in which the organic chemist goes about proving the structure of the molecules with which he works. Classical chemical methods for structure proof have served the organic chemist well, but modern electronic instrumentation provides more information of higher quality about organic molecules than can be obtained by classical analytical techniques and it does so rapidly and often nondestructively. Because modern organic chemistry relies heavily on various kinds of instrumentation, it is appropriate that this text provide an introduction to the most important of these techniques. We shall consider two of these, mass spectrometry and nuclear magnetic resonance spectroscopy, in this chapter, and will take up other instrumental methods later.

MASS SPECTROMETRY

In mass spectrometry ions are produced from a compound by bombarding the molecules of the material with a beam of high-energy electrons. Positive ions can be produced if the incident beam of electrons strikes a molecule in such a way that some of its electrons are expelled. Other organic molecules may absorb electrons from this beam giving negative ions. In conventional mass spectrometry the positively charged ions are usually the ones of interest.

The mass spectrometer is an instrument which generates ions from the molecule under investigation and then separates these ions according to their mass-to-charge ratio. The instrument produces a plot of the amount of a given ion that is formed against the mass-to-charge ratio of the ion. Usually one is dealing with singly charged positive ions so that the charge is 1+ and the mass-to-charge ratio of an ion is simply the mass of the ion. Plots of this kind are called *mass spectra*. A schematic diagram of a mass spectrometer is shown in Fig. 5.1 and an experimental mass spectrum is found in Fig. 5.2.

The term *low resolution* is applied to those instruments which can resolve ions whose masses differ by 1 atomic mass unit. *High resolution* mass spectrometers are capable of discriminating a mass difference of 1 part in 10,000. A high-resolution instrument could thus distinguish between the ion $C_2H_5^+$ (m/e = 30.0469) and the ion CH_2O^+ (m/e = 30.0105), while a low-resolution instrument would see no difference between them.

Generally, the ions of high stability that are produced in a mass spectrometer are those that reach the detector. However, the cations produced in a mass spectrometric experiment with organic compounds can be high-energy,

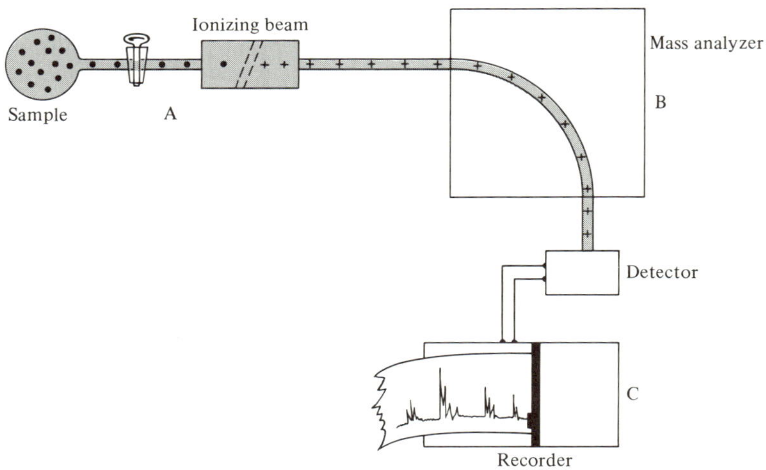

Fig. 5.1. A schematic representation of a mass spectrometer. Ions are generated at point A from the gaseous sample contained in the bulb. The ions are separated by their mass-to-charge ratio in the mass analyzer (B) and the abundance of each ion is detected and plotted (C). The mass analyzer may utilize magnetic or electrical fields to accomplish the separation of ions; consult an elementary physics book for the principles involved.

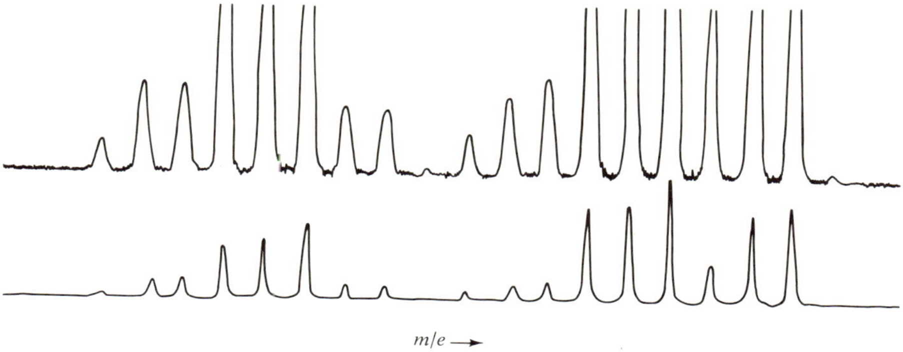

Fig. 5.2. A mass spectrum produced by a typical mass spectrometer. Each trace corresponds to a different gain setting so that very large and very small peaks can be recorded simultaneously.

excited species and can undergo chemical reactions in very short times—times so small that the reactions may be over before the ions enter the mass analyzer. The mass spectrum of an organic compound, therefore, does not simply consist of a single peak at the mass corresponding to the molecular weight of the compound, but also indicates the presence of other ions derived from the first-formed or parent ion by chemical reactions. The concentration of ions is very low in a mass spectrometer so that bimolecular reactions of the various ions formed is not likely. However, unimolecular reactions such as fragmentation and rearrangement do take place. In fragmentation, a molecular ion breaks down into smaller ions by simple breaking of a covalent bond in the molecule, while with rearrangement a reorganization (structural isomerization) of the molecular ion takes place prior to fragmentation. The weakest covalent bonds in a structure react most rapidly in fragmentation reactions so that ions derived by rupture of these bonds are the most abundant in the mass spectrum. Rearrangement reactions usually proceed in the manner expected for carbonium ion rearrangements.

Clearly a mass spectrum can yield a number of important clues regarding the structure of an organic molecule. The mass of the molecular ion formed by initial ionization of the material provides the molecular weight. Each of the additional peaks that appear in the spectrum corresponds to a part of the original molecular structure that can be identified by its mass. Like pieces of a jigsaw puzzle, these pieces of structural information can often be assembled to prove the structure of the molecule, or at least provide a strong indication of this structure. As an example, we can consider the mass spectrum of benzyl chloride, shown in Fig. 5.3. Initial bombardment of this molecule with electrons in the mass spectrometer causes ionization to the cation shown by loss of an electron from

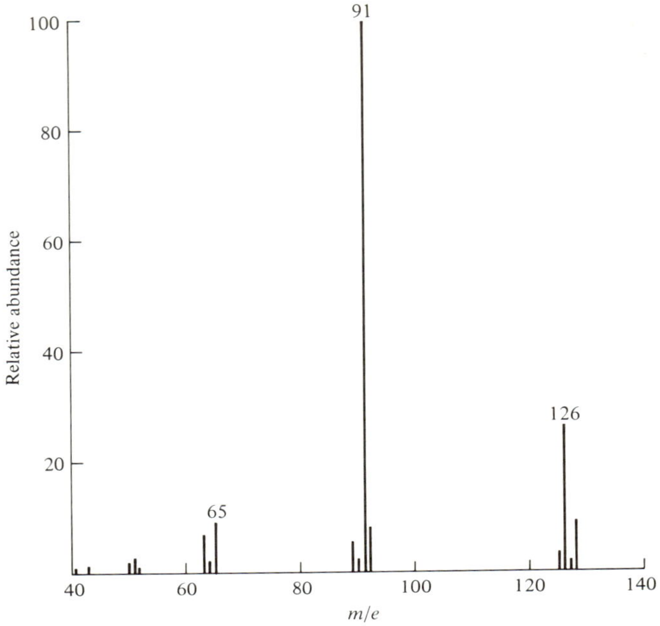

Fig. 5.3. A mass spectrum of benzyl chloride. (Low intensity peaks have been omitted.)

the structure. The mass of this ion, the molecular ion, is 126 ($C_7H_7Cl^+$). The molecular ion is not the most abundant ion in the spectrum, indicating that chemical reactions have taken place.

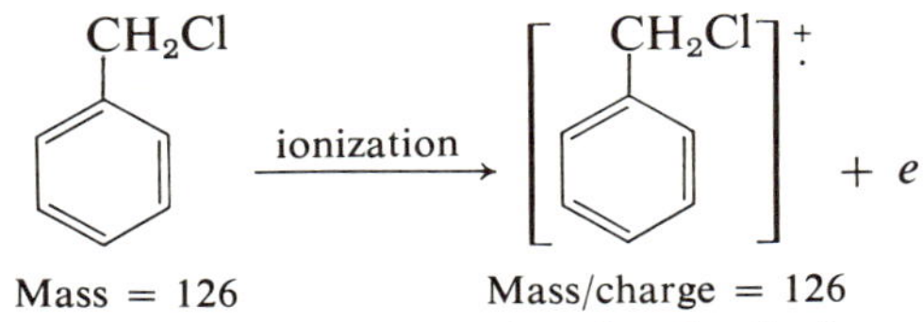

The carbon–chlorine bond is by far the weakest covalent bond in the molecule so that fragmentation at this bond to give $C_7H_7^+$ is observed. Rearrangement and fragmentation of the benzylic cation formed in this latter reaction account for the remaining major peaks in the spectrum at lower mass numbers.

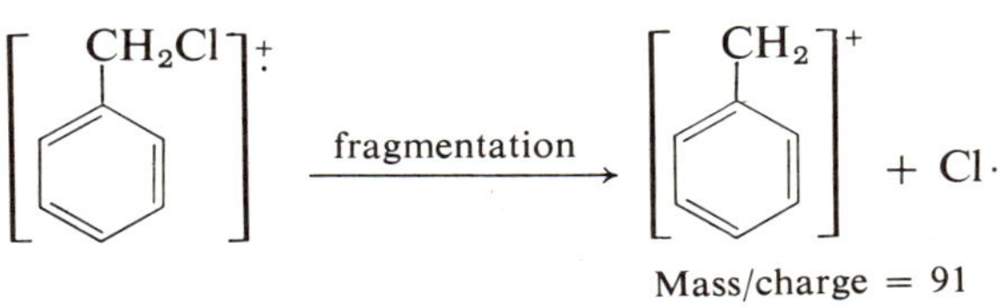

Great improvements in the instrumentation used for mass spectrometry have been made in recent years and the digital computer has been enlisted to aid in the interpretation of the spectra that these instruments produce.

NUCLEAR MAGNETIC RESONANCE SPECTROSCOPY

Nuclear magnetic resonance spectroscopy, commonly abbreviated nmr, is a technique that can be used to examine the environment of the nuclei of the atoms which make up a molecule. A given nuclear environment depends strongly upon the nature of the chemical bonds which hold the atom to the molecule and is also dependent on the number and types of other atomic nuclei in the immediate vicinity. A powerful diagnosis of molecular structure can, therefore, be obtained from an nmr experiment.

Nuclear Magnetism

The nuclei of many atoms have a special property called nuclear spin and nuclei with this property appear to rotate about an axis. The movement of electrical charges generates a magnetic field and it is found that nuclei that "spin" generate characteristic magnetic fields. The magnetic fields that come about in this manner are called nuclear magnetic moments. It is sufficient for our purposes to think of those nuclei that spin as behaving like small bar magnets with north and south poles. Important atomic nuclei which behave as bar magnets are ^{1}H (protons), ^{13}C, ^{19}F, and ^{31}P. Nuclei that do not have this magnetic property are ^{12}C and ^{16}O.*

To continue with the bar magnet analogy, let us suppose that we have a tray that contains a number of small bar magnets. The magnets will be assumed to be far enough from each other so that they do not interact appreciably with one another (Fig. 5.4). If the tray is vigorously shaken, the bar magnets will be randomly oriented and distributed on the tray. This crude example is analogous to the protons (hydrogen nuclei) of an organic molecule in a solution; the shaking of the tray corresponds to the thermal motions of the system which continuously agitate the molecules and thereby change the orientation of the molecules with respect to one another.

---

* Nuclei with even charge *and* even mass numbers have *no* spin property. Nuclei that have even mass but odd charge such as ^{14}N and ^{2}H *do* possess a nuclear spin but it does not result in a magnetic field that behaves as a simple bar magnet.

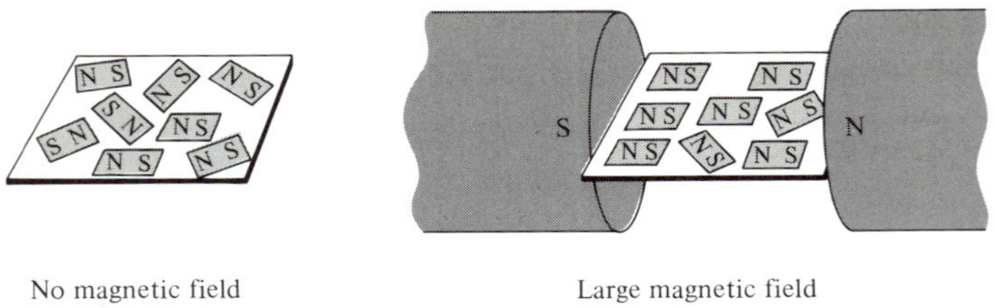

Fig. 5.4. The behavior of a tray of bar magnets in the absence and presence of a magnetic field.

Now suppose that the shaking tray of bar magnets is placed between the poles of a very powerful electromagnet. Because of the interaction of the magnetic fields of the bar magnets with the large external magnetic field, there will be a tendency for the bar magnets to become aligned along the axis of the external field, in spite of the fact that the tray holding them is in motion.

If the effects of tray-shaking are neglected temporarily, we see that there are two ways in which a bar magnet could be oriented in the presence of an external magnetic field (Fig. 5.5); the field direction of the bar magnet can be collinear (parallel) with the field direction of the large external field or the bar magnet can take the opposite (antiparallel) orientation with respect to the applied field. Like magnetic poles tend to repel one another; unlike magnetic poles attract. Thus one of these possible orientations (A) for the bar magnet in the external field is of higher energy than the other and there is a strong tendency for the bar magnet in this high energy state to flip (to reorient itself) into the lower energy state (B). However, the orienting effect of the electromagnet can be largely cancelled out by vigorous shaking of the tray.

Much the same thing happens in molecules except that with each nuclear "magnet" there are only two allowed energy states. The interaction energy between a nuclear spin and an external magnetic field is quantized in quite the same sense that the electronic energy of an atom is quantized. The thermal

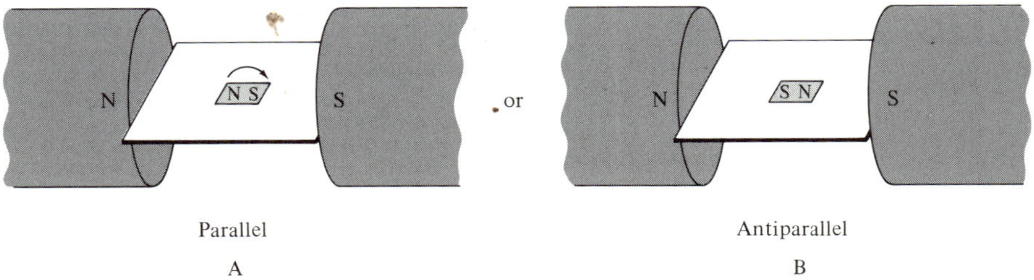

Fig. 5.5. The orientations of a bar magnet in a large magnetic field.

motions of molecules provide mechanisms whereby a nuclear spin can be transformed from one energy state to another and an equilibrium between the nuclear spin states accessible to the molecule is maintained. The number of nuclei in the spin state of lowest energy is slightly greater than the number in the higher energy state; it is the excess number of nuclei in the low energy state that gives rise to the nmr effect.

If we place a test tube containing trichloromethane into a large magnetic field, the proton in each molecule of the sample must be in one of the two allowed nuclear spin states. One of these states, corresponding to the favorable orientation B of Fig. 5.2, is lower in energy than the energy when the magnetic field is absent. The other state is higher in energy than when the magnetic field is absent.

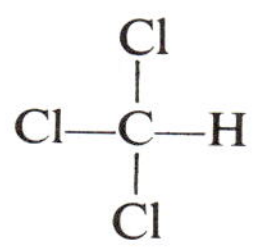

Trichloromethane
(chloroform)

Theory shows that E, the energy difference between these two states is given by the equation

$$\Delta E = \alpha H_0 \tag{5.1}$$

where α is a constant that is characteristic of the type of nucleus involved (here, a proton) and H_0 is the magnitude of the external magnetic field. Equation (5.1) is quite reasonable, as we would expect the magnitude of the interaction between the nuclear magnetic field and the external field to depend on the strength of the external field.

Interaction of Molecules with Light

We must digress at this point in order to note that atoms and molecules can interact with light in ways that can lead to the absorption of the light; the photochemical isomerization of olefins (Chapter 3) is one consequence of light absorption by organic molecules. Electromagnetic radiation can be thought of as a stream of small energy packets, called photons. The relation between the energy of a photon and the wavelength of the light is given by

$$E = \frac{hc}{\lambda} \tag{5.2}$$

where c is the velocity of light, λ is the wavelength of the light, and h is a proportionality constant, usually called Planck's constant. The quantity c/λ is the frequency characteristic of the light and is symbolized ν. The energies of several photons that will be important for this and later discussions are recorded in Table 5.1. It should be noted that the portion of the total electromagnetic radiation spectrum to which the human eye responds (visible light) is rather small.

Atoms and molecules can absorb or release energy in the form of photons if the size of the photon involved is exactly equal to the energy difference between two energy levels of the atom or molecule. Since there are only certain allowed energy levels in a particular atom or molecule, only photons of certain energies will be absorbed or emitted. Thus, only electromagnetic radiation of specific wavelengths is absorbed by a particular system. Since the spacings between energy levels change from molecule to molecule and are unique for each molecule, a study of the frequencies of light absorbed or emitted by a molecule can, in principle, provide a great deal of information that is significant for structural determinations. Molecular spectroscopy, the study of absorption of electromagnetic radiation by molecules, enables the organic chemist, in effect, to "photograph" the molecule under investigation.

Table 5.1
The Electromagnetic Spectrum[a]

Type of radiation	*Energy (cal/mole)*	*Frequency (Hz)*	*Wavelength (cm)*
X-rays	10^7–10^8	10^{17}–10^{18}	10^{-7}–10^{-6}
Ultraviolet light	10^5–10^6	6×10^{14}–10^{16}	10^{-6}–5×10^{-4}
Visible light	6×10^4–8×10^4	3×10^{14}–6×10^{14}	2×10^{-4}–5×10^{-4}
Infrared light	10^2–6×10^4	10^{12}–10^{14}	10^{-4}–10^{-2}
Microwave	1–10^2	10^{10}–10^{12}	10^{-1}–1.0
Radiowaves	10^{-4}–10^{-1}	10^6–10^9	10–10^4

[a] For a more complete tabulation consult J. R. Dyer, "Applications of Absorption Spectroscopy of Organic Compounds," p. 3. Prentice-Hall, Englewood Cliffs, New Jersey, 1965.

The energy difference between two nuclear spin states is very small (of the order of hundredths of calories per mole) so that the size of the photon that is absorbed in making a transition from the lower energy state to the upper energy state corresponds to electromagnetic radiation in the radiofrequency region (Table 5.1).

Nuclear Magnetic Resonance Spectra

There are two possible energy states for the proton nuclear spin of chloroform when the sample is in a magnetic field. If radiation corresponding exactly to the energy difference (ΔE) between these states impinges upon the sample, a photon can be absorbed. The result of the absorption of a photon is to excite a particular spin from the low energy state to the higher energy state. That is, by absorbing the energy of a photon, a nucleus can be transformed from the favorable orientation relative to the magnetic field to the high-energy, unfavorable orientation.

The condition that must be met for this absorption of energy is

$$\Delta E = h\nu = \alpha H_0 \tag{5.3}$$

To determine the values of ν and H_0 that correspond to the absorption of energy one of these quantities is held constant and the other is varied until an absorption of energy takes place. The electronic apparatus necessary to perform this operation experimentally is called a nuclear magnetic resonance spectrometer; the essential components of this instrument are sketched in Fig. 5.6. Typically, instruments of this type provide a plot of a quantity proportional to the amount of energy absorbed, as determined by the detector, against the value of the applied magnetic field. Such a plot defines the *nmr spectrum* of the compound in the sample tube.

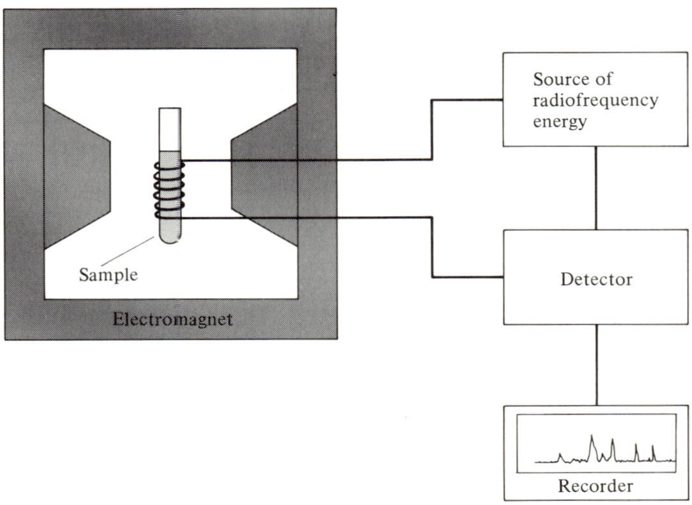

Fig. 5.6. A schematic representation of an nmr spectrometer.

By far the most common application of nmr spectroscopy is the study of the hydrogen nuclei of organic compounds. When used in this way the technique is often called pmr (proton magnetic resonance) spectroscopy. Proton nmr signals are the easiest to detect of the common nuclei mentioned above and commercial instruments have evolved to the point where pmr spectra are relatively easy to obtain. As we hope to show below, pmr spectroscopy has become an indispensable technique for the study of the structure of organic molecules.

The Chemical Shift

If the hydrogen nuclei of all organic compounds had exactly the same response to an external magnetic field, all pmr spectra would be identical, and this spectroscopic technique would be of little value. Nature has been kind to the organic chemist in this regard because protons that are chemically different give rise to nmr absorption signals that occur at slightly different values of the external field, H_0. The pmr spectrum obtained from a mixture of chloroform and cyclohexane, for example, illustrates this effect (Fig. 5.7). These slight changes in the value of H_0 that are necessary for resonance are called *chemical shifts*.

It is difficult to measure and interpret chemical shifts on an absolute basis and it is customary to include a small amount of a reference compound in the samples used for pmr spectroscopy. The most common reference material is tetramethylsilane (TMS); it is used since its pmr signal occurs at higher H_0 than most other proton nmr signals.

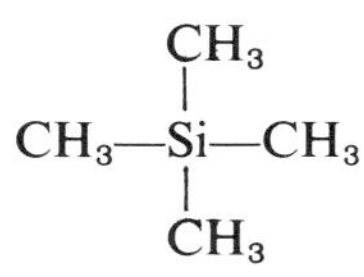

Tetramethylsilane
(TMS)

Chemical shifts are measured relative to the position of the reference signal in units of hertz (Hz) or cycles per second (cps). In order to convert these shifts into a dimensionless number, the separation of the signal of the sample from the signal of the reference compound, measured in Hz, is divided by the radiofrequency used for the experiment. That is,

$$\delta = \frac{\Delta}{\nu}$$

where Δ is the difference between the signal position for the reference com-

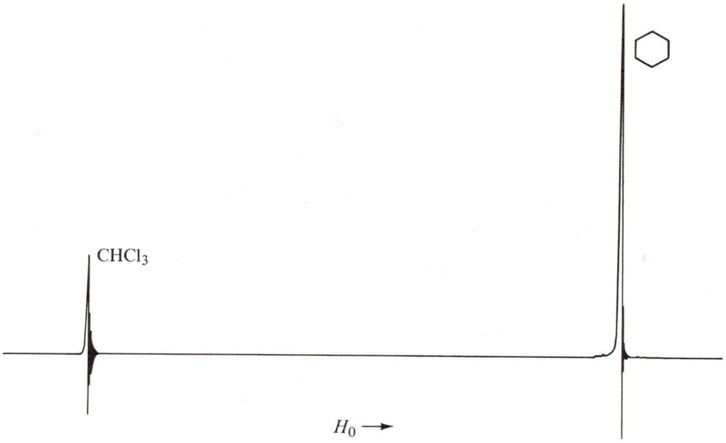

Fig. 5.7. The pmr spectrum of chloroform and cyclohexane.

pound and the signal of interest, expressed in Hz, and ν is the radiofrequency. Typically, δ will turn out to be between 0 and 10 ppm for proton magnetic resonance spectroscopy.* The chemical shift parameter, δ, is inherently a negative quantity if TMS is used as the reference compound. However, by convention the sign is tacitly neglected since nearly all values of δ for organic compounds are negative.

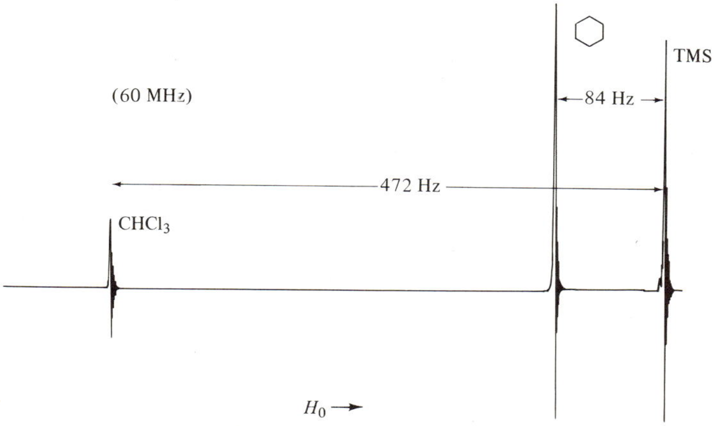

Fig. 5.8. The pmr spectrum of the sample used to obtain the spectrum in Fig. 5.7 with 10% TMS added. The separations between the signals and TMS are given in Hz (cps); the spectrometer operating frequency was 60 MHz.

* In most instruments for pmr spectroscopy, ν is between 60×10^6 Hz (60 MHz) and 300×10^6 Hz.

If a small amount of TMS is included in the chloroform–cyclohexane sample used to obtain the spectra in Fig. 5.7, the pmr spectrum shown in Fig. 5.8 is obtained. With the data given there, one calculates that the chemical shift parameter for cyclohexane is 1.40 ppm (84 Hz/(60 $\times$ 10^6)), while that for chloroform is 7.87 ppm.

Hydrogen atoms are bound to organic structures predominantly by the interaction of their 1*s* atomic orbitals with other orbitals of the molecule. If we restrict our attention only to a single hydrogen atom and consider only its associated 1*s* orbital, we have to some degree an isolated proton (hydrogen nucleus) surrounded by a nearly spherically symmetric distribution of electronic charge (Fig. 5.9). When the distribution of electronic charge is placed in a magnetic field, it circulates about the nucleus in paths defined approximately by the

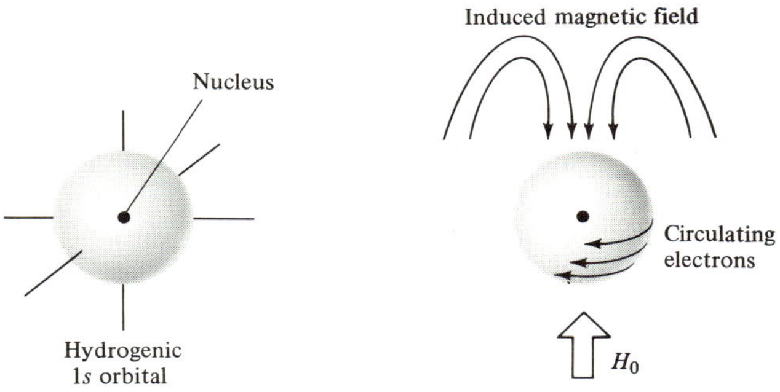

Fig. 5.9. The circulation of electrons in a hydrogen 1*s* orbital and the chemical shift effect.

atomic orbital. But circulating electrical charges generate magnetic fields and, in this case, the circulation is such that the generated magnetic field opposes (points in the opposite direction) the applied magnetic field. As a result, the nucleus of this atom experiences a total effective magnetic field that is not the same as the applied field. The magnitude of the induced magnetic field depends on the density of electrons in the orbital(s) around the hydrogen atom and the magnitude of the externally applied field. The magnetic field experienced by a given nucleus is thus dependent upon where the nucleus is in a molecule and how it is chemically bound to the molecule. The conditions of field and frequency that must be satisfied in order to have energy absorption occur remain the same, however, so that small adjustments in one of these operating parameters must be made to cause absorption by each nucleus in the molecule. If frequency is held constant, then the magnetic field required for absorption to occur will be different for each chemically unique nucleus in the molecule.

Table 5.2
Some Chemical Shifts for Protons in Organic Molecules

Type of proton	δ (*ppm*)[a]
TETRAMETHYLSILANE (reference)	0.0
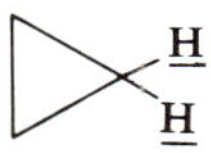	0.2
$R{-}C\underline{H}_3$	1.0
$R{-}C\underline{H}_2{-}R'$	1.3
$R{-}C(R')(R''){-}\underline{H}$	1.5
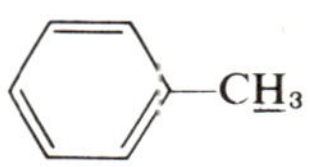	2.4
$-S{-}C\underline{H}_3$	2.5
$-C{\equiv}C{-}\underline{H}$	2.5
$-O{-}C\underline{H}_3$	3.6
$R{-}O\underline{H}$	3.0–5.0[b]
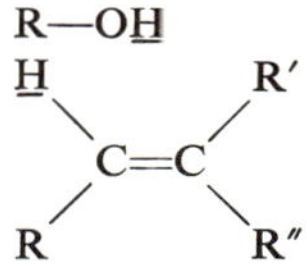	5.0–5.5
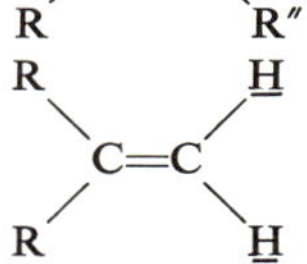	4.5–5.0
AROMATIC PROTONS	6.6–8.0
$R{-}\overset{+}{N}\underline{H}_3$	7.5
$R{-}C({=}O){-}\underline{H}$	9-10
$RCOO\underline{H}$	11–12[b]

[a] TMS was the reference sample; all signals quoted above appear downfield from TMS. *The δ values above are intended to indicate an approximate position for the proton resonance under consideration. Particular cases may deviate considerably from these values.*

[b] These signals are highly concentration and temperature-dependent.

The magnitude of the induced field will depend upon the electron density in the 1*s* orbital surrounding the proton. The chemical shift parameter will thus depend on the electronegativity of other atoms in the molecule to the extent that these atoms determine the distribution of electrons in the molecular orbital system of the molecule. There are several other effects that derive from the manner in which electrons circulate in molecules and which can contribute to the chemical shift of a given nucleus. We shall not consider these in further detail here; the most important point for our present consideration is that structurally different nuclei are usually distinguishable in the nmr spectrum by means of the chemical shift phenomenon. Enough work has been done over the past fifteen years to permit empirical correlations of chemical shifts with chemical type. The usual vehicle for presenting such correlations is a table, such as Table 5.2. More extensive correlations are available from the references cited at the end of this chapter.

Quantitative Analysis

The intensity of a signal in an nmr spectrum is directly related to the number of nuclei that absorb energy. The nmr spectrometer can thus be a quantitative analytical tool. However, only relative intensities of peaks can be determined with accuracy and for this purpose most commercial spectrometers include electronic circuits for integrating the area under an nmr signal. These areas are related to the number of nuclei present, and the relative areas of the various peaks in a pmr spectrum of an organic compound provide an indication of the relative abundance of each chemical type present. The pmr spectrum of toluene is shown in Fig. 5.10. This spectrum tells us that there are two distinct classes of protons in toluene, and with the aid of the chemical shifts measured from this spectra and of Table 5.2, we surmise that there are aromatic protons and the protons of a methyl group present in the molecule. The integration curve shows that the relative abundance of these two kinds of protons is 5:3. These data are quite consistent with the known structure of toluene.

Spin-Coupling Constants

There is usually more information about the structural features of organic molecules in an nmr spectrum than is provided by chemical shift data alone. The pmr spectrum of ethyl bromide, CH_3CH_2Br, is shown in Fig. 5.11. This spectrum is more complex than we might have expected, although integration of the peak

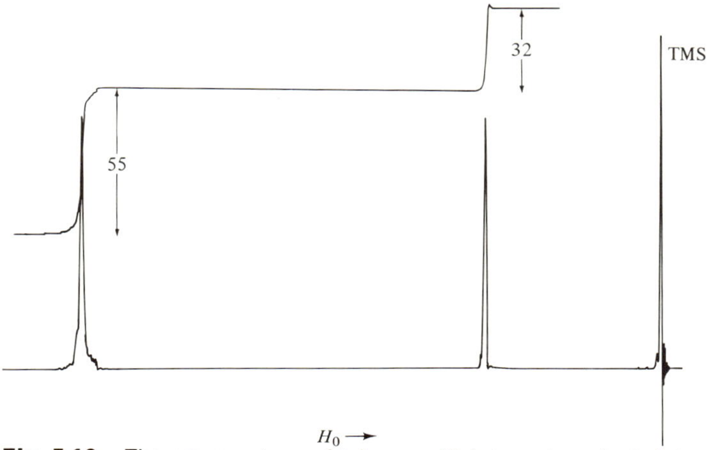

Fig. 5.10. The pmr spectrum of toluene with integration; the height of each step in the top curve is proportional to the area under the absorption peaks beneath it.

areas for the two groups of signals suggests that these multiplets correspond to the —CH_2— group (signal at 3.15 ppm, relative area = 2) and the —CH_3— group (1.72 ppm, relative area = 3). The fine structure in this spectrum is due to an additional magnetic interaction between nuclei known as *spin–spin splitting* or *spin–spin coupling*. A complete description of this effect is well beyond the scope of this discussion, and we shall simply state what is known as the *first-order* rule, in operational form. This generalization comes in two parts: (a)

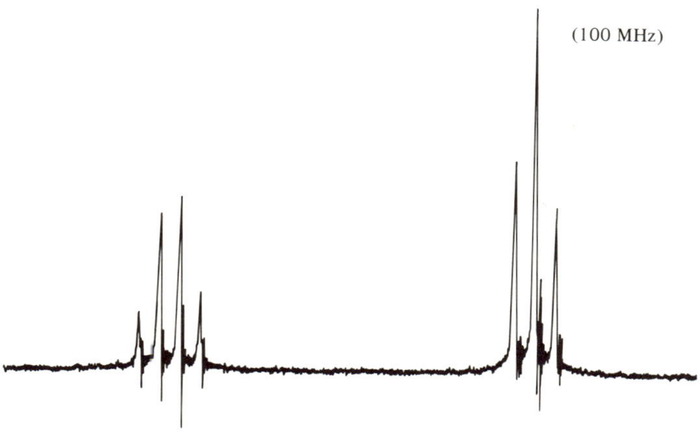

Fig. 5.11. The pmr spectrum of ethyl bromide at 100 MHz. The triplet signal is 172 Hz from TMS, while the quartet is 315 Hz from the reference peak.

Table 5.3
Some Proton Spin–Spin Coupling Constants

Type of structure	*J (Hz)*[a]
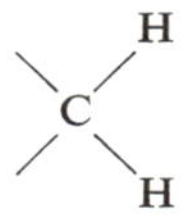	$J_{12} = -12$ to -15
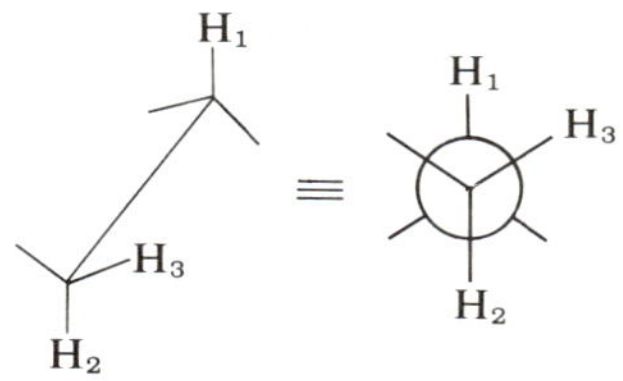	$J_{12} = 10$ to 18 $J_{13} = 1$ to 3.5 (note dihedral angle dependence)
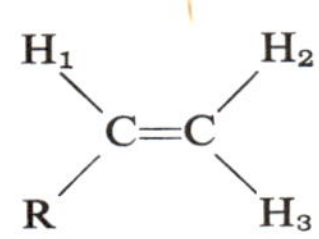	$J_{12} = 5$ to 14 $J_{13} = 11$ to 18 $J_{23} = -4$ to 0
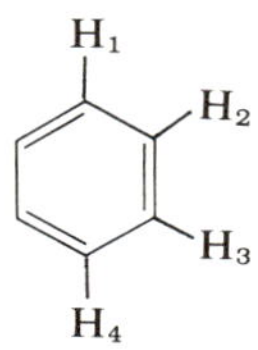	J_{12} (*ortho*) = 8 to 10 J_{13} (*meta*) = 2 to 4 J_{14} (*para*) = 0 to 2

[a] Although signed numbers are given here, it is only the absolute value that is important in the interpretation of first-order nmr spectra.

nuclei which have the same chemical shift do not show any observable spin–spin coupling, and (b) when there are *n* equivalent protons near a specific proton and the chemical shift difference between this proton and the *n* equivalent protons is large, the pmr signal for this proton will be split into $n + 1$ lines. The second part of this rule is illustrated by the pmr spectrum of ethyl bromide. The signals assigned to the methyl group ($\delta = 1.72$) appear as a triplet because of spin–spin coupling to the two adjacent —CH_2— protons. Similarly, a quartet is observed for the methylene (—CH_2—) protons because there are three adjacent, chemically equivalent protons in the methyl (—CH_3) group ($n + 1 = 4$). The spacing between the lines in each of these two multiplets is the same and is called the coupling constant. Coupling constants are usually given the symbol *J* and are recorded in units of Hz. Although chemical shifts depend upon the particular magnetic field and radiofrequency used to determine a pmr spectrum, the values of spin-coupling constants do not.

The magnitude of a spin-coupling interaction depends upon the number of covalent bonds that intervene between the interacting nuclei; if there are more than four covalent bonds separating these nuclei the coupling effect is usually negligibly small. The size of a spin–spin coupling constant also is a sensitive function of the geometrical properties of the structure which holds the interacting nuclei. These points are illustrated by the data in Table 5.3, which records typical values of J for several common situations. We should caution at this point that first-order rules for spin–spin splitting apply only when chemical shifts (δ) are large relative to J. When this condition is not satisfied, the splitting patterns become much more complex and a more complete theoretical treatment is necessary.

An example of the power of pmr techniques in the elucidation and confirmation of organic structures is afforded by the three structurally isomeric trichlorobenzenes. Each of these substances has the empirical formula $C_6H_3Cl_3$ and to distinguish between them by purely chemical studies would be tedious. The pmr spectrum of each substance shown in Fig. 5.12 identifies and establishes the

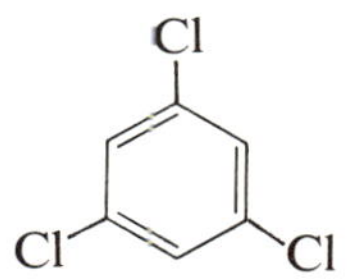

1,3,5-Trichlorobenzene

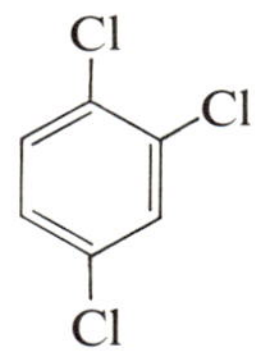

1,2,4-Trichlorobenzene

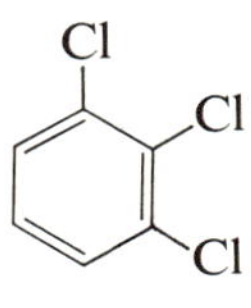

1,2,3-Trichlorobenzene

structure of each isomer easily. Spectrum (a), a single line, must correspond to 1,3,5-trichlorobenzene, since all of the protons in this molecule are chemically equivalent and have the same chemical shift. Spectrum (b), a doublet and a triplet ($J = \sim 7.5$ Hz) and relative areas 1 and 2 are assigned to 1,2,3-trichlorobenzene; the signals for the equivalent protons at positions 4 and 6 are split into doublets by spin–spin interaction with the proton at position 5, which, in turn, appears as a triplet because of the same interaction. Spectrum (c) is assigned to 1,2,4-trichlorobenzene in a similar way.

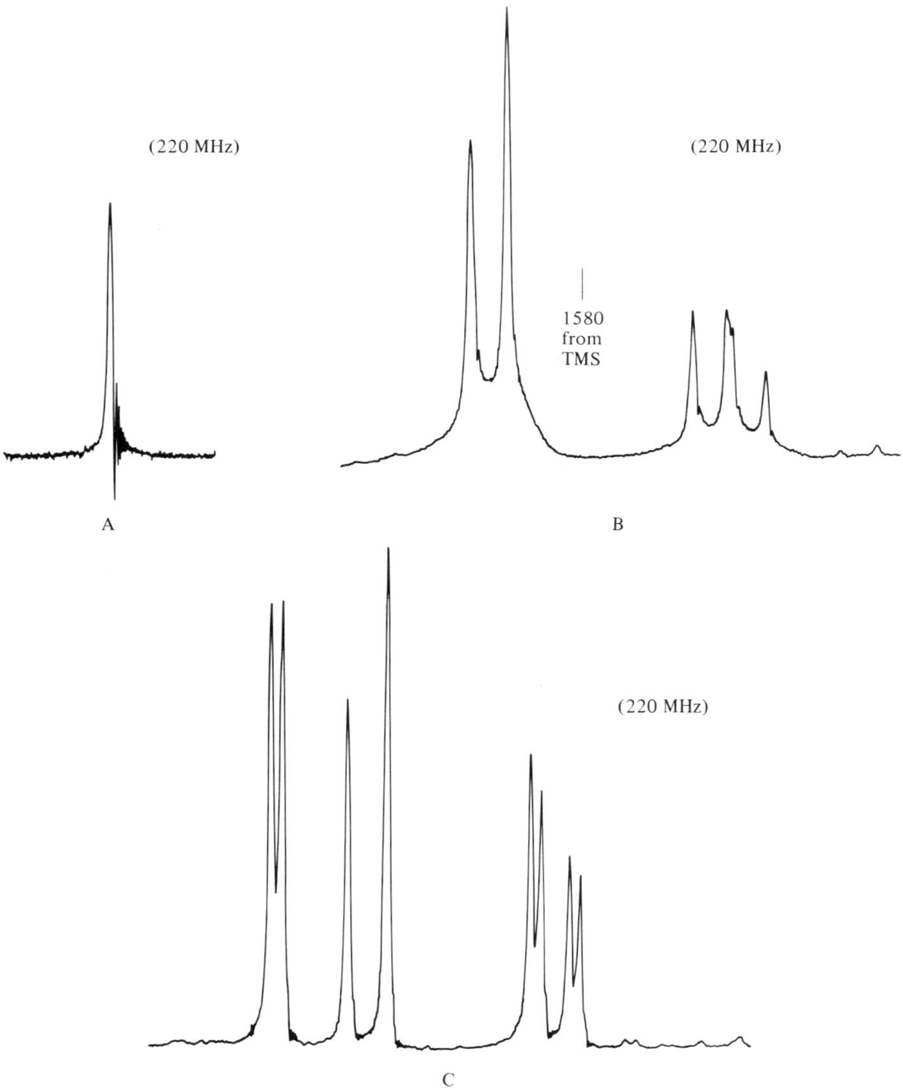

Fig. 5.12. Pmr spectra of three isomeric trichlorobenzenes recorded at 220 MHz.

PROBLEMS

1. For each of the following compounds state how many chemically unique types of protons are present, i.e., how many different chemical shift parameters, δ, are needed to characterize the spectrum of the compound.

(a) methane
(b) benzene
(c) ethane
(d) 1-hydroxypropane
(e) 1-bromo-2,4-dinitrobenzene
(f) naphthalene
(g) cumene
(h) 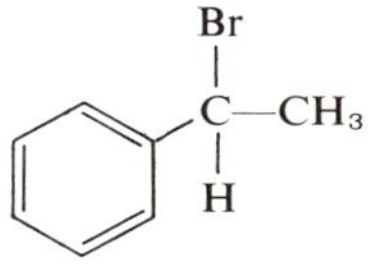

(i) 2-hydroxypropane
(j) 1-chloro-4-*t*-butylcyclohexane

2. In a certain sample the pmr signal for chloroform comes at 733.0 Hz from TMS. What is the chemical shift, δ, for chloroform in this sample if the spectrometer frequency is 100 MHz?

3. The mass spectrum of *n*-octane is shown below. Account for the structures of the ions at m/e = 114, 85, 71, 57, 43, and 39. What is the cause of the smaller peaks near these major ones? Hint: See Table 1.2.

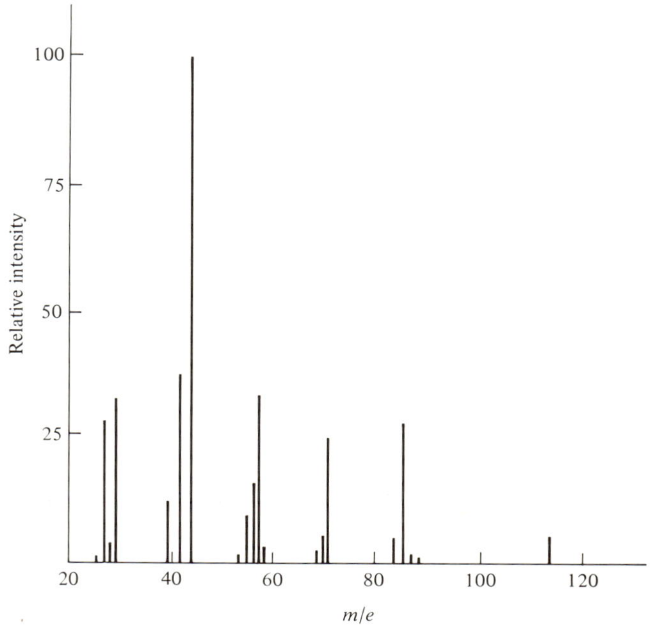

4. A mixture of methanol (CH_3OH) and methyl iodide the (CH_3I) gives pmr spectrum shown below. Rationalize the assignments of the three signals downfield from TMS.

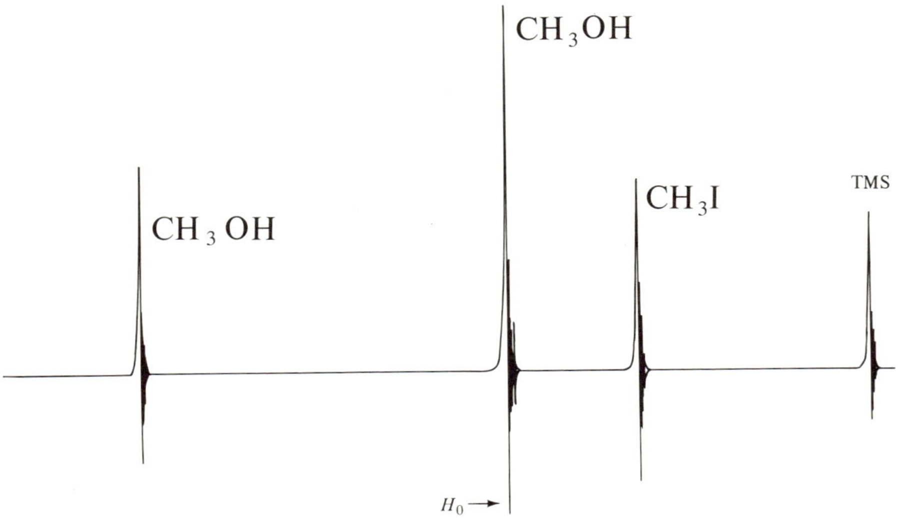

5. Predict the pmr spectrum of the following molecules using first-order considerations and the data in Tables 5.2 and 5.3.

(a) Methylacetylene
(b) Methyl iodide
(c) 1,1-Dimethylethene
(d) Vinyl bromide
(e) *trans*-1,2-Dichloroethylene
(f) Styrene
(g) CH_3—S—CH_3
(h) 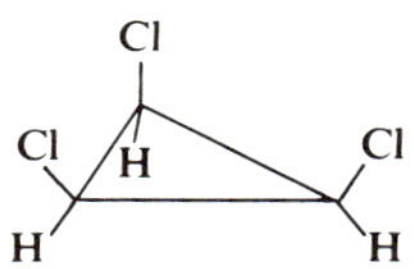

(i) 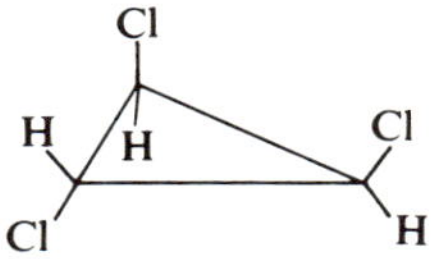

6. Predict the mass spectrum of iodobenzene. How will this spectrum differ from that of fluorobenzene?

7. Deduce the structures of the organic molecules which are characterized by the molecular formulas and spectra given on p. 129.

(a)

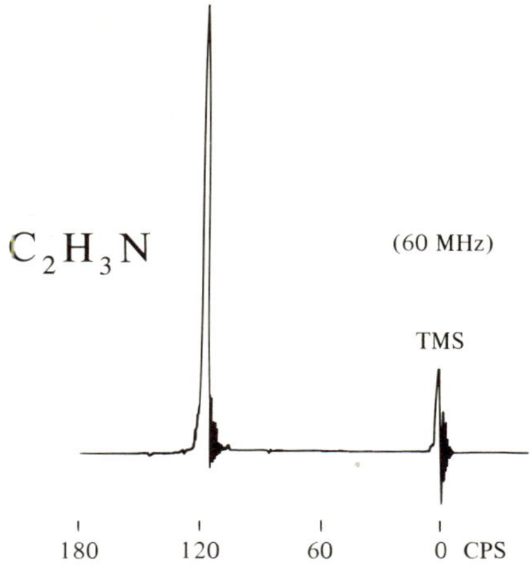
C_2H_3N
(60 MHz)
TMS
180
120
60
0 CPS

(b)

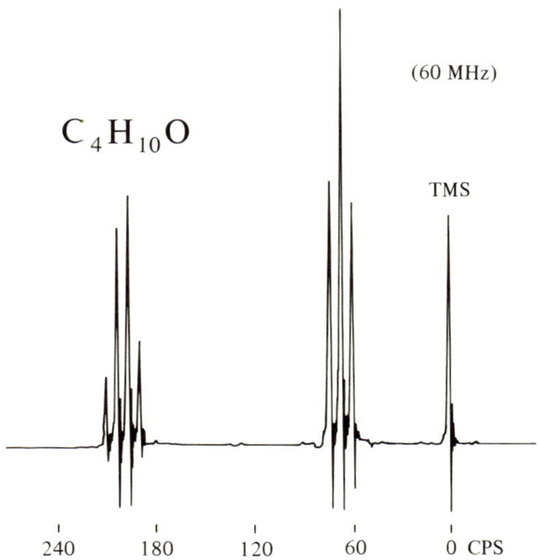
$C_4H_{10}O$
(60 MHz)
TMS
240
180
120
60
0 CPS

(c)

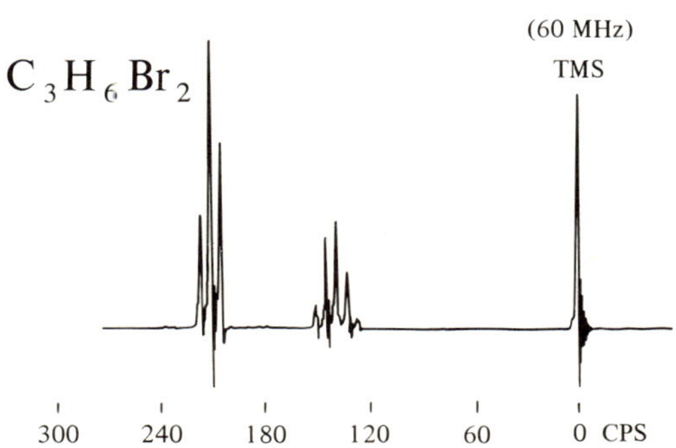
$C_3H_6Br_2$
(60 MHz)
TMS
300
240
180
120
60
0 CPS

SUGGESTED READINGS

Mass Spectrometry

H. Budzikiewicz, C. Djerassi, and D. H. Williams, "Mass Spectrometry of Organic Compounds." Holden Day, San Francisco, California, 1967.

R. W. Kiser, "Introduction to Mass Spectrometry." Prentice-Hall, Englewood Cliffs, N.J., 1965.

F. W. McLafferty, "Interpretation of Mass Spectra." Benjamin, New York, 1966.

Nmr Spectroscopy

E. D. Becker, "High-Resolution N.M.R. Spectroscopy." Acacemic Press, New York, 1969.

J. R. Dyer, "Applications of Absorption Spectroscopy of Organic Compounds." Prentice-Hall, Englewood Cliffs, N.J., 1965.

J. W. Emsley, J. Feeney, and L. H. Sutcliffe, "High-Resolution Nuclear Magnetic Resonance Spectroscopy." Pergamon, New York, 1966.

J. D. Roberts, "Nuclear Magnetic Resonance." McGraw-Hill, New York, 1959.

6

Compounds Derived from Water and Hydrogen Sulfide

"Water, water everywhere . . ." begins a familiar stanza and, indeed, no fluid is more abundant on the earth's surface than this ubiquitous liquid. Living systems are largely water and virtually all biologically significant reactions occur in an aqueous environment. Water is a good solvent for many organic and inorganic materials, coming close to earning its designation as the "universal solvent." The unique properties of water are unexpected; Table 6.1 compares the vaporization behavior of water to other hydrogen derivatives of adjacent elements in the periodic chart. The high boiling point of water and its attendantly large heat of vaporization (the energy required to convert 1 mole of liquid into 1 mole of gas) are clearly out of line and indicate that some unexpected interaction between water molecules must take place in the liquid state.

Table 6.1
Properties of Hydrides of Elements near Oxygen

Formula	H_4C	H_3N	H_2O	HF	(Ne)
Boiling point (°C)	−162	−33	100	20	−246
Heat of vaporization (kcal/mole)	2.0	2.6	9.7	1.8	0.43

By considering the geometrical properties of the $2p$ orbitals on the central oxygen atom, it has been predicted that the H—O—H bond angle in water should be about 90°; this prediction agrees fairly well with the experimental value of 104.5°. The bent nature of the water molecule has two important implications. First, the remaining electrons (two unshared pairs) on the oxygen

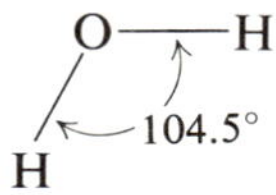

Bond angle in water

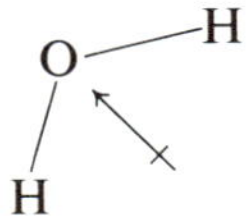

Expected dipole moment of water

atom are relatively exposed so that reactions with water and its analogs that utilize these electrons should not be particularly hindered by the presence of the hydrogen atoms. Second, a molecule of this shape with a highly electronegative atom like oxygen would be expected to be strongly dipolar.

These structural features of the water molecule largely account for the anomalous chemical and physical properties of this material. The intermolecular interaction between water molecules called hydrogen bonding is due to the dipolar nature of the water molecule and the exposed electron density on the oxygen atom. This interaction does not lead to the formation of a bond of the same strength as a typical covalent bond, but rather is weaker, with a bond energy of 2 to 10 kcal/mole. Hydrogen bonding accounts for the high heat of vaporization of water since an extensively hydrogen-bonded network must be disrupted before a water molecule can escape the liquid phase to enter the gas phase.

Since the water molecule is dipolar in nature, it can interact electrostatically with materials that are dissolved in it. Water is an especially good solvent for inorganic salts because the water molecules can orient themselves around cations and anions and electrostatically interact with them. The interaction of a solute and solvent is called solvation and the layer of solvent molecules most intimately involved with the solute is called the solvation shell.

In 1968, B. V. Deryagin and his co-workers in the Soviet Union described what they believed to be a new form of water, now usually called polywater or anomalous water. This material is obtained by condensation of water into very fine glass capillary tubes and has a higher density, boiling point, and viscosity than "normal" water. The discovery of "polywater" caused great excitement because of its theoretical and biological implications. Theorists quickly produced calculations to explain why new forms of water are possible and biologists

reflected upon the relevance of polywater to the cytoplasmic composition of living cells. Further experimentation showed that polywater contains high concentrations of the sodium and potassium salts of sulfate, nitrate, chloride, silicates, and borates, probably leached from the glass capillary, as well as appreciable concentrations of lipids and phospholipids (fatty materials), contaminants that plausibly could arise from human sweat. At the present, the high concentrations of impurities in polywater make it unlikely that a new form of water per se has been discovered, but rather that the properties of this material are a result of these impurities.

In this chapter we shall consider organic molecules which are derived from water by the substitution of one or both hydrogens by an alkyl or aryl radical. Many of the properties of these compounds are directly the result of the electronegativity of oxygen. The bonding pair of electrons in a carbon–oxygen bond is strongly attracted to the oxygen atom; the carbon–oxygen bond is said to be polarized. As indicated below, this polarization leaves the carbon atom electron-deficient. It can thus function as an electrophile. The excess of electronic charge

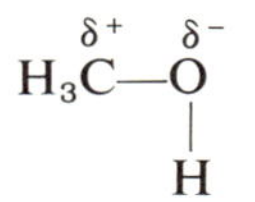

Polarization of a carbon–oxygen bond; the symbols δ^- and δ^+ mean "an excess of negative charge" and "a deficiency of negative charge," respectively

on the oxygen atom coupled with the already available electron pairs permits the oxygen atom to behave as a base.

ALCOHOLS

Substitution of one of the hydrogens of water by an alkyl radical gives a class of compounds known as alcohols. The —OH portion of an alcohol is referred to as a *hydroxyl* group and the carbon which bears the hydroxyl group is called the *carbinol* carbon. Some examples of alcohols are shown below.

Methanol is an important industrial chemical and is also used as nonpermanent antifreeze. It is extremely poisonous to humans. Ethyl alcohol is obtained commercially by hydration of ethylene or by fermentation reactions. It is the only simple alcohol that is not poisonous and has been a social drug since almost the beginning of mankind; its physiological effects need not be elaborated

CH_3OH

Methanol
methyl alcohol
(wood alcohol)

CH_3CH_2OH

Ethanol
ethyl alcohol
(grain alcohol)

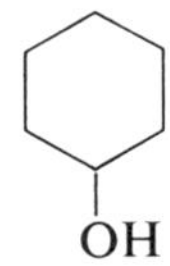

Cyclohexanol

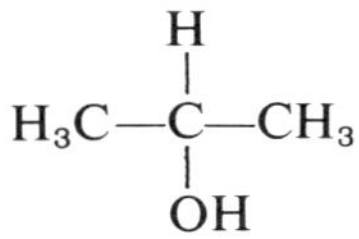

Isopropanol
isopropyl alcohol
(rubbing alcohol)

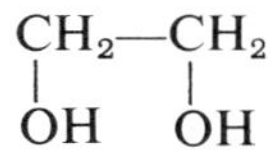

Ethylene glycol

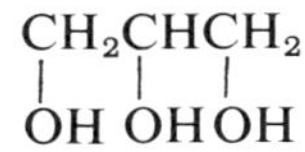

Glycerol
(glycerine)

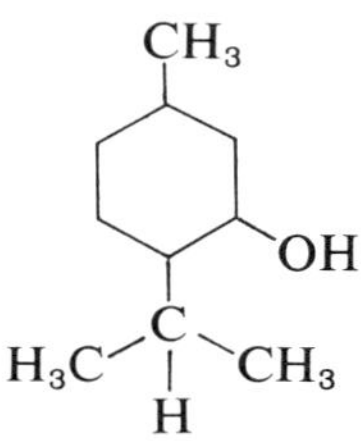

Menthol

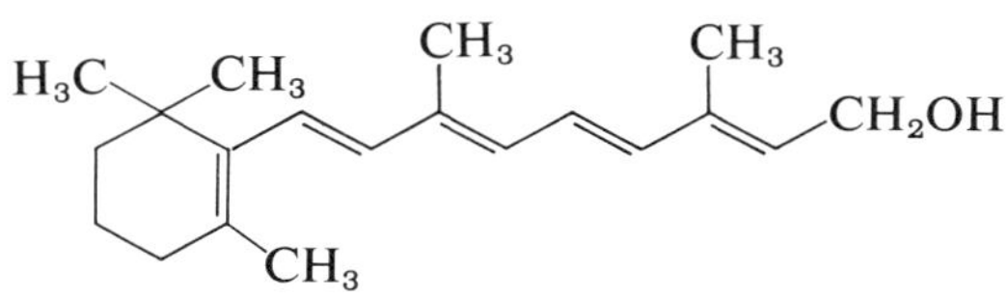

Vitamin A

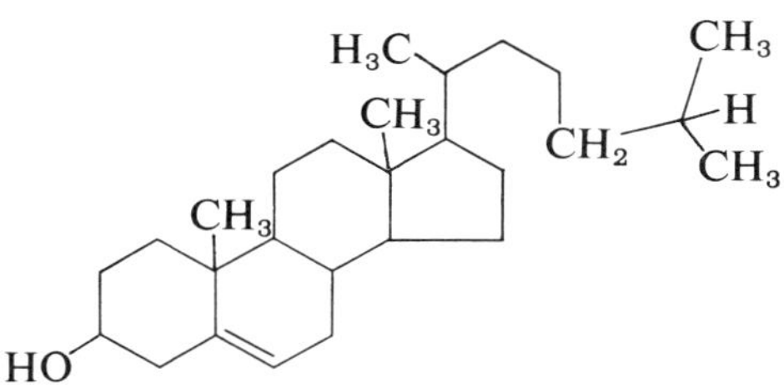

Cholesterol

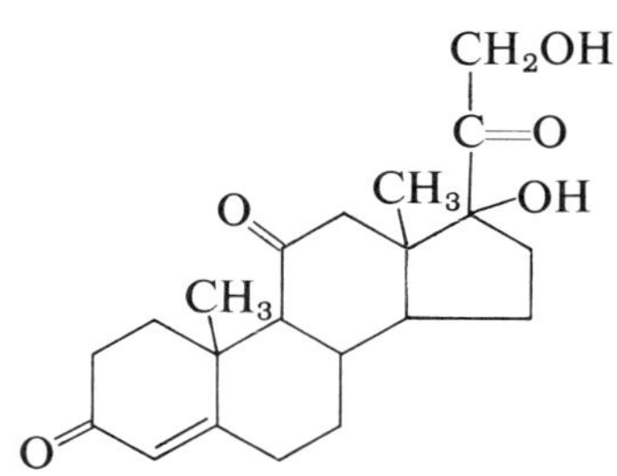

Cortisone

here. The concentration of ethanol in beverages is usually expressed as "proof" which is twice the volume percent of ethanol present in the solution. Menthol, a pungently odiferous compound in which the carbinol carbon is incorporated into a ring structure, is used in patent medicine, and is involved in the formulation of some tobacco products to give them "a breath of springtime." Cholesterol, a steroid, is widely distributed in mammals, being an important component of mammalian cell membranes. Cortisone is one of 28 hormones isolated from the adrenal cortex. It has a multitude of regulating functions in the human body and has been used in the treatment of rheumatoid arthritis. This steroid is characterized by a number of functional groups including two alcohol functions.

Nomenclature

Simple alcohols may be named by prefixing the appropriate name of the alkyl group to the word *alcohol*. In the IUPAC system, the longest chain containing the carbon bearing the hydroxyl group is chosen as the parent chain. The longest carbon chain must carry the hydroxyl group and the carbon attached to the hydroxyl group is given the smallest location number possible. The suffix *-ol* is then substituted for the *-e* in the name of the corresponding alkane. If a double or triple bond is present, the suffix *-enol* or *-ynol* is used with the proper locating numbers. When two or more hydroxyls are present the endings *-diol*, *-triol* are used. Alternatively, molecules which possess two hydroxyl groups on separate carbon atoms are often referred to as glycols. Some examples of these nomenclature systems appear below. The systematic name is given first, with the more common name in parentheses.

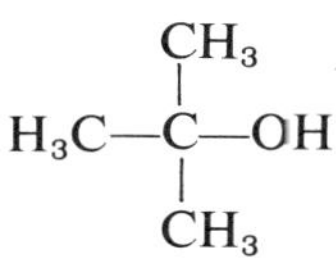

1,1-Dimethylethanol
(*t*-butyl alcohol)

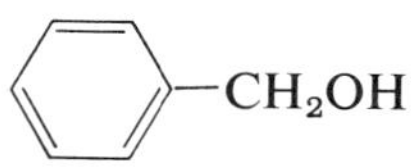

Phenylmethanol
(benzyl alcohol)

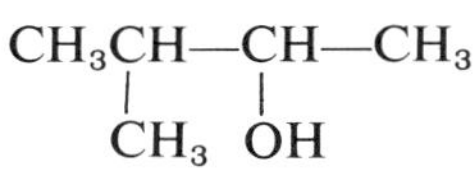

3-Methyl-2-butanol

$CH_2{=}CHCH_2OH$

2-Propenol
(allyl alcohol)

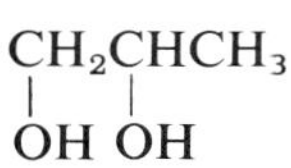

1,2-Propanediol
(propylene glycol)

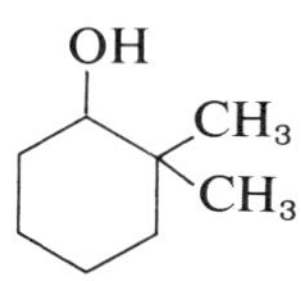

2,2-Dimethylcyclohexanol

It is often useful to classify alcohols according to the number of substituents attached to the carbon atom bearing the hydroxyl group. If there is one substituent attached to the carbinol carbon the alcohol is categorized as *primary*; if there are two, as *secondary*; and if three substituents are attached it is classified as *tertiary*. Thus, methanol and ethanol are examples of primary alcohols, isopropanol is a secondary alcohol, and *t*-butyl alcohol is a tertiary alcohol.

PHENOLS

Replacement of one of the hydrogens of water by a benzene ring gives a class of compounds known generically as phenols. These compounds may be named as benzene derivatives with the hydroxyl group assumed to have the first location number.

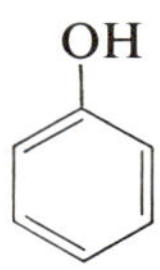

Phenol

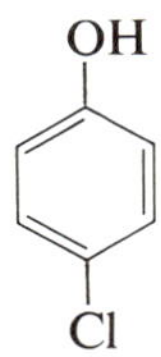

4-Chlorophenol
p-Chlorophenol

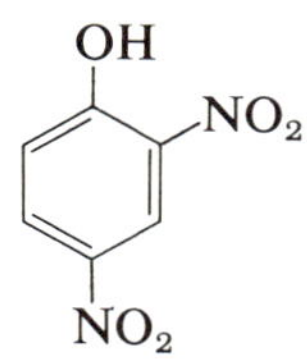

2,4-Dinitrophenol

The corresponding compounds derived from naphthalene are called naphthols.

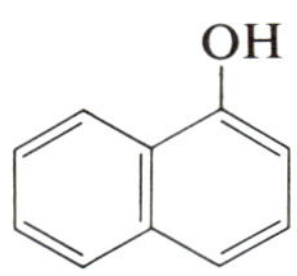

1-Naphthol
α-Naphthol

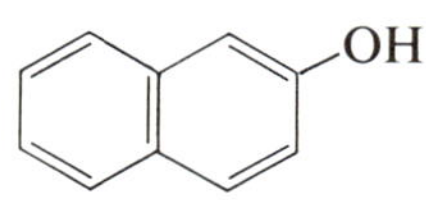

2-Naphthol
β-Naphthol

PHYSICAL PROPERTIES

In general, alcohols have higher boiling points and melting points and are more soluble in water than hydrocarbons of comparable molecular weight. This is not entirely a result of the higher polarity of alcohols. Although the oxygen atom of an alcohol is not very basic it can attract the hydroxyl hydrogen of another alcohol molecule so that the type of interaction previously termed hydrogen bonding can be set up. Hydrogen bonding will stabilize the liquid and solid forms of alcohols, thus accounting for the increased boiling and melting

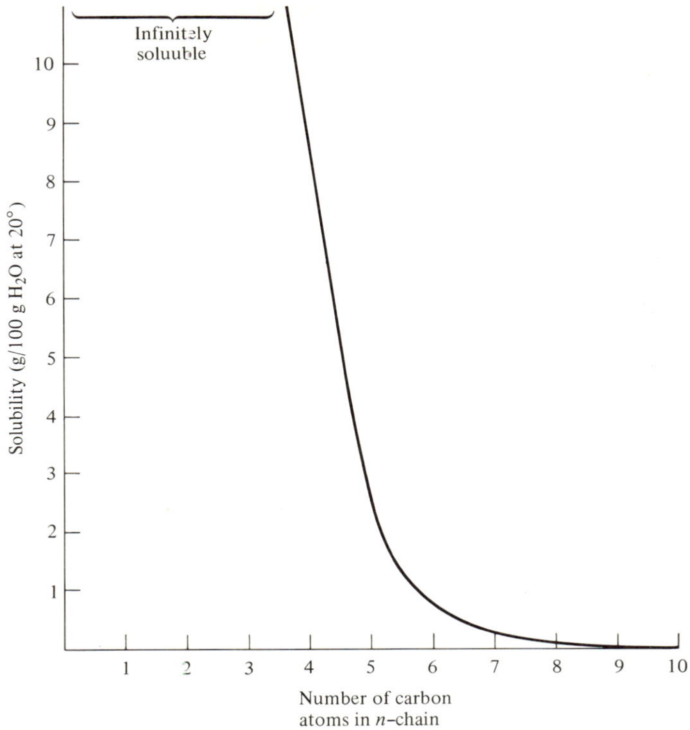

Fig. 6.1. The solubility of normal chain alcohols in water as a function of chain length.

points. Hydrogen bonding can also take place with solvent water molecules and increases the solubility of the alcohols; this effect makes the low molecular weight alcohols infinitely miscible with water in all proportions. As the chart in Fig. 6.1 indicates, the solubility of alcohols decreases as the amount of hydrocarbon character of the molecule increases. Alcohols with greater than nine carbon atoms are essentially no more soluble in water than hydrocarbons of similar molecular weights.

REACTIONS

Alcohols as Acids and Bases

Like water, an alcohol molecule can behave either as an acid or a base. In the autoprotolysis equilibrium one water molecule functions as an acid (proton donor), while the second one acts as a base (proton acceptor). The equilibrium

$$\mathrm{HOH + HOH \rightleftharpoons H_3O^+ + OH^+}$$

constant for this system is 10^{-14} (pK = 14). The corresponding equilibrium in methanol can be written

$$\mathrm{CH_3OH + CH_3OH \rightleftharpoons CH_3\overset{+}{O}H_2 + CH_3O^-}$$

and is characterized by an equilibrium constant of 10^{-16}; if water is the proton acceptor, $K \cong 10^{-18}$ (pK = 18). Comparing these data to that in Table 1.5

$$\mathrm{CH_3OH + H_2O \rightleftharpoons CH_3O^- + H_3O^+}$$

shows aliphatic alcohols to be very weak acids. The acidity of alcohols decreases slightly with increasing alkyl group size.

Phenols are much stronger acids than aliphatic alcohols, although they must still be regarded as weak when compared to hydrochloric acid, for example. The ionization constant for phenol is 10^{-10}, meaning that phenol is about a hundred million times stronger acid than methanol. Electron delocalization in the phenoxide anion accounts for this increased acidity. Phenol will be stabilized by

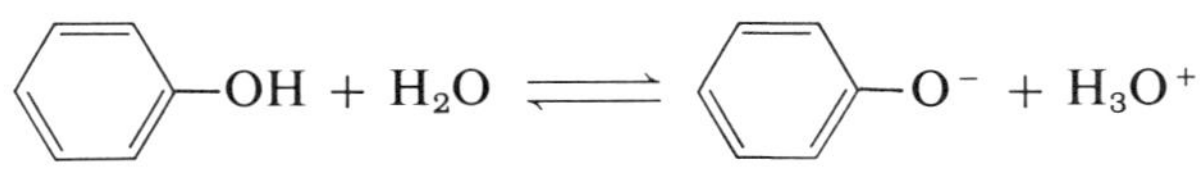

Phenol Phenoxide anion

resonance and some delocalization of the lone pairs on the hydroxyl group may take place. The resonance structures shown below indicate that this type of delocalization will not contribute much to the stability of this molecule since these forms involve a high-energy separation of formal electrical charges.

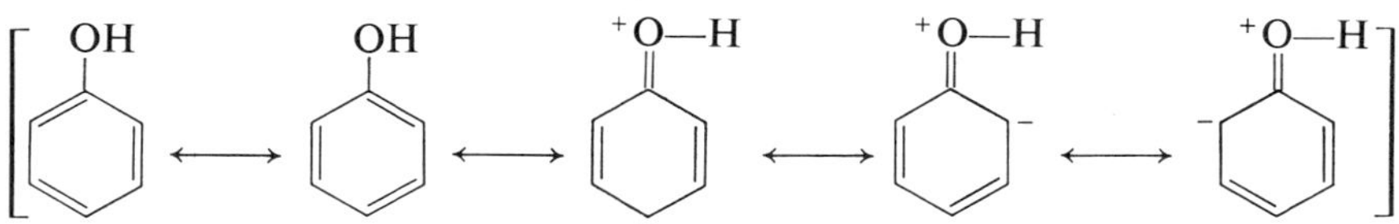

Declocalization of oxygen electrons

Resonance structures for phenol

However, in the phenoxide anion extensive delocalization of the extra electron can occur with no separation of charge. The energy of the phenoxide ion is

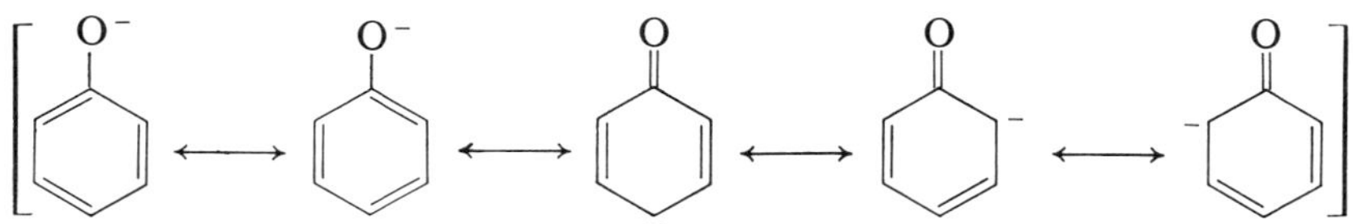

Delocalization of oxygen electrons

Resonance structures for phenoxide ion

lowered by resonance more than that of the starting phenol thereby making the free energy difference between phenol and phenoxide less than expected and shifting the acid–base equilibrium in favor of the phenoxide.

The conjugate bases of alcohol molecules are called alkoxide ions (RO^-). These anions are stronger bases than hydroxide (OH^-). Since the acidity of alcohols decreases in the order *primary* > *secondary* > *tertiary*, it may be deduced that the base strength of alkoxides follows the order of *tertiary* > *secondary* > *primary*.

Sodium alkoxides are conveniently prepared in the laboratory by reaction of the corresponding alcohol with sodium metal or sodium hydride.

$$CH_3OH + Na \longrightarrow \underset{\text{Sodium methoxide}}{CH_3O^-Na^+} + \tfrac{1}{2}H_2$$

$$(CH_3)_3COH + NaH \longrightarrow \underset{\text{Sodium } t\text{-butoxide}}{(CH_3)_3CO^-Na^+} + H_2$$

Alkoxides are useful when a strong base is needed for a reaction which must be performed in the absence of water; in this case the corresponding alcohol is used as a solvent for the reaction.

Oxidation of Alcohols

Oxidation of alcohols produces a class of materials known as carbonyl compounds which are characterized by a carbon attached to oxygen by a double bond; these compounds will be discussed in detail in subsequent chapters. There are three types of carbonyl compounds that may be obtained from the oxidation of alcohols: aldehydes (RCHO), ketones (R_2CO), and carboxylic acids (RCO_2H).

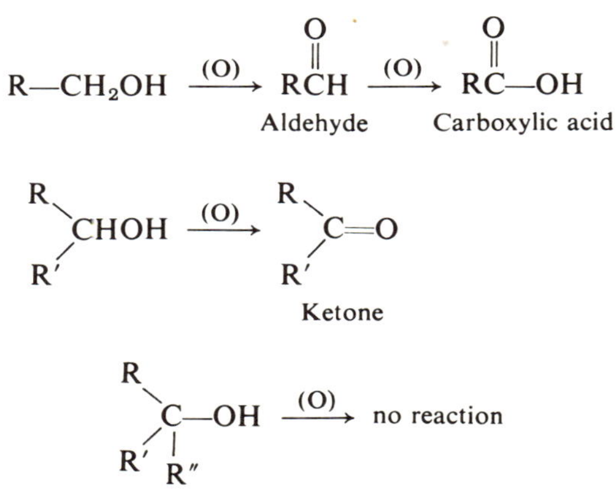

A primary alcohol may be oxidized to give an aldehyde which by further reaction gives a carboxylic acid. Often the rate of oxidation of the aldehyde is faster than its rate of formation, so that the product usually isolated upon oxidation of primary alcohol is the carboxylic acid. Careful choice of the oxidizing agent must be made to stop the oxidation at the aldehyde level.

$$\underset{\text{Ethanol}}{CH_3CH_2OH} \xrightarrow[200\text{–}300°C]{Cu} \underset{\text{Acetaldehyde}}{CH_3\overset{\overset{O}{\|}}{C}H}$$

$$\underset{n\text{-Butanol}}{CH_3CH_2CH_2CH_2OH} \xrightarrow[2.\ H^+]{1.\ KMnO_4,\ OH^-} \underset{n\text{-Butyric acid}}{CH_3CH_2CH_2\overset{\overset{O}{\|}}{C}OH}$$

If the carbinol carbon is attached to an sp^2- or sp-hybridized carbon, the C—OH is said to be "activated" and is much more susceptible toward oxidation. Milder oxidants may be used and no difficulties are encountered in stopping the oxidation of these primary alcohols at the aldehyde level. It should be remembered that these oxidizing agents (MnO_2, chloranil) are only effective with "activated" alcohols.

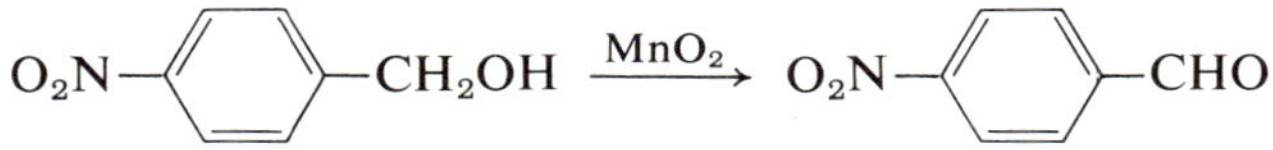

p-Nitrobenzyl alcohol *p*-Nitrobenzaldehyde

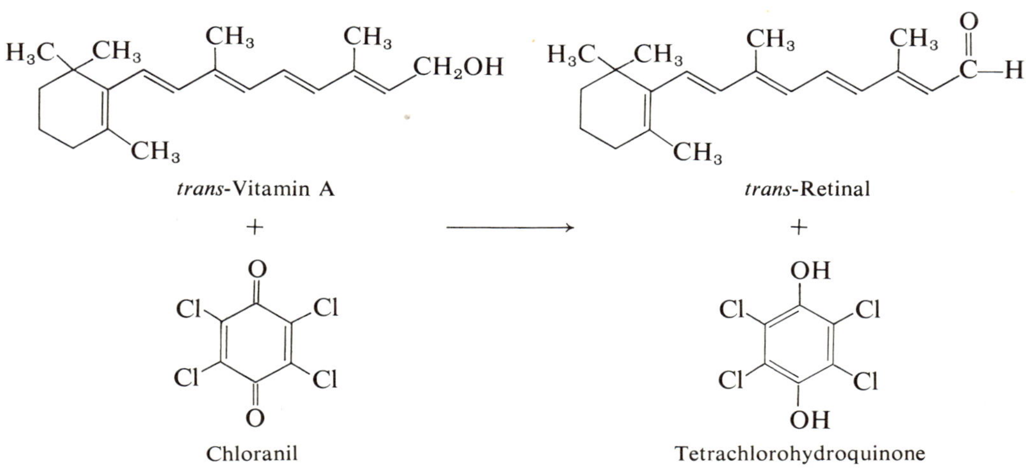

When a secondary alcohol is oxidized, a ketone is obtained. Ketones are resistant to further oxidation.

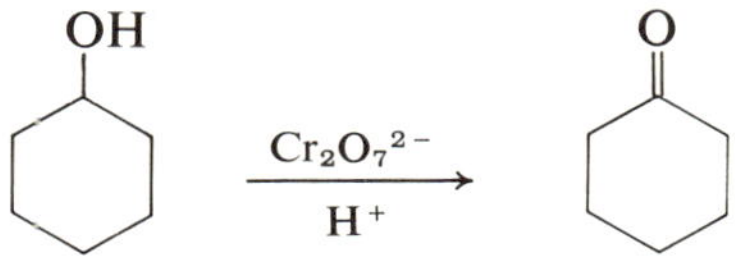

Cyclohexanol Cyclohexanone

$$\underset{\text{2-Pentanol}}{\underset{\displaystyle\;\;\;|\atop \displaystyle OH}{CH_3CH}CH_2CH_2CH_3} \xrightarrow[H^+]{Cr_2O_7^{2-}} \underset{\text{2-Pentanone}}{\underset{\displaystyle\;\;\;\|\atop \displaystyle O}{CH_3C}CH_2CH_2CH_3}$$

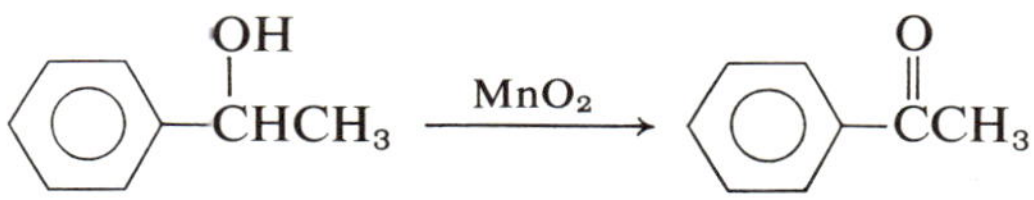

1-Phenylethanol Acetophenone

Formation of Alkyl Halides

Alcohols can be converted to alkyl halides by treatment with concentrated solutions of the corresponding hydrogen halide. This reaction involves the substitution of a hydroxyl group by a halogen; it does not occur with hydrogen fluoride.

$$\underset{n\text{-Butanol}}{CH_3CH_2CH_2CH_2OH} + HBr \longrightarrow \underset{n\text{-Butyl bromide}}{CH_3CH_2CH_2CH_2Br} + H_2O$$

$$\underset{t\text{-Butanol}}{CH_3\!-\!\overset{\displaystyle CH_3 \atop |}{\underset{| \atop \displaystyle CH_3}{C}}\!-\!OH} + HCl \longrightarrow \underset{t\text{-Butyl chloride}}{CH_3\!-\!\overset{\displaystyle CH_3 \atop |}{\underset{| \atop \displaystyle CH_3}{C}}\!-\!Cl} + H_2O$$

There is strong evidence that tertiary alcohols are converted to tertiary alkyl halides in this reaction by a mechanism that is different from the mechanism followed in the reaction of primary alcohols. These two important mechanistic types will be outlined here and discussed more fully in Chapter 8.

The first mechanism is called an S_N1 mechanism. This nomenclature signifies that the reaction involves substitution (S) by a nucleophile (N) and that the transition state for the slow step of the reaction contains only one (1) reactant molecule. Alcohols can function as bases; and in the presence of a strong acid, we can expect an equilibrium of the following type.

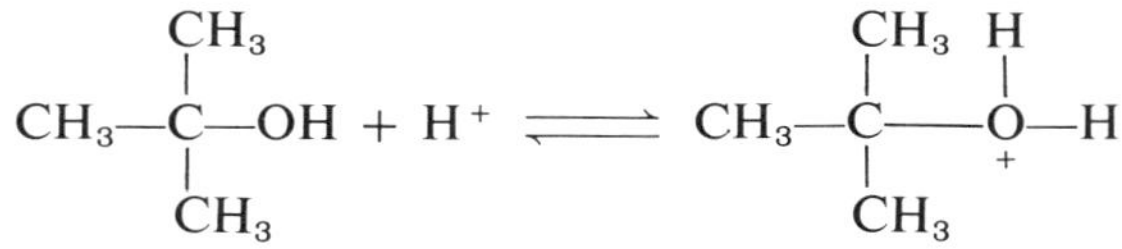

When the oxygen atom of an alcohol is protonated, the electrons involved in the C—O bond are even more attracted toward the oxygen atom than before and, as a result, the carbinol carbon of a protonated alcohol bears a much larger partial positive charge than its conjugate base. In many cases, it is energetically favorable for the C—O bond of the protonated alcohol to rupture, giving a positively charged alkyl group (carbonium ion) and water. The carbonium ion formed is electron-deficient (electrophilic) and has a great tendency to react with an electron-rich or nucleophilic species. Since concentrated acids are used, the concentration of halide ions present will be large and these nucleophilic

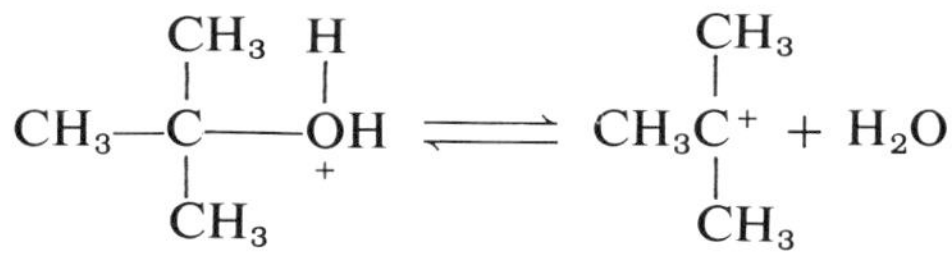

t-Butyl
carbonium ion

species can be captured by the carbonium ion to form *t*-alkyl halides. The formation of the carbonium ion is the slow step in this reaction and (except for

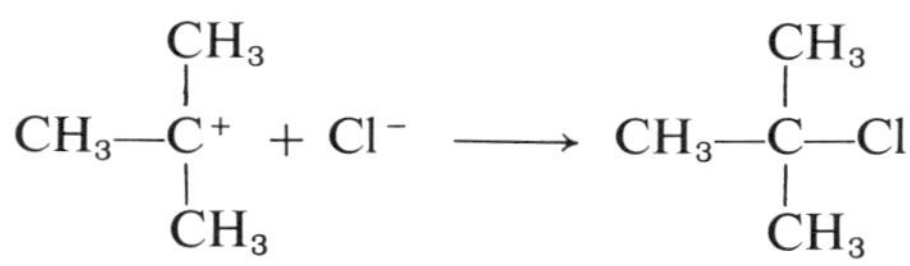

solvent molecules) only the alkyl halide participates in forming the transition state for the reaction.

The ease with which protonated alcohols will expel the water as a leaving group is related to the stability of the carbonium ions which are developed in the process. It has been noted before that tertiary carbonium ions are more stable than secondary ions which are more stable than primary alkyl cations. Primary alcohols, therefore, do not readily undergo S_N1 reactions.

Primary alcohols react with hydrogen halides by an S_N2 reaction. This process still involves substitution (S) by a nucleophile (N) but now occurs by a bimolecular mechanism; that is, there are two (2) molecules involved in forming the transition state for the reaction. The formation of an equilibrium concentration of protonated alcohol is again presumed to precede the rate-determining step of the reaction, but now a direct attack by the incoming or substituting nucleophile (halide ion in this case) occurs. The reaction proceeds through a

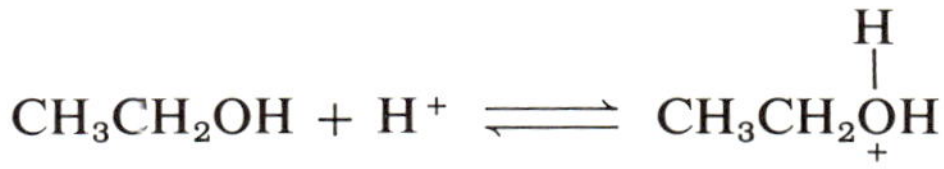

transition state in which a carbon atom becomes temporarily *pentavalent*; as the new bond between carbon and the approaching nucleophile is being formed, the C—O bond is being broken. Bond-making and bond breaking occur

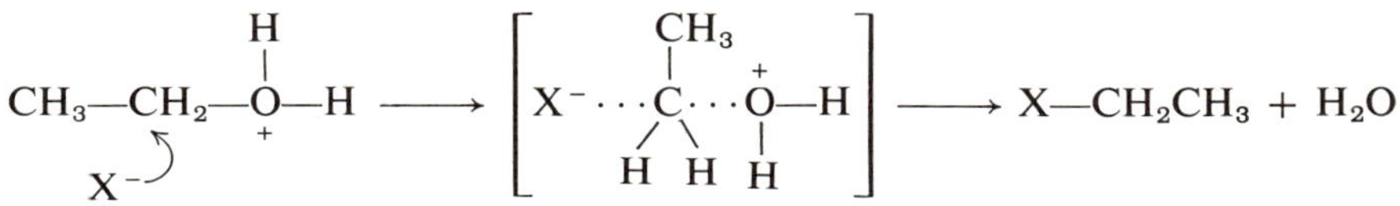

Transition state Products

synchronously in the S_N2 reaction but are separate steps in the S_N1 mechanism. Since two reactant molecules must come together to form the transition state for the slow step of the S_N2 reaction, the reaction rate will depend upon the concentrations of both species. The backside attack of the incoming nucleophile on the carbon bearing the leaving group would be hindered if this carbon were highly substituted with alkyl groups so that the ease of S_N2 displacements follows the order *primary* > *secondary* > *tertiary*.

Elimination Reactions

Alcohols which have a hydrogen atom attached to a carbon directly adjacent to the carbinol carbon may also undergo dehydration to give an alkene when treated with an acid. Reactions of this type are called elimination reactions and

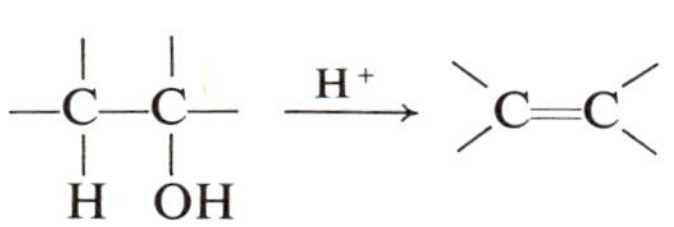

may be regarded as the reverse of the hydration reaction of alkenes mentioned in Chapter 3. Generally, the ease of elimination follows the order tertiary alcohols > secondary alcohols > primary alcohols. When the formation of more than one alkene is possible by elimination of water, the one in which the greatest number of alkyl groups is attached to the resulting double bond is usually the dominant product. The mechanisms of elimination reactions bear a strong resemblance to the S_N1 and S_N2 displacement reactions and often compete with these latter reactions. Elimination can be favored by using an

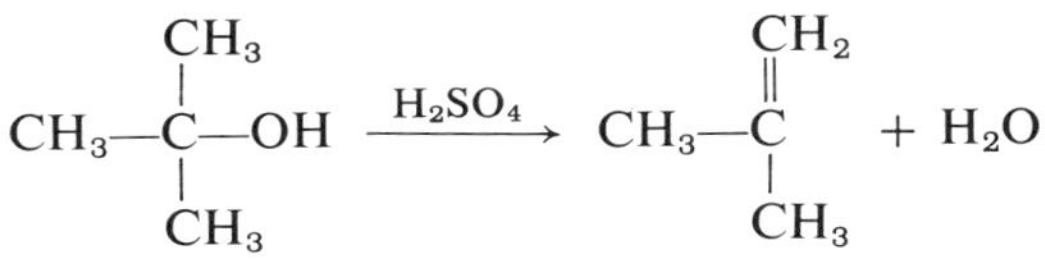

Isobutylene

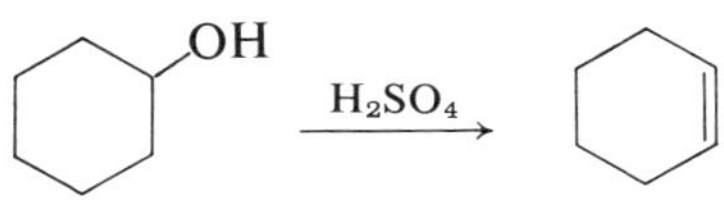

Cyclohexanol Cyclohexene

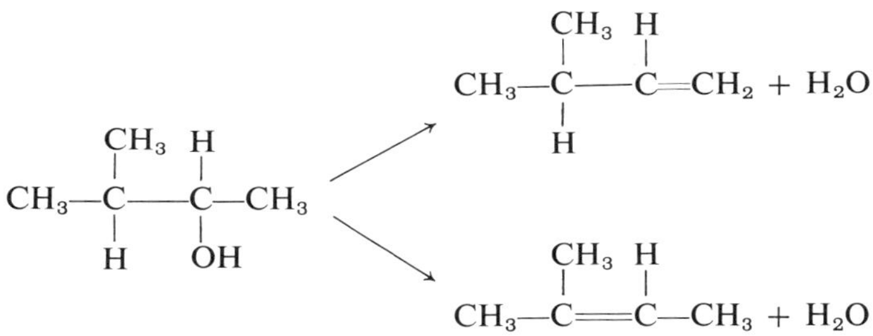

(major product)

acid with a poorly nucleophilic anion component such as sulfuric acid or phosphoric acid. Careful control of temperature and reaction conditions may be necessary to ensure that only elimination is the principal mode of reaction.

ETHERS

Substitution of both hydrogens of water for hydrocarbon groups gives a class of compounds known as *ethers*. The hydrocarbon radicals may be alkyl, cycloalkyl, or aromatic. Ethers are named by prefixing the names of the two groups attached to oxygen (in alphabetical order) to the word *ether*. If the two groups are identical, the prefix *di-* may be used but is not necessary.

The characteristic odor of ethyl ether still permeates modern hospitals where it finds use as an anesthetic. Dioxane is a cyclic ether and is a useful industrial solvent. Colchicine, a drug capable of arresting cell division and also used in the treatment of gout, is characterized by the presence of four methyl ether linkages at various points in the molecule.

$CH_3OCH_2CH_3$

Ethyl methyl ether

$CH_3CH_2OCH_2CH_3$

Diethyl ether
(ethyl ether)

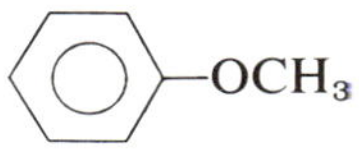

Methyl phenyl ether
(anisole)

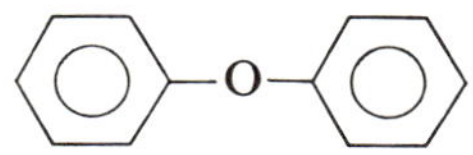

Diphenyl ether
(phenyl ether)

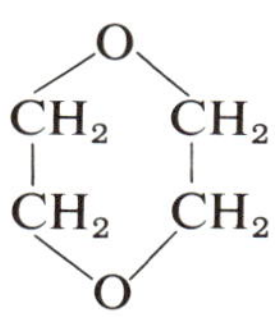

Dioxane

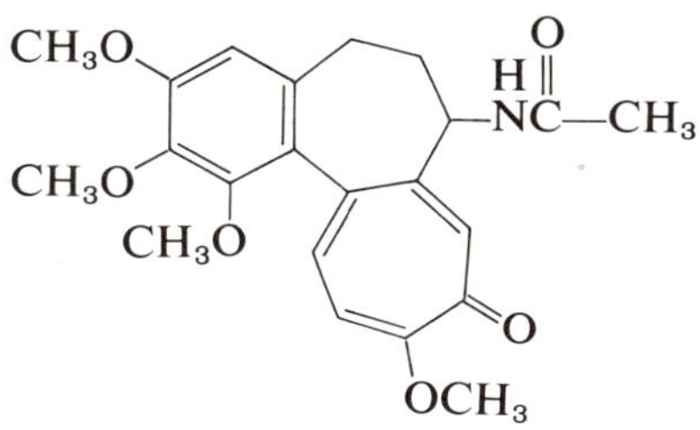

Colchicine

Ethers have no acidic hydrogen atoms and do not participate in the formation of hydrogen-bonded networks as do the alcohols. The boiling points of these compounds are lower than those of alcohols of comparable molecular weight, and approach the boiling points of the corresponding alkenes. Ethers are normally inert to basic reagents and, because of their low reactivity and high solubilizing ability for nonpolar organic molecules, they are often employed as solvents for organic reactions. Ethers can form dangerously explosive peroxides (see below) by a radical reaction with oxygen from the air and care should always be taken in the handling of these materials. Commercial ethers usually contain an additive to inhibit the formation of peroxides.

Reactions of Ethers

Like alcohols, ethers are weakly basic by virtue of the lone pairs of electrons on the oxygen atom and may react with strong acids to give *dialkyloxonium salts*. In some cases these salts have been isolated as reasonably stable compounds. Protonation of the oxygen atoms of ethers (a) makes the adjacent atoms much more electrophilic and thus susceptible to substitution and elimination reactions, and (b) provides a better leaving group for carbon–oxygen bond cleavages. For the above reasons, cleavage reactions of ethers occur in

$$CH_3CH_2{-}O{-}CH_2CH_3 + HCl_{(gas)} \longrightarrow CH_3CH_2{-}\overset{\displaystyle H}{\overset{|}{\underset{+}{O}}}{-}CH_2CH_3 \quad Cl^-$$

Diethyloxonium chloride
mp −92°C

acidic media. Hydrogen bromide is often used for this reaction and, in this case, the reaction affords an alkyl bromide and an alcohol as initial products of the cleavage reaction. If excess hydrogen halide is present, the alcohol is converted to the corresponding alkyl halide.

$$\underset{n\text{-Propyl ether}}{CH_3CH_2CH_2{-}O{-}CH_2CH_2CH_3} + HBr \longrightarrow \underset{n\text{-Propyl bromide}}{CH_3CH_2CH_2Br} + \underset{n\text{-Propanol}}{HOCH_2CH_2CH_3}$$

Phenols are better leaving groups than primary alcohols. Acid cleavage of alkyl phenyl ethers proceeds more rapidly than the corresponding reaction of dialkyl ethers; the products obtained are the phenol and an alkyl halide.

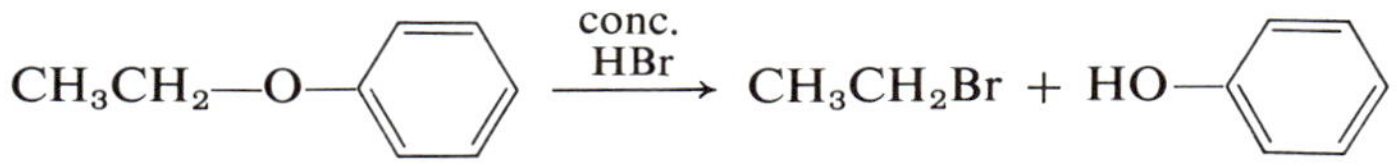

Dialkyl or alkyl aryl ethers may be prepared in the laboratory from a primary or secondary alkyl halide and an alkoxide or phenoxide anion. The reaction is known as the Williamson synthesis and proceeds by an S_N2 mechanism.

$$\underset{\text{Methyl iodide}}{CH_3I} + \underset{\substack{n\text{-Propoxide}\\ \text{(an alkoxide anion)}}}{{}^-OCH_2CH_2CH_3} \longrightarrow \underset{\text{Methyl } n\text{-propyl ether}}{CH_3OCH_2CH_2CH_3}$$

or

$$\underset{\text{Methoxide}}{CH_3O^-} + \underset{n\text{-Propyl bromide}}{BrCH_2CH_2CH_3} \longrightarrow CH_3OCH_2CH_2CH_3$$

In order to avoid elimination reactions in the synthesis of unsymmetrical dialkyl ethers, the alkyl halide used should be the one corresponding to the least branched partner in the reaction. Thus, in the preparation of *t*-butyl ethyl ether, the preferred reaction would be

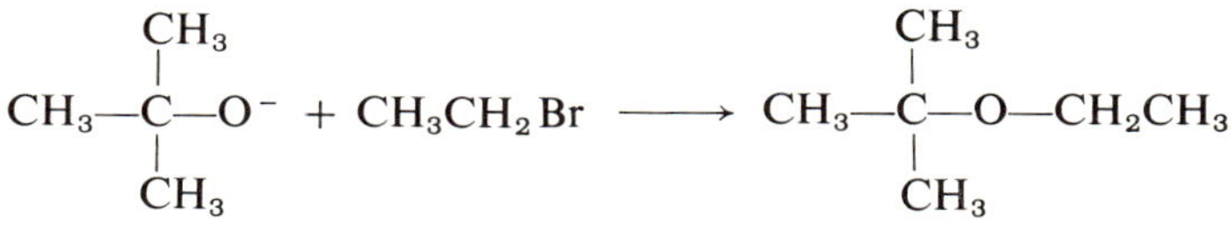

t-Butoxide *t*-Butyl ethyl ether

rather than the reaction between *t*-butyl bromide and ethoxide. In this latter case, ethoxide would promote elimination of HBr from the tertiary halide in competition with the formation of the desired ether because S_N2 reactions with tertiary halides are so unfavorable.

To prepare alkyl aryl ethers, the reaction of the appropriate phenoxide ion with an alkyl halide must be used.

p-Chloroanisole

Symmetrical dialkyl ethers may also be prepared by an acid-catalyzed dehydration of alcohols, preferably primary alcohols. Again, elimination is a serious side reaction if more highly branched alcohols are used. It should be noted that this reaction is simply the reverse process of the acid-catalyzed

$$2\,CH_3CH_2OH \underset{}{\overset{H^+}{\rightleftharpoons}} CH_3CH_2OCH_2CH_3 + H_2O$$

hydrolysis of ethers. The mechanism of dehydration involves formation of an alkyloxonium ion followed by nucleophilic displacement of water by the alcohol.

Recent research has provided methods for the synthesis of large cyclic ethers such as the one shown below. Rings of this

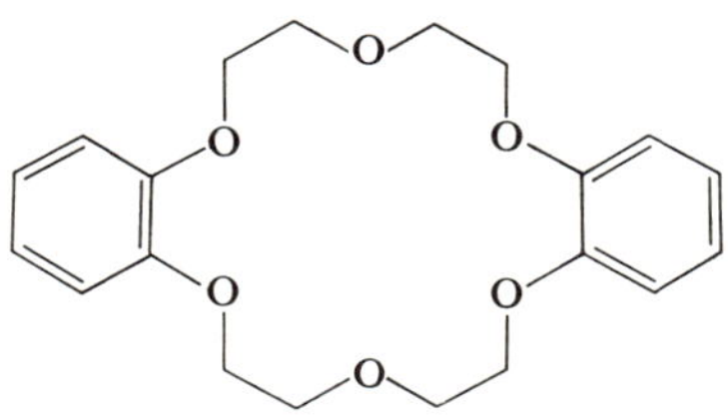

A "crown" ether

size are referred to as macrocycles. Molecular models of these novel macrocyclic ethers resemble a crown and they thus have

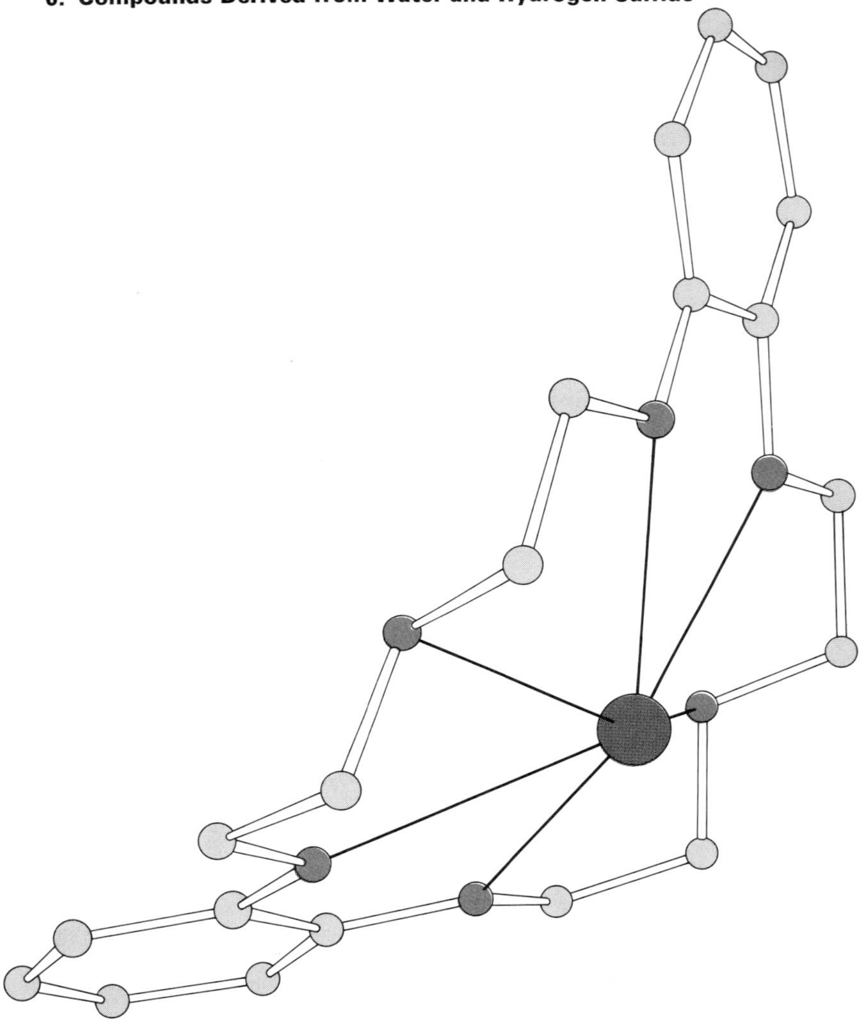

Fig. 6.2. The structure of the dibenzo-18-*crown*-6-ether–rubidium complex. [From D. Bright and M. R. Truter, *J. Chem. Soc.* (*B*), 1544 (1970).]

become known as "crown ethers" in common parlance. The cavity inside the "crown" in a sense is a ready-built solvation shell and these compounds form very firm complexes with alkali metal cations. The structure of one of these complexes is shown in Fig. 6.2. Because the exterior of the crown is rich in hydrocarbon character, these cation–ether complexes are quite soluble in nonpolar organic solvents. The possibility of carrying out reactions with ionic materials in non-polar solvents is raised by the availability of these macrocycles and is now under active investigation by a number of laboratories.

PEROXIDES

Peroxides are compounds which contain the —O—O— bond system. They may be thought of as derivatives of water in which one hydrogen atom has been replaced by an alkoxyl group. Alternatively, peroxides may be regarded as derivatives of hydrogen peroxide (H—O—O—H), a reagent occasionally used as an antiseptic and for the lightening of hair color when Nature has failed to provide the right shade genetically. Compounds with only one alkyl or aryl substituent are generically referred to as hydroperoxides; they are named by mentioning the name of the substituent followed by the word *hydroperoxide*.

$$CH_3O{-}OH$$

Methyl hydroperoxide

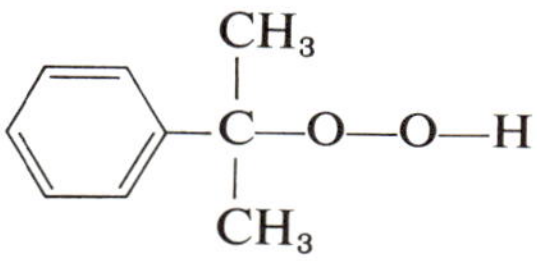

Cumene hydroperoxide
or cumyl hydroperoxide

The nomenclature of disubstituted peroxides is similar to the system used for ethers; the substituents are named in alphabetical order followed by the word peroxide. Peroxides and hydroperoxides are good oxidizing agents. The

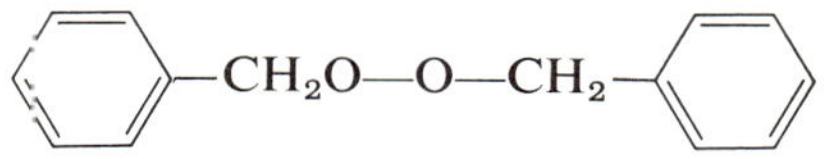

Dibenzyl peroxide

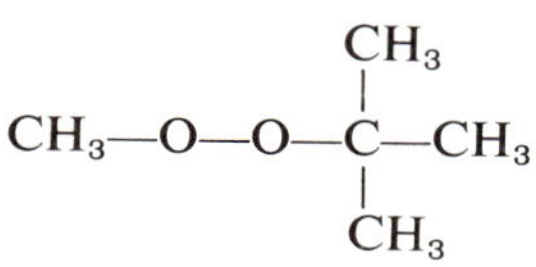

t-Butyl methyl peroxide

oxygen–oxygen bond is not a very strong one and these compounds may decompose into alkoxy radicals. The stability of organic peroxides to decomposition varies considerably with the nature of the substituent groups; some peroxides are dangerously explosive. Generally, stability increases as the proportion of peroxide atoms (O—O) to hydrocarbon atoms in the molecule is decreased.

DERIVATIVES OF HYDROGEN SULFIDE

Thiols

Thiols are derivatives of hydrogen sulfide (H_2S) in the same sense that alcohols are derivatives of water. Thiols are alternatively often referred to as mercaptans. The IUPAC or systematic names for simple thiols are formed by attaching to the name of the parent hydrocarbon the syllable *-thiol.* Common names are usually based on the name *mercaptan* and are generated by the same procedure as used for the common names of simple alcohols. Some examples of the nomenclature of thiols appear below. The sulfur-containing analog of phenol is called thiophenol.

CH_3SH — Methanethiol (methyl mercaptan)

$CH_3CH_2CH_2CH_2SH$ — 1-Butanethiol (*n*-butyl mercaptan)

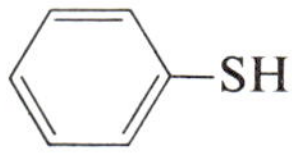

Thiophenol

Hydrogen sulfide is familiar to most as the odor of rotten eggs. The thiols maintain this family tradition by also possessing disagreeable odors. Butanethiol is a principal component in the secretion of the skunk, l-propanethiol is found in onions, and methanethiol occurs in radishes. The aroma of methanethiol is also evident in the urine after ingestion of asparagus.

Thiols are weak acids but are stronger acids than the corresponding alcohols. They can be converted to salts by treatment with strong bases and the resulting anions participate smoothly in S_N2 displacement reactions analogous to those of alcohols and alkoxide ions as indicated by the sequence below.

$$H_2S + NaOH \rightleftharpoons \overset{+}{Na}\overset{-}{SH} + H_2O$$

Sodium hydrosulfide

$$CH_3CH_2CH_2Br + NaSH \longrightarrow CH_3CH_2CH_2SH + NaBr$$

1-Propanethiol

$$CH_3CH_2CHSH + NaOH \rightleftharpoons CH_3CH_2CH_2\overset{-}{S}\overset{+}{Na} + H_2O$$

$$CH_3CH_2CH_2Br + Na^+ \ \overset{-}{S}CH_2CH_2CH_3 \longrightarrow CH_3CH_2CH_2{-}S{-}CH_2CH_2CH_3 + NaBr$$

Di-*n*-propyl sulfide

Sulfides

Sulfides are the sulfur analogs of ethers and the nomenclature of these compounds is similar to that of ethers. They can be formed by S_N2 displacement reactions between salts of thiols and alkyl halides or by addition of thiols to alkenes. The odors of sulfides are greatly attenuated relative to those of the corresponding mercaptans.

$CH_3S—CH_3$

Dimethyl sulfide

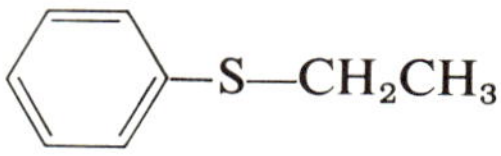

Ethyl phenyl sulfide

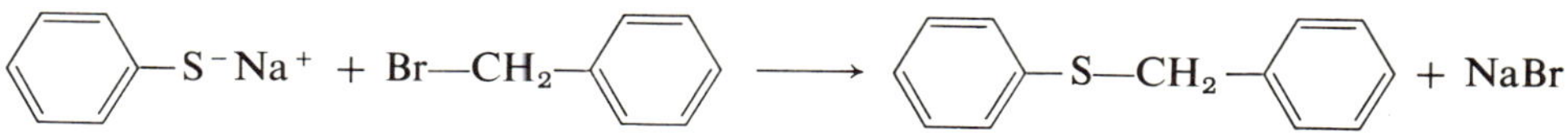

Sodium salt of thiophenol — Benzyl bromide — Benzyl phenyl sulfide

Disulfides

When a thiol is treated with a mild oxidizing agent (oxygen from the air will often do), an oxidation–reduction reaction occurs to generate a disulfide. These reactions likely involve radical species. Stronger oxidizing agents give a variety of products that arise by further oxidation of the disulfide to compounds which

$$CH_3SH + [O] \longrightarrow CH_3S\cdot + HO$$

$$2\ CH_3S\cdot \longrightarrow CH_3S—S—CH_3$$

Dimethyl disulfide

contain oxygen atoms directly attached to sulfur. A particularly important reaction involves the sulfur-containing amino acid cysteine. In the presence of an oxidizing agent this compound is converted to another amino acid, cystine, which is characterized by the presence of a disulfide bond system. These

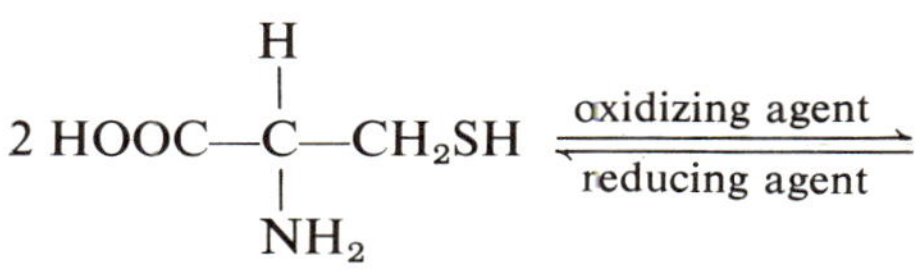

Cysteine

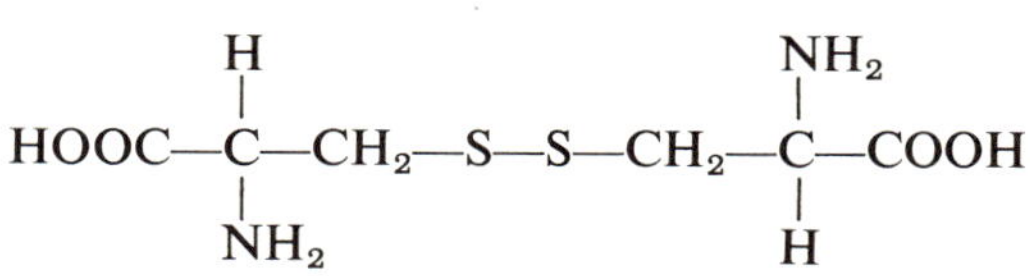

Cystine

two amino acids can be interconverted by treatment with an oxidizing agent to form the disulfide or with a reducing agent to regenerate the thiol. This interconversion is an important aspect of the chemistry of proteins, a subject that will be taken up later.

PROBLEMS

1. Write the structures that correspond to the following names:
   (a) Cyclohexyl methyl ether
   (b) *sec*-Butyl alcohol
   (c) Menthol *n*-propyl ether
   (d) *p*-Fluorobenzyl alcohol
   (e) 3,3,5-Trimethylcyclohexanol
   (f) Cyclopropanol
   (g) Dioxane
   (h) 1-Chloro-2-naphthol
   (i) *t*-Butyl hydroperoxide
   (j) 2-Iodo-3-cyclopentylphenol
   (k) *n*-Pentyl mercaptan
   (l) Potassium *t*-butoxide
   (m) *o*-Chlorothiophenol
   (n) Diphenyl peroxide
   (o) *m*-Bromoanisole
   (p) Dicyclobutyl sulfide
   (q) Dichloromethyl sulfide

2. Suggest a synthesis for each of the following molecules starting with *n*-butyl bromide.
   (a) *n*-Butanol
   (b) 1-Butanethiol
   (c) 1-Butene
   (d) *n*-Butyl methyl ether
   (e) Butyraldehyde ($CH_3CH_2CH_2\overset{\overset{\displaystyle O}{\|}}{C}—H$)
   (f) 2-Butanol
   (g) Di-*n*-butyl sulfide
   (h) 1,2-Dibromobutane
   (i) Lithium *n*-butoxide

3. Give an appropriate name for each of the following compounds:

(a) 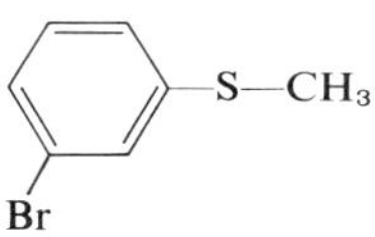

(b) 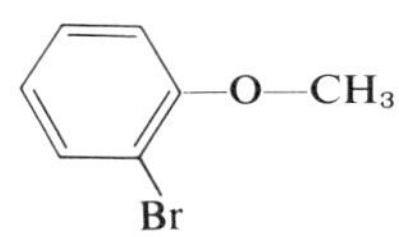

(c) 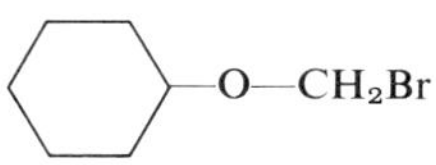

(d) $CH_3CH_2—S—S—CH_3$

(e) 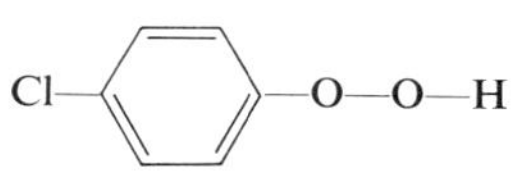

(f) 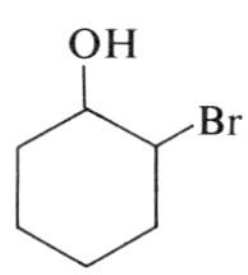

(g) 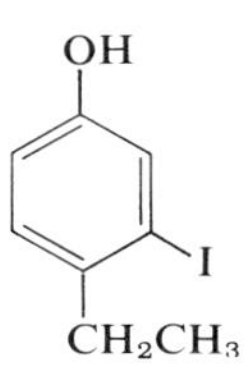

(h) 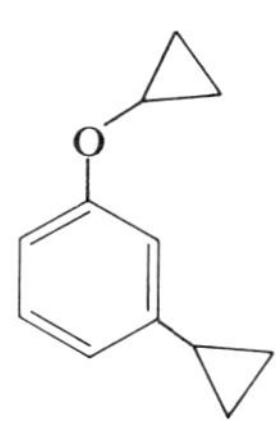

(i) $HOCH_2CH{=}CHCH_2OH$

(j) 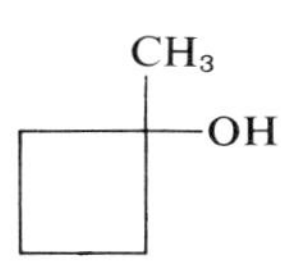

(k) 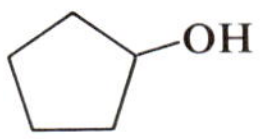

(l) 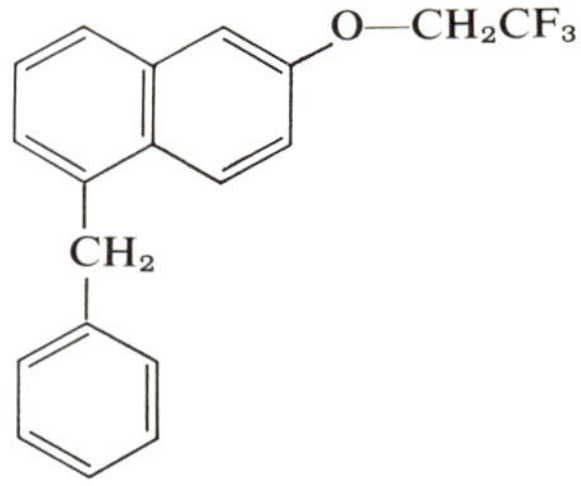

4. Most inorganic salts that are very soluble in water are only slightly soluble in methanol. Account for this difference in terms of the physical and chemical properties of water and methanol.

5. Δ′-Tetrahydrocannabinol (THC), shown below, is the major psychotomimetical component of marijuana. Predict the products that would be obtained by treating Δ′-THC with (a) Br_2 in carbon tetrachloride, (b) conc. HBr, (c) refluxing potassium permanganate, (d) methyl iodide.

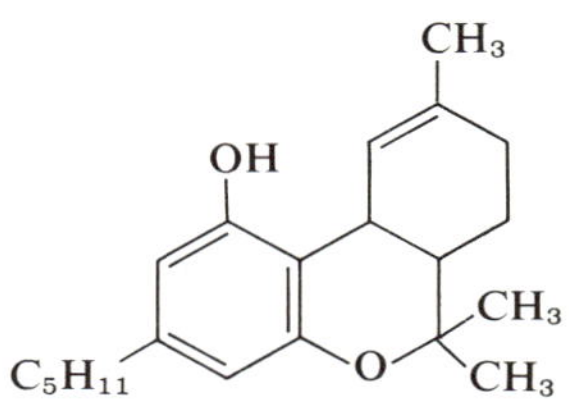

Δ′-THC

6. 2,4,6-Trinitrophenol (picric acid) has a pK_a of 1. How much stronger an acid is this compound than methanol? Than phenol? Suggest a reason why 2,4,6-trinitrophenol is such a strong acid.

7. Compound A (C_3H_8O) exhibited the pmr spectrum shown below. When A was oxidized with potassium dichromate, it was changed to compound B. The pmr spectrum of B is shown. What are A and B? How are the spectral data consistent with the structures you have proposed?

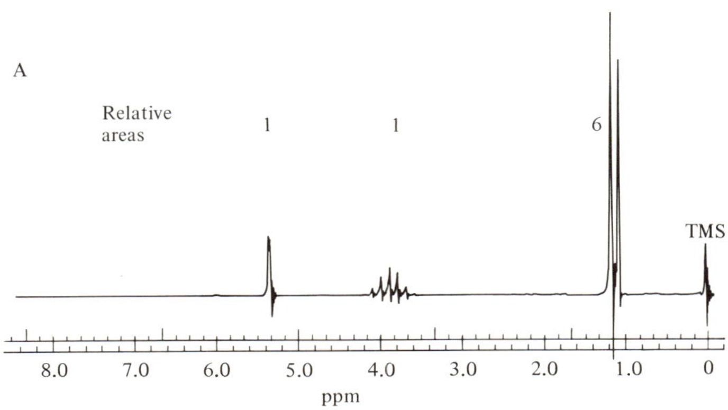

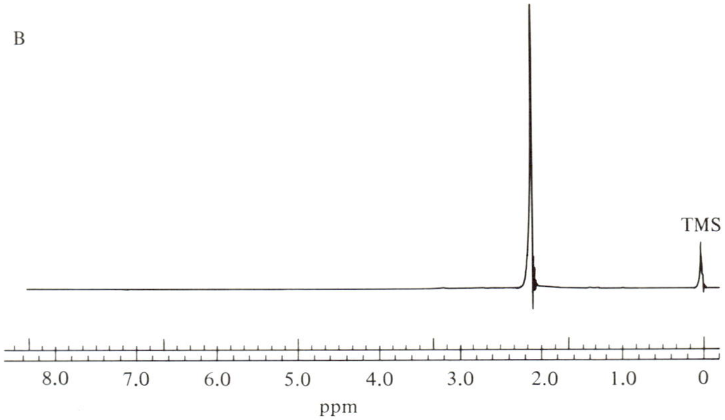

SUGGESTED READINGS

F. Challenger, "Aspects of the Organic Chemistry of Sulphur." Academic Press, New York, 1959.

J. J. Christensen, J. O. Hill, and R. M. Izatt, *Science* **174**, 459 (1971). (Macrocycles.)

R. E. Davis, D. L. Rousseau, and R. D. Board, *Science* **171**, 167 (1971). (Polywater.)

J. English, H. G. Cassidy, and R. L. Baird, "Principles of Organic Chemistry," 4th ed. McGraw-Hill, New York, 1971.

R. M. Johnson and I. W. Siddigir, "The Determination of Organic Peroxide." Pergamon, Oxford, 1970.

S. Patai, ed., "The Chemistry of the Ether Linkage." Wiley (Interscience), New York, 1967.

D. Swern, ed., "Organic Peroxides." Wiley (Interscience), New York, 1970.

7

Compounds Derived from Ammonia; Heterocyclic Molecules

A walk along the seashore provides an almost overwhelming number of sensory impacts. The sparkle of the sunlight as it is reflected from the water and sand and the crash of the surf as it smashes to the beach all delight one's aural and visual mechanisms. Equally apparent is the characteristic odor of the waterfront, especially near a fishing harbor. The pungent, ammoniacal, sometimes unpleasant aroma of these places derives largely from decaying sea life; it also provides a primitive if unrecognized introduction to the properties of a class of organic compounds called amines.

AMINES

The previous chapter discussed some of the organic compounds which can be thought of as being derived from water or hydrogen sulfide by replacement of hydrogen atoms in these parent compounds with alkyl, cycloalkyl, or aromatic radicals. Amines are derivatives of ammonia (NH_3) in which one or more of the N—H bonds is replaced by a bond to an organic group. If one hydrogen atom is removed from ammonia and an organic radical substituted, a primary amine is the generic type of organic compound which is the result. Replacement of two hydrogen atoms gives a secondary amine, while a tertiary amine is

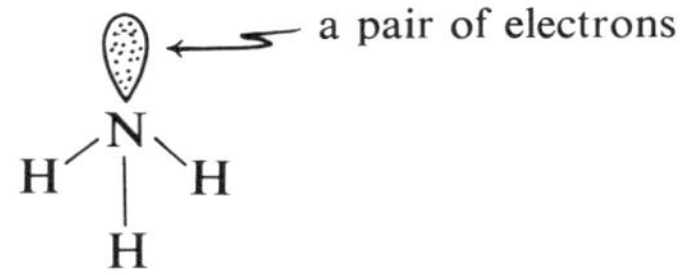

Ammonia

produced when all three bonds at the nitrogen atom are terminated with organic groups.

Primary amines	*Secondary amines*	*Tertiary amines*
CH_3NH_2	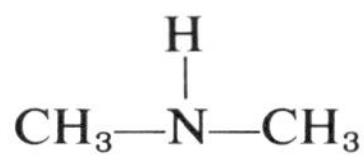	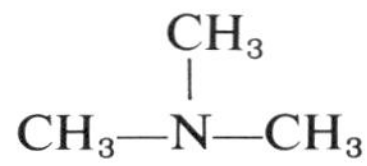
Methylamine	Dimethylamine	Trimethylamine
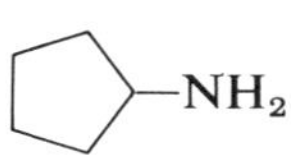	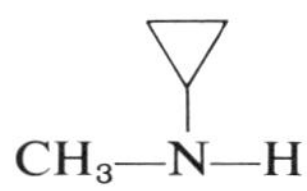	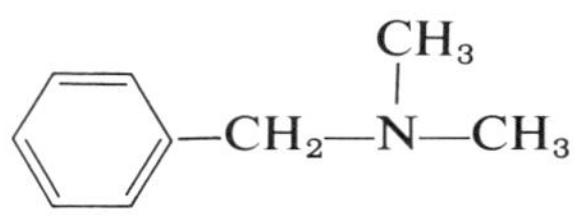
Cyclopentylamine	Cyclopropylmethylamine	*N*,*N*-Dimethylbenzylamine

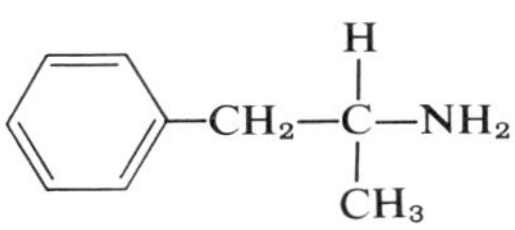

Benzedrine
(amphetamine)

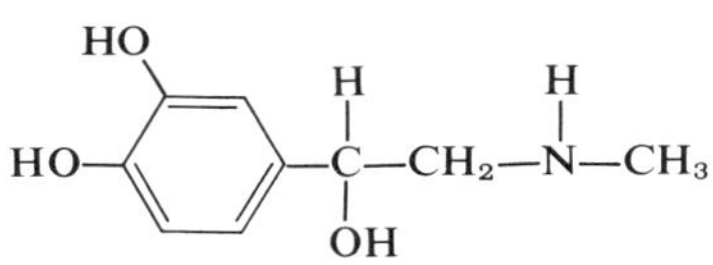

Epinephrine
(adrenaline)

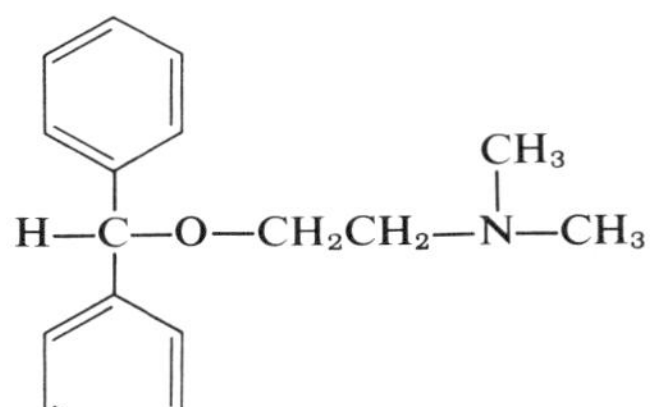

Benadryl

Amines, both as free materials and in chemical combination with other molecules, are widely distributed in nature, being especially prevalent in the nervous systems of vertebrates. The adrenaline which is rapidly released in the body during times of stress provides an example of a physiologically important secondary amine. It is also a polyhydroxylic substance. Benzedrine ("bennies," "red jackets," "speed") is a synthetic primary amine which can mimic the chemical and physical properties of other, naturally occurring amines in the nervous system to produce frenetic "highs," while benadryl, a tertiary amine, is useful in inducing sleep and in treating the symptoms of allergies. An ether linkage is also contained in the latter compound.

The —NH_2 group and derivatives of this group are known as amino groups.

It is possible for a single organic molecule to contain more than one amino group in its structure. If two are present, the compound is generically known as a diamine; if three are in evidence, the compound is a triamine, and so forth.

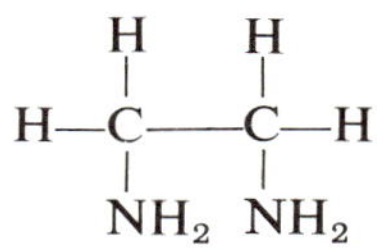

Ethylene diamine

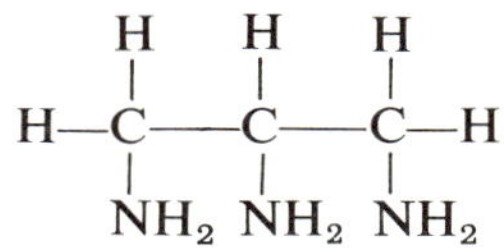

Propylene triamine

It should be recalled that ammonia is basic and that the reaction of an acid with ammonia affords the ammonium salt of the acid.

$$NH_3 + HCl \longrightarrow NH_4^+ + Cl^-$$

The ammonium ion is a tetrahedrally symmetric ion having a structure quite similar to that of methane. It is possible to replace one or more of the N—H bonds in the ammonium ion with linkages to organic radicals. The alkyl-

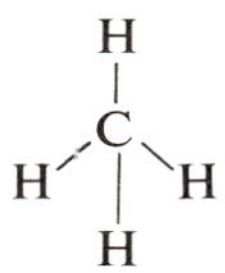

Methane

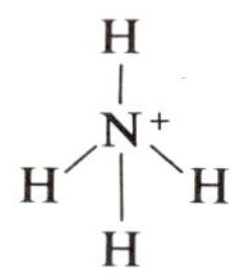

Ammonium ion

ammonium salts that result are soluble in water because they contain a "built-in" ionic charge that interacts strongly with water molecules. These salts retain the basic tetrahedral molecular arrangement found in the ammonium ion. It

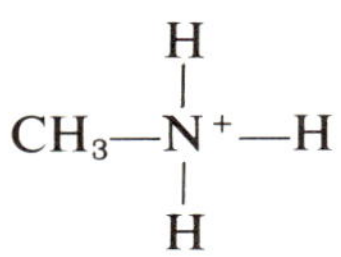

Methylammonium ion

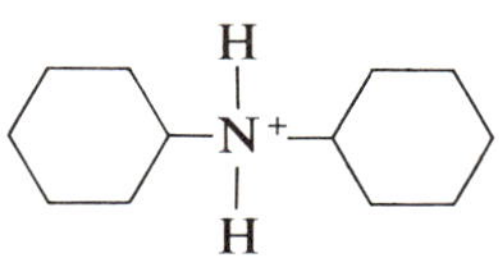

Dicyclohexyl ammonium ion

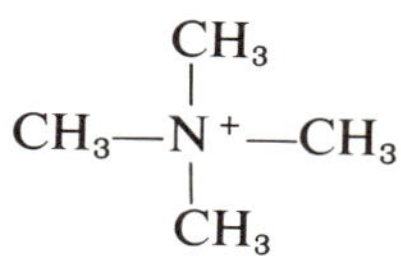

Tetramethylammonium ion

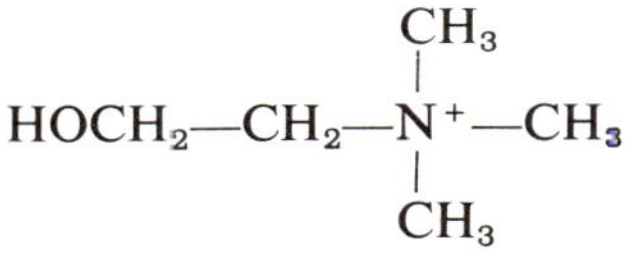

Choline

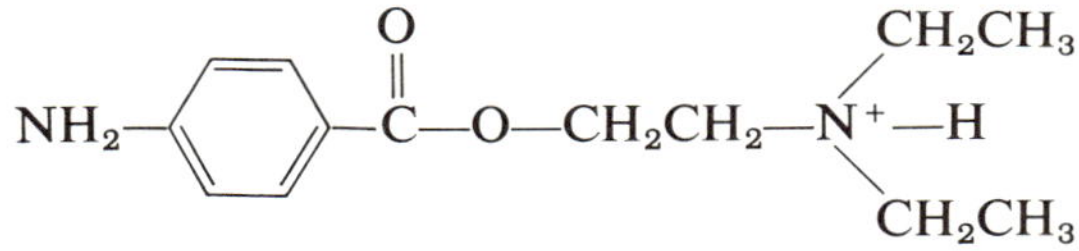

Novocaine

should be remembered that an anion must be present with these materials to balance the positive charge of each ammonium ion present.

The ammonia molecule has a pyramidal shape, with the nitrogen atom occupying the apex of the pyramid. A lone pair of electrons in ammonia and derivative amines is pointed away from the nitrogen atom, well-exposed and readily available for reactions. This availability of the lone pair of electrons on ammonia or an amine gives these compounds a basic nature and makes them potent nucleophiles. In the ionic ammonium or alkylammonium compounds this lone pair of electrons is busy binding a fourth group to the nitrogen atom, so that these latter materials are neither bases nor nucleophiles.

Nomenclature

Simple amines are named by denoting the substituents at the nitrogen atom by the appropriate radical names, combining these and appending the word *-amine*. Some examples of this approach are shown below. For more complex structures, the primary amine defined by the most complicated substituent and the amino nitrogen is taken as the parent molecule. The other groups attached to the nitrogen are indicated by their radical names, prefixed by the designator *N*-. Amino groups may be located by position numbers in systematic names when more than one functional group is present.

$CH_3CH_2NH_2$

Ethylamine

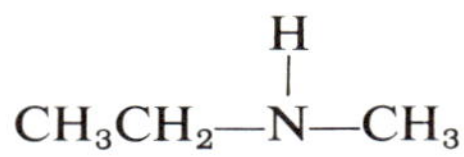

Methylethylamine or *N*-methylethylamine

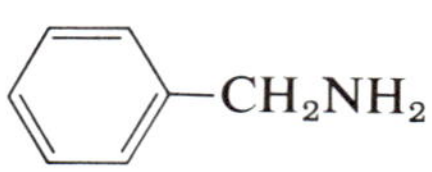

Benzylamine

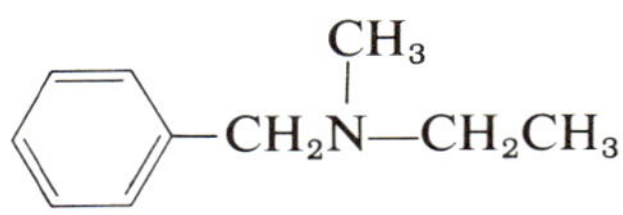

N-Methyl-*N*-ethylbenzylamine

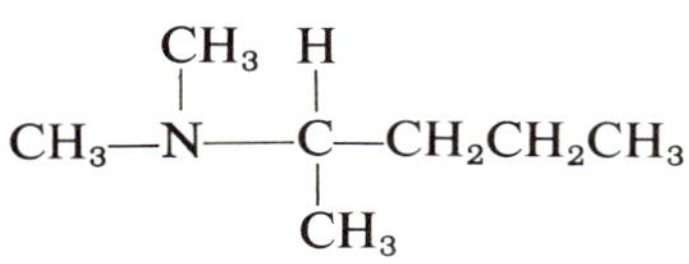

2-(*N*,*N*-Dimethylamino)pentane

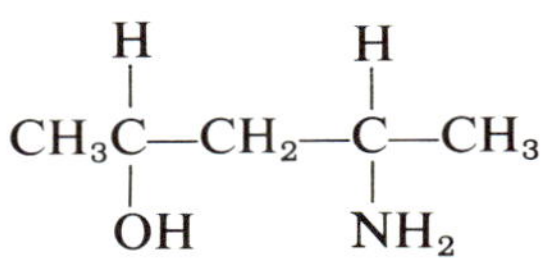

4-Amino-2-pentanol

Aromatic amines have the amino group directly attached to an aromatic ring. The simplest aromatic amine is *aniline*. Substituted anilines are named as derivatives of aniline as indicated in the examples.

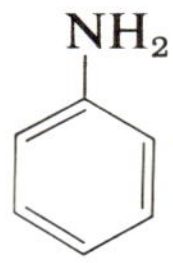

Aniline

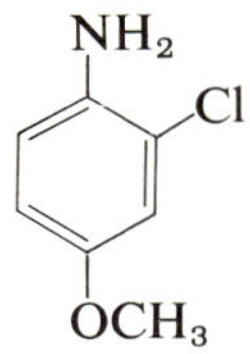

2-Chloro-4-methoxyaniline

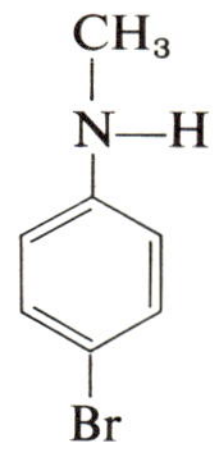

4-Bromo-*N*-methylaniline

Physical Properties

Simple amines are polar molecules and can form hydrogen bonds to water and other hydroxylic solvents. Low molecular weight amines (less than six carbon atoms) are soluble in water, whereas larger aliphatic and aromatic amines behave more as hydrocarbons, being sparingly soluble in water but easily soluble in organic solvents such as benzene, ethanol, and chloroform. Molecules of primary and secondary amines can hydrogen-bond to one another so that the boiling points of these materials are higher than those for nonpolar compounds of similar molecular weight.

Amines as Bases

As noted above, a characteristic property of amines is their basicity. The qualitative idea of basicity as the relative ease with which the structure will accept a proton can be made more quantitative by evaluating the position of an equilibrium such as that set up between an amine and water. The value of the

$$CH_3\text{—}\ddot{N}H_2 + H_2O \rightleftharpoons CH_3NH_3^+ + OH^-$$

equilibrium constant (K_b) for this system is a measure of the ability of an amine (or other base) to accept a proton (H^+) from a reference acid, in this case, water. (Since the concentration of water in the above expression is essentially constant, it is incorporated into the equilibrium constant, K_b.) A list of some K_b values is given in Table 7.1. As these data indicate the nature of the organic substituents at the amino group can have an appreciable influence on the basicity of the amine. The most striking change occurs when an aromatic ring is substituted; methylamine is about a millionfold stronger base than aniline, and diphenylamine is a much weaker base than aniline. The reason for this drastic change in amine basicity is largely the resonance interaction which can take

Table 7.1
Basicity of Some Amines

Name	K_b	pK_b[a]
Ammonia	1.6×10^{-5}	4.8
Methylamine	4.0×10^{-4}	3.4
Dimethylamine	5.0×10^{-4}	3.3
Trimethylamine	6.3×10^{-5}	4.2
Aniline	4.0×10^{-10}	9.4
N-Methylaniline	6.3×10^{-10}	9.2
N,N-Dimethylaniline	1.0×10^{-9}	9.0
Diphenylamine	6.3×10^{-14}	13.2
p-Anisidine (*p*-methoxyaniline)	2.0×10^{-9}	8.7
p-Nitroaniline	1.0×10^{-13}	13.0

[a] $pK_b = -\log(K_b)$. The larger the pK_b, the weaker is the base.

place in the free amine between the unshared pair of electrons on the nitrogen atom and the π-electronic system of the aromatic ring. This resonance delocalization lowers the free energy of the free amine relative to the protonated form

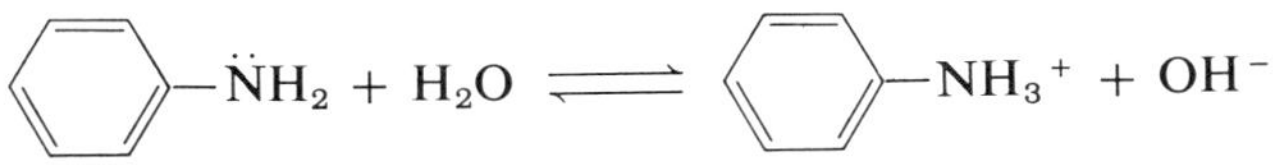

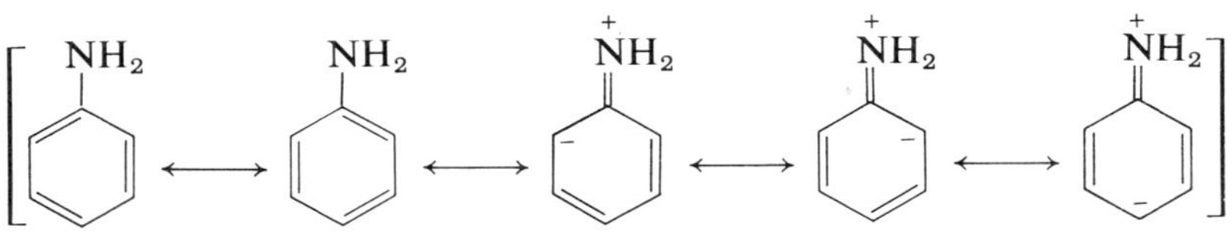

Some resonance structures for aniline

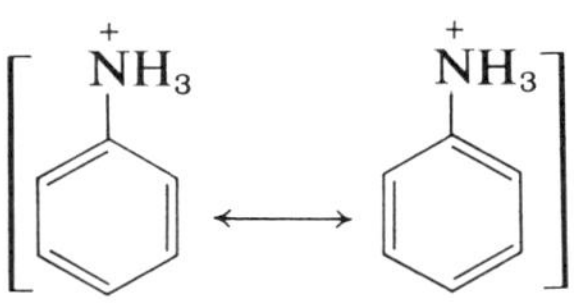

Some resonance structures for protonated aniline

so that ΔG° for the equilibrium involving aniline is larger than ΔG° for the equilibrium between methylamine and water. Resonance interaction makes the free aniline molecule more stable than would be expected so that it has a greater reluctance to accept a proton. It is noteworthy that diphenylamine is a much weaker base than aniline, which is very likely a reflection of the fact that the second aromatic ring can also interact with the amino group. Diphenylamine is such a weak base that it is soluble only in concentrated acids.

SOME REACTIONS OF AMINES

Nucleophilic Displacements

The same pair of electrons on the nitrogen atom that makes amines behave as bases or proton acceptors also endows these compounds with strong nucleophilic tendencies. Thus, amines, especially alkyl-substituted amines, can displace halogens from alkyl halides by an S_N2 mechanism.

$$\underset{\text{Ethyl chloride}}{CH_3CH_2Cl} + NH_3 \longrightarrow \underset{\text{Ethylamine}}{CH_3CH_2NH_2} + HCl$$

$$CH_3CH_2NH_2 + CH_3CH_2Cl \longrightarrow \underset{\text{Diethylamine}}{CH_3CH_2{-}\overset{\displaystyle H}{\overset{|}{N}}{-}CH_2CH_3} + HCl$$

$$CH_3CH_2{-}\overset{\displaystyle H}{\overset{|}{N}}{-}CH_2CH_3 + CH_3CH_2Cl \longrightarrow \underset{\text{Triethylamine}}{(CH_3CH_2)_3N} + HCl$$

$$(CH_3CH_2)_3N + CH_3CH_2Cl \longrightarrow \underset{\text{Tetraethylammonium chloride}}{(CH_3CH_2)_4N^+Cl^-}$$

The primary amine formed when ammonia reacts with an alkyl halide is a somewhat stronger base and a better nucleophile than ammonia itself. The reaction of primary amines with alkyl halides is, therefore, more rapid than their formation from ammonia. Because multiple reactions of the type indicated above can occur in displacement reactions of amines with alkyl halides, the reaction conditions must be carefully controlled to obtain the desired product. Although aromatic amines are less reactive, displacement reactions using these compounds also occur, especially if the aromatic ring is so substituted that the electrons at the nucleophilic nitrogen atom are made more available for reaction.

Nitrous Acid

Both aromatic and aliphatic amines undergo reaction with nitrous acid, HNO_2.

Primary Aliphatic Amines. When a primary aliphatic amine is treated with nitrous acid, a vigorous reaction ensues which generates products that suggest the intermediacy of a carbonium ion. A diazonium ion (RN_2^+) is an intermediate in this reaction. Because the nitrogen molecule is thermodynamically

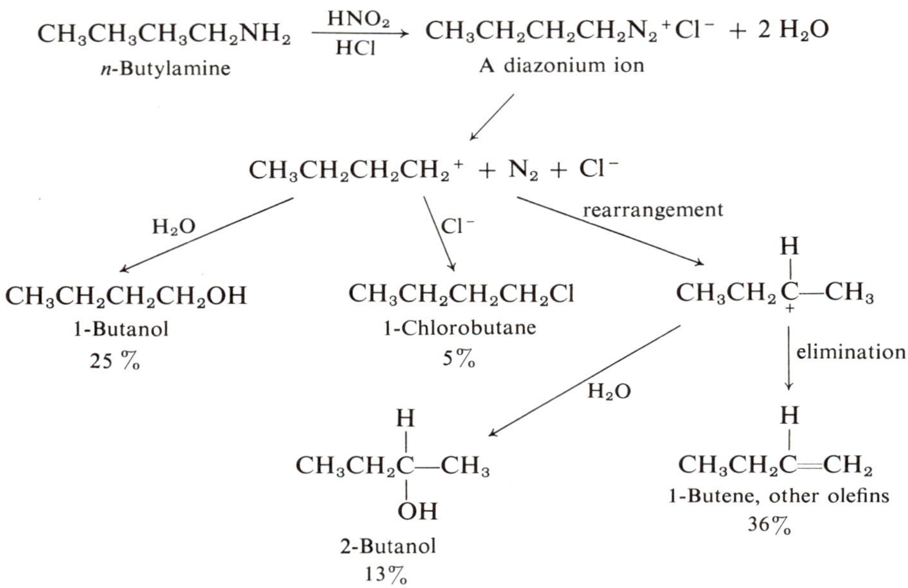

Reaction of a primary amine with nitrous acid

very stable, the loss of N_2 from diazonium ions is usually a strongly favored process unless the nature of the starting amine is such that the ion can be stabilized against further reaction.

Aliphatic diazonium ions (those derived from primary alkyl amines) are very unstable and have been isolated in only a few instances. They decompose rapidly to primary carbonium ions which then react rapidly with solvent or other nucleophiles in the reaction mixture. The primary ion may also rearrange its structure to a more stable one and, finally, olefin-forming elimination reactions can occur. The decomposition of alkyl-substituted diazonium ions is clearly not a very clean reaction. The evolution of nitrogen gas during the process is quantitative, however, and the reaction provides, if nothing else, a convenient method for determining the presence of primary amino groups.

Primary Aromatic Amines. With primary aromatic amines, the diazonium cations formed by treatment of the amines with nitrous acid are relatively stable and often they can be isolated in solid form. The greater stability of aromatic diazonium species is the result of resonance interaction between the nitrogen function and the aromatic ring.

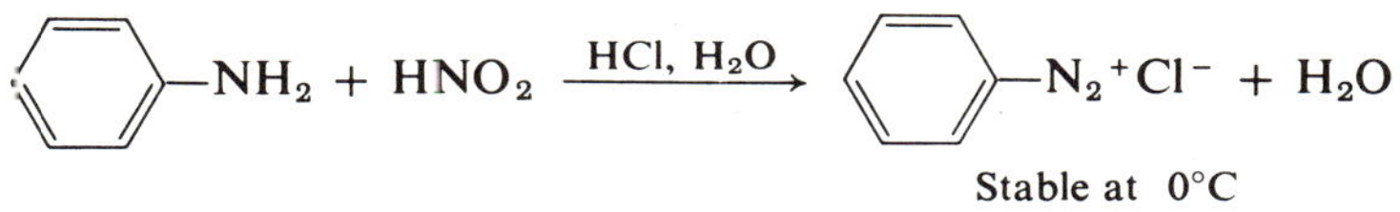

When aqueous solutions of aromatic diazonium compounds are warmed, the displacement of a nitrogen molecule by water and subsequent formation of a hydroxyl compound occurs to give a phenol.

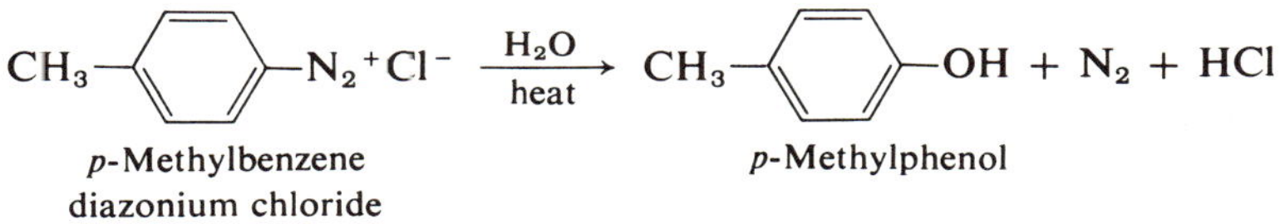

Diazonium compounds may be reduced with either loss of nitrogen to give the parent hydrocarbon or to form a new class of molecules known as hydrazines; these will be considered in more detail later in this chapter.

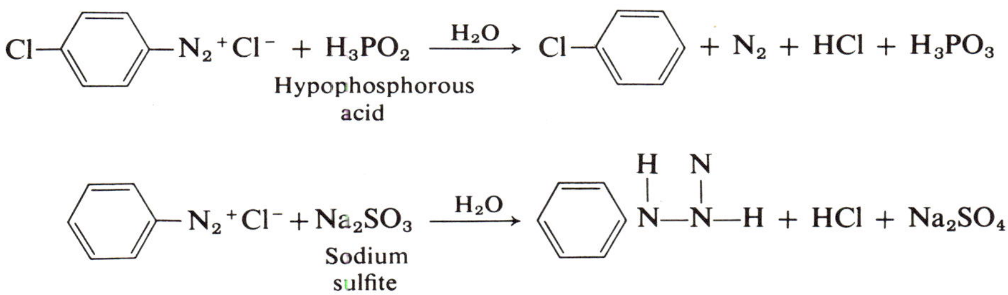

Reductions of aromatic diazonium compounds

Aromatic diazonium cations are electron-deficient and thus should be capable of undergoing electrophilic reactions. A variety of coupling reactions involving the reaction of an aromatic diazonium ion with an activated aromatic ring system

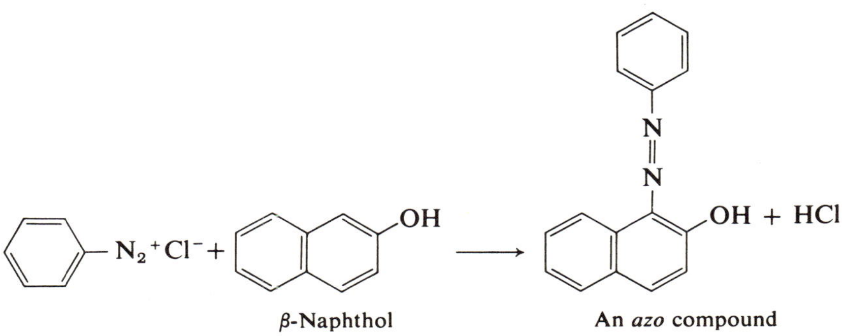

occur to give compounds in which the nitrogen–nitrogen bond in the diazonium system remains intact. The new class of compounds formed in this way is important commercially since the presence of the nitrogen–nitrogen double bond or *azo* linkage usually imparts a strong color to the organic compound which contains it; compounds of this nature are used as dyes for fabrics. By appropriate choice of substituents at each end of the *azo* linkage, the color of the organic compound can be varied over the entire range of the visible spectrum.

Secondary and Tertiary Amines. Secondary amines, whether aromatic or aliphatic, form *N*-nitroso derivatives when treated with nitrous acid. Tertiary amines with all aliphatic substituents at nitrogen undergo no detectable reaction with nitrous acid. If one or more of the substituents at nitrogen is aromatic, a

$$CH_3{-}\overset{\overset{\displaystyle H}{|}}{N}{-}CH_2CH_3 + HNO_2 \xrightarrow{HCl} CH_3{-}\overset{\overset{\displaystyle NO}{|}}{N}{-}CH_2CH_3 + H_2O$$

An *N*-nitroso amine

reaction of a tertiary amine with nitrous acid does occur, but it is an electrophilic substitution reaction at the *ortho* or *para* position of the aromatic ring by the nitrosonium ion (NO^+), the amino function serving only to activate the aromatic ring for this reaction.

$$CH_3{-}\overset{\overset{\displaystyle CH_3}{|}}{N}{-}CH_2CH_2CH_3 + HNO_2 \longrightarrow \text{no reaction}$$

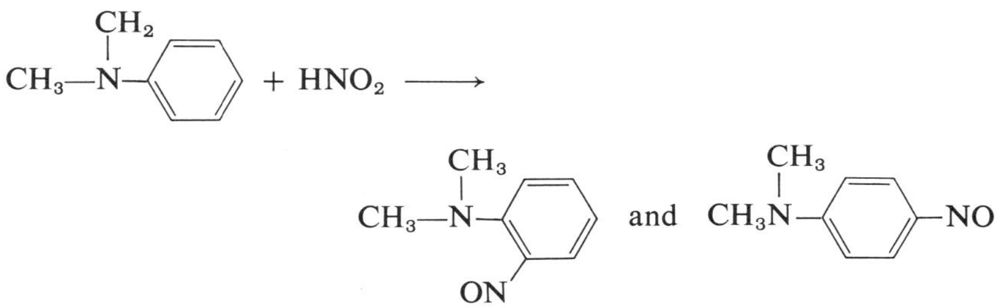

Electrophilic substitution products

AZO COMPOUNDS AND HYDRAZINES

It has previously been noted that compounds containing the *azo* bond system (—N=N—) can be prepared by diazonium coupling reactions when the substituents at each end of this bond system are aromatic in character. A wide variety of *azo* compounds is known in which one or both of the substituents at the *azo* linkage are aliphatic in nature. In these cases, the compounds are less stable than when both of the substituents are aromatic. Aromatic substituents

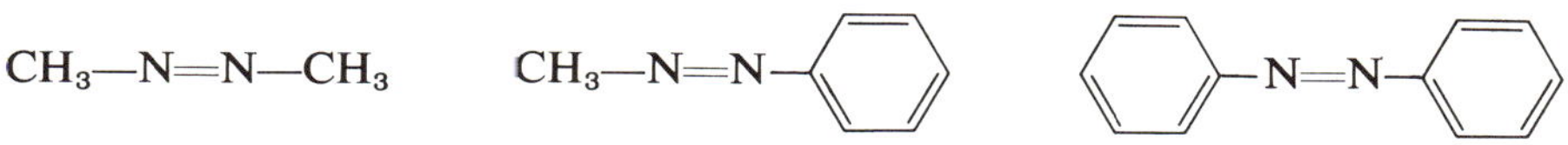

Azomethane Methylazobenzene Azobenzene

interact by resonance with the nitrogen–nitrogen double-bond system to stabilize the molecule against decomposition. With aliphatic substituents present, a molecule of nitrogen is relatively easily extruded from an *azo* compound to produce, initially, two organic radicals. These compounds thus can serve as initiators for radical polymerization reactions. When one of the substituents at

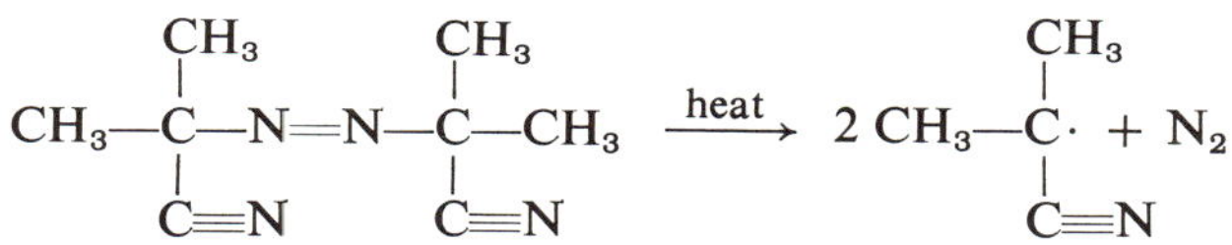

Azobisisobutyronitrile (AIBN)

the *azo* linkage is a hydrogen atom, the resulting compounds are quite unstable and decompose within a few hours after formation to give the hydrocarbon derived by collapse of the two ends of the molecule.

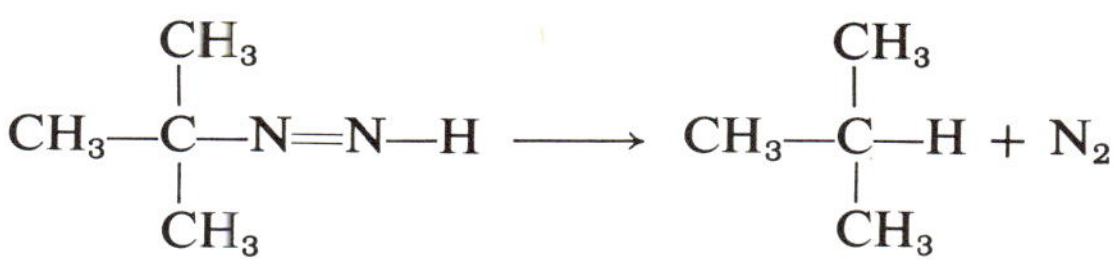

In analogy with the reduction of alkenes, the nitrogen–nitrogen double bond in *azo* compounds can be reduced to give another class of organic compounds known as hydrazines. These materials can be regarded as two amines back to back. They are strongly basic compounds and are powerful nucleophiles. The

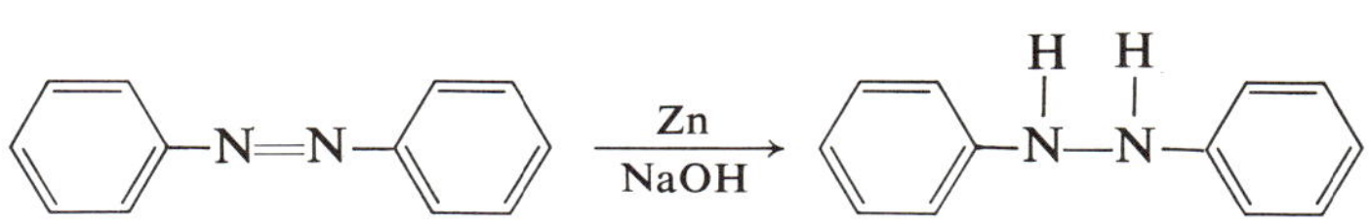

Azobenzene Hydrazobenzene (*sym*-diphenylhdrazine)

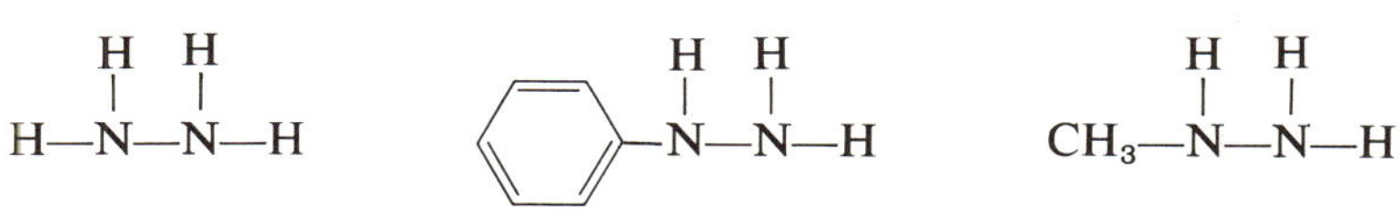

Hydrazine Phenylhydrazine Methyl hydrazine

parent compound, hydrazine (N_2H_4), is used as a rocket fuel. Further reduction can break the bond between the nitrogen atoms in the hydrazines to form the corresponding amines.

HETEROCYCLES

Heterocycles are organic molecules that have cyclic structures in which one or more atoms in the ring is an element other than carbon. The elements of the ring which are not carbon atoms are called *heteroatoms* and, most commonly, these heteroatoms are oxygen, nitrogen, and sulfur. For example, the cyclic ether shown below is a heterocycle, with the heteroatom being oxygen.

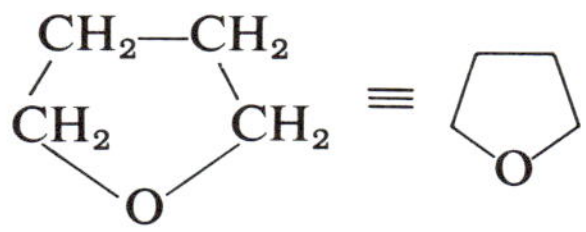

Tetrahydrofuran

Although the nomenclature systems for heterocycles are honeycombed with trivial names, the syllables *az-*, *ox-*, and *thi-* indicate that the heteroatoms in a structure are nitrogen, oxygen, or sulfur, respectively. The atoms of a heterocyclic ring are numbered so that the heteroatom(s) have the smallest possible numbers; the oxygen atom of tetrahydrofuran is position 1 for nomenclature purposes.

Oxygen Heterocyles

The saturated, oxygen-containing heterocycles shown below have essentially the same chemistry as the noncyclic ethers. The larger rings are good solvents for organic molecules and are, therefore, useful as media for organic reactions. Ethylene oxide can be polymerized to give long-chain polyethers.

Ethylene oxide

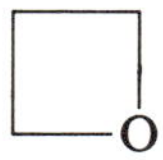

Oxetane

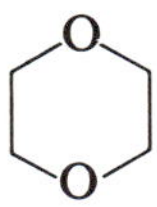

1,4-Dioxane

A number of oxygen containing heterocycles are also characterized by the presence of carbon–carbon double bonds in the ring structure. Several of the more common of these compounds are sketched on p. 167.

Furan
(oxole)

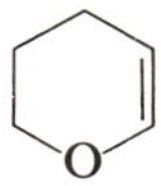

Dihydropyran

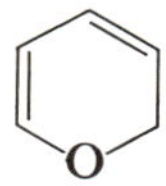

α-Pyran

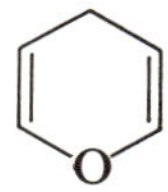

β-Pyran

The six-membered oxygen-containing heterocycle called pyran is the parent compound for a variety of plant pigments of which cyanidin chloride is an example. These compounds usually occur in combination with sugars and metal ions and take on colors ranging throughout the visible spectrum from red to blue.

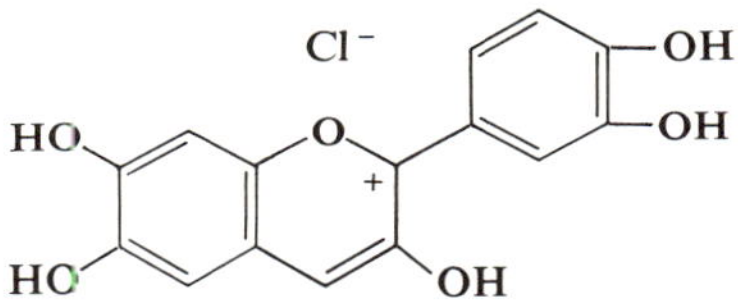

Cyanidin chloride
(blue pigment of the cornflower)

Nitrogen Heterocycles

Saturated heterocyclic amines are obtained by replacing the oxygen atoms in the series of compounds shown earlier with nitrogen atoms. These materials are secondary, aliphatic amines and are both strong bases and good nucleophiles. Aziridine and azetidine are unstable in strongly acid solutions and will undergo ring-opening reactions under these conditions.

Ethylene imine
or aziridine

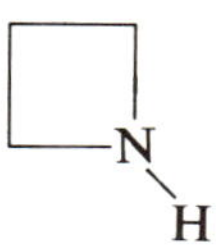

Azetidine

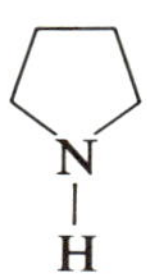

Pyrrolidine

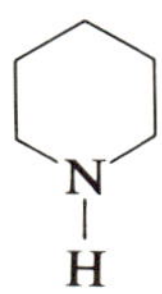

Piperidine

The venom of the red fire ant which is found in the United States from North Carolina to Texas provides an example of a naturally occurring piperidine. The active ingredient in this venom is 2-methyl-6-*n*-undecylpiperdine, shown on p. 168.

Alkylated piperidines are rarely found in animals but have been isolated from a number of plants.

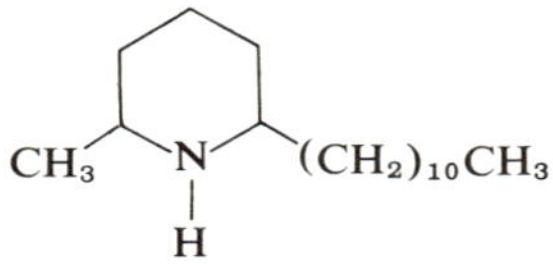

2-Methyl-6-*n*-undecylpiperidine

Several important nitrogen heterocycles which contain sp^2 carbon atoms are pyrrole, imidazole, pyridine, and pyrimidine. The electronic structures of pyrrole and furan have much in common. Remembering that there is an "extra" pair

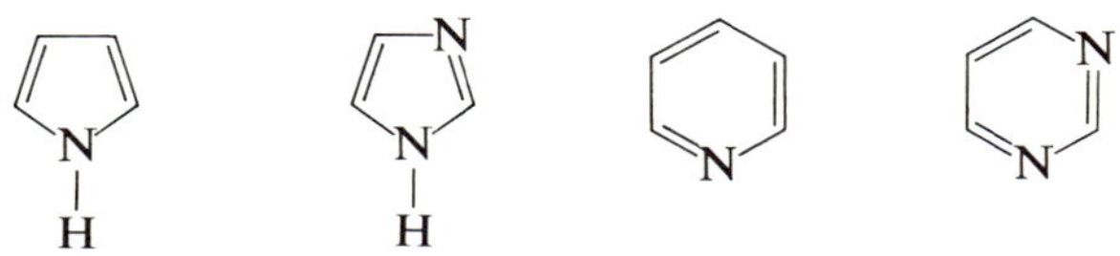

Pyrrole Imidazole Pyridine Pyridmidine

of electrons on the heteroatom in each of these compounds, it can be seen that the π-electronic systems of both molecules are closed rings of $2p$ orbitals (Fig. 7.1). Like benzene, another molecule we have encountered where this is true, an appreciable amount of electron delocalization is possible in furan and pyrrole. We anticipate that these molecules will be unusually stable and will take part in many reactions during which the integrity of the rings is maintained. The resonance structures shown below provide a way to represent the delocalization of π electrons in these molecules.

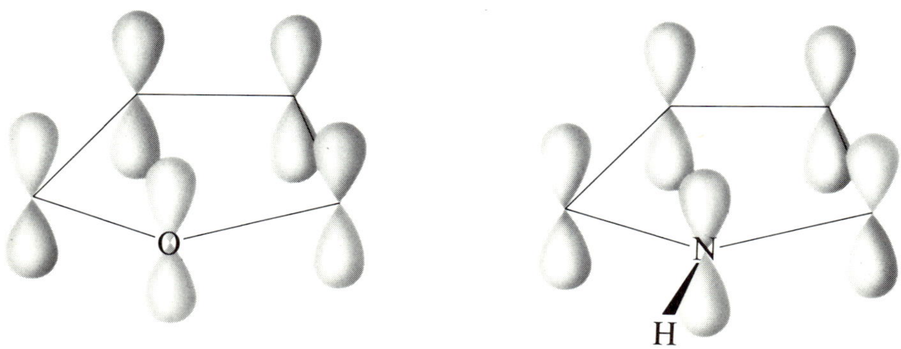

Fig. 7.1. Representations of the π-electronic systems in furan and pyrrole.

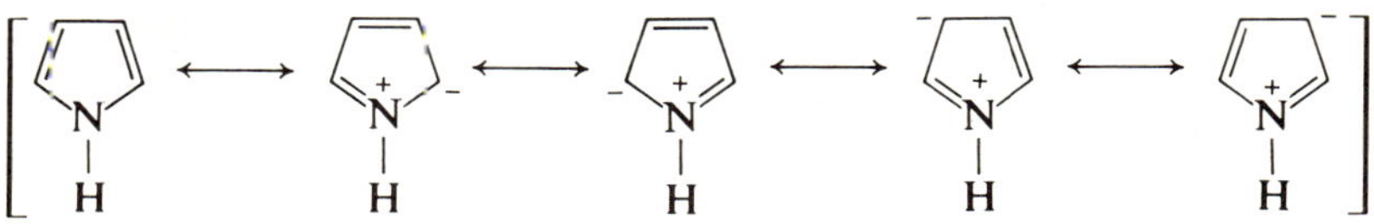

Resonance structures for pyrrole

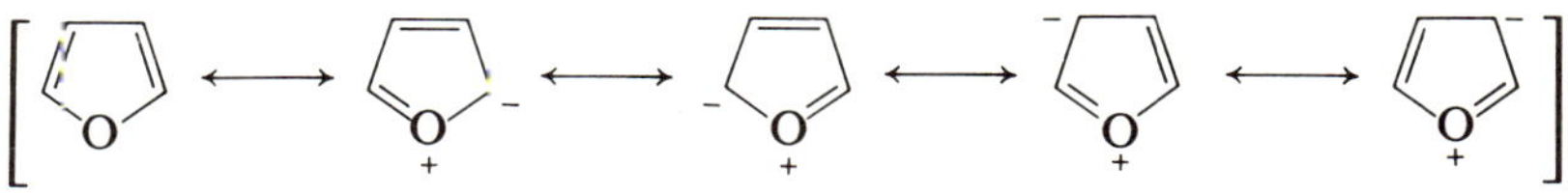

Resonance structures for furan

An immediate consequence of the aromatic nature of pyrrole is the greatly decreased basicity of the molecule. The pair of electrons at the nitrogen atom which would normally be used to bind an incoming proton is now delocalized over the entire ring so that the attraction of a proton to the nitrogen is not as strong as it is in pyrrolidine, for example. More importantly, formation of the ammonium salt of the pyrrole destroys the π-electronic system of the molecule, thereby raising the free energy of the protonated form. Unfortunately, an accurate measurement of the K_b for pyrrole is not possible since the molecule decomposes in acidic solutions.

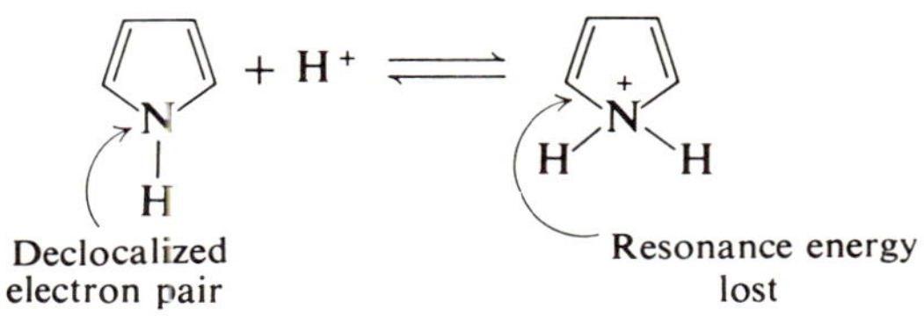

Pyrrole as a base

Because the resulting anion is stabilized by resonance, pyrrole is a reasonably strong acid. The dissociation constant in this case is 3×10^{-14}, about the same as that of water. Pyrrole is at least 10^{18} times stronger as an acid than is pyrrolidine.

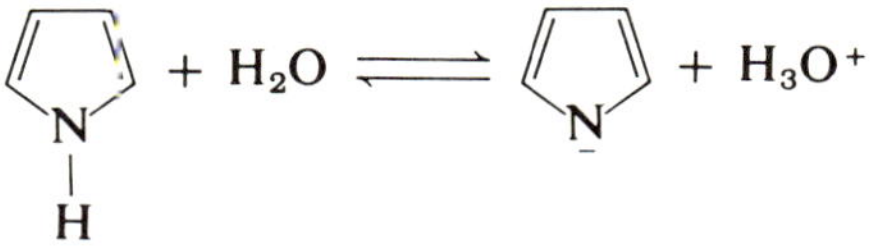

Pyrrole as an acid

Imidazole is a heterocycle that incorporates an acidic and a basic nitrogen into the same ring. As the electronic structure depicted in Fig. 7.2 indicates, one nitrogen atom of imidazole is similar to the one found in pyrrole, while the other

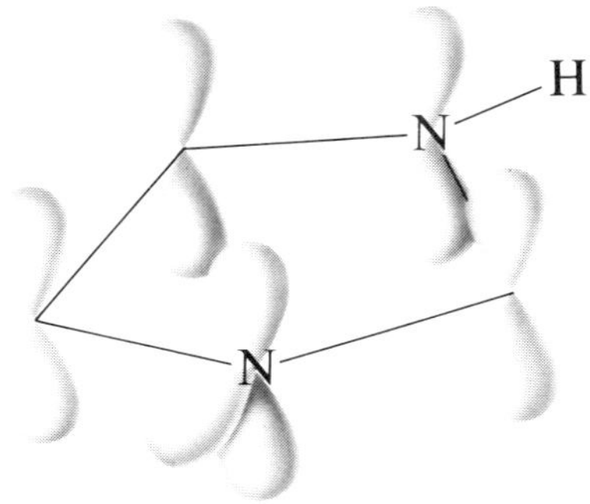

Fig. 7.2. The electronic structure of imidazole.

nitrogen features an unshared pair of electrons that projects away from the ring. It is this latter nitrogen that is the basic site of the molecule. Imidazole does not appear as a free molecule in nature, but derivatives of imidazole are important parts of the catalytic proteins known as enzymes; these catalysts often utilize the ability of imidazole to act as an acid or a base.

Pyridine is structurally very similar to benzene, but the resonance energy of pyridine (40 kcal/mole) is slightly more than that of benzene. The nitrogen atom in the ring is similar to the basic site in imidazole so that pyridine is moderately basic with $K_b = 1.7 \times 10^{-9}$.

The pyridine ring system and reduced forms of the pyridine ring systems are common features of organic compounds synthesized by plants. An important example is nicotinamide which is related structurally to nicotine, a poisonous component of tobacco smoke. Nicotine provides an example of both types of nitrogen heterocycle with a tertiary amine-like pyrrolidine ring attached as a side chain to an aromatic pyridine system. Nicotinamide, found in combination with other organic fragments, takes part in oxidation–reduction reactions in living organisms. This pyridine derivative cannot be synthesized *in vivo* by animals and must, therefore, be ingested in order to maintain life.

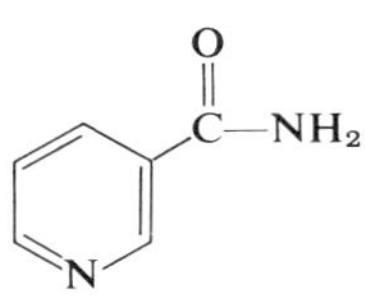

Nicotinamide

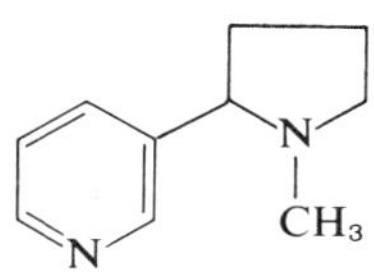

Nicotine

Reactions of Heterocycles

The delocalized π-electronic structure of pyrrole, furan, and thiophene, the sulfur analog of furan, enables these molecules to participate in electrophilic substitution reactions in much the same way as benzene. Electrophilic substitution takes place at carbon 2 as the following examples indicate.

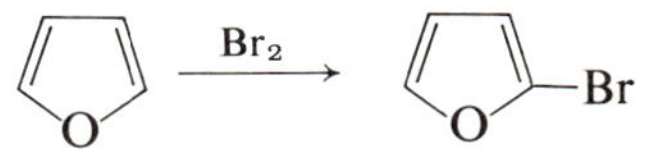

2-Bromofuran

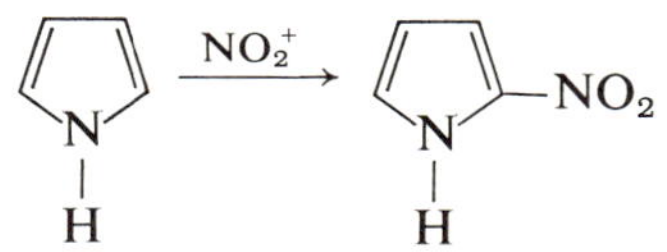

2-Nitropyrrole

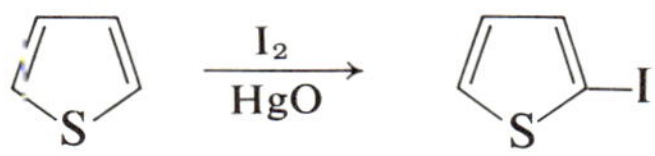

Thiophene 2-Iodothiophene

Pyridine takes part in electrophilic substitution reactions; in these reactions it behaves much like nitrobenzene in that the electrophile attacks at the carbon *meta* to the nitrogen atom and the rates of the reactions are much less than those of the corresponding reactions with benzene. Vigorous conditions are usually required for electrophilic substitution on pyridine.

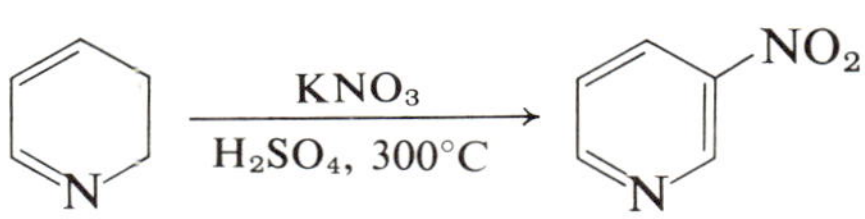

3-Nitropyridine

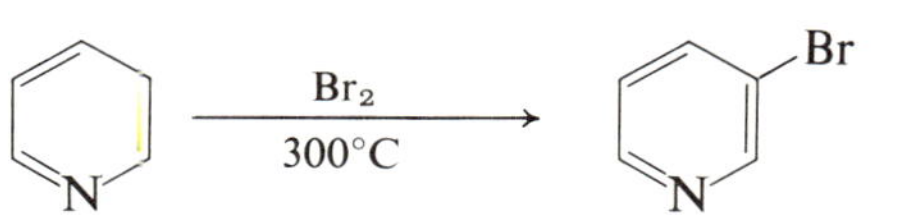

3-Bromopyridine

Electrophilic substitution with pyridine

Pyridine is more susceptible to substitution by nucleophiles than by electrophiles. With hydroxide or amide ions, substitution at the carbon adjacent to the heteroatom takes place. The hydroxyl-substituted pyridine provides us with the first example of a phenomenon that commonly occurs in nitrogen-containing

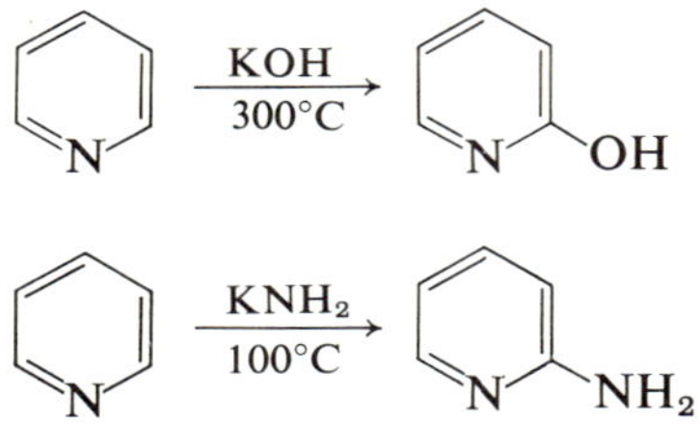

2-Aminopyridine

heterocycles with hydroxyl groups adjacent to the nitrogen: tautomerism to a carbonyl form can take place. Although this change appears to be unfavorable due to loss of resonance energy, the carbonyl form is favored in this equilibrium.

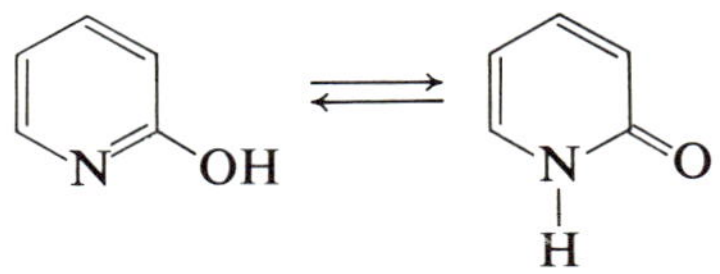

2-Pyridone

Tautomerism in 2-Pyridone

Alkaloids

Alkaloids are biologically important nitrogenous heterocycles that are synthesized by plants. The simplest natural alkaloids can be thought of as derivatives of pyridine. Coniine (2-*n*-propylpiperidine) is a constituent of hemlock and quite

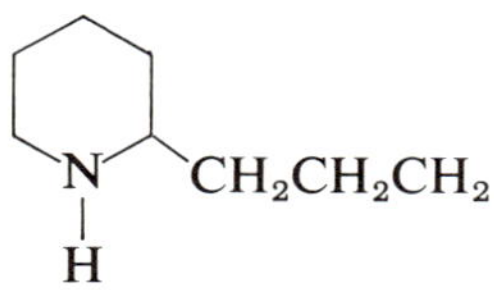

Coniine

toxic, as the story of Socrates amply indicates. It is structurally very similar to the venom component noted previously. The ring system in this compound is essentially a reduced pyridine (or piperidine) skeleton.

If one imagines fusing a benzene ring to the unsaturated nitrogen heterocycles discussed above, new classes of heterocycles can be generated that are of biological importance. Fusion of a benzene ring to pyrrole affords indole, while

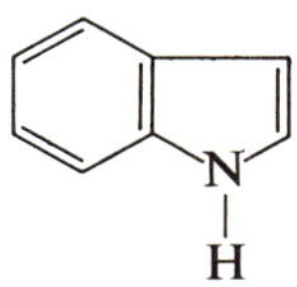

Indole

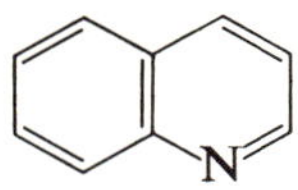

Quinoline

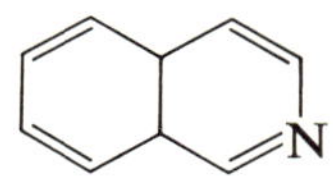

Isoquinoline

attachment of benzene to a pyridine ring can give quinoline or isoquinoline, depending upon how the attachment is made. The last two compounds can be regarded as analogs of naphthalene in the same sense that pyridine is an analog of benzene.

Alkaloids based upon the indole skeleton are widely distributed in nature. A natural plant growth regulator, 3-indolylacetic acid, is clearly structurally

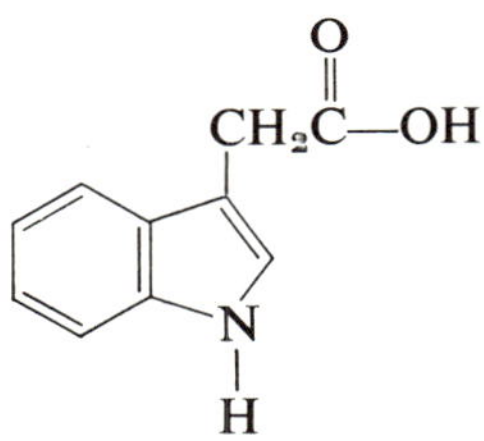

3-Indolylacetic acid

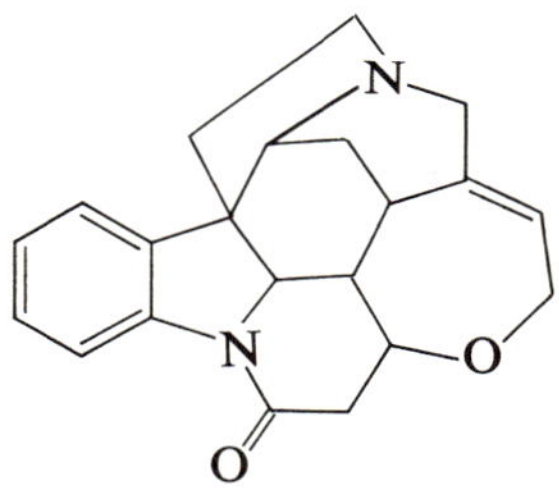

Strychnine

related to indole and the basic outline of the indole skeleton can be seen in the structure of strychnine, a potent poison often prominent in Sherlock Holmes novels. Many hallucinogenic drugs are derivatives of indole. A clue to the mechanism of the physiological action of these materials may rest in the fact that derivatives of indole play important roles in the chemistry of the brain. These include tryptamine and 5-hydroxytryptamine (serotonin), structures in which the indole ring is apparent.

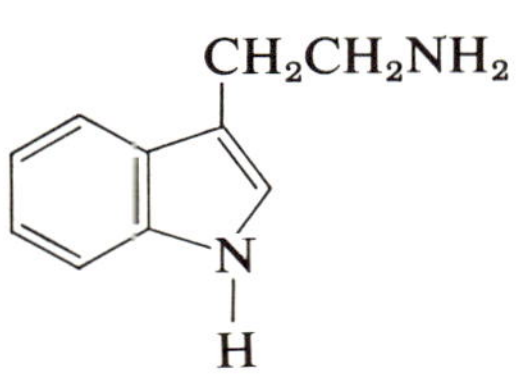

Tryptamine

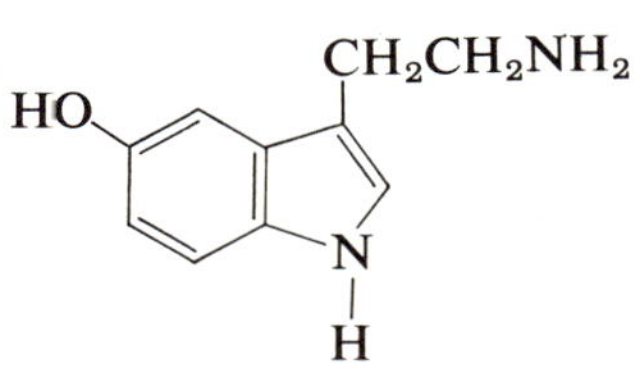

Serotonin

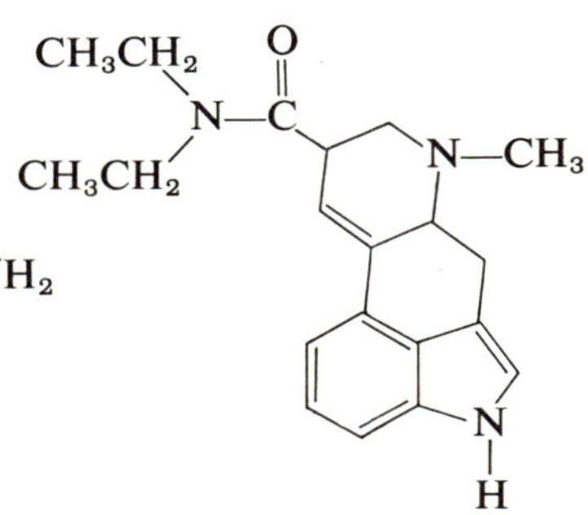

Lysergic acid diethylamide (LSD)

By contrast, naturally occurring materials based on the quinoline ring structure are not common. Quinine, which occurs in the bark of the Cinchona tree has been used for centuries for treatment of malaria. Synthetic drugs such as primaquinine have largely replaced quinine for this purpose. Both compounds contain the quinoline ring as an integral part of their structure.

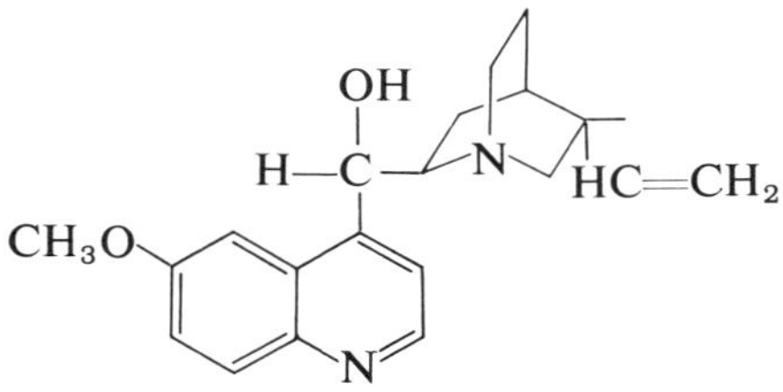

Quinine

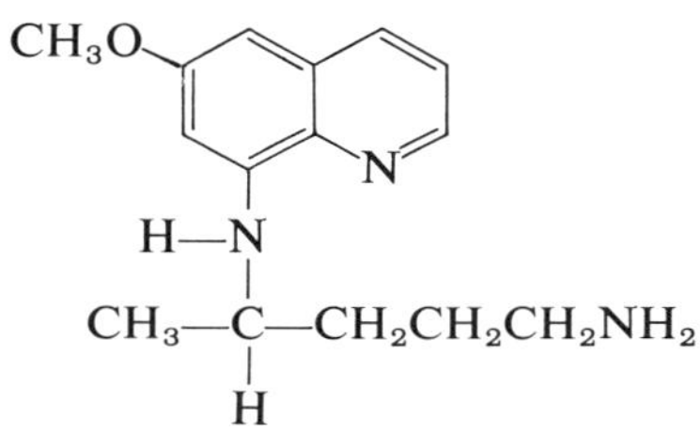

Primaquine

A large number of alkaloids are based upon the ring system of isoquinoline. For example, the liquid from the seed capsules of the opium poppy contain more than 20 alkaloids of this type including papaverine, which is useful as an antispasmodic drug, and morphine.

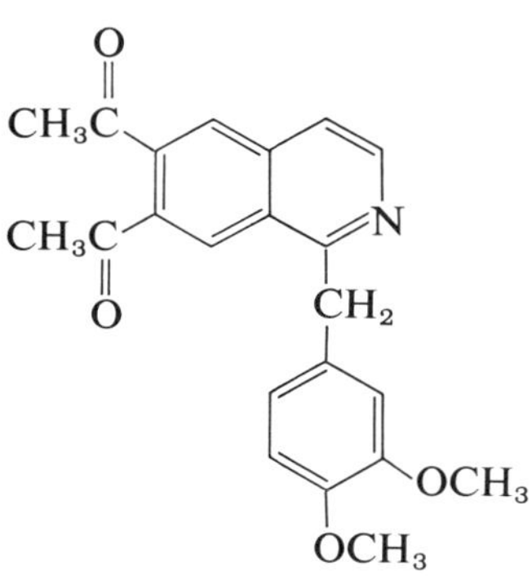

Papaverine

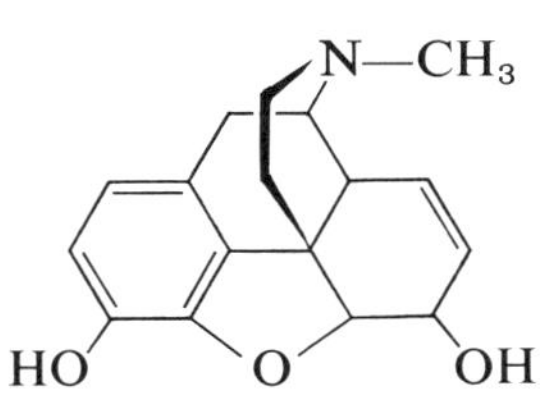

Morphine

The full literature of heterocyclic chemistry is staggeringly large and even an adequate summary is beyond the scope of this text. Discussion of the chemistry of heterocycles important in the transmission of genetic information will be made later, but the reader should realize that this introduction to heterocyclic compounds has been necessarily incomplete and brief.

PROBLEMS

1. Provide an acceptable name for each of the following compounds.

(a) 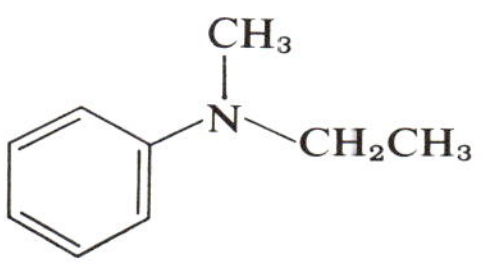

(b) $CF_3CH_2NH_2$

(c) 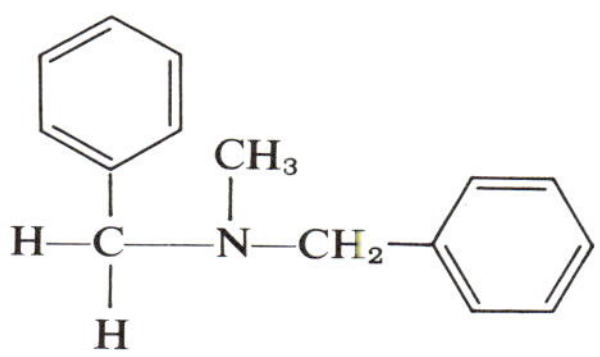

(d) 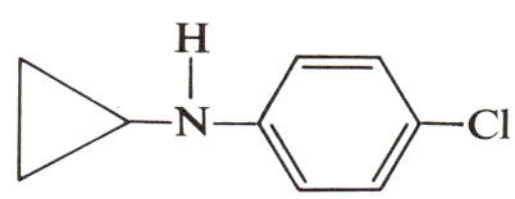

(e) 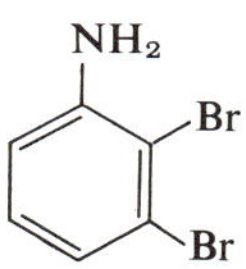

(f) $(CH_3CH_2CH_2)_4NCl$

(g) 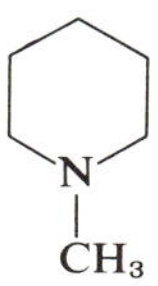

(h) 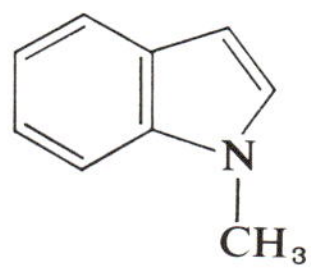

(i) 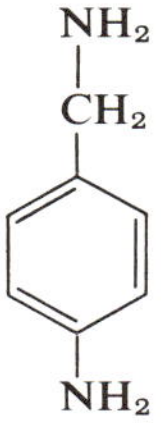

(j) 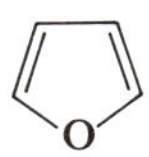

2. Give the structural formulae that correspond to the following names.
   (a) *p*-Chloroaniline

(b) *N*-Ethyl-*N*-methylhexylamine
(c) α-Naphthylamine
(d) *N*-Cyclopropyl-*N*-methylaniline
(e) 2-Amino-3-octanol
(f) Cyclohexylamine
(g) *N*-Chloropyrrolidine
(h) Aziridine
(i) Tetrabutylammonium fluoride
(j) Hexadecyltrimethylammonium bromide

3. Identify the functional groups in papaverine, quinine, and LSD that have been encountered so far.

4. *p*-Nitroaniline is a much weaker base than aniline while *p*-methoxylaniline is a stronger base than aniline. Rationalize these observations.

5. Predict the products that would be obtained by treating the following amines with nitrous acid.

(a) CH_3NH_2

(b) 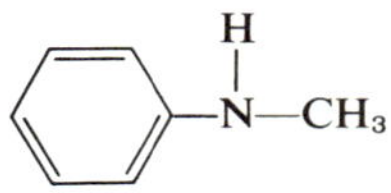

(c) 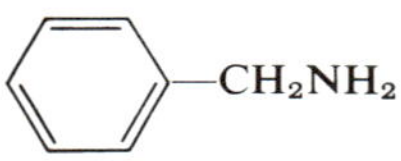

(d) 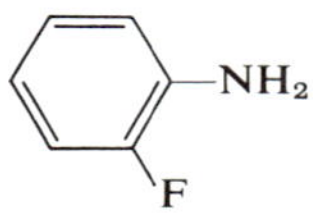

(e) 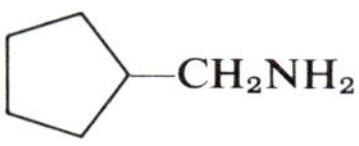

(f) 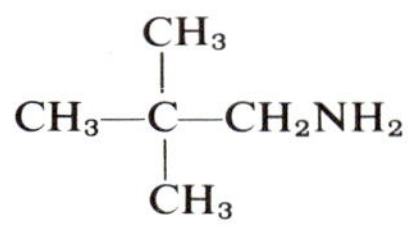

(g) 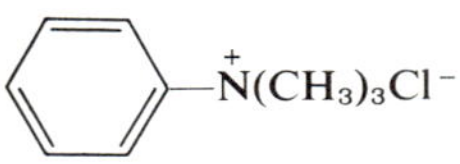

(h) 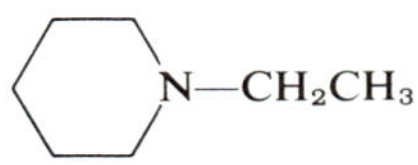

6. Predict which of the following pairs of amines will be most basic.

(a) 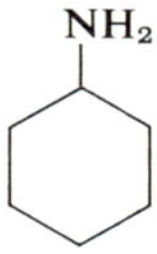 *or* 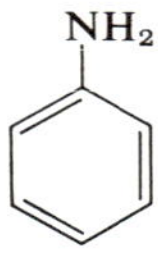

(b) 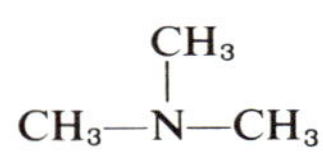 *or* 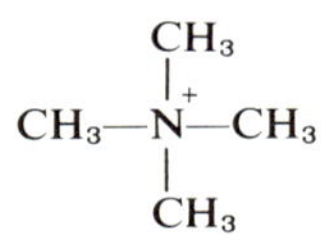

(c) 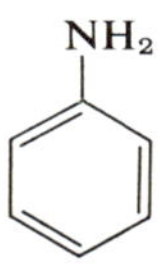 *or* 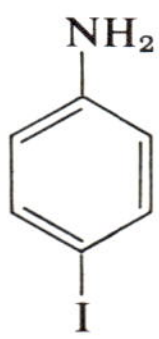

(d) NH_3 *or* NF_3

(e) 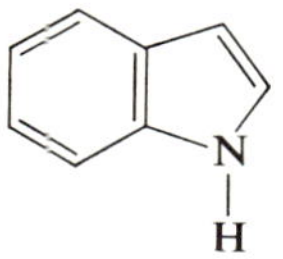 *or* 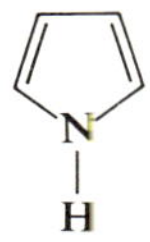

(f) $CH_3CH_2—N(CH_2CH_3)—CH_2CH_3$ *or* $CH_3—N(CH_3)—CH_3$

7. Suggest a mechanism for the reaction of benzenediazonium chloride ($C_6H_5N_2{}^+Cl^-$), with phenol. The product of the reaction is 4-hydroxyazobenzene.

8. Two geometrical isomers of azobenzene are known. Sketch their structures.

9. In a nonpolar solvent like benzene the heterocycle imidazole is associated into chains of about 20 imidazole units in length. These structures do not form in water solution. Explain these observations in terms of the hydrogen-bonding abilities of imidazole.

10. It was found that 48 ml of 0.0582 *N* HCl was required to neutralize 0.178 g of a substance known to be a monoamine. What is the molecular weight of this substance?

11. Predict the pmr spectrum of each of the following compounds.
   (a) Ethylamine
   (b) *N*,*N*-Dimethylaniline
   (c) 2,3,5,6-Tetrabromoaniline

SUGGESTED READINGS

R. M. Acheson, "An Introduction to the Chemistry of Heterocyclic Compounds." Wiley (Interscience), New York, 1960.

M. Calvin and M. J. Jorgenson, "Bioorganic Chemistry," pp. 167 and 179. Freeman, San Francisco, California, 1968.

R. Elderfield, ed., "Heterocyclic Compounds." Wiley, New York, 1950–1967. (A series of reference volumes.)

F. Franzen and K. Eysell, "Biologically Active Amines Found in Man." Pergamon, Oxford, 1969.

D. Ginsburg, "Concerning Amines." Pergamon, 1967.

A. R. Katritzky and J. M. Lagowski, "Heterocyclic Chemistry." Methuen, London, 1960.

S. Patai, ed., "The Chemistry of the Amino Group." Wiley (Interscience), 1968.

8

Organic Reaction Mechanisms: Displacement and Elimination Reactions

The mechanism of an organic reaction is a detailed description of how the atoms and electrons of reactant molecules become rearranged into the molecular configuration that defines the product molecules. We would like to have information about how each chemical bond in the reactants is perturbed as these molecules gain enough energy to attain the transition state required for transformation into products. For reactions in solution, an understanding of the interactions between solvent molecules and the reactants, transition states, and products would be useful. The ideal organic reaction mechanism could be likened to the individual frames of a motion picture film; each frame shows what is happening at a particular instant in time and, when viewed continuously, provides a coherent representation of a sequence of events. No organic reactions are understood well enough to permit refined mechanisms of this idealized type. However, as a result of efforts over the past half-century, the essential features of a number of organic reaction mechanisms are available. The reactions of alkenes and aromatic compounds with electrophiles have been discussed briefly in previous chapters; these mechanisms are well established. In this chapter we shall examine some other reactions and try to indicate some of the approaches the organic chemist uses in elucidating a reaction mechanism.

It should be emphasized that organic mechanisms in the real world are nonideal when compared to the standard outlined above. All reaction mechanisms are based upon experimental observations; the mechanism of a particular reaction must be chemically consistent with all the known facts regarding that reaction. As new techniques become available, cleverly designed experiments can provide sufficient new information that revision of a currently accepted mechanism is necessary.

TYPES OF ORGANIC REACTIONS

Organic reactions can usually be classified into one of the following categories:

(1) Additions
(2) Substitutions
(3) Eliminations
(4) Rearrangements

In an addition reaction, some or all of the atoms of one reactant are added to the structure of a second reactant. The formation of 1,2-dibromoethane from bromine and ethylene is an example of an addition reaction. When one functional group on a reactant has been displaced by another functional group, a

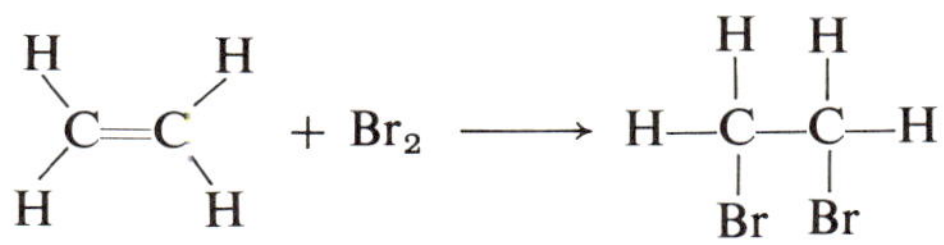

An addition reaction

substitution reaction has taken place. An example of a substitution reaction is the generation of methanol from hydroxide ion and methyl iodide. Elimination reactions can be regarded as the reverse of an addition reaction—one or more

$$CH_3I + OH^- \longrightarrow CH_3OH + I^-$$

A substitution reaction

groups of atoms are removed from the structure of a reactant molecule, as in the synthesis of cyclohexene from cyclohexanol. Finally, a molecular rearrangement is a structural reorganization within the reactant molecule. Rearrangement

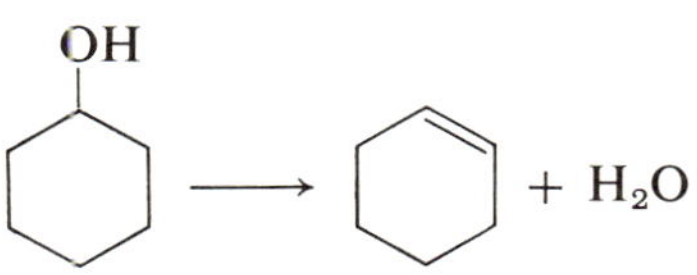

An elimination reaction

reactions sometimes accompany reactions of the other three types mentioned, or can occur separately. This classification of reactions carries no implication regarding the mechanism of the reactions but merely summarizes the outcome of a given transformation. Complicated reactions can often be broken down into a series of simple reactions of these types.

How Are Mechanisms Determined?

Product Studies. What kind of information does one gather in order to suggest a mechanism for an organic reaction? Clearly the first order of business is to ascertain exactly the nature of the products of the reaction. Many organic reactions do not proceed cleanly to give a single product but rather produce an array of products by several competing reactions. In addition to determining the number of products formed and their abundances, structural studies must be performed for each product including, if possible, details about the stereochemistry of the product. For example, the fact that *cis*-1,2-cyclohexandiol is formed when cyclohexene is treated with permanganate under mild conditions has important implications regarding the mechanism of this oxidation.

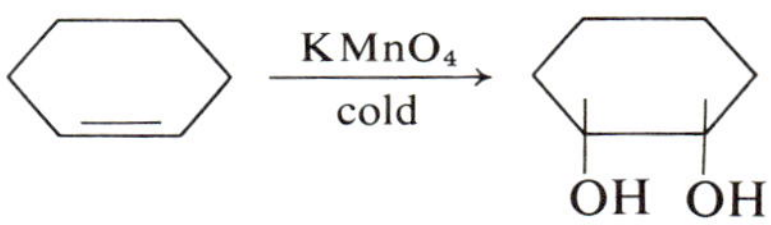

Rate Studies. The rate at which an organic reaction proceeds and the dependence of that rate on the initial concentrations of reactants provide pertinent information that can bear on the mechanism. For simple reactions, it is often found that the reaction rate will either depend upon (1) the concentration of a single reactant, or (2) the concentrations of two reactants. In the first case, the reaction follows a first-order rate law which can be expressed

$$\text{Rate} = k_1 \,[\text{reactant}]$$

where k_1 represents a proportionality constant and the square brackets mean "concentration of." If two reactant concentrations are important the rate law becomes

$$\text{Rate} = k_2 \,[\text{reactant}_1][\text{reactant}_2]$$

where k_2 represents another proportionality constant. This is a second-order rate law. Usually, a first-order rate law for a reaction implies that only a single reactant is involved in reaching the rate-determining transition state for the reaction. Most simply, a second-order rate law suggests that two distinct reactant molecules are simultaneously involved in forming the transition state for the rate-determining step of the reaction. Defining the rate law for a reaction will, thus, usually define the molecules involved in forming a transition state.

Perturbation Studies. Once the products of a reaction are known and the rate law for the reaction is determined, it is sometimes possible to suggest a mechanism for the reaction based upon precedent and intuition. Additional experimental work can then be directed toward substantiating the proposed mechanism. We know that structural changes in organic molecules can alter the distribution of electrons within the molecule and one possible substantiating experiment might be to introduce a substituent into one of the reactants, determine its effect on the reaction rate, and interpret the rate change in light of known properties of the substituent. For example, the rate of bromine addition to styrene should depend upon the electron density at the double bond since attack of a positively charged bromonium ion (Br^+) at this position is an important feature of the mechanism for this reaction. It would be predicted that a strongly electron-withdrawing substituent like a nitro group attached to the aromatic ring of styrene should lower the rate of addition of bromine.

Another occasionally useful perturbation of a reaction is a change in solvent. Important properties of solvents include polarity (as the result of a permanent dipole moment in the solvent molecule) and the ability to participate in special interactions such as hydrogen-bonding. Highly polar, interacting solvents like water and the alcohols are good solvents for polar reactants while solvents which are largely hydrocarbon in character (cyclohexane, benzene) will best dissolve nonpolar organic molecules. Solvent–solute interactions can affect the free energy of reactants and of transition states and, if the difference in polarity (electronic charge–separation) is very great between reactants and a transition state, the rate of the reaction may be strongly affected by a change in the polarity of the solvent for the reaction.

It was mentioned in Chapter 1 that isotopes of common elements are known to exist. Because of the difference in atomic mass of isotopes, molecules containing various isotopes of a given element react at slightly different rates. For example, in the radical halogenation of methane the substitution of deuterium (^2H) by a chlorine atom occurs at a slower rate than the substitution of a normal hydrogen (^1H). The rate of substitution (R_H) of a hydrogen atom (mass = 1) is several times faster than the rate of replacement (R_D) of a heavy hydrogen or deuterium atom (mass = 2). To a first approximation, large isotope effects will appear only when the chemical bond holding the isotopic atoms to the reacting molecule is ruptured during the course of the reaction. The observations of isotope effects on the rate of an organic reaction can help identify which bonds in the reactant molecule are primarily involved in forming the transition state for reaction.

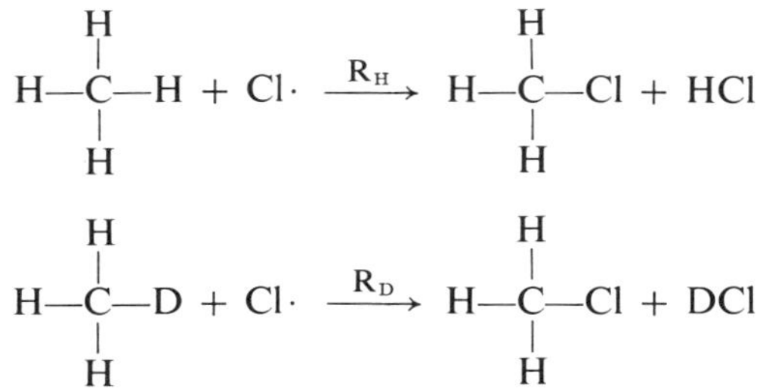

NUCLEOPHILIC SUBSTITUTION REACTIONS

We mentioned above in broad outline some experimental approaches that can be used to provide evidence upon which to base a mechanism for an organic reaction. These techniques and several others have been applied extensively to the study of nucleophilic displacement reactions in saturated compounds. These reactions involve the substitution of a nucleophile (**N**) for a leaving group (**L**) which is attached to a saturated organic molecule.

$$\mathbf{N} + \mathrm{R{-}}\mathbf{L} \longrightarrow \mathrm{R{-}}\mathbf{N} + \mathbf{L}$$

The leaving group **L** will depart from R—**L** with both electrons used initially to form the R—**L** bond, while the incoming **N** by virtue of its nucleophilic character will supply the electrons used to make the R—**N** bond. A number of common leaving groups are shown in Table 8.1. Groups which are weak bases in the free state tend to have good leaving abilities in displacement reactions.

Nucleophiles for aliphatic displacement reactions should possess a readily extensible region of high electron density. Strong bases should be good nucleophiles since R—**N** bond formation can be considered analogous to H—**N** bond formation during protonation of the base, **N**. However, base strength is not the whole story and the relative reactivities of nucleophiles will depend upon the reaction conditions (solvent, temperature, etc.) as well as the total electronic structure of the nucleophile. Nucleophilic reagents with a large number of polarizable electrons in their structure are better nucleophiles than might be anticipated on the grounds of basicity toward H^+ alone. Examples of such reagents are iodide ion (I^-), the mercaptide anions derived from thiols, and the thiophenoxide anions. The relative reactivities of some common nucleophiles

Table 8.1
Leaving Groups for Displacement and Elimination Reactions

Type of compound	*Bound group*	*Free group (after reaction)*
Chloride	~C—Cl	Cl^-
Bromide	~C—Br	Br^-
Iodide	~C—I	I^-
Carboxylic ester	~C—O—C(=O)—R	$^-$O—C(=O)—R
Sulfonate ester	~C—O—SO_2R	$^-SO_3R$
Phosphate ester	~C—O—P(=O)(OH)—O^-	$^-$O—P(=O)(OH)—O^-
Diphosphate ester	~C—O—P(=O)(OH)—O—P(=O)(O^-)—O^-	$^-$O—P(=O)(OH)—O—P(=O)(O^-)—O^-
Protonated alcohol	~C—O^+(H)—H	H—O—H
Protonated ether	~C—O^+(H)—R	H—O—R
Tetraalkylammonium ion	~C—N^+(R)(R)—R	R—N(R)—R

follow the qualitative order $C_6H_5S^- > CN^- > I^- > RNH_2 > RO^- > Br^- > Cl^- > NO_3^-$.

The major mechanisms for nucleophilic aliphatic substitution reactions have been delineated briefly in Chapter 6. In the S_N1 mechanism breaking of the bond to the leaving group and bond formation to the entering nucleophile are distinct steps in the reaction; the destruction of the covalent bond to the leaving group in the reactant molecule is the slow or rate-determining step in this reaction. For the S_N2 mechanism, two reactants (R—**L** and **N**) are simultaneously involved in generating the transition state for reaction. In this case both bond-making and bond-breaking proceed in synchrony. Tertiary aliphatic compounds react according to the S_N1 mechanism because the approach to the backside of the carbon bearing the leaving group is sterically hindered by the three alkyl groups attached to this carbon and, moreover, these alkyl groups exert a potent stabilizing effect on the carbonium ion formed at the central carbon.

S_N1 Reactions

The experimental evidence supporting the S_N1 mechanism is of several types. The reaction of *t*-butyl chloride with hydroxylic solvents is a good example of this reaction mechanism and shall be used in the discussion which follows. As has been described previously, the minimum number of steps in this reaction are (1) ionization of the alkyl halide to give a carbonium ion which (2) undergoes reaction with nucleophiles in the medium. Since solvent molecules far outnumber other nucleophiles in the system, the major product is *t*-butanol if the solvent is water, or a *t*-butyl ether if the solvent is an alcohol.

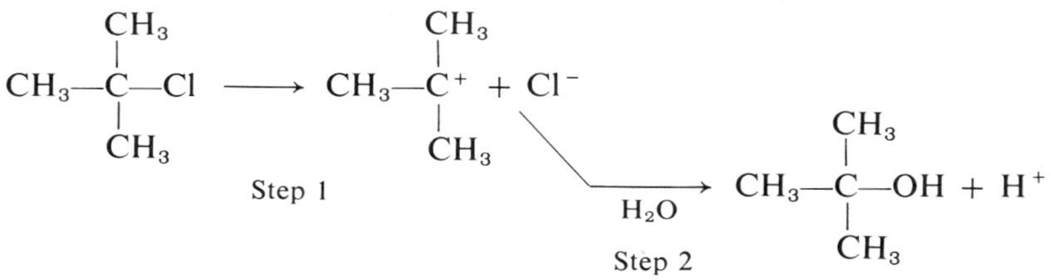

The rate law for the reaction of this alkyl halide in water is

$$\text{Rate} = k[t\text{-butyl chloride}]$$

and is consistent with the notion that only a single molecule (the chloride) is involved in the formation of the rate-determining transition state. Solvent molecules have been neglected in this formulation but clearly are important as the organic cation will strongly interact with the solvent. The importance of solvent for the S_N1 reaction is illustrated by the observation that the reaction of *t*-butyl chloride in water is faster than the corresponding reaction in ethanol. Ethanol is a less polar, more hydrocarbon-like solvent and is less favorable to the formation of ionic transition state.

It is found experimentally that *t*-alkyl halides react by the S_N1 mechanism about 100,000 times faster than secondary alkyl halides and at least a million

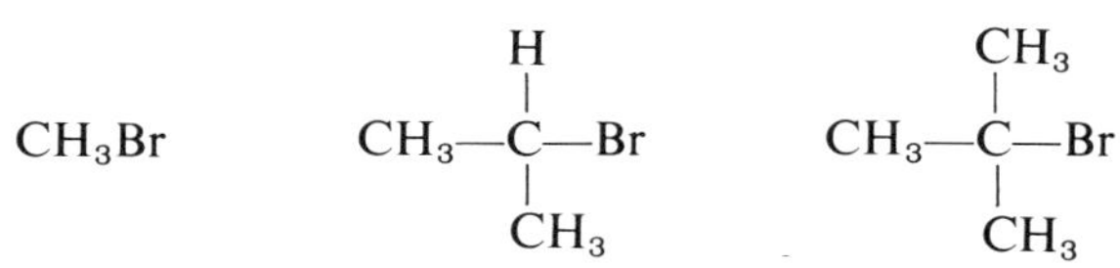

	Methyl bromide	Isopropyl bromide	*t*-Butyl bromide
Approximate relative rate for S_N1 hydrolysis	1	10	1,000,000

times faster than primary alkyl halides. Because the stability of the intermediate carbonium ion in the S_N1 reaction is strongly affected by the number and type of alkyl groups attached to the central carbon atom, it is expected that as the methyl groups of *t*-butyl chloride are successively removed a decrease in reaction rate will be observed.

Resonance effects also can stabilize carbonium ions and it has been found that benzyl chloride, a primary alkyl halide, reacts about as rapidly as isopropyl chloride in an S_N1 reaction. In the reaction of benzyl chloride, the 2*p*-orbital at the cationic center can overlap with the aromatic π-electronic system of the

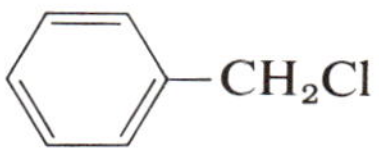

Benzyl chloride

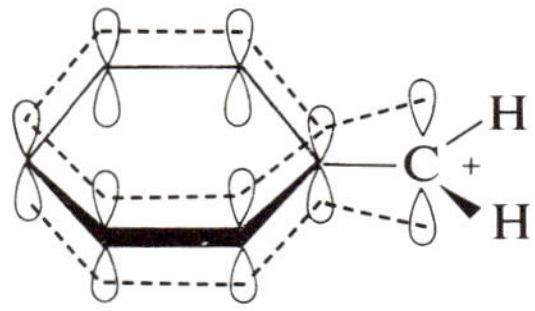

Benzyl cation

benzene ring. This interaction can be described by resonance structure of the type shown below.

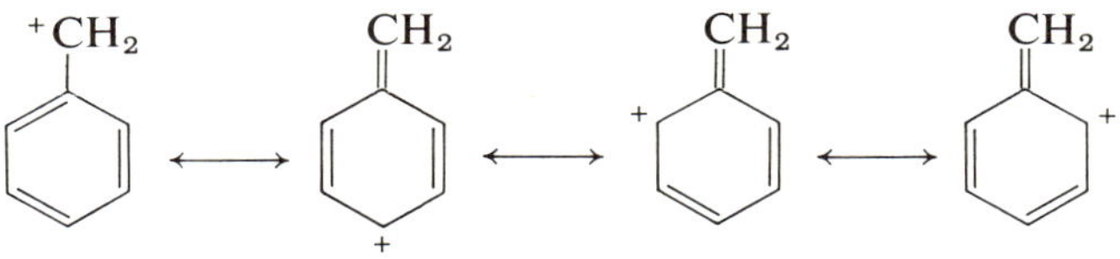

Some resonance structures for the benzyl carbonium ion

Thus, in this primary alkyl halide, one benzene ring appears to exert about as much stabilizing influence on the carbonium ion center as two methyl groups. Placing additional aromatic rings adjacent to the positive charge stabilizes the system even more; triphenylmethyl chloride is about 10^5 more reactive than *t*-butyl chloride.

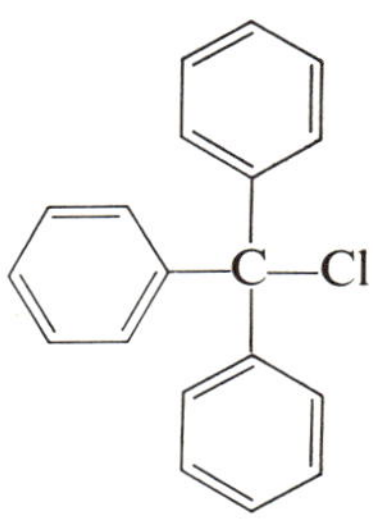

Triphenylmethyl chloride

Probably the most compelling evidence for the S_N1 mechanism comes from studies which relate the stereochemistry of the product molecules to the stereochemistry of the reactants. We will describe in more detail in Chapter 10 how organic molecules with four different substituents can be isomeric with and distinguishable from their mirror image form. This type of isomerism is illustrated by the two structures of 1-phenylethanol shown below. These two forms

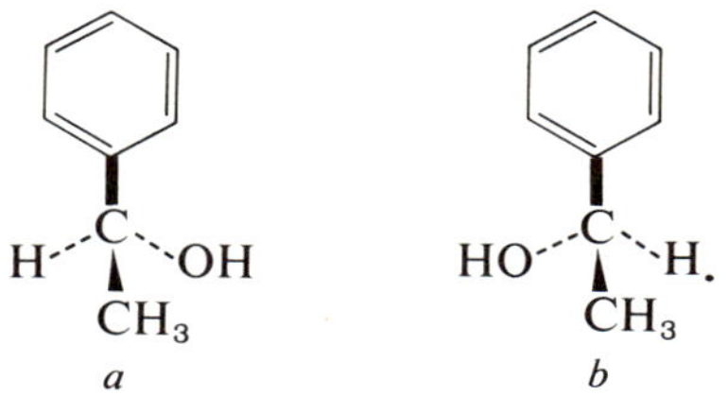

Mirror images of 1-phenylethanol

can be distinguished by appropriate experiments. When the 1-phenylethyl chloride corresponding to one of these isomers is solvolyzed in a water–acetone mixture, *both* isomers of the product alcohol are formed in essentially equal amounts. This observation can be accounted for best in terms of a reaction mechanism which features an intermediate with a plane of symmetry which reacts equally well from either side of the plane with solvent molecules to give the products. A planar benzylic carbonium ion fulfills these specifications.

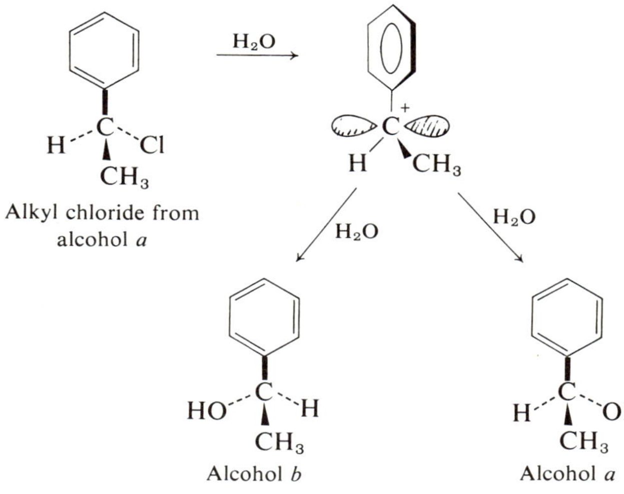

S_N2 Reactions

The S_N2 mechanism for nucleophilic substitution is followed by primary aliphatic compounds. The transition state for this mechanism contains both the incoming nucleophile (**N**) and the starting aliphatic material (R—**L**).

$$\mathbf{N} + \mathrm{R{-}}\mathbf{L} \longrightarrow [\mathbf{N}\cdots\mathrm{R}\cdots\mathbf{L}] \longrightarrow \mathbf{N}\mathrm{{-}R} + \mathbf{L}$$

In this transition state the new covalent bond from **N** is being formed at the same time the bond to the leaving group **L** is being broken; bond-making and bond-breaking steps in this mechanism occur simultaneously. The carbon atom to which **L** is attached is attacked from the rear by **N**. This carbon atom becomes temporarily pentavalent, although two of the bonds it makes are unusually long and, therefore, weak. Examples of reactions which proceed by S_N2 processes are the reactions of methyl chloride shown below.

		Nucleophile		Product
CH_3Cl Methyl chloride	+	$^-OCH_3$	$\longrightarrow$	$CH_3OCH_3 + Cl^-$ Methyl ether
	+	$^-SCH_3$	$\longrightarrow$	$CH_3SCH_3 + Cl^-$ Methyl sulfide
	+	^-CN	$\longrightarrow$	CH_3CN Acetonitrile (methyl cyanide)
	+	$N(CH_3)_3$	$\longrightarrow$	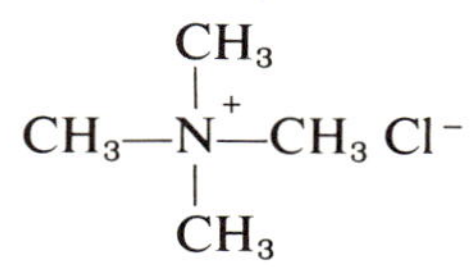 Tetramethylammonium chloride

The rate laws for each of these reactions is

$$\text{Rate} = k_2\,[\text{alkyl halide}][\text{nucleophile}]$$

The reaction rate depends upon the concentration of *both* reactants so that, for example, doubling either concentration would double the rate of the reaction. This rate law implies that the reactants must collide with one another for the reaction to take place, a conclusion consistent with the proposed transition state for the S_N2 reaction. Study of structural effects also suggests that the reactants must closely approach one another in that the rate of the reaction is very sensitive to the size of the alkyl groups attached to the carbon atom at which displacement is occurring. A comparison of the rate of S_N2 reaction of some primary halides is shown below. Adding bulky, space-filling organic groups to the primary carbon produces a very appreciable deceleration in reaction rate. The optimum transition state geometry with the incoming nucleophile and the carbon-leaving group bond collinear becomes increasingly

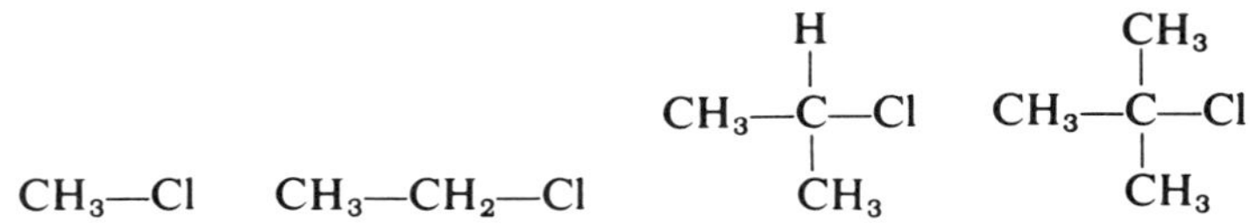

	Methyl chloride	Ethyl chloride	Isopropyl chloride	*t*-Butyl chloride
Approximate relative rates for S_N2 displacement reactions	1	0.03	0.001	0.000001

more difficult to attain as the substituent at the carbon atom gets larger.

Convincing evidence for the S_N2 mechanism comes from study of alkyl halides with mirror-image isomers. When 2-chlorobutane is treated with hydroxide ion, two possible alcohols, *c* and *d*, could be formed by nucleophilic displacement of the chlorine atom. It is found that only *one* of these, *d*, is

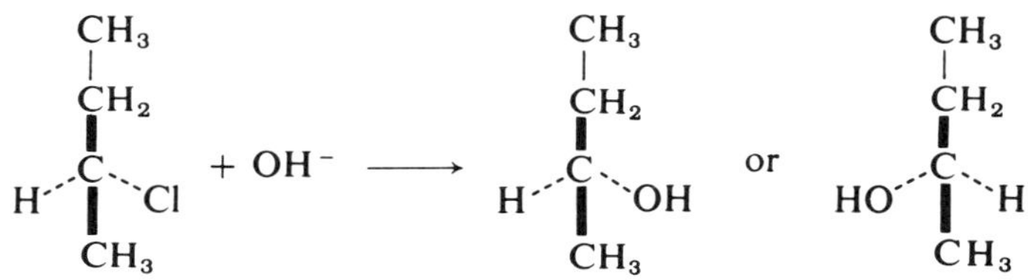

2-Chlorobutane *c* *d*

formed. This result is consistent with the bimolecular transition state for this reaction shown below; during the attack on the carbon bearing the chlorine atom a new covalent bond to hydroxyl is formed on the opposite side of the molecule.

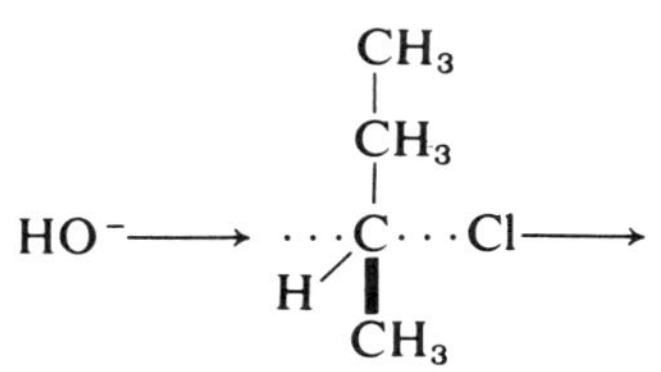

Approach to the transition state for the above S_N2 reaction

ELIMINATION REACTIONS

In an elimination, parts of the starting organic molecule are expelled to give the products. It is useful to classify elimination reactions according to the relative positions of the groups that are removed from the starting material. If both of these groups are on the same carbon atom as in the case of elimination of HCl from chloroform, a reaction to be discussed below, the process is referred to as an α-elimination. Alternatively, the eliminated groups can be on adjacent carbons in which case the process by which they are removed is called a β-elimination. β-elimination reactions are by far the more prevalent of these two; they can be regarded as the reverse of the various olefin addition reactions which were described in Chapter 3.*

α-Elimination

Chloroform is unexpectedly reactive toward hydroxide ion when one considers the fact that carbon tetrachloride is inert to these conditions and that dichloromethane reacts only sluggishly by an S_N2 process. A different kind of reaction is thus indicated for chloroform.

Isotope-tracer studies show that when chloroform is treated with a strong base like hydroxide a small concentration of trichloromethylide ion is produced. This latter species is characterized by a negatively charged carbon atom and is known as a *carbanion.*

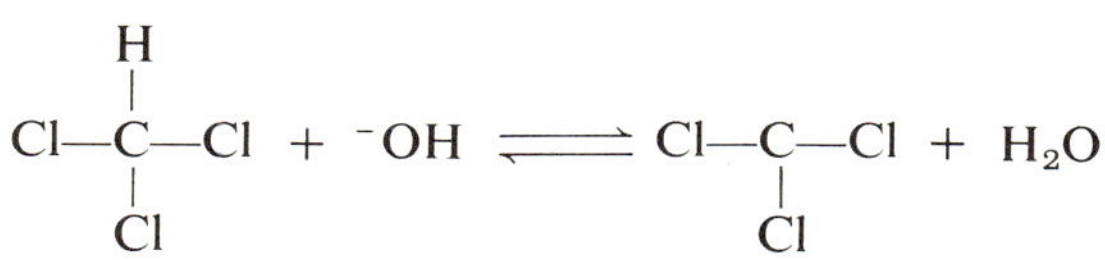

Trichloromethylide
(a carbanion)

The trichloromethyl anion can expel a chloride ion, leaving a divalent

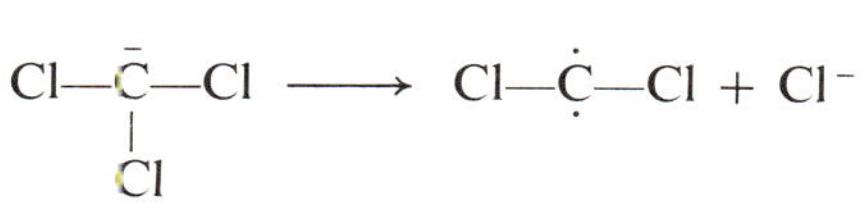

carbon species which is representative of a class of intermediates known as carbenes or methylenes. Carbenes can react with olefins to give cyclopropanes and the presence of dichlorocarbene in the α-elimination reaction of chloroform can be detected by adding cyclohexene to the reaction mixture. The added olefin

* Other elimination reactions are known in which the eliminated groups are more widely separated in the starting material. These are usually rather specialized reactions and will not be considered.

traps the transient carbene and a bicyclic molecule containing a cyclopropane ring is formed.*

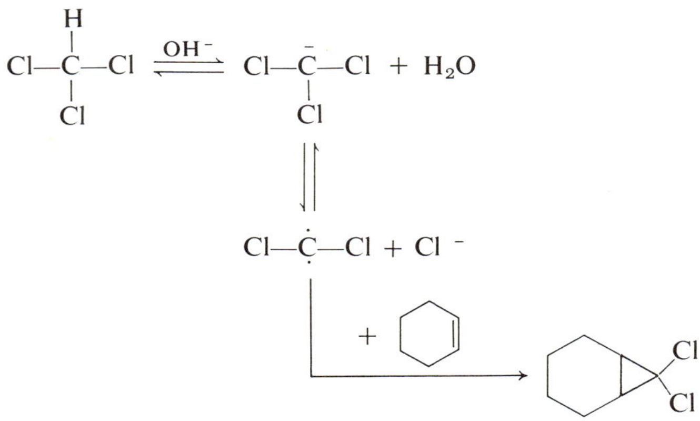

Detection of the carbene intermediate in an α-elimination

The various reaction steps in the α-elimination from chloroform are discrete in time, with the removal of the hydrogen atom preceding the expulsion of the chloride ion. In other α-eliminations, these two steps can merge into one. The reaction of chlorodifluoromethane apparently takes place by this type of concerted mechanism.

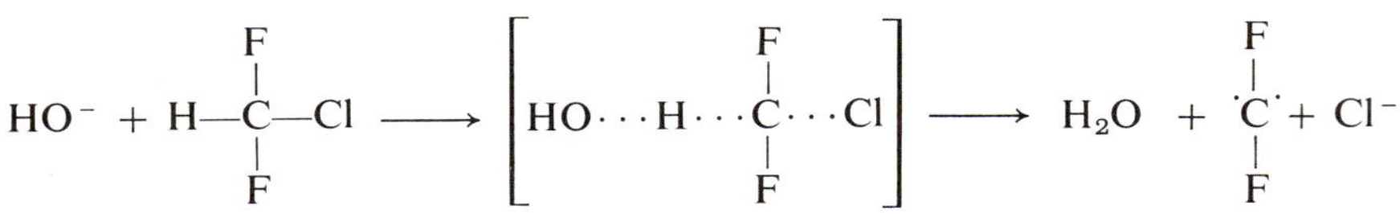

Transition state   Difluorocarbene

β-Elimination

Most research on β-elimination reactions has focused on the elimination of HX from alkanes where X is a good leaving group like halide or trimethylammonium. Substitution and elimination reactions usually go on concurrently and the major pathway for reaction of an alkyl derivative is determined by the reaction conditions (solvent, temperature, concentration), the nature of the leaving group, and the properties of the nucleophile or base used to bring about reaction.

---

* Trapping experiments of this nature are a great help in verifying the presence of suspected reaction intermediates, thereby helping to establish reaction mechanisms.

E2 Elimination Reactions. When ethyl bromide is heated with sodium ethoxide solution in ethanol, elimination of HBr occurs and ethylene is formed. An appreciable yield of diethyl ether is formed by the expected S_N2 reaction.

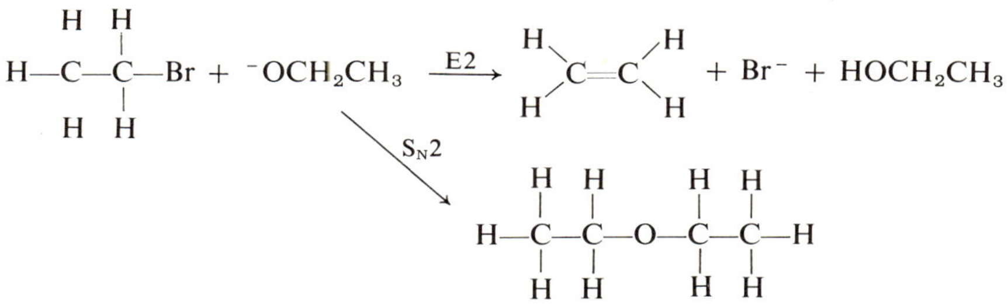

The rate law for the elimination reaction has been determined and, like the S_N2 process, it is found that

$$\text{Rate} = k_2 \text{ [ethyl bromide][ethoxide ion]}$$

Therefore, both reactants are involved in forming the transition state for elimination of HBr. The symbol E2 is used to indicate this reaction mechanism, meaning an elimination reaction (E) in which two reactant molecules are used to generate the rate-determining transition state. The transition state structure generally accepted for the E2 reaction is shown below; it attempts to portray nearly simultaneous abstraction of a hydrogen atom *beta* to the leaving group

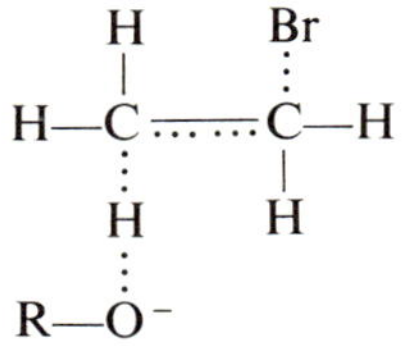

A transition state for E2 elimination

with partial breaking of the carbon-leaving group bond and incipient formation of the double bond. The degree to which these three features of the E2 mechanism vary depends somewhat upon the nature of the leaving group and the attacking base. In β-phenylethyl derivatives, for example, it appears that β-hydrogen abstraction runs well ahead of breaking the bond to the leaving group, probably because of the ability of the neighboring benzene ring to delocalize the forming negative charge. In any event, separate reaction steps are not involved in the E2 elimination, but rather a smooth transfer of electrons from the base molecule to the leaving group takes place.

The proposed transition state for E2 elimination is consistent with observations

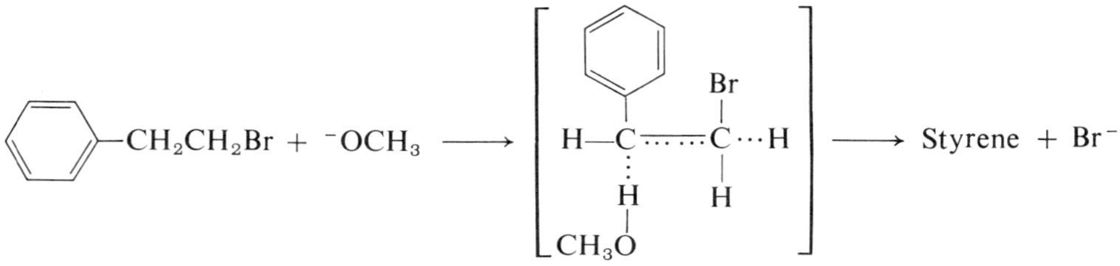

Transition state

that have been made regarding the stereochemistry of the alkenes formed in the reaction.

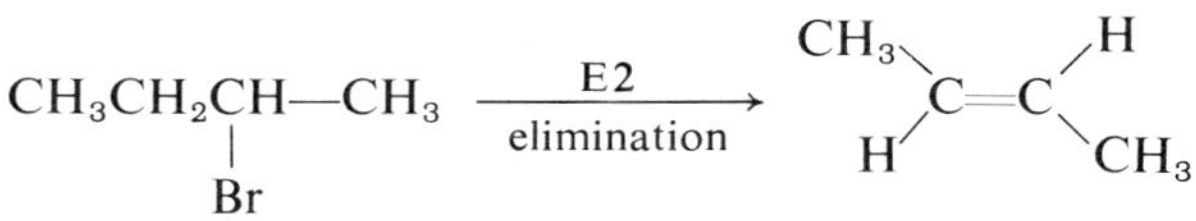

2-Bromobutane

+some *cis* isomer
+some 1-butene

When 2-bromobutane is treated with a strong base, the major elimination product is *trans*-2-butene. The three staggered conformational isomers of the starting material are represented below. Two of these (*e* and *g*) are set up nicely

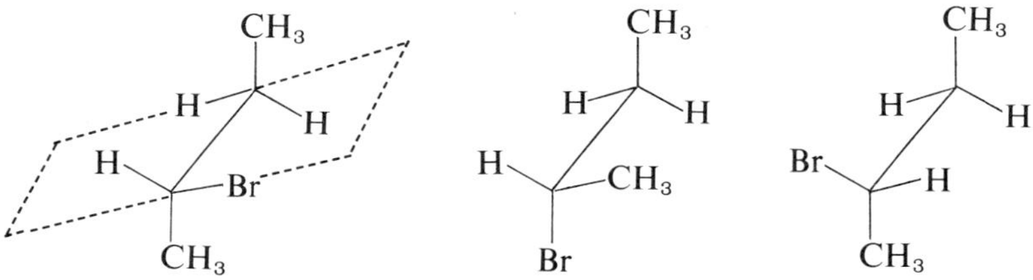

e *f* *g*

for the formation of an E2 transition state since the C—H and C—Br bond are essentially within the same plane, as indicated in the drawings. Elimination from conformation *f* should not be as facile as the reaction with the other conformers. To get a *cis*-olefin as the product of the reaction, a more eclipsed conformation (*h*) similar to *f* would have to be generated, at a considerable increase in the free energy of reaction and a concomitant decrease in reaction

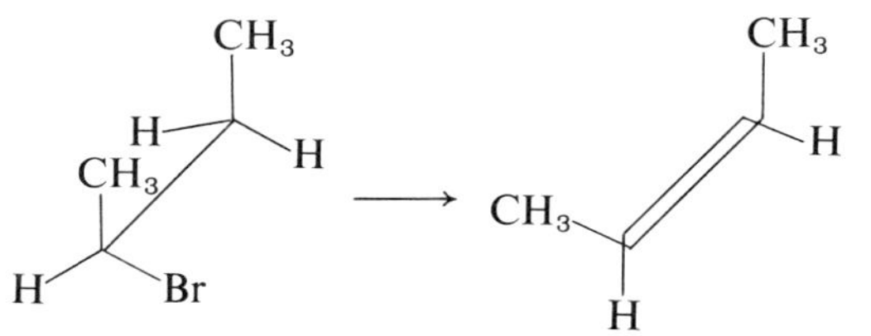

h *cis*-2-Butene

rate. The formation of *trans*-olefins is the dominant mode of reaction and the preponderance of the *trans* isomer in the product mixture increases as the size of the groups attached to the double bond is increased.

The formation of 1-butene in the reaction above illustrates an additional complication in these elimination reactions which has to do with the *direction* of elimination when two or more structural isomers can be formed. Generally the most highly substituted olefin will be formed as illustrated by the observation that 2-butene is the major product obtained by elimination of HBr from 2-bromobutane.

Strong bases are required for the E2 reaction and the use of high reaction temperatures and bases like the alkoxide ions or amide ion ($^-NH_2$) favors the formation of E2 elimination products in competition with S_N2 displacement products.

E1 Elimination Reactions. Elimination reactions also compete with displacement reactions when carbonium ion intermediates are involved; a carbonium ion with a β-hydrogen may expel this atom as a proton (H^+) to give a neutral alkene. The solvolysis of *t*-butyl derivatives in alcohol generally leads to about 80% of the *t*-butyl ether, the S_N1 substitution product, and about 20% of isobutylene, an elimination product. This elimination reaction has the

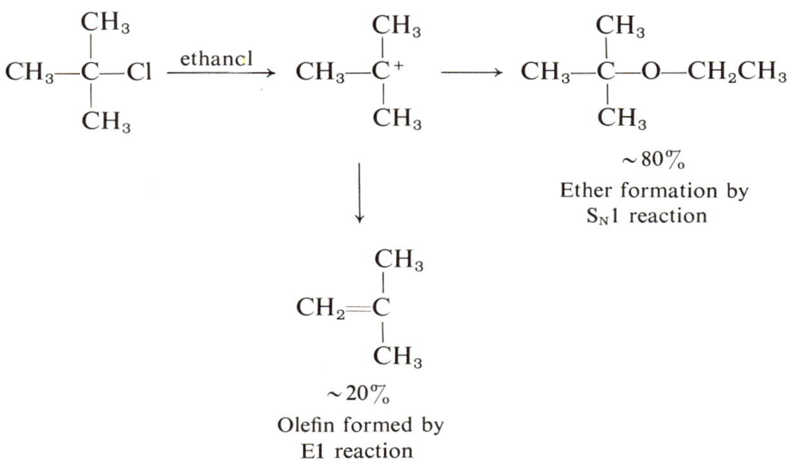

same rate law as the S_N1 substitution reaction and is believed to involve the same intermediate, the carbonium ion. The reaction is called an E1 reaction, the symbolism being an extension of that described above for the E2 mechanism. Structural influences on the rate of the E1 reaction are reminiscent of those found for S_N1 reactions; elimination from tertiary alkyl groups is more rapid than from secondary alkyl groups and primary groups are least reactive. *Trans* alkenes are favored when geometrical isomerism is possible in the products of an E1 elimination reaction, and the most highly substituted alkene is formed predominantly if the elimination can proceed in more than one direction. It is possible in both the E1 and S_N1 reactions for the intermediate carbonium ion to internally rearrange its structure in order to stabilize itself. In the dehydration of 3,3-dimethyl-*t*-butanol, the initially formed carbonium ion is a secondary ion. It rearranges to a tertiary ion before proton loss to give the observed olefinic product. Migrations involving alkyl groups, hydrogen atoms, and aromatic

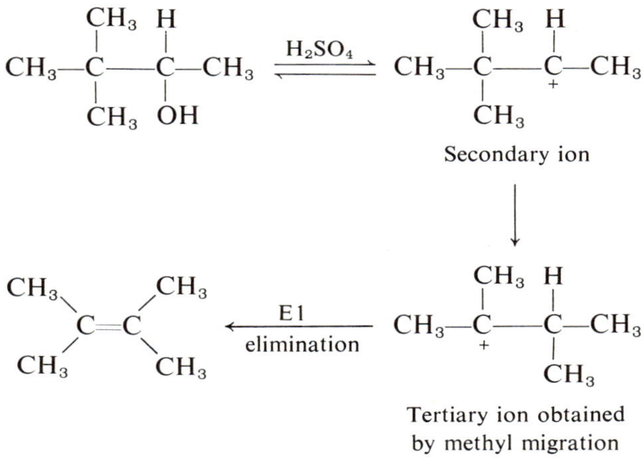

groups have been observed. Carbonium ion rearrangements are not limited to ions generated by hydrolytic or solvolytic reactions; carbonium ions formed in the deamination of amines, during Friedel–Crafts alkylation reactions or under electron impact in the mass spectrometer can also undergo rearrangements.

CATALYSIS

Thermodynamic considerations determine the relative energies of the products and the starting materials of a reaction system and, as we have seen, a free-energy barrier ($\Delta G^{\ddagger}$) separates these two states and must be surmounted by the reactant molecules in order that they can be converted to products. Any substance which can change the magnitude of the free-energy barrier that separates reactants and products is called a catalyst. Catalysts are chemical species that modify the rate of a reaction but can be recovered unchanged themselves at the end of the reaction. They change the transition state free energy for a reaction or open up an alternative mechanistic pathway for the conversion of reactants to products (Fig. 8.1). While we normally tend to think of catalysts as materials which accelerate a reaction, catalysts can also act in the negative sense to slow reaction rates. This latter type of catalyst is often called an inhibitor.

Many catalysts function in ways that are not yet completely understood, especially heterogeneous catalysts such as those used for catalytic hydrogenation.

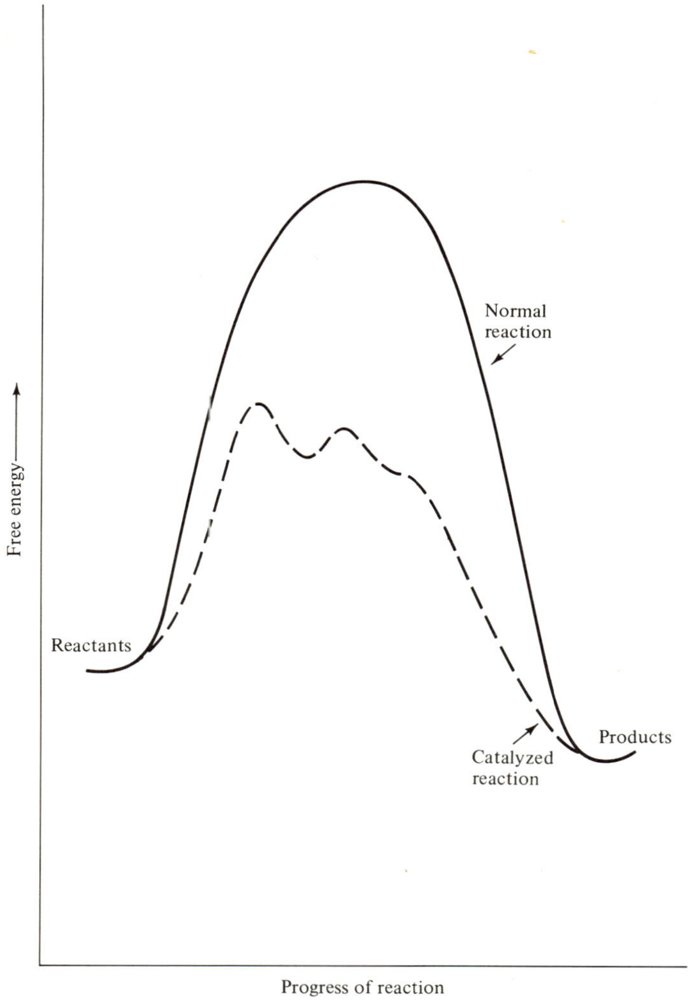

Fig. 8.1. The effect of a catalyst on the free-energy profile for a simple reaction.

However, many organic reactions that occur in homogeneous solutions are catalyzed by acids or bases and these reactions are susceptible to interpretation in terms of the expected acidity or basicity of the oganic molecules involved. We have already examined reactions in which the role of the acid catalyst is apparent. These include the acid-catalyzed addition of water to alkenes and the reverse reaction, the acid-catalyzed elimination of water from alcohols to give alkenes. It is clear that a catalyst for a reaction in one direction (e.g., alkene to alcohol) must also be a catalyst for the reverse reaction (alcohol to alkene) if the experimental conditions are the same. Catalysts only aid a system in reaching its equilibrium condition; they do not alter that position once reached.

Acid- and base-catalyzed reactions usually involve the making and breaking of bonds between oxygen, nitrogen, or carbon and a hydrogen atom. With the exception of proton removal from carbon, the rates of these proton transfer reactions are usually very rapid relative to the overall rate of the reaction being catalyzed.

Virtually all reactions in living systems are catalyzed by special molecules designed expressly for these purposes. These are protein molecules called enzymes. Their enormous catalytic power arises even though the materials are forced to operate at moderate temperatures and in aqueous solution at near neutrality. We shall have a chance to explore aspects of catalysis by enzymes in Chapter 13.

PROBLEMS

1. Sketch free energy–reaction coordinate profiles for a S_N1 and a S_N2 reaction using specific reactant molecules.

2. Predict the products of the following reactions. Be especially wary of possible carbonium ion rearrangements.

(a) Propyl alcohol + HBr $\longrightarrow$

(b) CH_3CH_2I + NaCN $\longrightarrow$

(c) 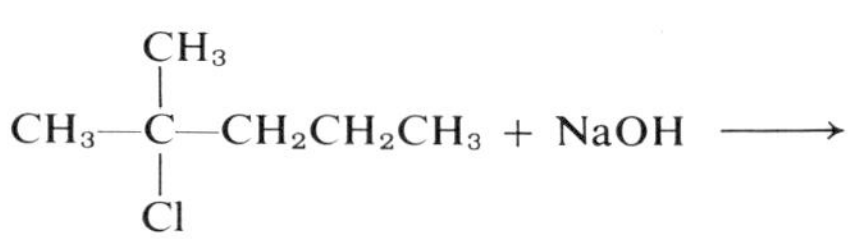

(d) 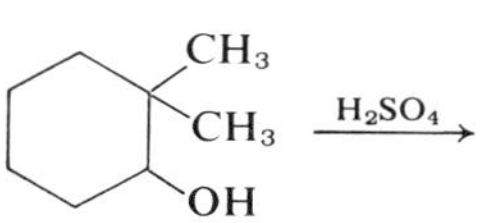

(e) *cis*-1-Chloro-2-methylcyclohexane + lithium methoxide ⟶

(f) 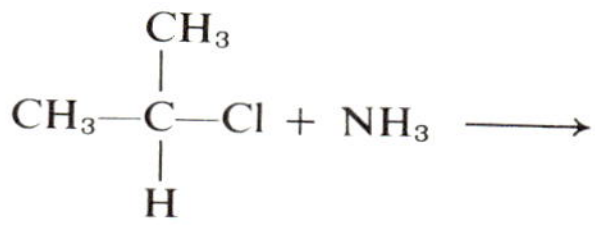

(g) 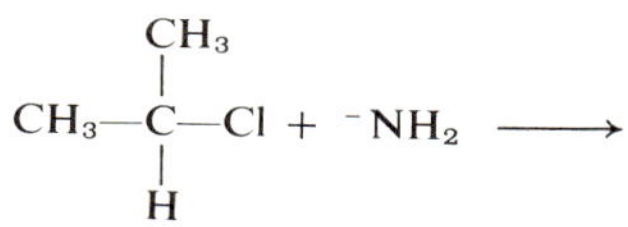

(h) 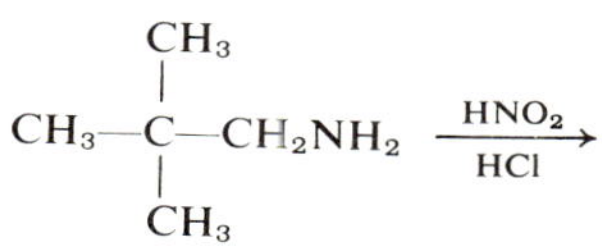

(i) 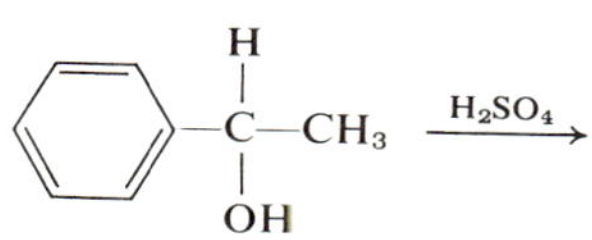

(j) 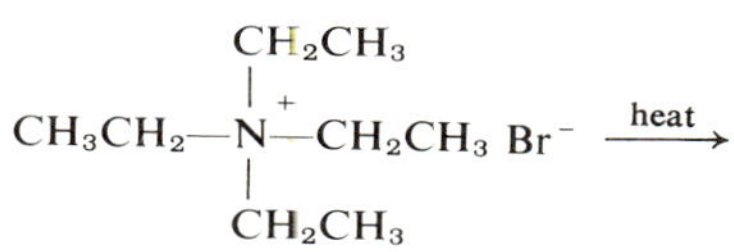

(k) 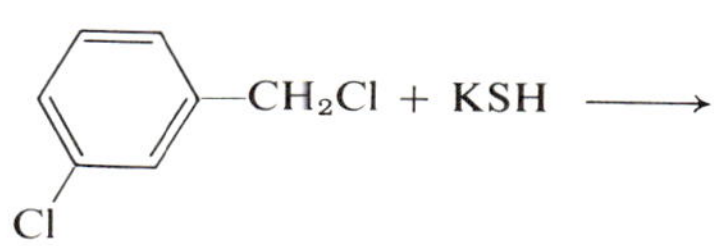

(l) 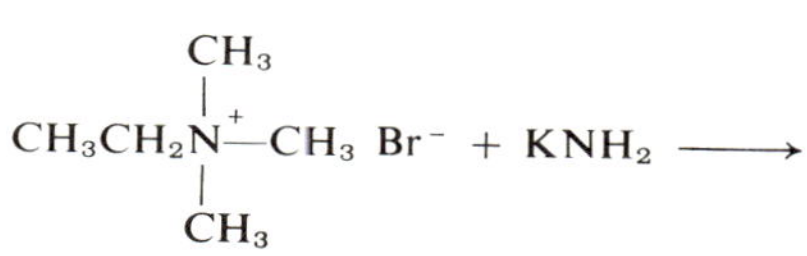

(m) 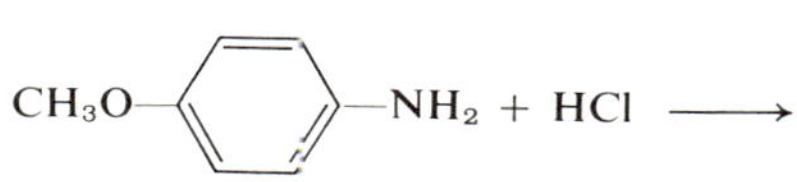

(n) 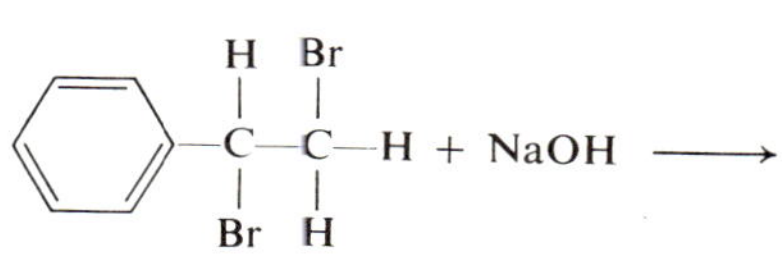

(o) 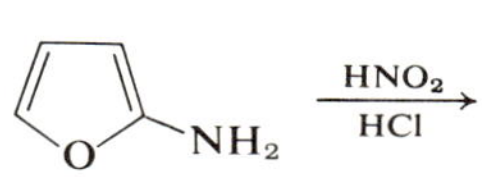

(p) 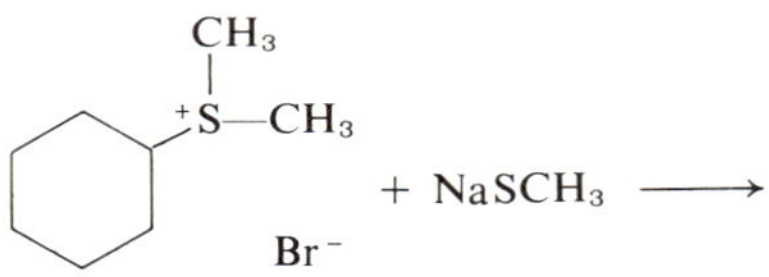

(q) 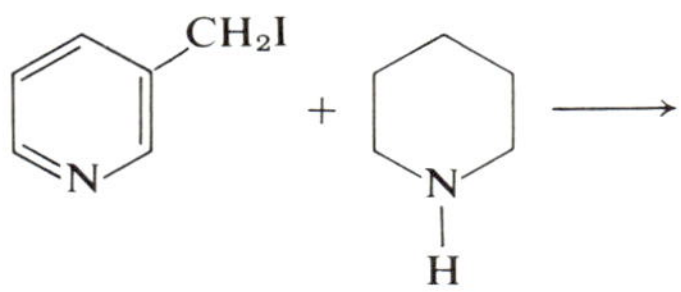

3. Predict the effect on rates and products of changing the solvent for the following reactions from 10% ethanol/90% water to pure ethanol.

(a) 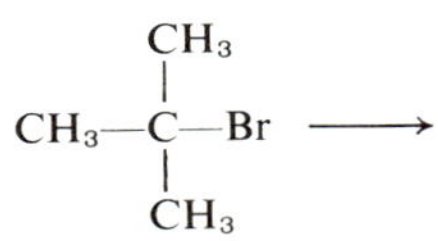

(b) 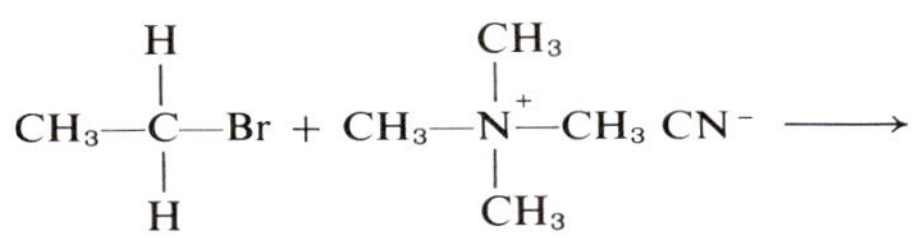

(c) $CH_3Br + NH_3 \longrightarrow$

SUGGESTED READINGS

D. BANTHORPE, "Elimination Reactions." Elsevier, Amsterdam, 1963.

R. BRESLOW, "Organic Reaction Mechanisms," 2nd ed. Benjamin, New York, 1969.

C. A. BUNTON, "Nucleophilic Substitution at a Saturated Carbon Atom." Elsevier, Amsterdam, 1963.

A. A. FROST AND R. G. PEARSON, "Kinetics and Mechanism." Wiley, New York, 1961.

E. S. GOULD, "Mechanism and Structure in Organic Chemistry." Holt, New York, 1959.

A. STREITWEISER, *Chem. Rev.* **56**, 571 (1956).

9

Aldehydes and Ketones

In 1969 radio astronomers obtained experimental evidence for the existence of formaldehyde (H_2CO) in the outer reaches of the universe. This discovery was of interest because this simple molecule can be utilized in reactions which lead to a variety of organic molecules and, thus, may have been important in the generation of the materials essential for the existence and definition of living organisms. Formaldehyde is the simplest example of an organic compound having a carbonyl group (C═O). It is useful to examine closely the electronic structure of this compound and to compare it to that which has previously been described for ethylene. In the latter molecule, it will be recalled, the double bond is formed by overlapping an sp^2-hybridized carbon atomic orbital on each carbon atom to give a σ bond. Simultaneously, overlap of the unhybridized $2p$ orbitals on each carbon produces a π bond that is the second component of this double bond. A similar electronic structure is found in formaldehyde (Fig. 9.1). The exact nature of the hybridization of the oxygen atom is not particularly important for the present discussion; one should remember, however, that there are two atomic orbitals at the oxygen that are doubly occupied by electrons but are not directly involved in bonding to the carbon. They function to make the carbonyl group somewhat basic toward proton donors or Lewis acids. Conclusions regarding the molecular structure of carbonyl-containing compounds can be made from the electronic arrangement

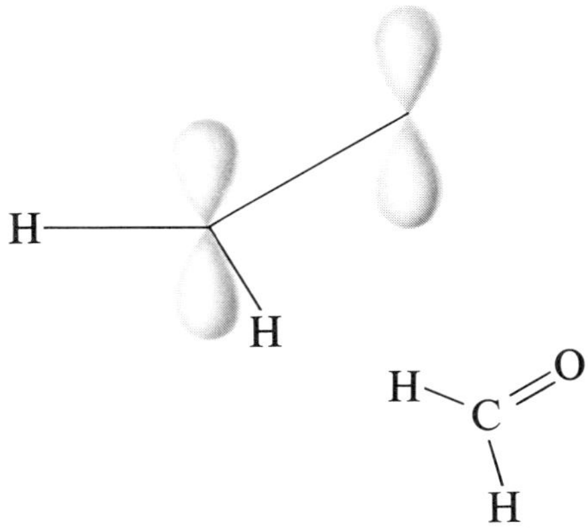

Fig. 9.1. The electronic structure of formaldehyde.

described above. The carbonyl group should be coplanar with the first atoms of the substituents attached to the carbon atom of this group. Moreover, the bond angle between the substituents should be about 120°. Experimentally, this bond angle is between 118° and 122°, in good agreement with this prediction.

There is an important distinction between the electronic structures of ethylene and formaldehyde. Since oxygen is more electronegative than carbon, the electrons in both the σ and π components of the carbonyl double bond will be considerably shifted toward the oxygen atom. That is, the C=O bond is polarized, with a partial buildup of electronic charge apparent at the oxygen atom and a partial decrease in electron density at the adjacent carbon. This situation can be represented by the notation shown below. Alternatively, a resonance structure which emphasizes the polarized nature of the carbonyl group can be

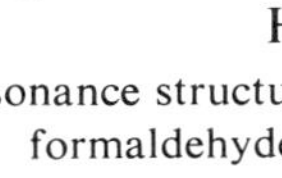

Electron polarization in the carbonyl group

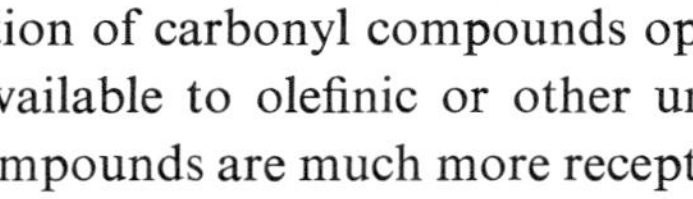

Resonance structures of formaldehyde

used. Electronic polarization of carbonyl compounds opens additional reaction pathways that are not available to olefinic or other unsaturated compounds. In particular, carbonyl compounds are much more receptive to attack by nucleophiles than are the alkenes.

CLASSES OF CARBONYL COMPOUNDS

There are a number of classes of organic compounds which contain the carbonyl group. The distinctions between these classes is made according to the nature of the substituent(s) attached to the carbonyl group at the positions analogous to those occupied by the hydrogen atoms of formaldehyde. When one of these positions is occupied by an aliphatic or aromatic hydrocarbon group, the resulting compound is called an *aldehyde*. Replacement of both of the

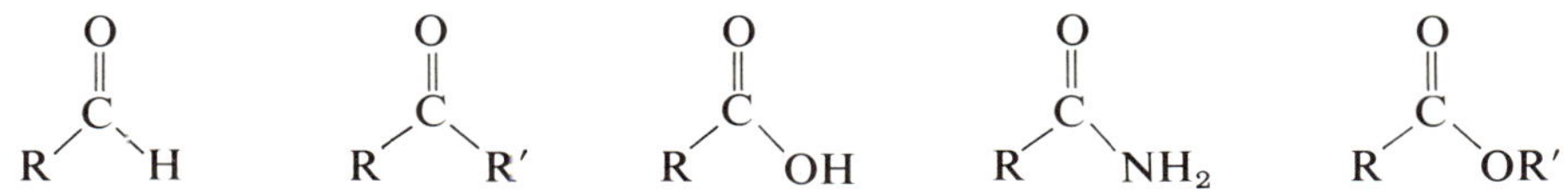

Aldehyde Ketone Carboxylic acid Amide Ester

hydrogen atoms by organic substituents generates a *ketone*. Substituents at the carbonyl group need not be strictly hydrocarbon in nature; hydroxyl, amino, or alkoxyl groups may also be attached to the carbonyl group to afford carboxylic acids, amides, and esters respectively. The presence of elements more electronegative than carbon at the carbonyl group drastically modifies the chemistry of the group; on the other hand, the reactions of aldehydes and ketones are quite similar to one another, but different from these other classes of compounds. We shall, therefore, consider the chemistry of aldehydes and ketones separately and the chemistry of acids, esters, amides, and other compounds with electronegative atoms directly attached to the carbonyl function will be examined later.

Nomenclature

Aldehydes. The systematic IUPAC name for an aldehyde is based upon the name for the corresponding hydrocarbon; for this purpose, an aldehyde is considered to be a derivative of the longest chain of carbon atoms in the molecule which contains the aldehyde functional group (—C(=O)—H). This chain is numbered by starting at the aldehyde carbon atom which is always designated carbon atom 1. The corresponding hydrocarbon is named and then the terminal *-e* in the hydrocarbon name is replaced by *-al* to indicate that the compound is an aldehyde. The application of these rules in generating systematic names for aldehydes is illustrated below.

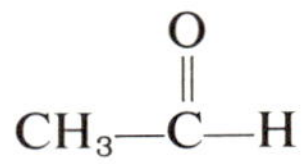	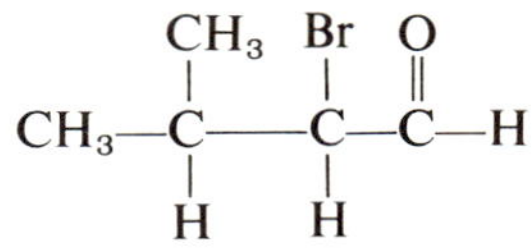
Ethanal (corresponding hydrocarbon is ethane)	2-Bromo-3-methylbutanal (corresponding hydrocarbon is 2-bromo-3-methylbutane)

Common names for aldehydes may also be obtained from the names of the corresponding carboxylic acids as indicated in the examples below. The aldehyde name is obtained by replacing the *-ic* (or *-oic*) of the acid name with *-aldehyde*. The presence and position of substituents along the backbone carbon chain are frequently indicated by Greek letters.

Carboxylic acid	Aldehyde
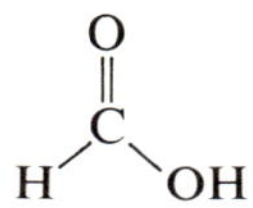 Formic acid	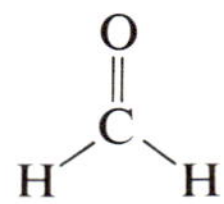 Formaldehyde
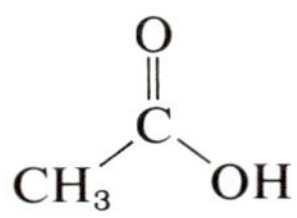 Acetic acid	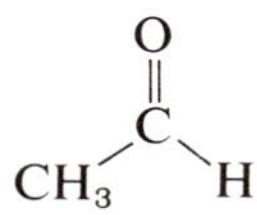 Acetaldehyde
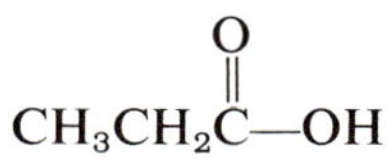 Propionic acid	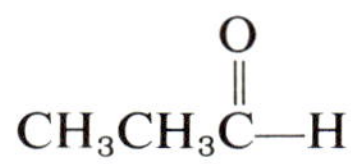 Propionaldehyde
 Butyric acid	 Butyraldehyde
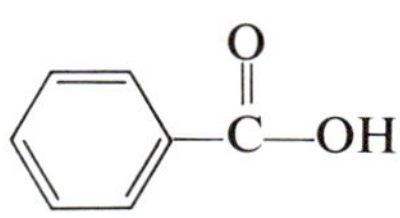 Benzoic acid	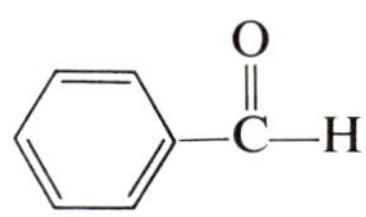 Benzaldehyde

Thus the carbon atom adjacent to the carbonyl group in an aldehyde is called the α-carbon, the next one along the chain is the β-carbon and so forth. The presence of a substituent is then denoted by attaching the appropriate Greek letter to the stem name for the substituent as the examples below indicate.

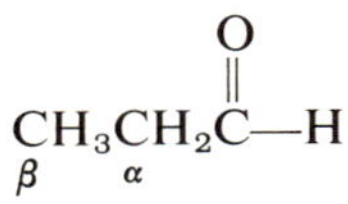

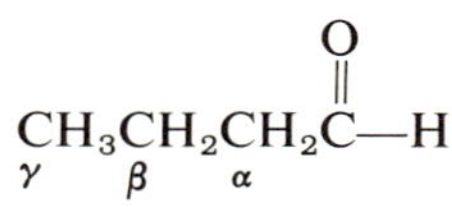

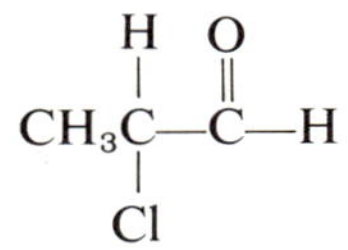

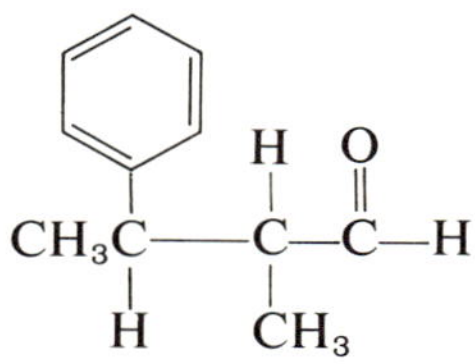

α-Chloropropionaldehyde α-Methyl-β-phenylbutyraldehyde

Ketones. The IUPAC nomenclature system for ketones utilizes the ending *-one* to denote the presence of "internal" carbonyl groups. To name a ketone systematically, one first locates the longest hydrocarbon chain which contains the carbonyl group and then numbers the contiguous carbon atoms in the chain in such a way that the carbonyl carbon has the lowest possible number. The name is formed by writing (1) the names of substituents (if any) and their position numbers, (2) the position number for the carbonyl (or keto) function, and (3) the name of the hydrocarbon corresponding to the backbone of the molecule. The terminal *-e* on the hydrocarbon name is replaced by *-one*. This procedure is illustrated below.

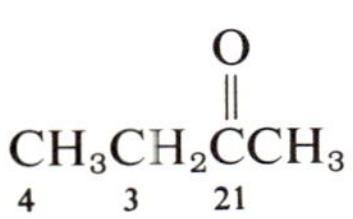

2-Butanone

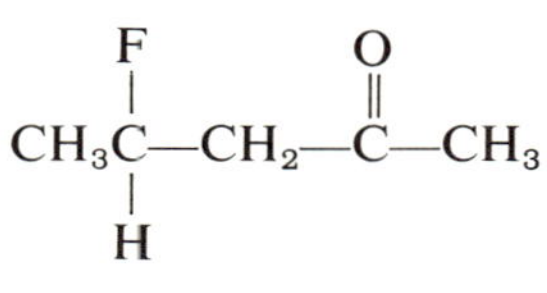

4-Fluoro-2-pentanone

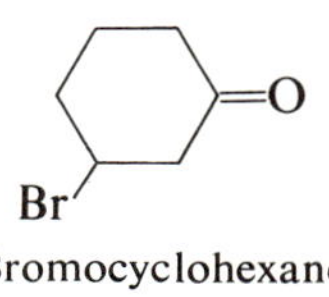

3-Bromocyclohexanone
(need not specify position of carbonyl group since it will always be carbon 1)

If two *keto* functions are present, their position-designating numbers are collected and the ending *-dione* is used.

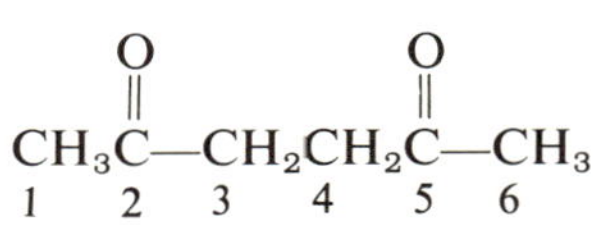

2,5-Hexanedione

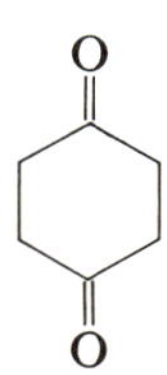

1,4-Cyclohexanedione

Common names also abound for a number of important ketones. These include acetone and acetophenone.

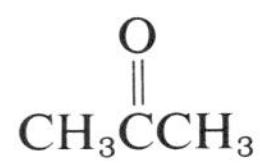

Acetone
(2-propanone)

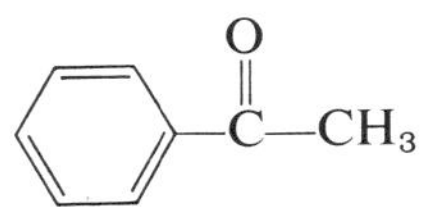

Acetophenone
(methyl phenyl ketone)

Also, a nomenclature reminiscent of that described for ethers is used with ketones. In this method the names of the two groups attached to the carbonyl group are written as separate words and the word *ketone* is appended. Some examples of common names obtained in this way are shown below.

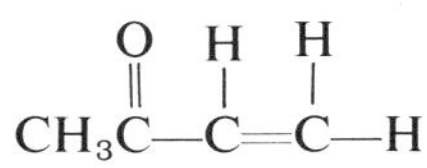

Methyl vinyl ketone

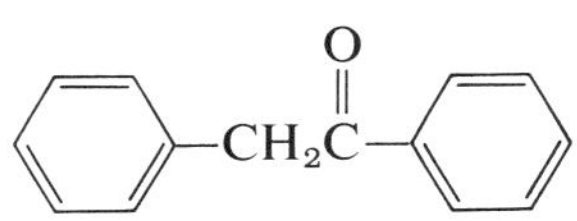

Benzyl phenyl ketone

Properties and Occurrence of Aldehydes and Ketones

As was noted above, the carbonyl function is quite polar with an excess of electronic charge accumulating at the electronegative oxygen atom. Aldehydes and ketones are, therefore, polar molecules. This polar character encourages favorable interactions between these molecules and polar solvents such as water and alcohols. Formaldehyde and acetone are infinitely soluble in water but as the hydrocarbon content of the aldehyde or ketone is increased, the solubility in water decreases. Butyraldehyde and benzaldehyde have only a limited solubility in water and acetophenone is essentially insoluble in water. Formaldehyde is a gas, while other aldehydes and ketones of molecular weight less than about 230 are liquids. The boiling points for these liquids are higher than those of the corresponding hydrocarbons. Larger molecules of this type are solids.

A characteristic property of many aldehydes is their distinctive aroma. Formalin, a 37% solution of formaldehyde in water, is familiar to most students of biology as the "pickling" or preservative solution for biological specimens.

Vanillin, a substituted benzaldehyde, gives vanilla flavoring its characteristic odor, and cinnamaldehyde is a constituent of cinnamon.

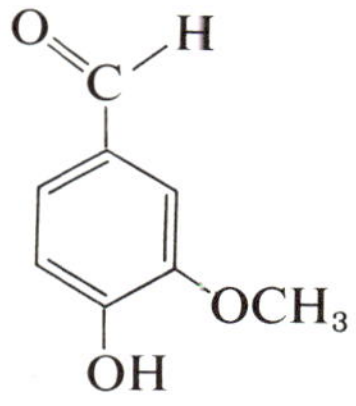

Vanillin

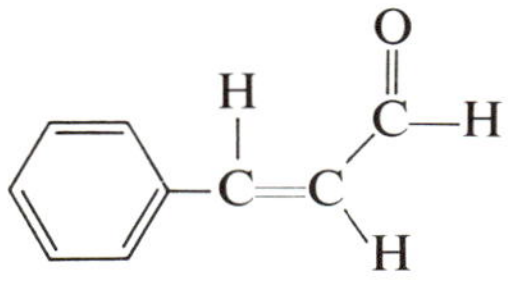

Cinnamaldehyde

A number of naturally occurring and synthetic ketones are used in perfumes and flavorings. These include civetone and muscone. These materials are sex pheromones, isolated from the civet cat and male musk deer, respectively, and are strain-free rings of 17 and 15 carbon atoms.

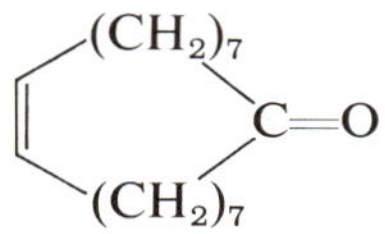

Civetone

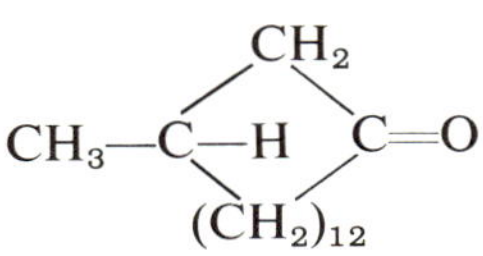

Muscone

Camphor, an interesting polycyclic molecule, has been used as a medicinal agent for centuries, possibly because of its irresistible scent. A dione, biacetyl, and acetoin give butter its delicate flavor; these compounds are usually added to margarine to make it "taste like the high-priced spread."

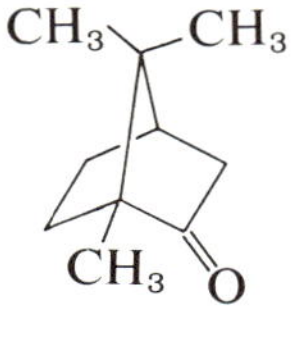

Camphor

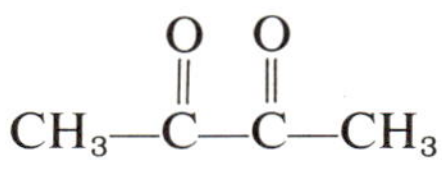

Biacetyl

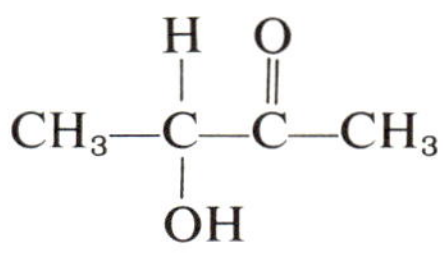

Acetoin

The ketone functional group is found in a wide variety of compounds that occur in nature, in structures that range from quercitin, a plant pigment responsible for the hue of red roses, to testosterone, a male hormone which leads to the development of secondary sex characteristics in man and other mammals.

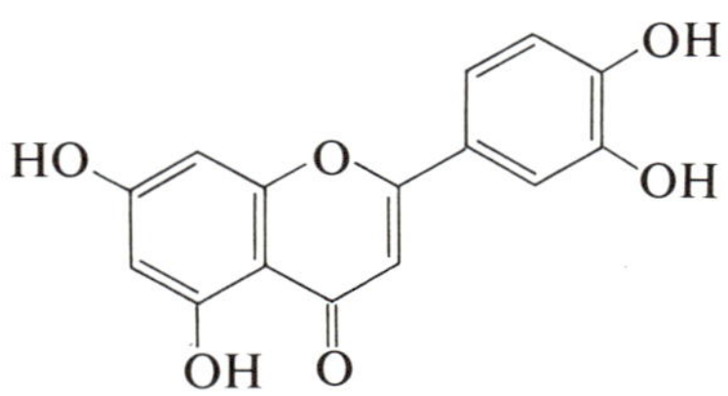

Quercitin

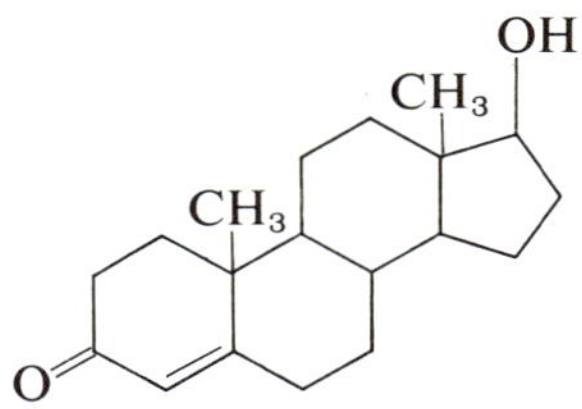

Testosterone

REACTIONS OF ALDEHYDES AND KETONES

In many important reactions, aldehydes and ketones behave in the same way and it is usually sufficient to focus our attention primarily upon the chemistry of the carbonyl group per se in these molecules. It should be recalled that the carbonyl function is strongly polarized with a deficiency of electronic charge developed at the carbonyl carbon atom. Being electron-deficient, this atom is relatively electrophilic and it will, therefore, be reactive toward nucleophilic (electron-rich) reagents.

Additions

Addition of Water (Hydration). Alcohols and water are nucleophiles and these molecules will attack and eventually add to the carbonyl group of an aldehyde or ketone. When an aldehyde is dissolved in water an equilibrium is reached between the aldehyde and the product formed by addition of water to the carbonyl group. The addition product is called a *hydrate* of the aldehyde or ketone. Its formation can be envisioned as an attack of the nucleophile water on the carbonyl carbon to give an intermediate which finally affords the carbonyl hydrate by proton reshuffling. It is important to note that a change in hybridiz-

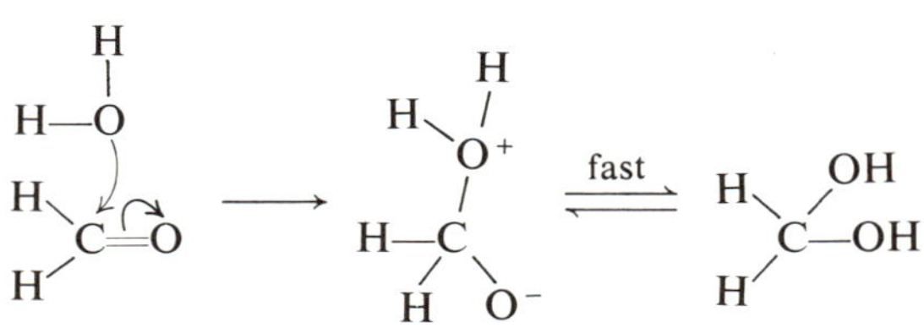

Formation of formaldehyde hydrate

ation of the carbonyl carbon atom occurs during this reaction—from an sp^2-hybridized state to a tetrahedral, sp^3 state. The equilibrium between an aldehyde (or ketone) and its hydrate is established rapidly in water solution. However, the position of the equilibrium, i.e., the relative amounts of carbonyl compound and hydrate present after equilibrium has been reached, is strongly dependent upon the nature of the alkyl or aromatic groups attached to the carbonyl carbon. Formaldehyde is nearly 100% hydrated in neutral water, but benzaldehyde forms an undetectable amount of the corresponding hydrated form.

Addition of Alcohols and Thiols. Alcohols will also add to the carbonyl group of an aldehyde or ketone in a reaction analogous to the hydration reaction of water. In alcohol solution an equilibrium between an aldehyde and its alcohol addition product, called a *hemiacetal*, is established; the corresponding reaction between an alcohol and a ketone gives a *hemiketal*. The hemiacetals or hemiketals are usually present in low amounts and are not isolatable since most

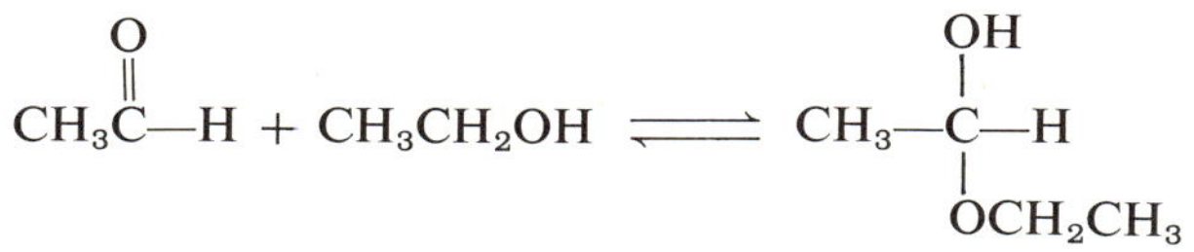

Acetaldehyde Ethanol A hemiacetal

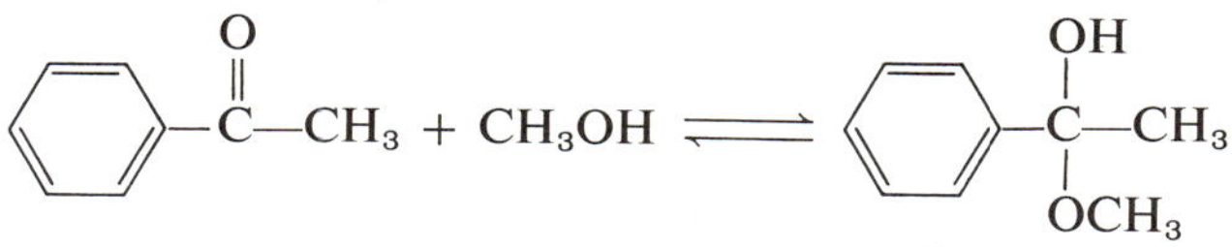

Acetophenone Methanol A hemiketal

isolation procedures shift the equilibrium in favor of the free carbonyl compound and the alcohol.

If an acid catalyst and an excess of alcohol are present, hemiacetals and hemiketals react with an additional molecule of alcohol to give *acetals* and *ketals*, respectively. The role of the acid catalyst is all-important in forming acetals and ketals and it is well to stop to consider how this catalytic effect arises. First of all, the presence of acid can facilitate the formation of hemiacetal or hemiketal. The carbonyl oxygen is a relatively basic site and if a proton (H^+) is accepted there, a full positive charge will develop on the carbonyl carbon, making it much more likely that it will be attacked by a nucleophile such as an alcohol. Loss of a proton from the structure gives the *hemi* compound. Since protons are not consumed in the reaction sequence as written but merely facilitate its course, the role of the acid (proton donor) is truly that of a catalyst. Both oxygen atoms on the hemiacetal are also basic—one is an alcohol oxygen,

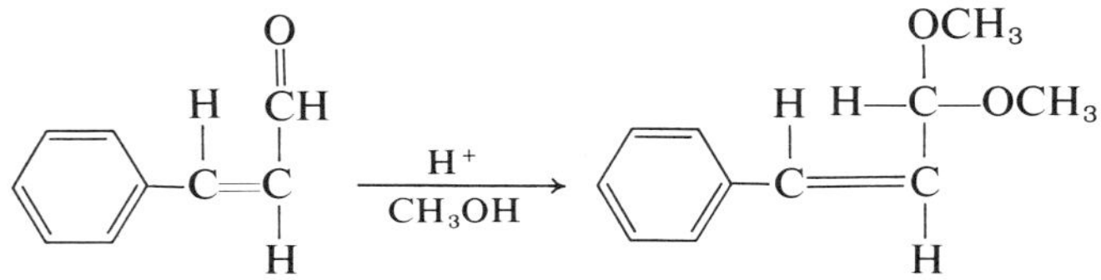

Cinnamaldehyde

Cinnamaldehyde dimethyl acetal (an acetal)

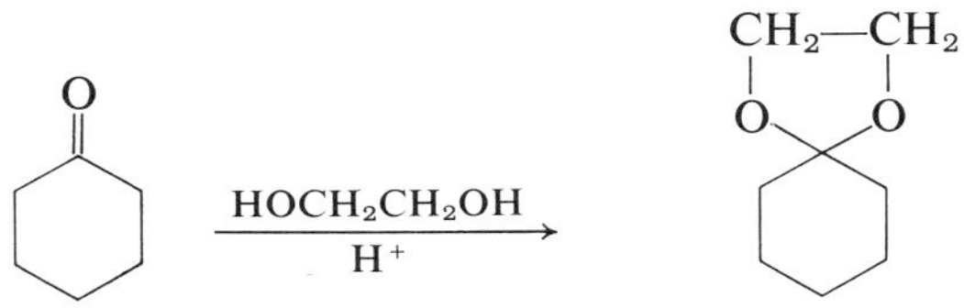

Cyclohexanone

A cyclic ketal formed with ethylene glycol (Here one molecule supplies both alcohol groups for the ketal formation)

while the other is an ether oxygen. If a proton is placed on the ether oxygen of the hemiacetal, an intermediate is formed in the reaction which can either expel the proton to give back the hemiacetal or return via the equilibria indicated to the starting aldehyde. When the hydroxyl oxygen is protonated a different intermediate results which can form a new carbonium ion by loss of water. This ion, by capturing an alcohol molecule, ultimately leads to the acetal. The intermediate is relatively stable since the ether oxygen can stabilize

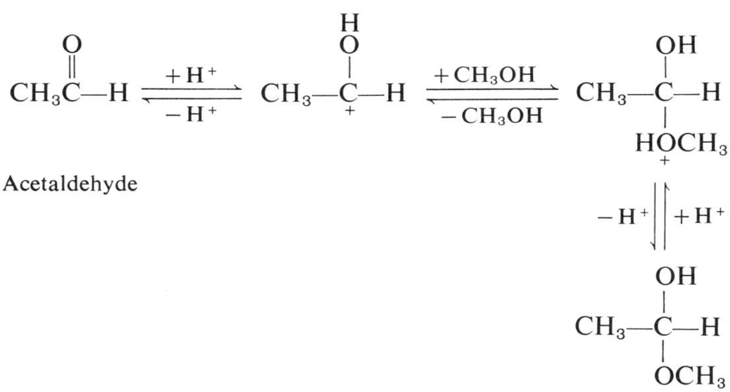

Acetaldehyde

A hemiacetal

Hemiacetal formation

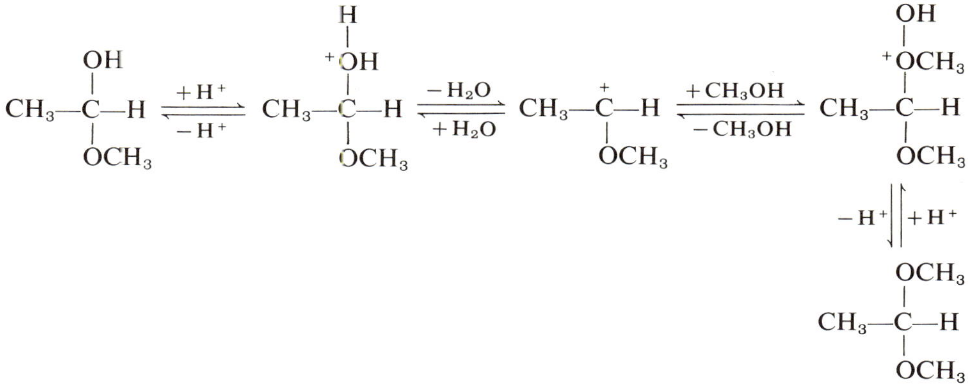

An acetal

Acetal formation

(The symbols at each equilibrium indicate the molecules added or subtracted in that step)

the adjacent positive charge by resonance interaction. All of these equilibria are present simultaneously in an acidic, alcoholic solution of an aldehyde, and the materials which can be isolated from the mixture will depend upon the relative stabilities of the species involved and the concentrations of the reactants. The formation of an acetal is usually accomplished by using a solution of the aldehyde in the alcohol, so that the concentration of alcohol is in large excess over the concentration of the aldehyde. By Le Chatelier's principle, this excess concentration of alcohol will increase the concentration of acetal present at equilibrium; the presence of the acid ensures that the equilibrium condition is established in a reasonable time.

Acetals are stable in neutral and basic solutions. Since they can be isolated and stored, they can be used in other organic reactions. But if an acetal is placed in dilute aqueous acid the equilibria described above are reestablished. Since water, rather than alcohol, would then be present in large excess, the aldehyde is regenerated. Thus the same reactions which lead to the formation of an acetal also result in its destruction. Acetals are hydrolyzed in aqueous acids to the original aldehyde and alcohol molecules.

The acetal formation reaction can be used to form polymers of aldehydes. Delrin, a plastic manufactured by du Pont, is a stabilized polymer of formaldehyde and consists of alternating methylene (CH_2) and oxygen groups. Another commercially important polymer which utilizes acetal linkages is poly(vinyl–butyral). In this case adjacent hydroxyl groups on polyvinyl alcohol are used to form a cyclic acetal with *n*-butyraldehyde. The resulting transparent polymer

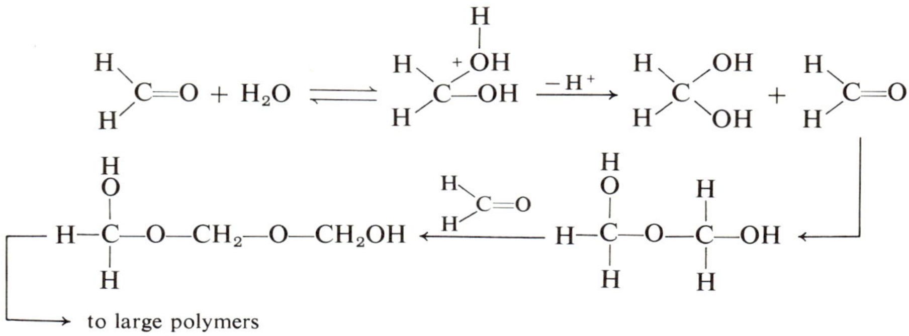

Polymerization of formaldehyde

is used as the central layer between two layers of plate glass to make "safety" glass such as that found in automobile windshields.

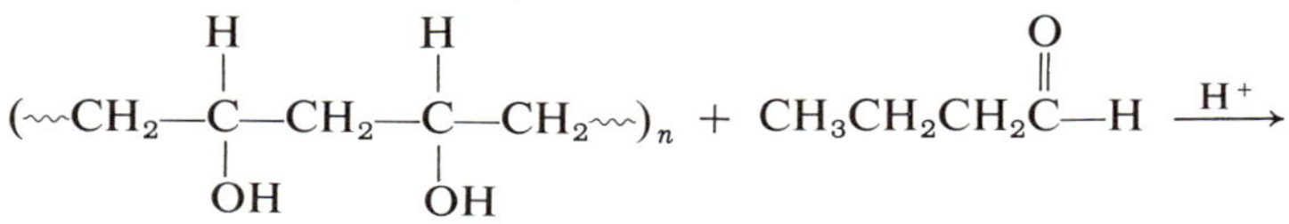

Polyvinyl alcohol n-Butyraldehyde

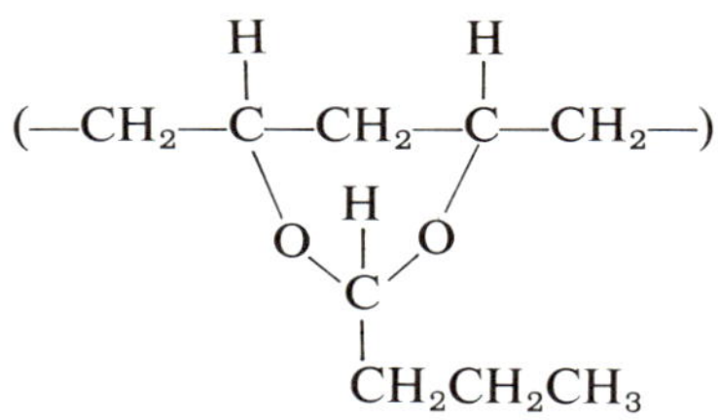

Poly(vinyl butyral)

Thiols react with aldehydes and ketones to form thioacetals and thioketals. These products are analogous to the compounds formed with alcohols but tend to be more stable than their oxygen analogs.

Addition of Bisulfite and Cyanide. Bisulfite and cyanide ions add to the carbonyl group of aldehydes and ketones. The amount of product that can be isolated from the reaction is a function of the structure of the carbonyl compound under study. Bisulfite addition products are often insoluble in concentrated sodium bisulfite solution and precipitate from the reaction mixture. The product obtained by addition of hydrogen cyanide to a carbonyl compound is called a *cyanohydrin.*

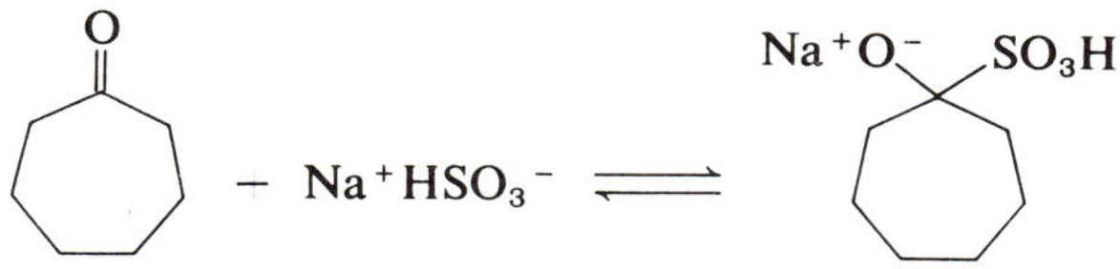

Cycloheptanone — Sodium bisulfite — Bisulfite addition compound

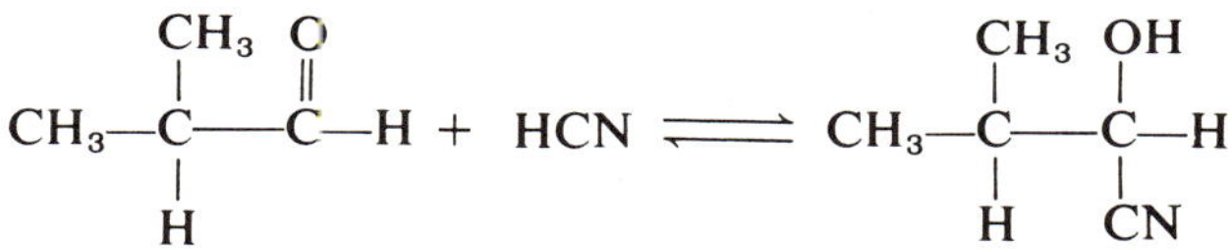

Isobutyraldehyde — Hydrogen cyanide — Isobutyraldehyde cyanohydrin

Addition of Organometallics. An important class of carbonyl addition reactions involves the use of organometallic reagents. An organometallic is an organic compound which contains a formal covalent bond between a carbon atom and a metal atom. The most common metals found in this situation are lithium, sodium, magnesium, zinc, and cadmium. Organolithium reagents can be prepared by treating an alkyl or aryl halide with lithium metal. An example is the synthesis of *n*-butyllithium.

$$\underset{n\text{-Butyl bromide}}{CH_3CH_2CH_2CH_2Br} + 2\ Li \longrightarrow \underset{n\text{-Butyllithium}}{CH_3CH_2CH_2CH_2Li} + LiBr$$

Organomagnesium compounds are usually called Grignard reagents, after the French chemist, Victor Grignard, who discovered these remarkable compounds. Alkyl and aryl halides (chlorides, bromides, iodides, but not fluorides) react with metallic magnesium to give alkyl or aryl magnesium halides. Since metals are less electronegative than carbon, the carbon–metal bond is polarized with

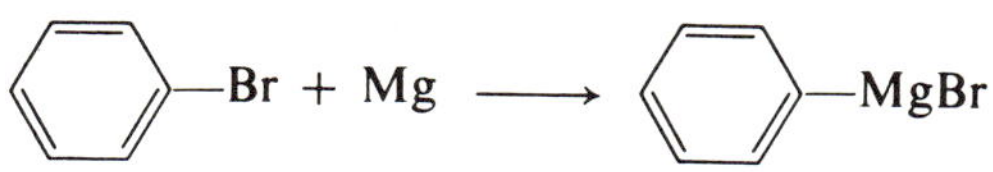

Bromobenzene — Phenylmagnesium bromide

the carbon atom acquiring an excess of electronic charge. That is, the carbon atom becomes carbanionic in nature and can function as a nucleophile.

The initial product formed when a Grignard reagent is added to an aldehyde or ketone is probably a magnesium alkoxide complex. This addition reaction leads to the formation of a new carbon–carbon bond. One should note that

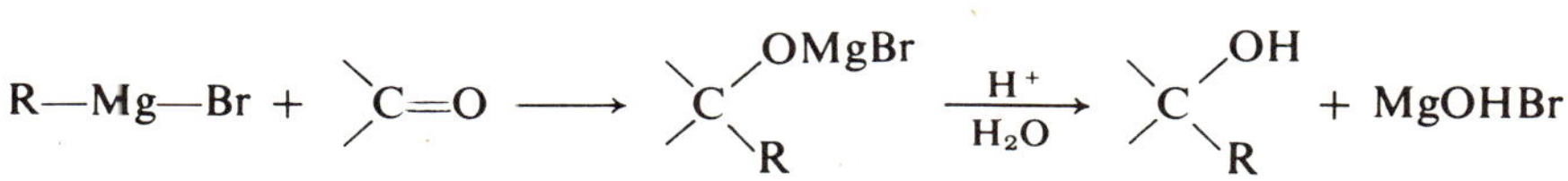

Reaction of a Grignard reagent with an aldehyde or ketone

reaction of a Grignard reagent with formaldehyde forms a primary alcohol, while a Grignard reagent with other aldehydes gives secondary alcohols. Addition of Grignard reagents to the carbonyl group of ketones affords tertiary alcohols. Since there is a rather limited number of reactions available for making

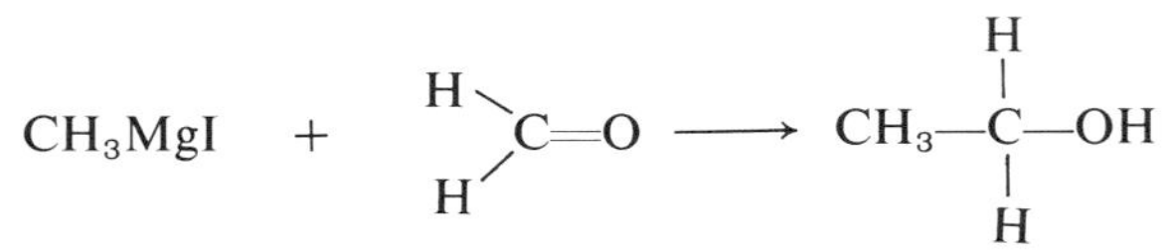

Methyl magnesium iodide — Formaldehyde — Ethanol

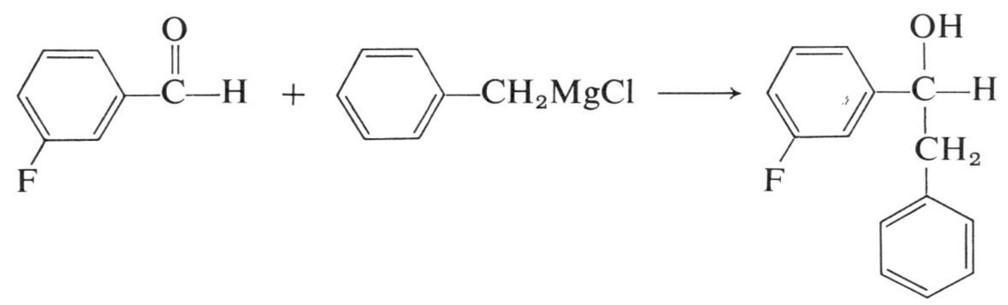

3-Fluorobenzaldehyde — Benzyl magnesium chloride — 1-(3-Fluorophenyl)-2-phenyl-ethanol

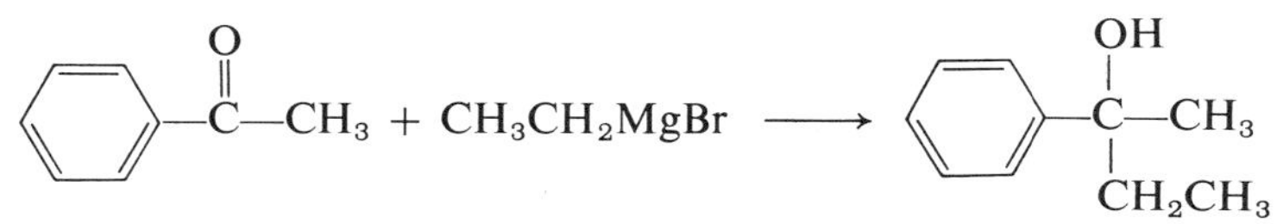

Acetophenone — 2-Phenyl-2-butanol

Alcohol synthesis with Grignard reagents

new carbon–carbon bonds, the Grignard reagent has an important place in synthetic organic chemistry.

Addition with Elimination

Most typical amines are strong bases and good nucleophiles. It is reasonable, therefore, to expect these organic compounds to react with aldehydes and ketones in the same manner as the other nucleophilic molecules mentioned above. Amines do add to the carbonyl group of aldehydes or ketones, but subsequent elimination reactions usually take place after this addition is completed.

The presence of an acid catalyst is important in effecting the addition of an amine to an aldehyde or ketone. As in the case of acetal formation, the acid probably acts to increase the electrophilicity of the carbonyl carbon atom so that attack by the nucleophilic amine to form a tetrahedral intermediate is facilitated. This intermediate loses the elements of water in an acid-catalyzed process to give a new compound which has a carbon–nitrogen double bond to the amine structure.

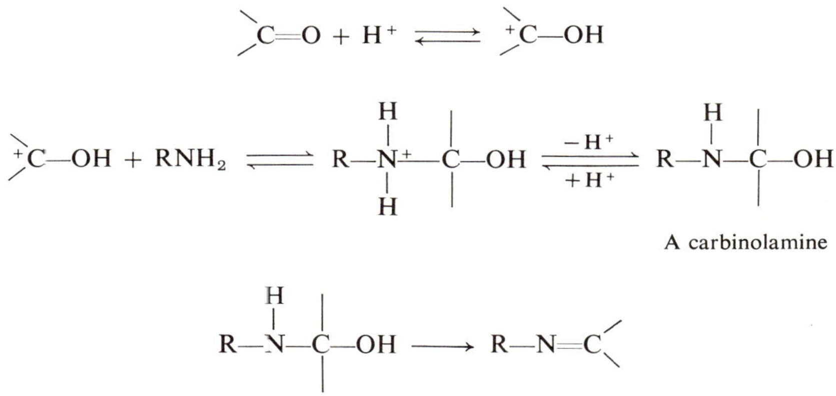

General sequence for the reaction of an amine with an aldehyde or ketone

The products of this sequence of addition–elimination reactions are called *aldimines* or *Schiff bases* when the carbonyl group is derived from an aldehyde, and *ketimines* when the carbonyl group is from a ketone.

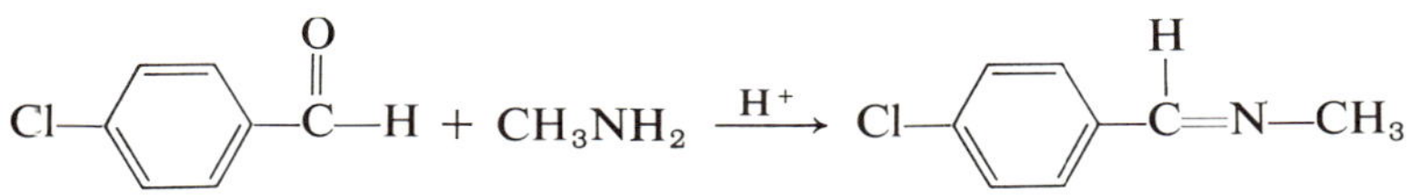

p-Chlorobenzaldehyde

An aldimine (Schiff base)

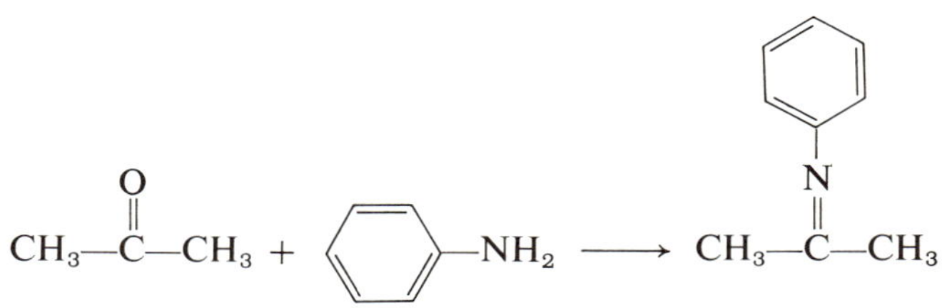

Acetone

A ketimine

The exact mechanism by which these imines are formed depends upon the basicity of the amine and upon the nature of the carbonyl compound. However, in most cases it is found that the reaction is highly sensitive to the pH of the reaction solution.

The substituent R on the general amine shown on p. 213 need not be limited to alkyl or aryl groups for the addition–elimination reaction with carbonyl compounds to occur. This group can also be a hydroxyl, amino, or substituted amino function. The structures and names for the products of the reactions of these substituted amines with aldehydes or ketones are given in Table 9.1.

Table 9.1
Carbonyl Addition–Elimination Reactions with Various Amine Derivatives

Amine derivative	*Product*
$R—NH_2$ R = alkyl, aryl An amine	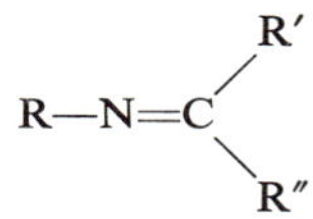 An imine
$HO—NH_2$ Hydroxylamine	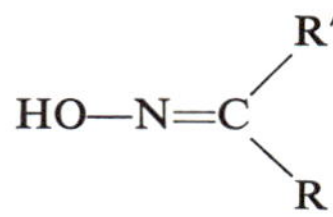 An oxime
$H_2N—NH_2$ Hydrazine	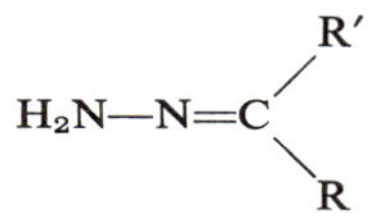 A hydrazone
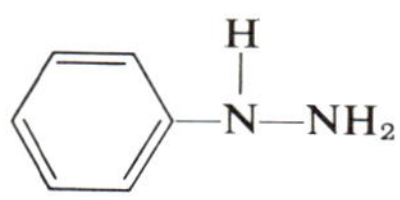 Phenylhydrazine	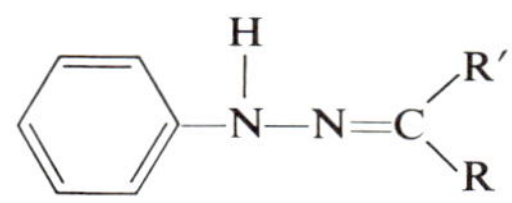 A phenylhydrazone
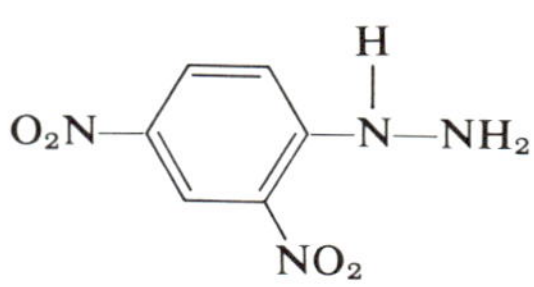 2,4-Dinitrophenylhydrazine	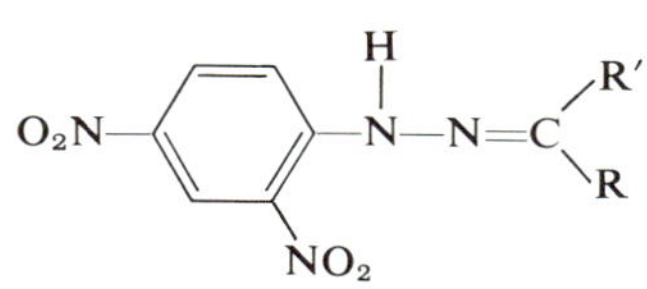 A 2,4-dinitrophenylhydrazone
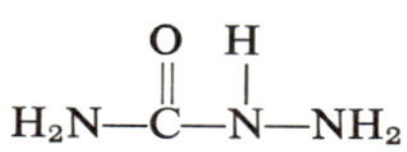 Semicarbazide	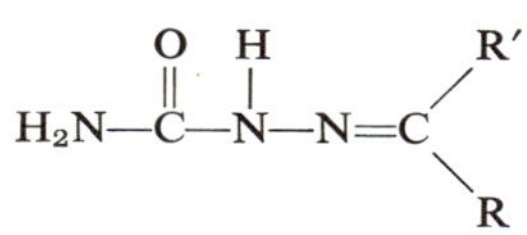 A semicarbazone

Reduction Reactions

Catalytic Reduction. The carbonyl group can be reduced to a lower oxidation state by a number of reactions. Catalytic hydrogenation of an aldehyde or ketone is a slower reaction than hydrogenation of an alkene or alkyne and it is

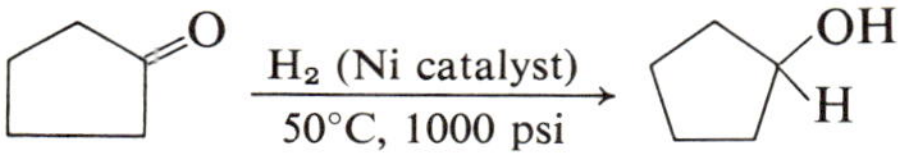

usually not possible to selectively reduce a carbonyl group in the presence of a double or triple bond under these conditions.

Metal Hydride Reduction. Metal hydride reagents offer the best method for reducing carbonyl groups under relatively mild conditions that do not affect carbon–carbon multiple bonds. Two metal hydrides in common use are lithium aluminum hydride ($LiAlH_4$) and sodium borohydride ($NaBH_4$). In these compounds the hydrogen atoms are directly attached to the aluminum and boron atoms, respectively. Hydrogen is more electronegative than either of these metals so that in these compounds it exists essentially as the hydride ion (H^-). It is possible to regard this nucleophile as being capable of adding to the carbonyl group of an aldehyde or ketone; a depiction of the reaction of $LiAlH_4$ with acetaldehyde is shown below. The alkoxyl-substituted complex ions of aluminum are not stable in aqueous acid so that hydrolysis of the reaction mixture liberates the reduction product; unreacted lithium aluminum hydride reacts with water to give hydrogen and various aluminum hydroxides.

Of the two metal hydride reagents mentioned, lithium aluminum hydride is by far the most potent reducing agent. It will reduce virtually any carbonyl

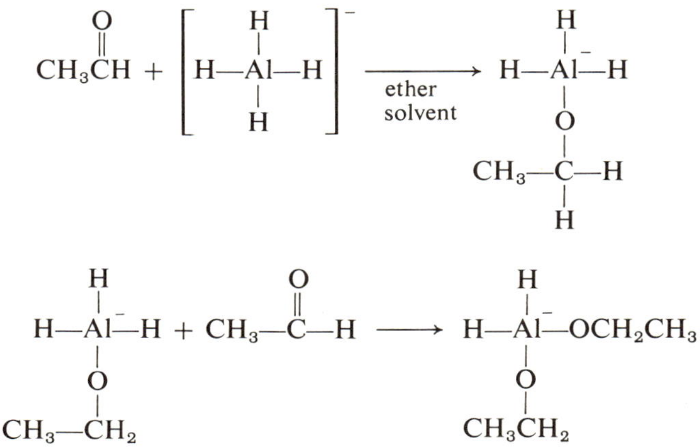

Similar reactions finally give

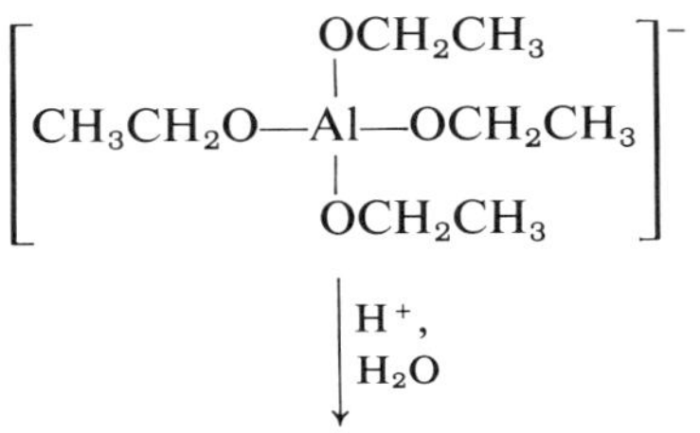

$$4\,CH_3CH_2OH + [AlOH_4]^-$$

Overall reaction:

$$4\,CH_3\overset{\overset{\displaystyle O}{\|}}{C}—H + [AlH_4]^- \longrightarrow 4\,CH_3CH_2OH + \text{aluminum salts}$$

group including those found in esters, acids, and amides, as well as aldehydes and ketones.

Sodium borohydride is a much milder reducing agent, is less sensitive to moisture, and can be used in hydroxylic solvents. It works well with aldehydes and ketones but will not reduce other carbonyl-containing functional groups. Neither of these reagents will reduce unsaturated carbon–carbon bonds, so that, combined with catalytic hydrogenation, appreciable selectivity in reducing reagents is available.

Oxidation of Aldehydes

Aldehydes can be readily oxidized to the corresponding carboxylic acids. This reaction takes place by a chain mechanism involving radical species; oxygen from the air is a sufficient source of oxidizing agent. Aldehydes should, therefore, be stored in tightly sealed containers and, if necessary, be purified before use in a reaction.

Complex ions of various metals can also oxidize aldehydes. In the process, the metal atom, originally present in a higher oxidation state, is reduced. Oxidation of aldehydes by the diamine silver ion ($Ag(NH_3)_2{}^+$) produces free silver metal which is deposited on the walls of the reaction vessel as a shiny mirror. This oxidant is known as *Tollen's reagent* and its reaction with aldehydes is a useful test for diagnosing the presence of the aldehyde (—CHO) functional group.

In the Fehling's and Benedict's reagents, the oxidizing agent is a complex ion of copper. As an aldehyde reacts with these solutions, insoluble copper(I) oxide is formed and can be recognized by its characteristic brick-red appearance.

These reagents have historically been of importance for detecting the presence of aldehydes in physiological fluids. Benedict's solution can detect as little as 0.01% concentration of some sugars.

Neither Tollen's reagent nor the Fehling's and Benedict's reagents are necessarily specific reagents for detecting the presence of an aldehyde; any easily

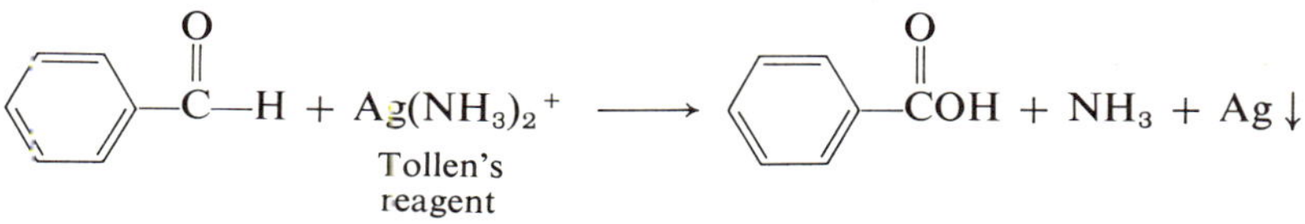

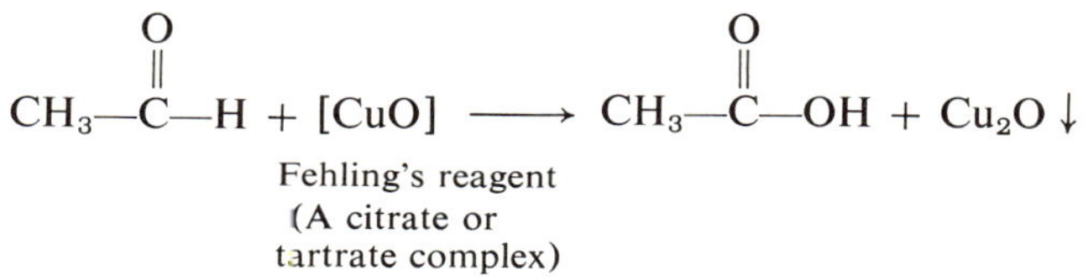

oxidized organic compound will react with these reagents and therefore give a positive test. In particular, ketones with a hydroxyl group on an α-carbon atom are oxidized by these reagents. If the presence of a carbonyl group is indicated by other evidence (for example, the formation of a 2,4-dinitrophenylhydrazone derivative), then Tollen's reagent can often be used to decide whether or not the compound is an aldehyde. Benedict's and Tollen's reagents may be used to distinguish aromatic from aliphatic aldehydes. Tollen's reagent will react with all aldehydes while Benedict's test is usually positive only for aliphatic aldehydes.

Condensation Reactions

The carbon–hydrogen bonds in alkanes or other hydrocarbons are quite stable in the presence of strong bases. The pK_a of methane has been estimated to be about 40, corresponding to a very weak acid indeed. However, the pK_a of acetone is about 20. The C—H bonds in acetone are more acidic because the carbanion formed upon ionization of one of the α-hydrogens is substantially stabilized by electron delocalization into the carbonyl group, as indicated by the resonance structures below.

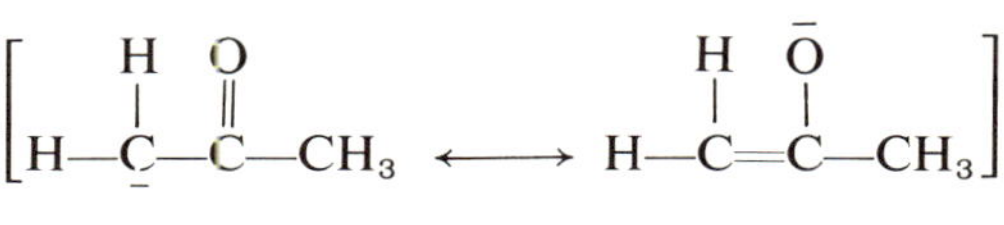

Appreciable concentrations of carbanions can be generated from aldehydes or ketones with α-hydrogens by treating the carbonyl compound with a base; the extent to which carbanion formation occurs will depend on the structure of the material and upon the strength of the base used to abstract the proton. Only α-hydrogen atoms in an aldehyde or ketone are in a position to interact with the carbonyl group in the manner described and hydrogen atoms more distant in a molecular structure behave essentially like normal hydrocarbon C—H bonds in that they have no marked acidity.

Carbanions generated from aldehydes or ketones with α-hydrogen atoms have many of the properties one would anticipate for a negatively charged ion. They are clearly strong bases and will revert back to starting material under conditions where protonation is possible. They are also nucleophilic species and, as such, can participate in carbonyl addition reactions. For example, when acetaldehyde is exposed to aqueous hydroxide ion, a carbanion is formed by proton abstraction from the α-carbon. This is an equilbrium situation in which both the starting aldehyde and the corresponding carbanion are present in the reaction mixture. The carbanion adds to the carbonyl group of a molecule of acetaldehyde, to produce aldol by formation of a new carbon–carbon bond. This sequence of equilibria is reversible, and for aldehydes, tends to favor the

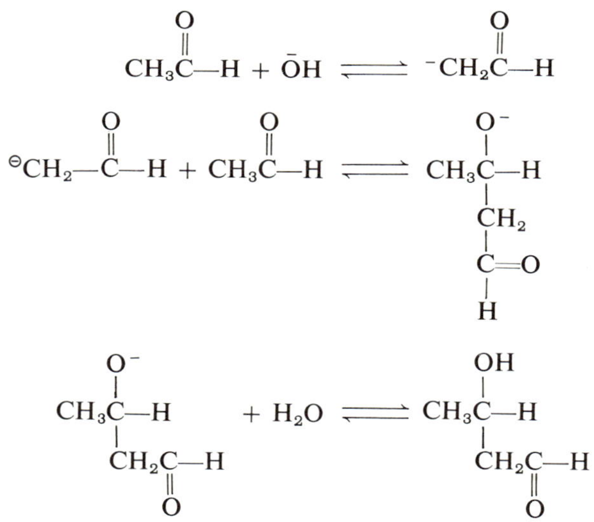

3-Hydroxybutanol
(aldol)

condensation product. It is a prototype for an entire class of reactions known as *aldol condensations*. β-Hydroxy carbonyl compounds are not particularly stable and may undergo elimination of water to give a double bond in conjugation with the carbonyl group. When elimination takes place, the acid–base equilibria in the condensation reaction are shifted in favor of product and the overall reaction proceeds to completion. The formation of an α,β-unsaturated

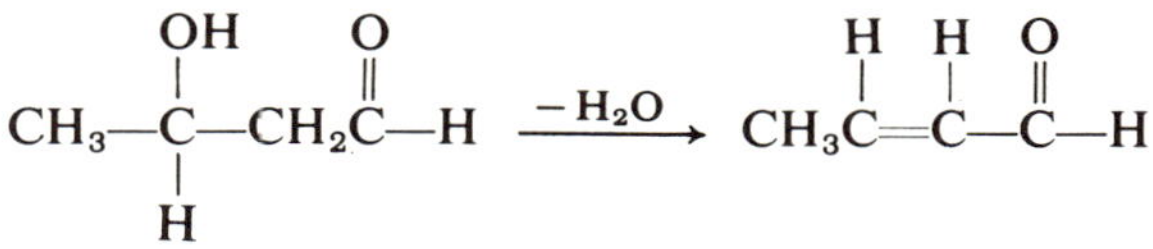

Aldol Crotonaldehyde

carbonyl compound in the aldol condensation is similar to the addition–elimination reactions which primary amines undergo with aldehydes or ketones. Both involve the attack of a nucleophile upon the carbonyl group followed by elimination of water to produce a new double bond to the former carbonyl carbon atom.

Aldol condensation reactions are most attractive from the viewpoint of organic synthesis when they lead cleanly to a single product. This means that if two different aldehydes (or ketones) are to be condensed by this reaction they should have appreciably different reactivities of either the α-hydrogens or the carbonyl groups. Particularly useful are reactions in which one of the carbonyl compounds does not have an α-hydrogen and therefore cannot react with itself.

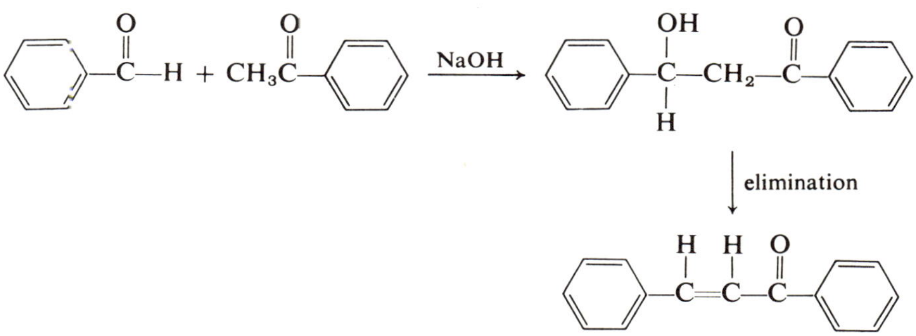

An aldol condensation

Halogenation

The α-hydrogen of a carbonyl compound can be replaced by halogen (chlorine, bromine, iodine) by treating the compound with the molecular halogen. The product isolated depends upon the acidity of the reaction mixture. At low pH (under acidic conditions), a single substitution of halogen for hydrogen occurs. In basic media, all α-hydrogen atoms are replaced by halogen. The base-catalyzed

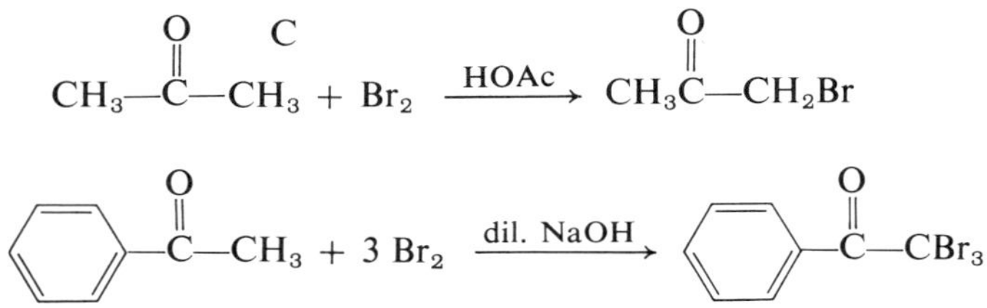

halogenation reaction very likely involves the generation of a carbanion at the α-position. The first halogen substitution greatly increases the stability of the α-carbanion so that subsequent reactions with halogens become more favorable.

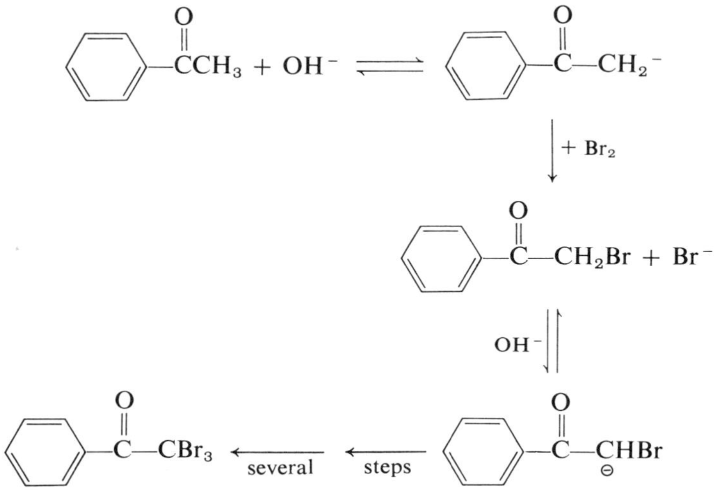

Bromination of acetophenone

When a methyl ketone, such as 2-butanone, is subjected to base-catalyzed iodination, the expected triiodonated compound is initially formed but it is

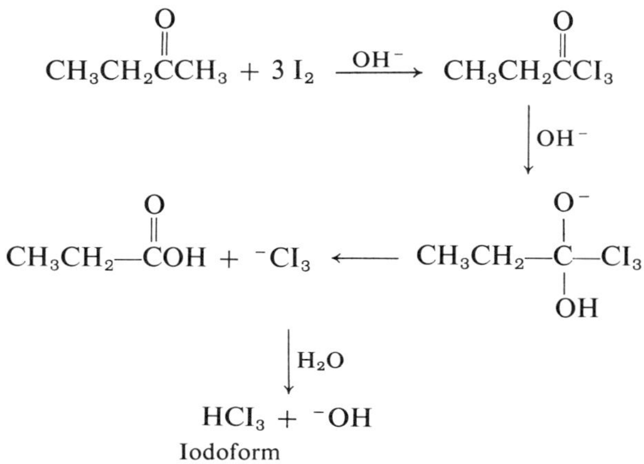

attacked by hydroxide to give a tetrahedral addition complex which disproportionates to a carboxylic acid and a triiodo carbanion (CI_3^-). Capture of a proton by this anion affords a yellow solid, iodoform, which can be recognized by its color and strong medicinal odor. The iodoform reaction is a means of detecting the presence of methyl ketones and, thus, provides an additional chemical test that can be used to characterize aldehydes and ketones. This reaction can also be used to prepare carboxylic acids from methyl ketones.

QUINONES

Quinones are cyclic diketones in which the two carbonyl groups are conjugated. They may be obtained by oxidation of 1,2-dihydroxy and 1,4-dihydroxy aromatic compounds or their amino-substituted analogs. *Meta* quinones are unknown. Quinones and their polyhydroxylic precursors are readily interconvertible by oxidation–reduction reactions, the ease of the interconversion being

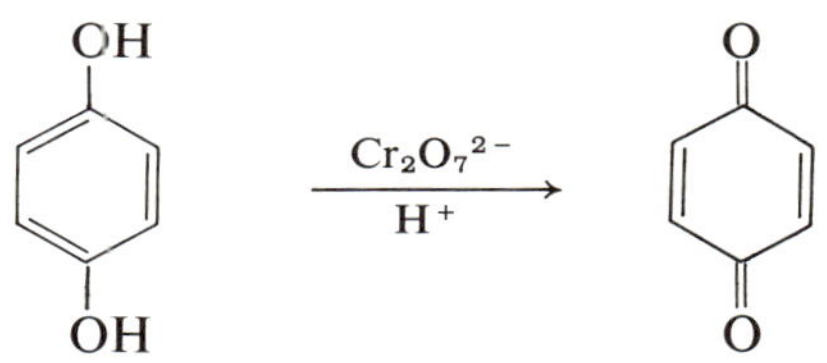

1,4-Dihydroxybenzene (hydroquinone) — 1,4-Benzoquinone (*p*-benzoquinone)

dependent upon the pH of the reaction system. Color is a characteristic property

1,2-Naphthoquinone

of quinones. Most simple *ortho* quinones, of which 1,2-naphthoquinone is an example, are red or orange, whereas the majority of the *para* quinones are yellow.

An appreciable number of pigments isolated from plants and, to a lesser extent, from animals have been characterized as quinones. Examples include perezone, polyphoric acid, and lawsone. Perezone is isolated from the root of the Mexican species of *Perezonia*. Polyphoric acid can be extracted from a fungus that grows on diseased oak trees. Testing has shown that this material is an active antitumor agent. Lawsone, which is extracted from the leaves of the tropical shrub, henna, dyes silk and wool an orange hue and preparations of henna leaves have been used to tint hair red. It is said that Mohammed dyed his beard with such a concoction.

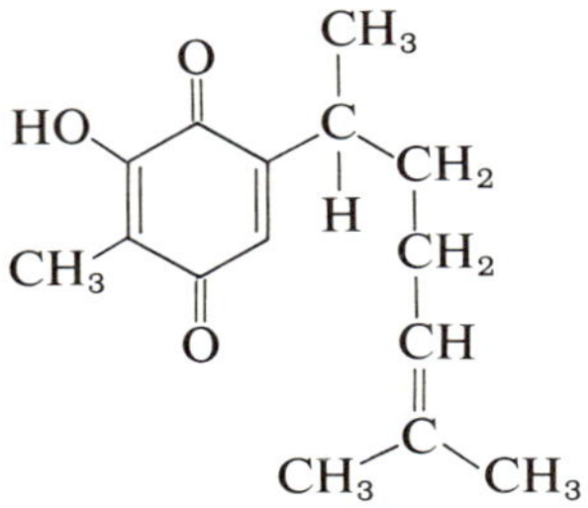

Perezone
(orange)

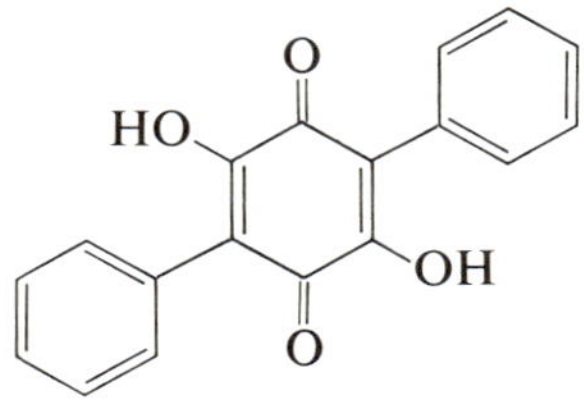

Polyphoric acid
(brown-violet)

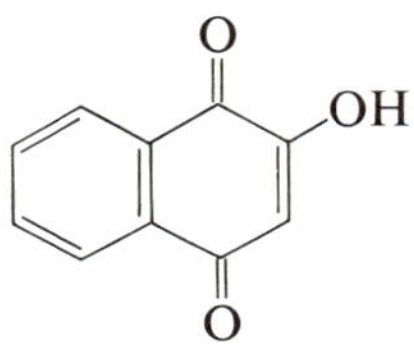

Lawsone
(yellow)

A group of structurally related quinones, known collectively as coenzyme Q, is involved in the transport of electrons during cellular respiration. In the structure of these coenzymes, the number (n) of isoprene $\left(-CH_2\overset{H}{\overset{|}{C}}=\overset{CH_3}{\overset{|}{C}}-CH_2-\right)$ units in the side chain varies up to 10. A shorthand designation for these molecules is CoQ_n, where n indicates the number of isoprene units present. The ability of quinones to undergo facile oxidation–reduction reactions is central to

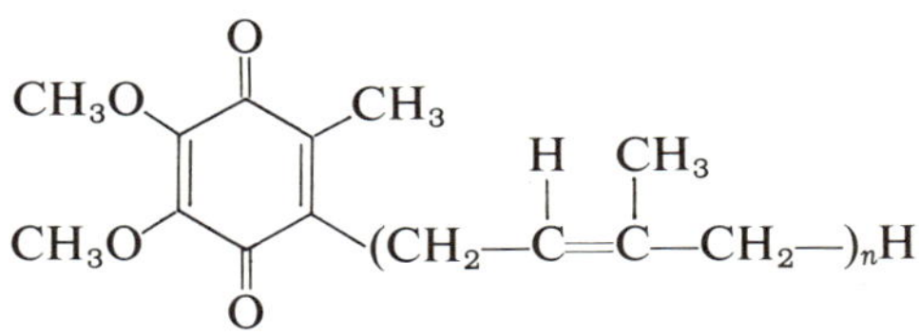

Coenzyme Q

the role of CoQ in electron transport. Most mitochondria also contain vitamin K and related naphthoquinones; the functional importance of these molecules is not yet clear.

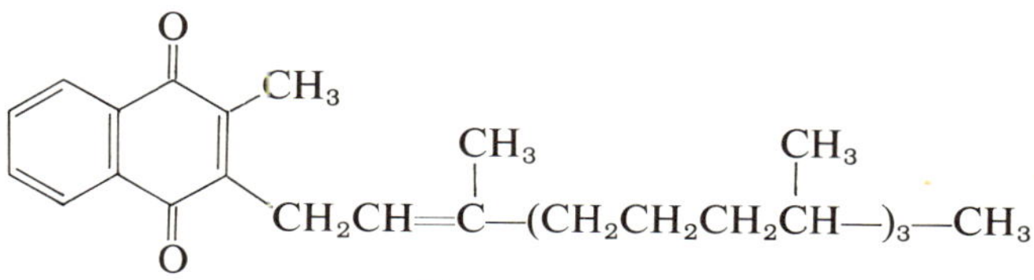

Vitamin K_1

ANTIFERTILITY DRUGS

The rapid rate of increase of the earth's population has become a matter of considerable concern. Although any open discussion of human sexual activity is incredibly complicated by the social matrix of the participants, there has been an increasing sense of uneasiness about this problem and some countries have now embarked on large-scale programs to consciously limit their populations. The past fifteen years have seen several major breakthroughs in birth control techniques; one of these is the development of orally administered contraceptive agents. The chemistry of these materials and their interactions with human biochemical systems are complex. The synthesis and structural chemistry of oral contraceptives provide a number of examples of ketone or aldehye chemistry and some of these are discussed briefly below.

Human sex hormones are steroids and are based structurally upon the cyclopentanoperhydrophenanthrene skeleton. For review, this skeleton is reproduced below, and the position-numbering scheme indicated. Most steroids have methyl groups attached to carbon atoms 13 and 10 of this structure; the carbon atoms of these methyl groups are numbered 18 and 19, respectively. The steroidal skeleton is essentially a planar one and the methyl groups mentioned above appear on top of the mean plane through the molecule. When representing the stereochemical features of a steroid it is usual to use dotted lines to represent bonds to substituents that project below the plane of the molecule and solid lines to indicate bonds to groups that point above it.

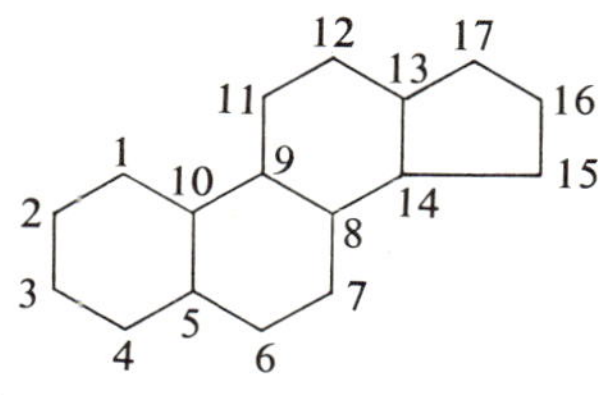

The cyclopentanoperhydrophenanthrene skeleton

The naturally occurring female hormone progesterone has often been called "nature's contraceptive" since one of its effects is to prohibit ovulation during pregnancy. This compound has no biological activity when taken orally, possibly

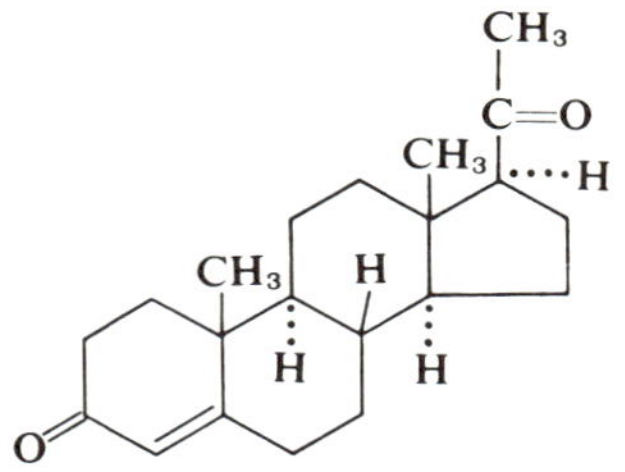

Progesterone

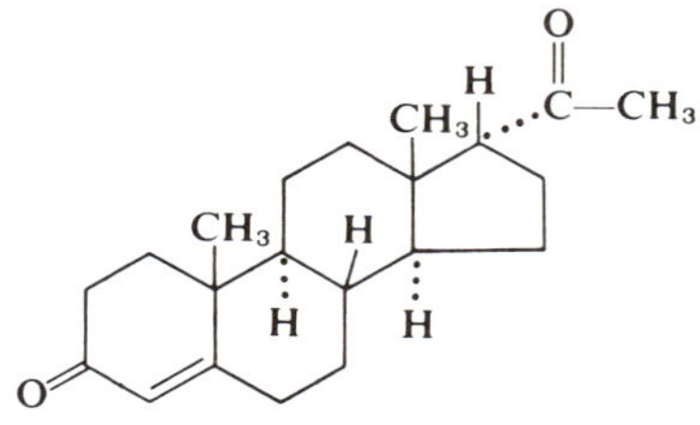

17-Isoprogesterone

because it is destroyed by the gastric juices of the stomach. It has long been assumed that the biological activity of this molecule was very specifically related to its structure since very similar molecules including the stereochemical isomer 17-isoprogesterone had no such activity. It was subsequently found that removal of a methyl group (carbon 19, attached to carbon 10 of the skeleton) and inversion of the stereochemistry at carbon 14 gave a steroid that behaved biologically nearly identically to progesterone. This molecule is called 14-*iso*-17-isonorprogesterone. It soon became clear that the methyl

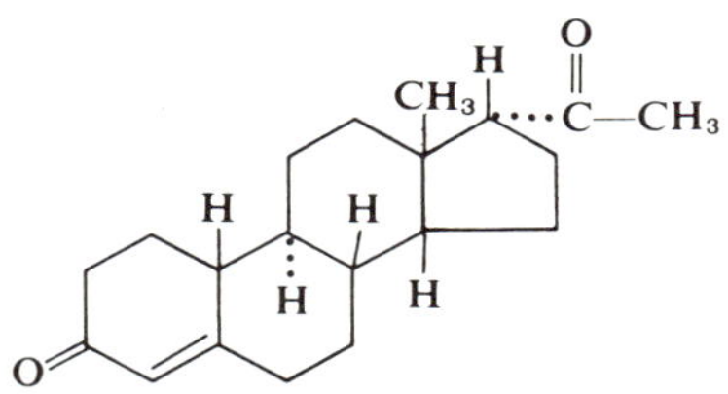

14-Iso-17-isonorprogesterone

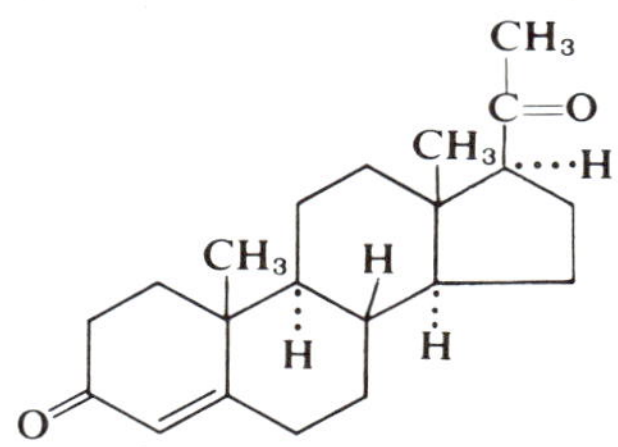

19-Norprogesterone

group attached to carbon 10 (carbon 19) was somehow associated with progestational behavior. It was reasoned that removal of the corresponding angular methyl group in progesterone to give 19-norprogesterone should increase the activity of this natural hormone and the new substance was indeed found to have 4 to 8 times the activity observed for progesterone when injected parenterally. Clinical trials with a similar compound, 19-nor-17α-ethynyltestosterone, showed it to have very high potency as a progestational agent when administered orally. This compound is now marketed in combination with other steroids as an oral contraceptive, i.e., "the pill." The commercial production of Enovid and Norlutin, two isomeric oral contraceptives, involves, in part, the sequence of reactions shown below. The carbonyl group of the starting material is protected by the formation of a cyclic ketal at this position. Oxidation of the alcohol function at carbon-17 gives another ketone which is then treated with the organometallic reagent sodium ethynylide ($Na^{+}\,{}^{-}C{\equiv}C{-}H$) which adds to this carbonyl group much in the manner that Grignard reagents add to this functional group. Acid-catalyzed hydrolysis of the ketal gives the two isomeric steroids.

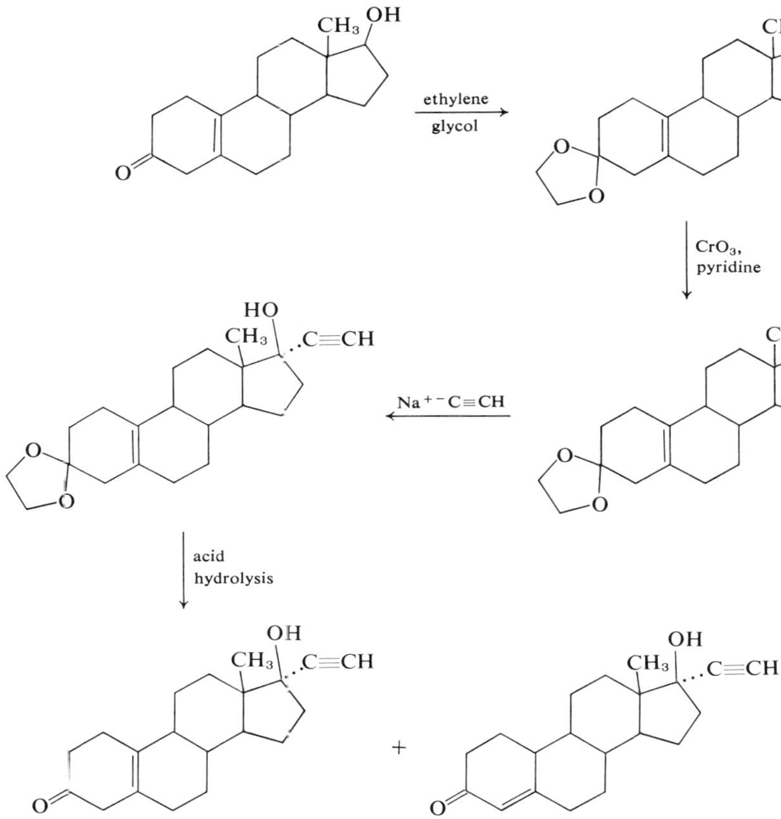

Enovid

Norlutin
(19-nor-17α-ethynyltestosterone)

PROBLEMS

1. Write the structures of the organic molecules which correspond to the following names.

(a) α-Bromopropionaldehyde
(b) Methyl hydrazine
(c) 3-Iodobenzaldehyde
(d) Heptanal
(e) Cyclobutanedione
(f) Cyclopentanone
(g) Ethylene glycol
(h) Cyclopentyl magnesium chloride

(i) Ethyl isopropyl ketone
(j) α-Fluoroacetophenone
(k) 1,1,1-Trichloropentan-3-one
(l) *cis*-3-Hepten-2-one

2. The following aldehyde is obtained by ozonolysis of natural rubber. Name the aldehyde. What is the structure of natural rubber?

$$H{-}\overset{O}{\overset{\|}{C}}{-}CH_2CH_2{-}\overset{O}{\overset{\|}{C}}{-}CH_3$$

3. Write the structure for the major products of the following reactions.

(a) Testosterone + sodium borohydride $\longrightarrow$

(b) $CH_3\overset{O}{\overset{\|}{C}}{-}CH_3$ + hydroxylamine $\longrightarrow$

(c) $Cl{-}CCl_2{-}\overset{O}{\overset{\|}{C}}{-}H + H_2O \longrightarrow$

(d) 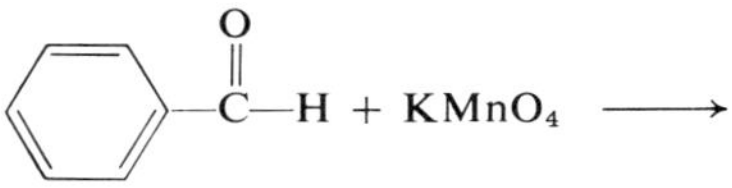

(e) 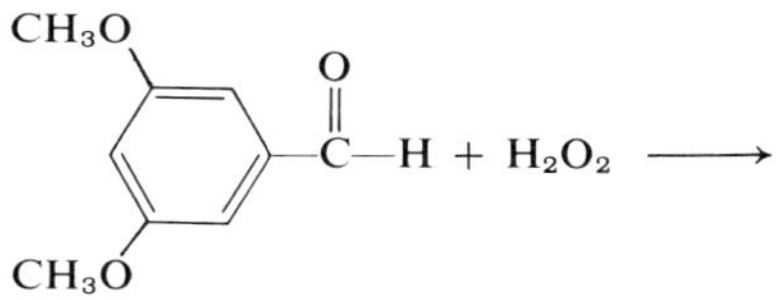

(f) 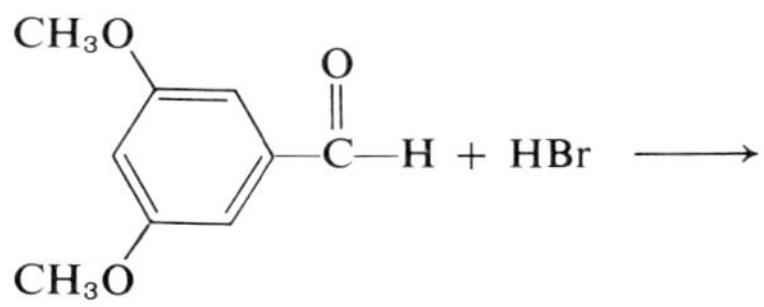

(g) 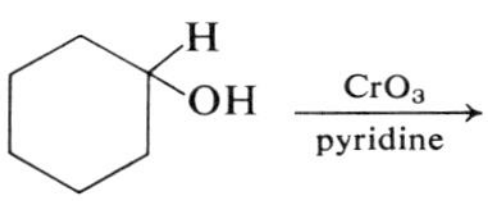

(h) 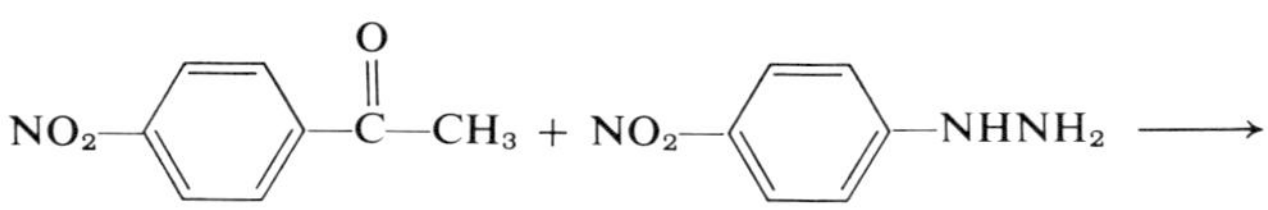

(i) 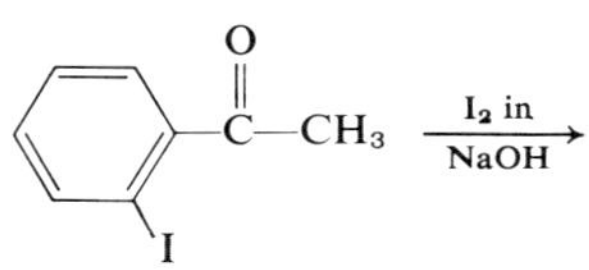

(j) Acetaldehyde + formaldehyde $\xrightarrow{\text{NaOH}}$

(k) 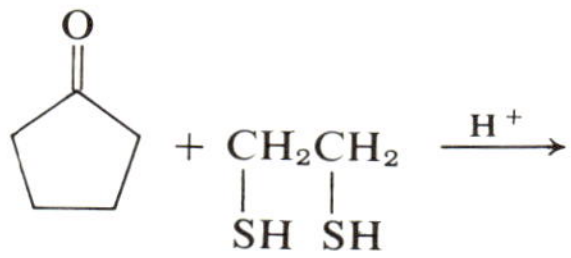

(l) 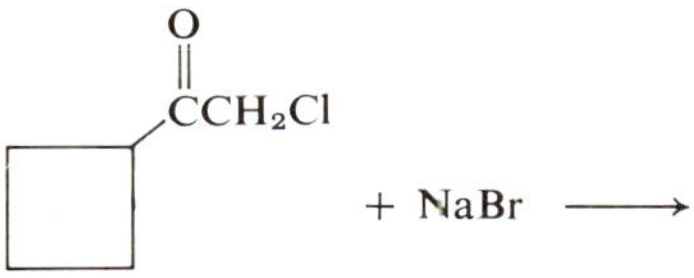

(m) Cinnamaldehyde + lithium aluminum hydride $\longrightarrow$

(n) Cinnamaldehyde $\xrightarrow[\text{Pt catalyst}]{H_2}$

4. Devise a series of reactions that could be used to synthesize the following molecules from the starting material indicated.

	Starting material	Desired product
(a)	$CH_3\overset{\overset{\Large O}{\|}}{C}CH_3$	$CH_3-\underset{CH_3}{\overset{CH_3}{C}}-OH$
(b)	$CH_3\overset{\overset{\Large O}{\|}}{C}CH_3$	$CH_3-\underset{CH_3}{\overset{H}{C}}-OH$
(c)	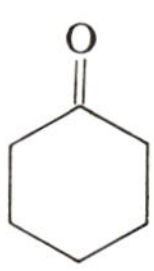	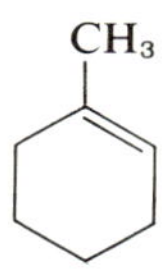
(d)	CH_3CH_2OH	$CH_3-\overset{H}{C}=\overset{H}{C}-\overset{O}{\overset{\|}{C}}-H$
(e)	CH_3CH_2OH	$CH_3CH_2CH_2\overset{O}{\overset{\|}{C}}-OH$

5. The heterocycle paraformaldehyde is used as a snail bait. How is this molecule related to formaldehyde?

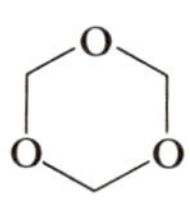

paraformaldehyde

6. Predict the pmr spectrum of each of the following molecules.

(a) 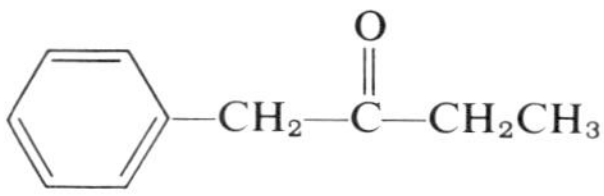

(b) $CH_3\overset{\overset{\textstyle O}{\|}}{C}—H$

(c) $ClCH_2\overset{\overset{\textstyle O}{\|}}{C}—H$

7. Suggest reactions to affect the reduction of 3-cyclohexenone to cyclohexanone and to cyclohexanol.

8. Organometallic reagents are very strong bases. What will be the products of the following reactions?

(a) $CH_3Na + H_2O \longrightarrow$

(b) $CH_3MgBr + CH_3COOH \longrightarrow$

(c) 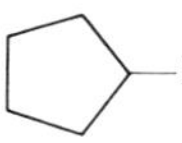 $—MgBr + H_2O \longrightarrow$

9. The dependence on acidity of the rate of an addition–elimination reaction between an aldehyde or ketone and an amine is shown below. Rationalize this acidity dependence in terms of the mechanism for this reaction discussed earlier.

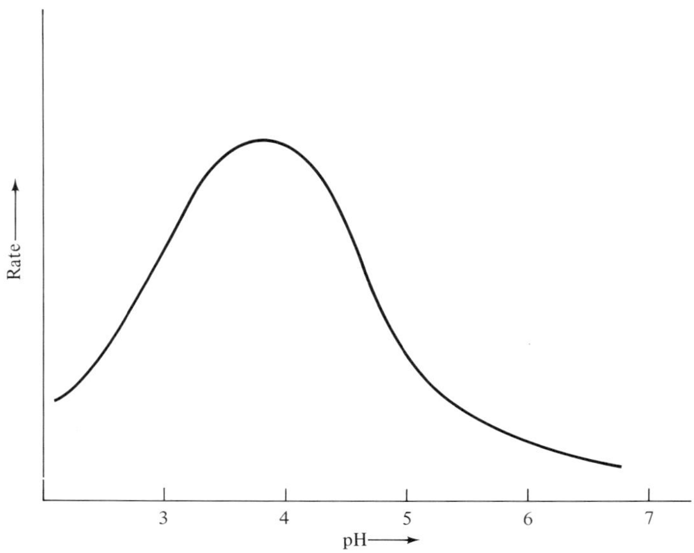

SUGGESTED READINGS

C. DJERASSI, *Science* **153**, 493 (1966). (Steroidal oral contraceptives.)

L. F. FIESER, Steroids. *In* "Bioorganic Chemistry" (M. Calvin and M. J. Jorgenson, eds.). Freeman, San Francisco, California, 1968.

C. D. GUTSCHE, "The Chemistry of Carbonyl Compounds." Prentice-Hall, Englewood Cliffs, New Jersey, 1967.

K. E. MAXWELL, "Chemicals and Life." Dickenson, Belmont, California, 1971.

S. PATAI, ED., "The Chemistry of the Carbonyl Group." Wiley (Interscience), London, 1966.

R. H. THOMSON, "Naturally Occurring Quinones." Academic Press, New York, 1971.

10

Carbohydrates

The most abundant class of organic compounds in nature is the carbohydrates; carbohydrates play a central role in the utilization of radiant energy from the sun by living organisms on earth. Plants convert carbon dioxide and water to carbohydrates and oxygen with the aid of sunlight and these carbohydrates are used by the animal kingdom as food or energy sources. In the process, they are converted back to carbon dioxide, water, and energy.

These substances derive their name from the fact that their empirical formulas can be broken down into a certain number of carbon atoms (carbo-) and water molecules (-hydrates). A high percentage of wood and plant fibers is made up of carbohydrates, as are the starchy components of cereal grains and tubers, such as potatoes. The compounds commonly regarded as sugars are carbohydrate in nature. Experimental study of carbohydrates is complicated by the fact that these organic compounds are often difficult to work with in the laboratory since they are reluctant to crystallize. Moreover, another kind of isomerism that we have not yet considered makes possible a large number of isomers for a single compound. In this chapter, we shall first examine this isomerism and then some aspects of carbohydrate chemistry.

OPTICAL ISOMERISM

We have already seen how some organic compounds can have the same elemental composition and yet be different compounds. The first major type of isomerism encountered was that of *structural* or *constitutional* isomerism. Structural isomers are a set of molecules that differ by the nature and sequence of the chemical bonds which hold the constituent atoms together. Some structural isomers of C_3H_5Cl are shown here. When the number and type of the

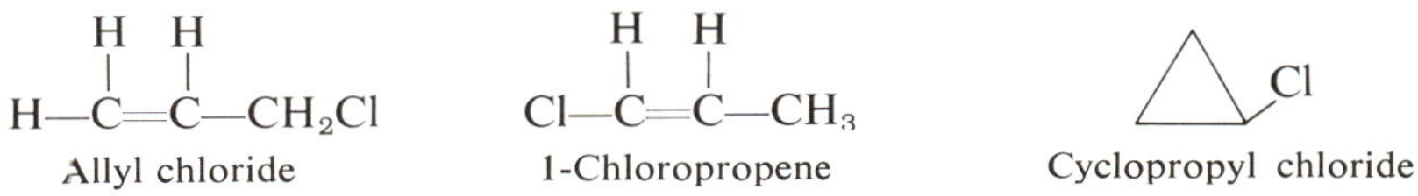

Structural isomers of C_3H_5Cl

bonds in the components are the same, but the isomers differ in the way in which the atoms are arranged in space, the molecules are *stereoisomers*. The conformational isomerism in butanes or substituted cyclohexanes has already been noted; in these cases the rate of interconversion of the stereoisomers (conformers) is rapid. In other molecules, such as *cis*- and *trans*-2-butene, certain features make interconversion of the stereoisomers a slow process and the various isomeric forms can be isolated. This latter type of stereoisomerism has previously been termed geometrical isomerism; geometrical isomers usually differ by the orientation of certain groups relative to a certain plane through the molecule.

A different kind of stereoisomerism arises when two molecules are mirror images of one another but are not identical molecules. The nonidentity of mirror images is not uncommon in the human experience. A person's right hand, for example, is similar in every way to his left in terms of structural features and yet the two hands are not identical. To be truly identical, it would be necessary to be able to superimpose every point of one hand on the corresponding point of the other. This is not possible with human hands and it is not possible with some sets of molecules. The lesson to be learned from our hands is that the mirror image of an object is *not* necessarily identical with the object itself. Molecules which are not identical with their mirror images must, therefore, be isomeric; this type of isomerism is called optical isomerism.

Asymmetric Carbon Atoms

What are the structural requirements of optical isomers? There cannot be optical isomers of methane, since the mirror image form of methane is identical to the original image. But if one considers the secondary alcohol, 2-butanol, it

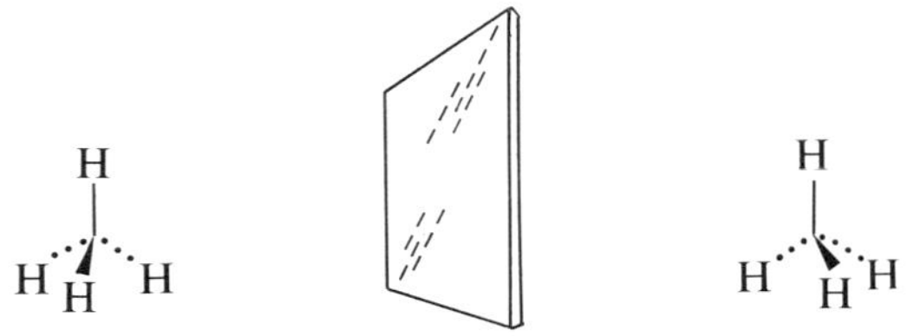

Original image of methane   Mirror   Mirror image

is seen that in this case it is possible to generate a nonidentical mirror image form. This molecule illustrates the fact that having four different substituents at a

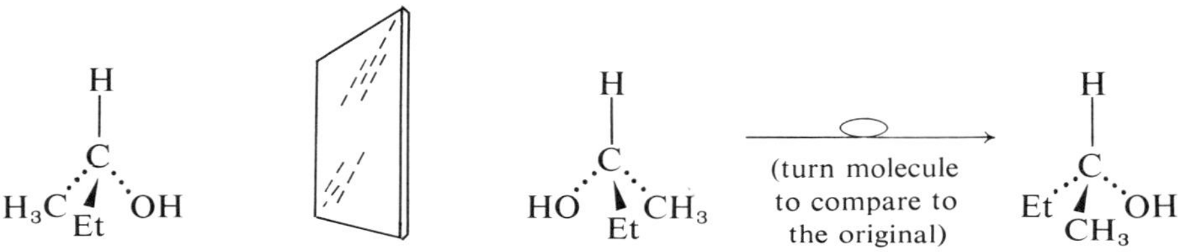

Original image   Mirror   Mirror image

(Et is used for C_2H_5, the ethyl group)

carbon atom is usually a sufficient condition for the generation of optical isomers. A carbon atom of this kind is called an *asymmetric* carbon. The two, nonidentical mirror image forms of a molecule are referred to as *enantiomers*.

Meso Compounds

The presence of asymmetric carbon atoms in a molecular structure does not guarantee that the molecule can exist in enantiomeric forms. The halogenated ethane shown below in the eclipsed configuration is not asymmetric even though there are two asymmetric carbon atoms, as defined above, in its structure. In this structure, called a *meso* form, the two elements of molecular asymmetry combine to create a symmetric structure that does not have a nonidentical mirror-image form. When considering whether or not a molecule is identical with its mirror image, the various conformations that are possible are ignored if their interconversion is rapid at laboratory temperatures. If one can find a single conformation that generates a molecule that is the same as its mirror image, the molecule is not regarded as having an enantiomer. Thus, the halogenated ethane depicted on p. 233 has no enantiomer even though some conformations obtained by rotation about the carbon–carbon bond do have mirror image forms that are not superimposable on the original molecule.

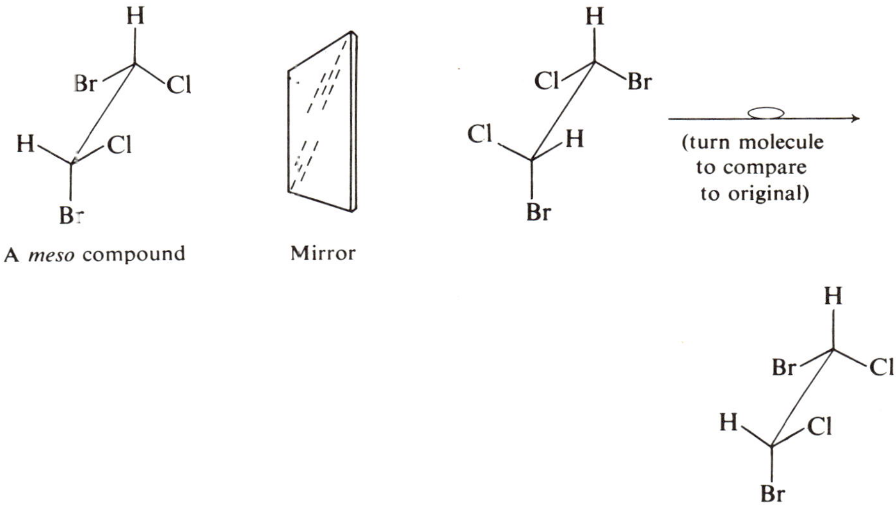

Optical Rotation

It takes a rather subtle experimental probe to determine which of two enantiomers one has in hand. The two forms of glyceraldehyde are sketched below. Note that this molecule is both an aldehyde and a dihydroxy compound. Each of these two forms has exactly the same number and type of chemical bonds. All typical physical properties with which we are familiar—boiling point,

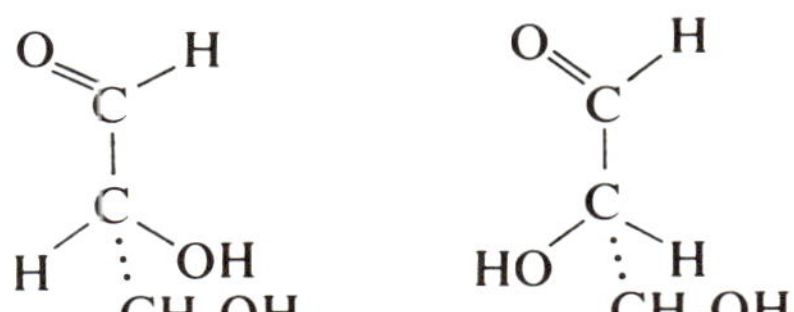

D(+) L(−)

Optical isomers of glyceraldehyde

melting point, nmr spectra, solubility, to name a few—depend upon the nature of these bonds and, therefore, all the physical properties (except one) of the molecule are identical. Similarly, the chemical reactions that have been mentioned for aldehydes or alcohols will occur in the same manner with both enantiomers. The only way to distinguish between two molecules that are enantiomeric is to place them in an environment that is, in some sense, asymmetric. One configuration of the molecule will interact with this environment in a different way than will the molecule of the opposite configuration.* A method for doing this is to expose the set of enantiomeric molecules to a biological system. As we shall see in the ensuing chapters, biochemical reaction systems in nature are very highly asymmetric. Enzymes, for example, will react with only one molecule in a pair of enantiomers; the enzyme provides an asymmetric environment for the enantiomers and preferentially interacts with one of them. The sense of taste is based on a biological system; one member of a pair of enantiomers often has a different taste than the other. (Tasting is not a recommended experimental method for distinguishing between enantiomers, however.) Finally, recent work has shown that enantiomers can have different odors; the compounds largely responsible for the characteristic odors of spearmint and caraway are enantiomers.

Another technique for distinguishing enantiomers makes use of the one physical property that is not the same for both isomers: the way in which each interacts with plane-polarized light. Ordinary light, such as that in this room, is made up of a collection of electromagnetic waves of various wavelengths, each being propagated relatively incoherently from the source to the object being viewed. There are available experimental methods for "sorting-out" a collection of light rays so that only light waves traveling in a single plane are allowed to pass through the separation device. Polaroid sun glasses are a familiar utensil for doing this; borrow a friend's set and try turning one eyepiece 90° relative to another. The light plane selected by the first is perpendicular to that which would be allowed to pass by the second so that no light gets through the set of two lenses.

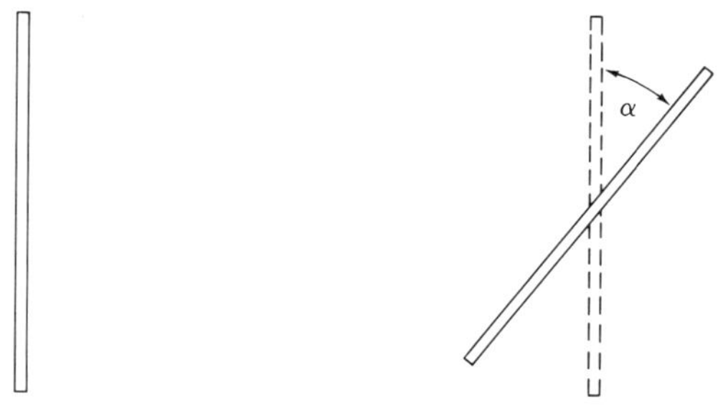

Original plane of polarization of light beam

Plane of polarization after passing through an asymmetric molecule

* An anology to this approach for distinguishing between enantiomers is the fit of one's left or right hand into a set of gloves. Each member of a pair of gloves provides an asymmetric environment for one hand and the left hand interacts differently with one glove in the pair as compared to the other.

A selected plane of light rays is called plane-polarized light. If the light beam is observed after it passes through an enantiomeric molecule, the plane of polarization will appear to be tilted relative to its initial orientation. The angle, α, by which the emergent light beam has been rotated relative to the initial plane of polarization is called the optical rotation, and molecules which can turn the plane of polarization in this manner are said to be optically active. The direction that an optically active compound rotates plane-polarized light is often indicated before the name of the compound. Those materials which rotate polarized light to the right are *dextrorotatory*, indicated by the symbol (+), while compounds which have the opposite effect are called *levorotatory* with the symbol (−) being prefixed to their name. Enantiomers rotate plane-polarized light in opposite directions but the absolute value of α is the same for both isomers. While there is no necessary correlation between the sign of an optical rotation and the configuration of the substituents about a given asymmetric carbon atom, optical rotation is a convenient way of identifying and characterizing asymmetric molecules.

Absolute Configuration

All of the optically active compounds that will be of interest to us in this text derive their symmetry from the presence of four different substituents at the asymmetric carbon atom. Although one can obtain experimentally the direction and magnitude of the optical rotation, a primary concern remains finding out which of the two possible mirror image forms one has at hand. This is known as determining the *absolute configuration* of the molecule.

The hydroxyaldehyde, glyceraldehyde, has one asymmetric carbon atom. There are, therefore, two possible stereoisomers of glyceraldehyde that are nonidentical mirror images. One of them rotates polarized light to the left while the other isomer to an equal but opposite extent is dextrorotary. Somewhere in the early history of organic chemistry the decision was made to arbitrarily assign the configuration shown earlier to the dextrorotatory isomer. The levorotatory form is, therefore, taken to be the enantiomer of this molecule.

The absolute configurations of other optically active compounds are determined by relating the configuration of their asymmetric carbons to one of these glyceraldehyde isomers.

Specifying the Configuration of Asymmetric Atoms

If an asymmetric molecule has a configuration of substituents at the asymmetric carbon that can be directly related to (+)-glyceraldehyde, it is said to be the D isomer, regardless of the sign or magnitude of the molecule's optical rotation. Molecules that are configurationally relatable to (−)-glyceraldehyde are designated L isomers. The symbols D- and L- have no particular significance but rather are merely designations for families of absolute configurations. It should be emphasized again that the sign of optical rotation (+ or −) of a molecule has no necessary relation to the absolute D or L configuration of the molecule. One only can be sure that the optical rotation of an L enantiomer will be opposite that of the corresponding D enantiomer.

Alternatively the configuration of an asymmetric carbon can be specified by the Cahn–Ingold–Prelog convention. This system for the designation of configuration is more flexible and more widely applicable than the D, L system described above.

In applying this convention, priorities are assigned to each of the four substituents attached to the asymmetric carbon. The priorities are determined in the same manner as used for the *E–Z* nomenclature for *cis–trans* olefin isomers described in Chapter 3; priorities are determined by the atomic number of the substituents with the highest priority assigned to the atom of largest atomic number. When groups are bound to the asymmetric carbon by the same element, the second atom in the group is used to determine the ranking of the group. The substituent with the lowest priority is placed pointing away from the observer. He looks along the bond from the asymmetric carbon toward this group and examines the order of the three remaining substituents. If their priorities are arranged so that they decrease in a clockwise sense, the configuration is said to be the *R* configuration. If the arrangement of priorities has a counterclockwise sense, the configuration is designated by *S*.

Racemization

A mixture of exactly equal amounts of D and L isomers is called a *racemic mixture*. Although each component of the mixture rotates plane-polarized light, the total rotation of the sample is zero since the equal but opposite rotations from the two enantiomers cancel. *Racemization* is the conversion of a single optical isomer into a racemic mixture; racemization occurs whenever, during the course of a chemical reaction, an asymmetric carbon atom assumes a symmetrical configuration even though the lifetime of the symmetrical form is short. An example of a racemizing reaction is the hydrolysis of α-methylphenethyl chloride (Fig. 10.1). This reaction proceeds almost entirely by an S_N1 mechanism and during the reaction the original asymmetric halide is converted into a carbonium ion which has a plane of symmetry. Capture of solvent water molecules can take place with equal probability from either side of this plane so that equal amounts of the D and the L isomers of the corresponding

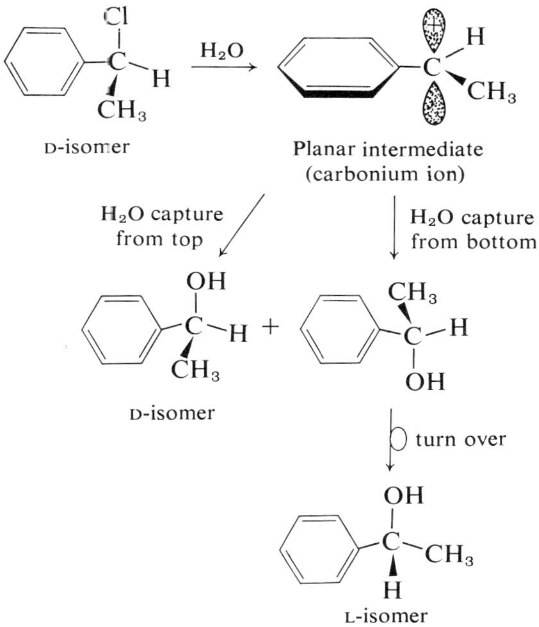

Fig. 10.1. The stereochemical course of the S_N1 reaction.

alcohol are formed. Thus, in this case a racemic product results from reaction of an optically active starting material.

Racemization of the optically active ketone shown in Fig. 10.2 would occur through the intermediacy of a carbanion; capture of a proton from the solvent can take place equally well from the top or bottom of its symmetry plane, giving a racemic ketone.

Most laboratory reactions with organic compounds produce racemic mixtures if the products are asymmetric. For example, the addition of benzyl magnesium bromide to acetophenone would give a product alcohol that has an asymmetric carbon atom but equal amounts of the two enantiomers are isolated (a racemic mixture) since there is no reason that one isomer should be formed preferentially.

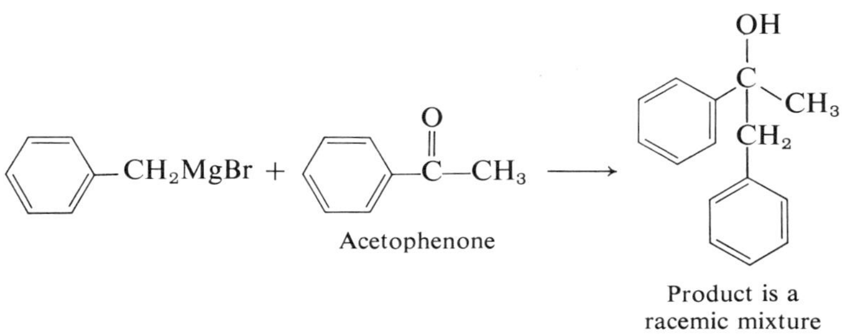

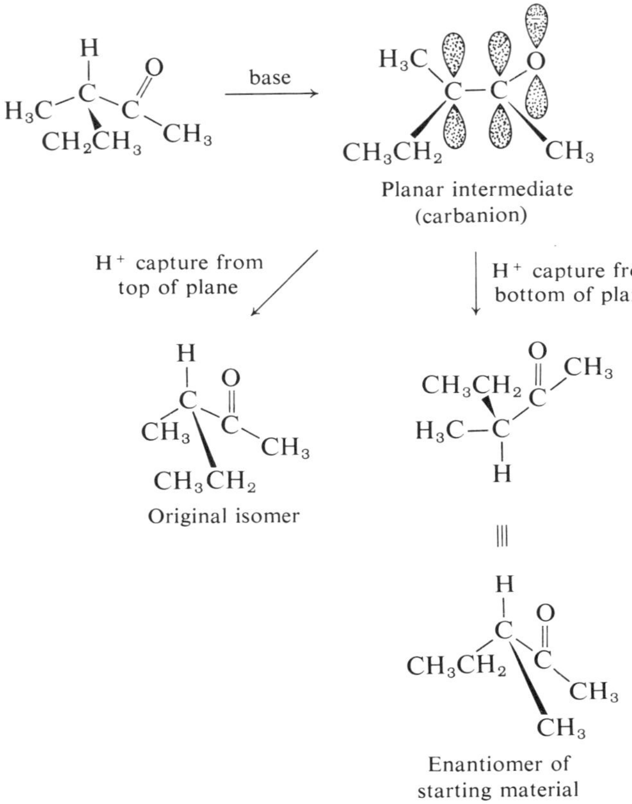

Fig. 10.2. Racemization of an optically active ketone.

Diastereomers

When there are two asymmetric carbon atoms in a molecule, four possible optical isomers can be formed. In this set of molecules, there are two pairs of enantiomers. The physical and chemical properties of each member of a pair will be identical within the limitations discussed above. Those molecules in this set of four that are not enantiomers of one another are still optically isomeric and are called diastereomers. The intramolecular interactions in a set of diastereomers are different and, therefore, the physical properties of diastereomers are different. Consider molecules A and B on p. 239. These species are enantiomeric and each of these molecules has the same number and type of interactions between the atoms within their structures. As a result, their physical and chemi-

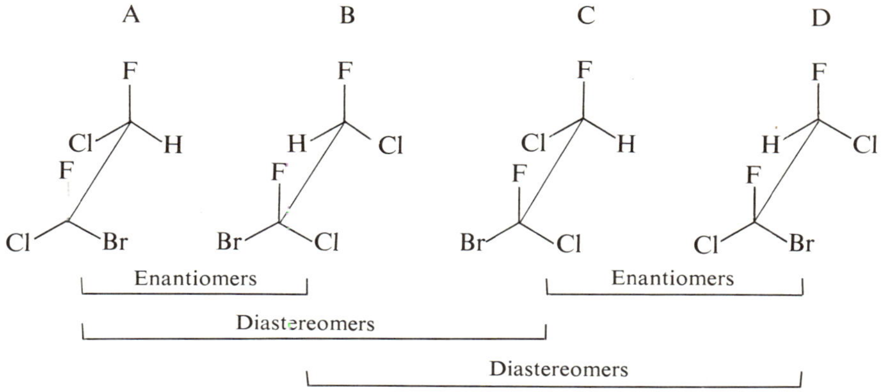

cal properties are identical. However, when comparing A to molecule C, it is found that there is a different set of interactions between the atoms in A than are found in C. For example, A has a Cl—Cl interaction that is missing (or at least different) in C. Molecules A and C are diastereomeric and they have different physical properties which can be used to distinguish between them.*

Resolution

The separation of a racemic mixture into the two optically active components is called *resolution*. One can take advantage of the chemical and physical differences between diastereomers in order to effect the separation of a racemic mixture into its two optically active components. In this method the racemic mixture is treated with an optically pure (100% one isomer) compound, usually

* When both groups of substituents on the two asymmetric carbons are identical, a pair of (identical) *meso* compounds is found among the four possible optical isomers. The physical properties of the *meso* compound are different from the properties of the other two molecules in the set.

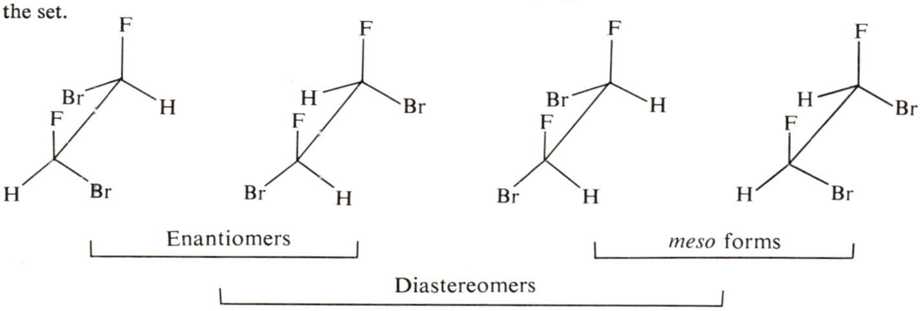

(The mirror image is identical to original molecule)

obtained from natural sources. For separation of the mixture of racemic tertiary alcohols formed in the Grignard reaction above, one could form a set of diastereomers by treating the racemic mixture with D-lactic acid, a hydroxycarboxylic acid found in muscle tissue. (The compound formed is an ester, a functional group to be taken up in the next chapter.) This reaction produces two compounds that contain two asymmetric carbon atoms (one carbon each

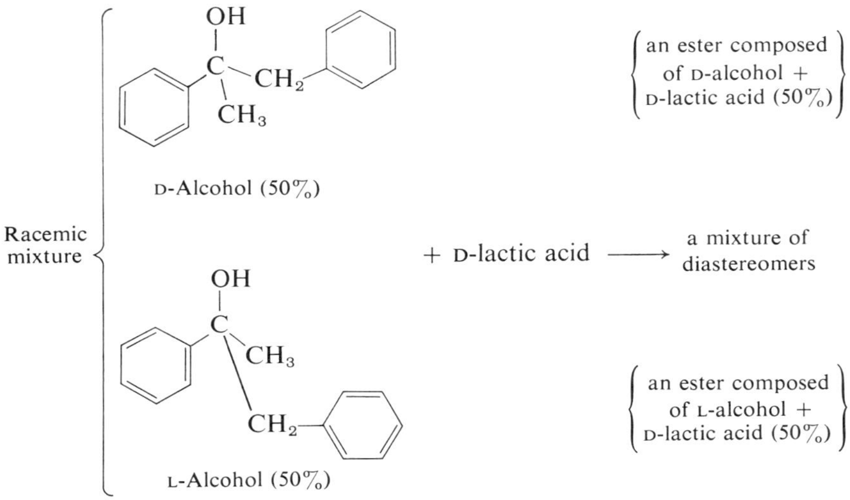

from the alcohol and the acid). They are stereoisomeric materials but are not enantiomers; they are diasteriomers. In the case described here, fractional crystallization gives the two diastereomeric esters, each of which can be decomposed in aqueous base to regenerate the lactic acid and one enantiomer of the starting alcohol. Removal of the acid from the hydrolysis reaction gives the resolved, optically active alcohol.

MONOSACCHARIDES

Although there is a staggering variety of carbohydrate structures found in nature, they are quite similar in their main structural features. The simplest carbohydrates are called monosaccharides. These compounds are polyhydroxy aldehydes or polyhydroxy ketones.

Glucose, also known as dextrose, is by far the most abundant organic compound found in nature and occurs in both the free state (fruit juices, honey) and as the monomeric unit in polymers (cellulose). Additional classification of monosaccharides includes the use of the term *aldose* if the sugar contains an aldehyde group or *ketose* if a ketone function is present. The number of carbon atoms present in the structure can also be designated by generic names like triose, tetrose, pentose, and hexose for three, four, five, and six carbon atoms in the skeleton. These names can also be combined so that glucose, for example, would be regarded as an aldohexose. Most common sugars are pentoses or hexoses.

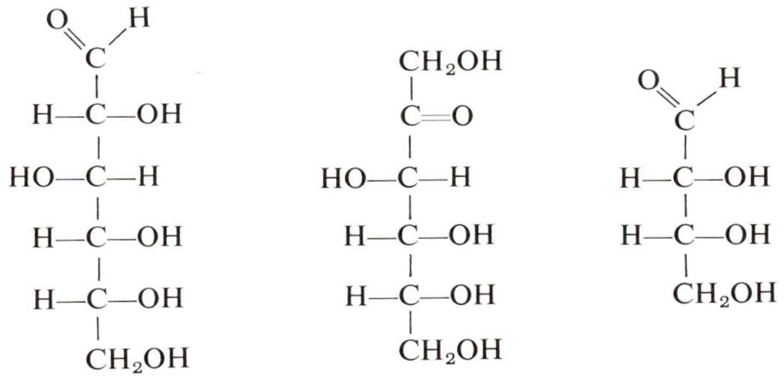

Glucose Fructose Erythrose

Some monosaccharides

In considering the examples of monosaccharides depicted above, a number of structural features are apparent. First, an aldehydic or ketonic carbonyl group is present as well as a number of alcohol (hydroxyl) groups. Second, the majority of the carbon atoms in the molecules are asymmetric in that each one of these has four different groups attached to itself. Each asymmetric carbon has two possible configurations so that $2^4 = 16$ stereoisomers are possible for an aldohexose, for example. One difficulty with this rampant isomerism is representing it in two dimensions on paper. A convention developed by Emil Fischer, a pioneer in sugar chemistry, is normally used in this regard. In the Fischer convention, the carbon chain of the sugar is drawn vertically with the carbonyl carbon at the top. The atoms of the chain are numbered consecutively in such a way that the carbonyl carbon has the lowest number. At each asymmetric carbon those bonds drawn horizontally are understood to project above a mean plane through the molecule, while those drawn vertically correspond to bonds which project below it. In this process each asymmetric carbon is viewed from the side of the mean plane on which it resides. A ball-and-stick model of glucose is shown in Fig. 10.3; this illustration should be compared to the planar representation of glucose above, which was drawn according to this convention. In understanding the Fischer projection technique it will be helpful to closely examine a molecular model.

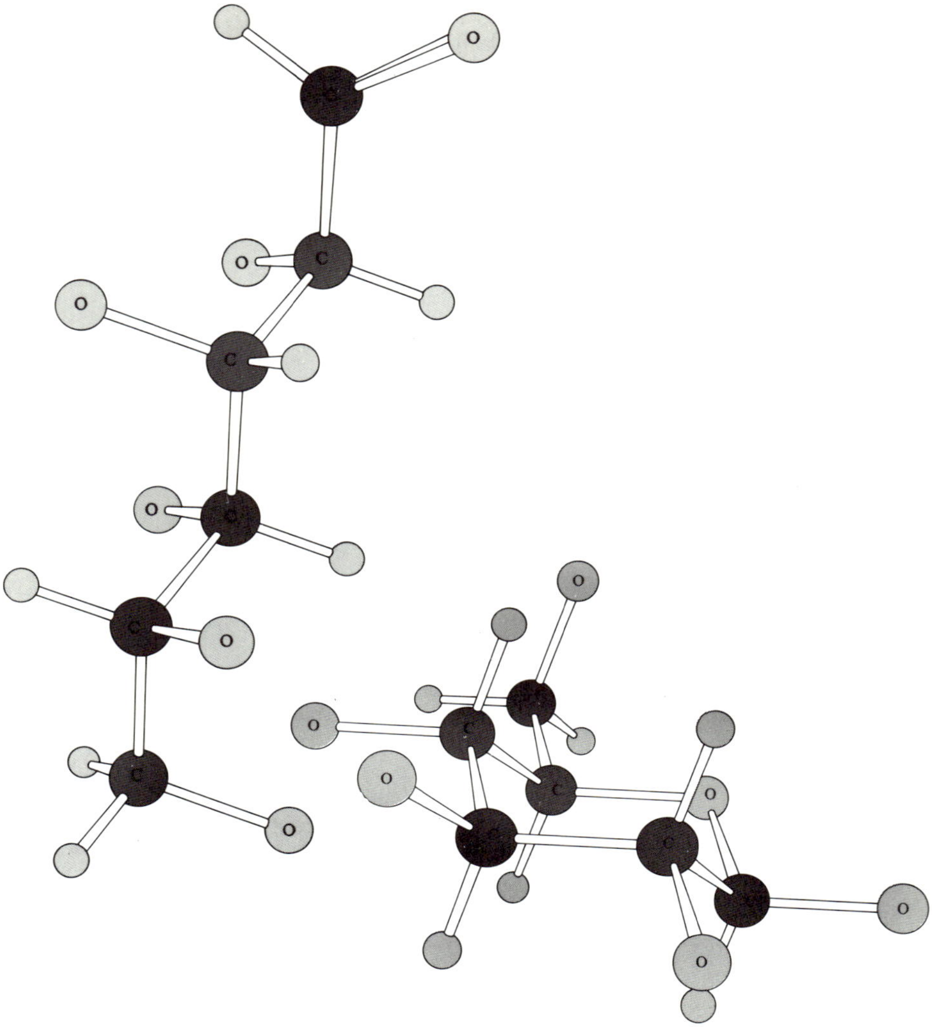

Fig. 10.3. Three-dimensional drawings of glucose in the linear and cyclic forms.

Reactions of Monosaccharides

Many, but not all, of the reactions of the carbonyl group that were described in the previous chapter on aldehydes and ketones are observed with the carbonyl groups in carbohydrates. Aldoses are easily oxidized by Tollen's reagent or Fehling's solution. Although ketones are normally inert to these reagents, the presence of a hydroxyl group at an α-carbon atom changes the reactivities of both groups sufficiently so that oxidation of α-hydroxy ketones with these reagents is possible. As a result, ketoses with hydroxyl groups *alpha* to the carbonyl function also react with these oxidizing agents. Sugars which react with Tollen's or Fehling's solutions are known as *reducing sugars* while those that do not are called *nonreducing sugars*.

Both ketoses and aldoses can be reduced with sodium borohydride. The reactions are carried out in aqueous solution, but conditions of pH and temperature must be carefully controlled when reducing some sugars since racemization

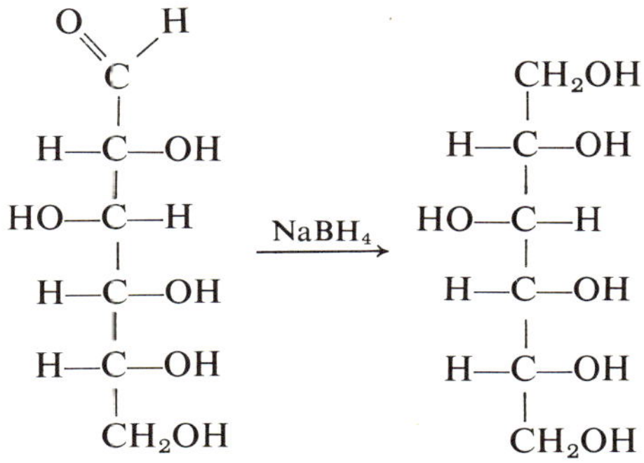

D-Glucose Sorbitol

and other undesirable side reactions can take place. Typically, borohydride reductions of monosaccharides are carried out at pH 8 in a borate buffer. Catalytic hydrogenation using platinum or nickel catalysts has also been employed for the reduction of sugars.

We have already noted that complex metal reagents (Tollen's and Fehling's solutions) can oxidize sugars. These reactions are not very useful for preparation

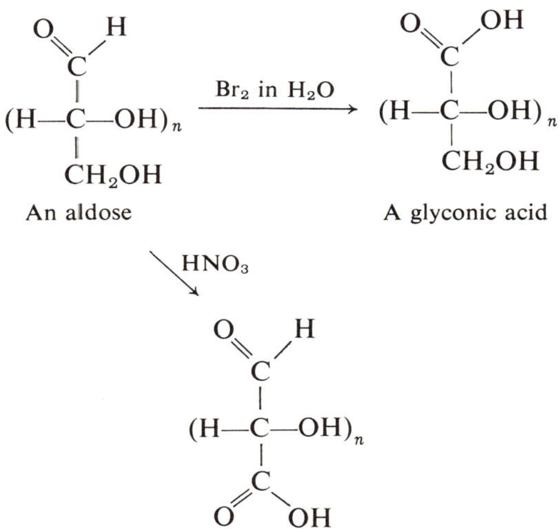

An aldose A glyconic acid

A glycaric acid

of oxidized sugars since the basic solutions in which they are used can cause isomerization or decomposition of the saccharide. Aldoses can be gently oxidized with bromine in water to monocarboxylic acids. The use of more drastic oxidizing agents such as nitric acid leads to the formation of dicarboxylic acids, since the terminal hydroxymethyl group (—CH_2OH) is also oxidized under these conditions. Ketoses are not oxidized by bromine water and stronger oxidants break the carbon chain adjacent to the carbonyl group.

Many of the carbonyl addition reactions that occur with nucleophilic reagents and aldehydes or ketones have counterparts in the chemistry of sugars. Hydrogen cyanide adds to the carbonyl groups of sugars to give cyanohydrins and the formation of oximes by an addition–elimination reaction takes place. Phenylhydrazine, another reagent which reacts with carbonyl groups by an addition elimination sequence of reactions, will react in this manner with an aldose or ketose if the concentrations of phenylhydrazine and the sugar are carefully controlled. With an excess of phenylhydrazine a more complex reaction leading to the formation of a compound containing two phenylhydrazine residues per sugar molecule takes place. These products are called osazone derivatives and are usually solid, easily crystallizable products. In osazone formation three

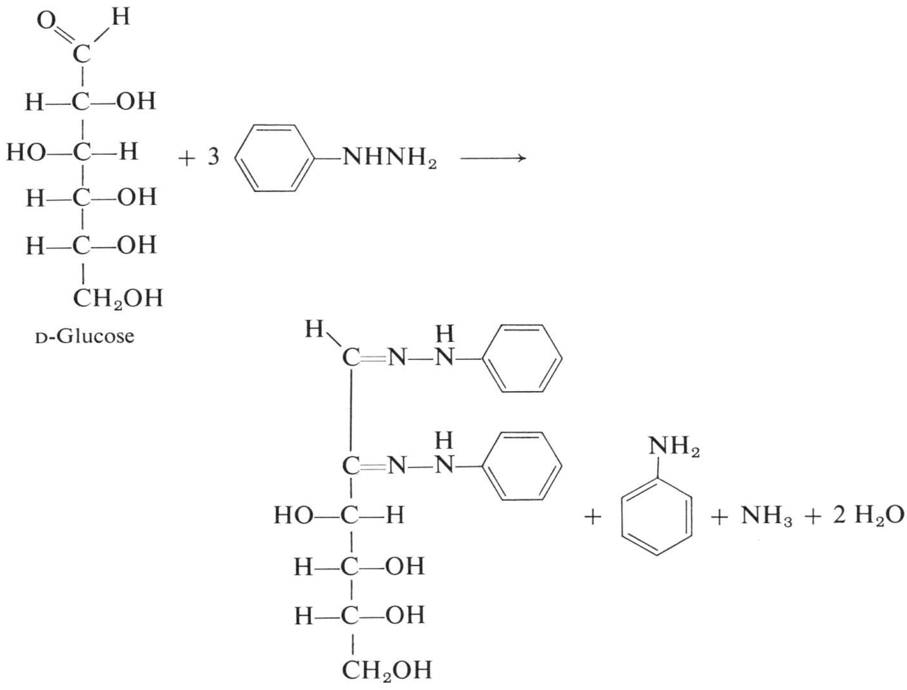

Osazone derivative of glucose

molecules of phenylhydrazine and one molecule of sugar eventually react to give the product. During this process an oxidation–reduction reaction that produces aniline and ammonia takes place.

One carbon atom can be removed from the chain of an aldose by several methods. One of these is known as the Wohl degradation and, as illustrated in the reaction sequence below, involves the formation of an oxime and dehydration of the oxime to a nitrile function. The reversibility of the addition of hydrogen cyanide to an aldehyde allows this material to be converted to an aldose which has one less carbon atom than the starting material.

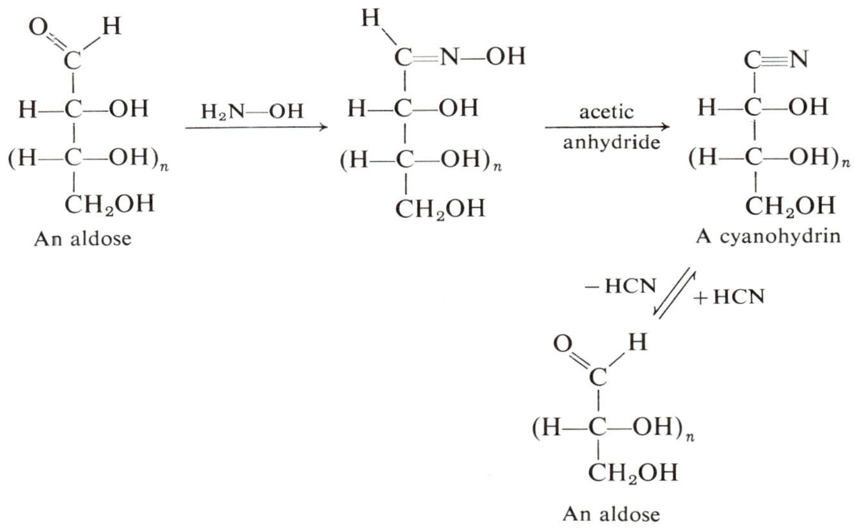

The Wohl degradation

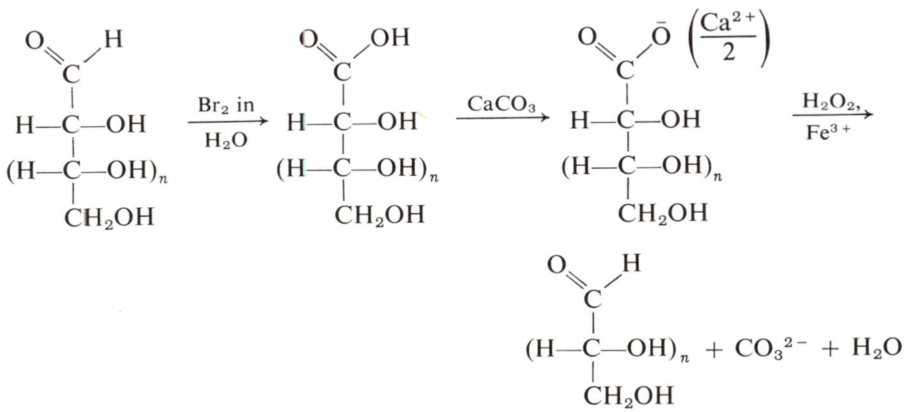

The Ruff degradation

Another method, known as the Ruff degradation, involves oxidation of the calcium salt of the acid corresponding to the starting aldose. The acid is typically prepared by the bromine–water oxidation reaction.

With these degradation reactions and other transformations of sugars it is possible to experimentally approach the question of the configuration at each asymmetric carbon atom in a sugar molecule.

Cyclic Forms of Glucose

There are a number of observations which suggest that the structure of glucose we have been writing up to this point is not totally correct. One of these is that it is possible to isolate two crystalline forms of D-glucose which have different melting points and optical rotations. The most common form, α-D-glucose, has an initial optical rotation in water of +113°, but the rotation slowly changes to a steady value of +52.5°. The other form, β-D-glucose, has an initial rotation of +19°, but this rotation also changes to +52.5° as the solution stands. These materials are probably isomeric, but it is difficult to see how the open-chain structure that has been drawn for glucose can be consistent with this. Moreover, attempts to form a dimethyl acetal of glucose by treating the compound with methanol and an acid catalyst are unsuccessful. This reaction gives two isomeric products which do not react with Fehling's or Tollen's reagents, thus suggesting that the carbonyl function is no longer present; pmr spectroscopy, however, shows that each isomer contains only *one* methoxyl group (CH_3O—) rather than two, as would be expected if an acetal had been formed.

The name glycoside is given to the products of reactions between alcohols (or phenols) and sugars. It has been shown that the two glycosides formed from glucose and methanol have cyclic, cyclohexane-like structures rather than an open-chain form. These structures are shown below. One should note that the

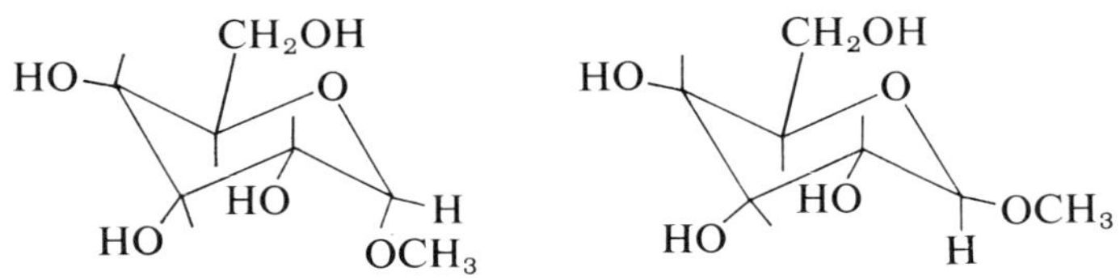

Methyl α-D-glycoside   Methyl β-D-glycoside

rings are oxygen-containing heterocyclic rings and they are, in fact, acetals. One alcohol group linked to what was formerly the aldehyde carbon atom of glucose comes from the glucose molecule itself, while the second alcoholic group derives from the reactant alcohol. The formation of glycosides introduces a new asymmetric center into the glucose molecule, and it is the stereochemistry at this

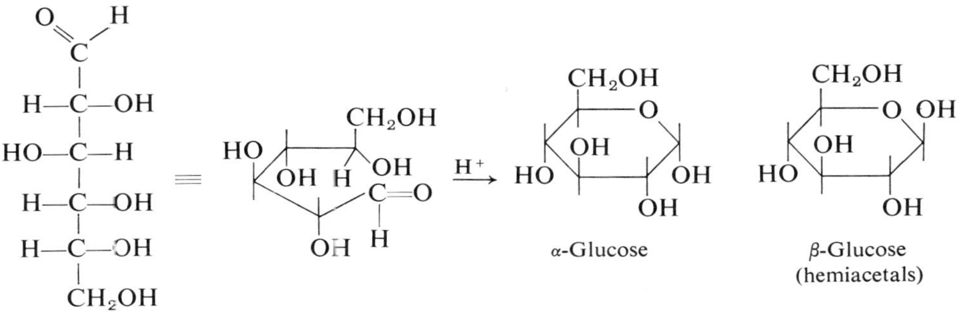

Open-chain representation of glucose

α-Glucose

β-Glucose (hemiacetals)

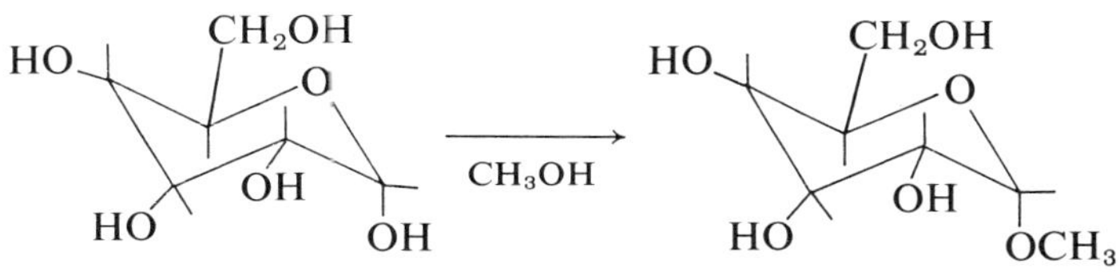

α-Glucose

Methyl α-D-glycoside (acetal) (The β isomer is also formed)

carbon atom (formerly the aldehyde carbon) that differentiates the α and β isomers of the methyl D-glycosides. The new asymmetric carbon is called the anomeric carbon. The six-membered glycoside ring formed from glucose can "ring-flip" in the same manner as cyclohexane. However, the ring conformation which places all of the substituents except, perhaps, the one at the anomeric carbon, into equatorial environments is overwhelmingly preferred. Regarding the glucose ring as fixed in this conformation, we can see that the isomeric methyl glycosides differ only in whether or not the methoxyl group is axial or equatorial at the anomeric carbon. The methyl glycosides, like other acetals, are stable in basic or neutral solutions but undergo hydrolysis in dilute acid.

These observations regarding the glycosides of glucose suggest the reason for the apparent anomalies that were mentioned above. Glucose exists as a cyclic hemiacetal and the two isomeric forms observed differ only in the arrangement of the hydroxyl group at the anomeric carbon. The changing optical rotation that is observed when either of the pure isomeric forms is dissolved in water indicates that the two forms are in equilibrium. The rate of change of the rotation that is observed experimentally must be a reflection of the rate at which equilibrium between the α and β forms is established and the rotation of $+52.5°$ that is observed when equilibrium is achieved must be due to a mixture of the α and β isomers of the sugar. This process of equilibration is called *mutarotation*; it involves breaking and reforming the hemiacetal linkage. The open-chain form of glucose that we have been drawing until this time is present in only small amounts in solution. However, because these equilibria are readily reversible in acid solution, glucose gives many of the reactions to be expected of an aldehyde. Like D-glucose, other monosaccharides can exist as anomeric isomers (or anomers) that are in equilibrium with each other via corresponding open-chain forms.

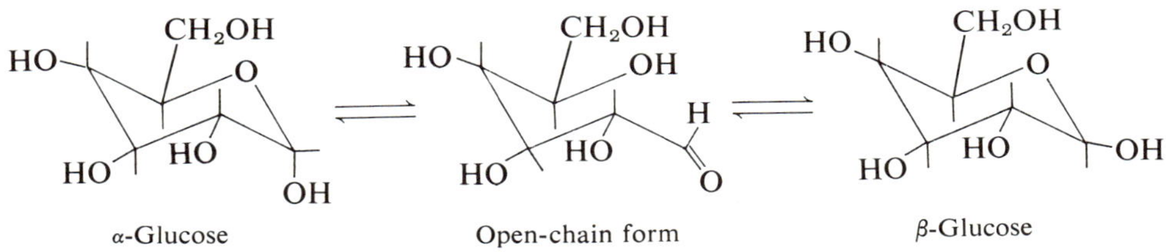

Mutarotation of glucose

Experiments have shown that most monosaccharides exist as either a five-membered or six-membered hemiacetal ring. D-Ribose, for example, is found in cyclopentane-like ring structure as is the ketose, fructose. The latter molecule

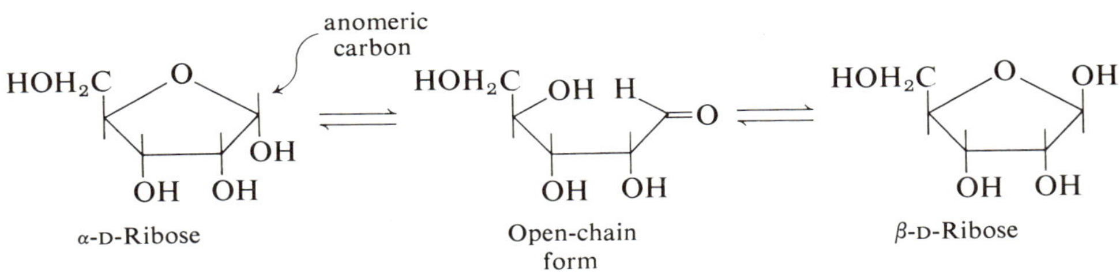

Mutarotation in ribose

also is in equilibrium with a form which is characterized by a six-membered hemiacetal ring (Fig. 10.4). An early and influential worker in the field of carbohydrate chemistry, W. N. Haworth, suggested that the size of the hemiacetal ring formed by a monosaccharide be designated with reference to the two oxygen-containing heterocycles, furan and pyran. The sugars with six-membered rings are, therefore, said to be *pyranoses* while those with five-membered rings are *furanoses*. These names can be used, when necessary, to indicate that a cyclic

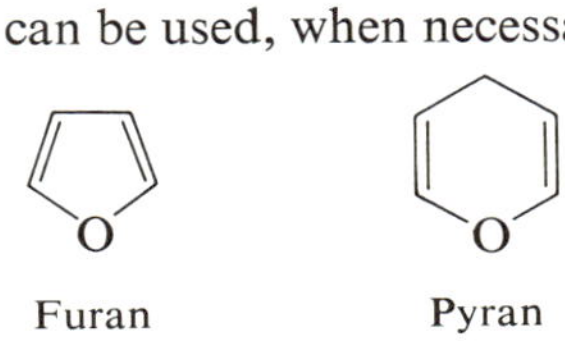

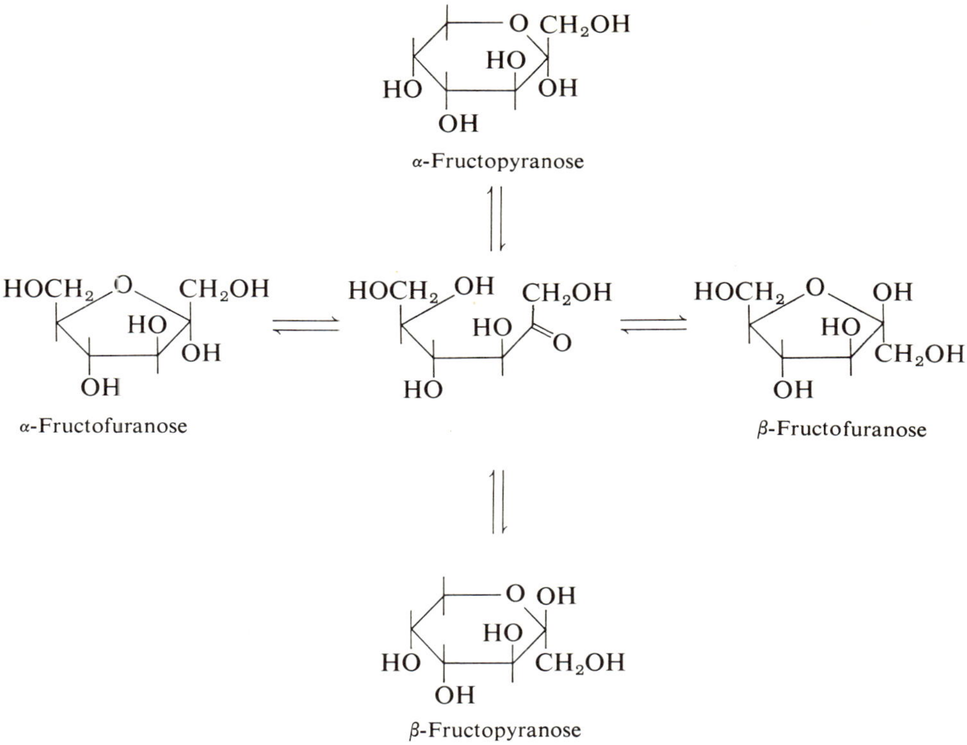

Fig. 10.4. Anomeric equilibria in D-fructose.

system is present. Thus, α-D-glucose in the cyclic form could be called α-D-glucopyranose. Similarly, α-D-ribose would be referred to as α-D-ribofuranose.

The discovery that ascorbic acid (vitamin C) could prevent the dreaded disease scurvy was made more than 200 years ago. This vitamin has received new prominence recently as the result of suggestions that this substance, in large doses, can prevent colds in human beings and provide relief from some of

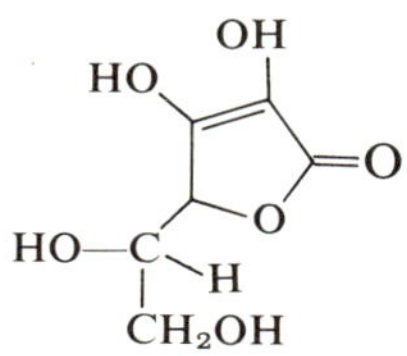

Ascorbic acid

the symptoms of arthritis. Although the subject is highly controversial, many continue to take the drug, no doubt chagrined that medical science is capable of elaborate organ transplantation but still has not found a cure for the common cold. Ascorbic acid resembles a carbohydrate but is a carboxylic acid derivative rather than an aldehyde or ketone. The heterocyclic ring in the molecule is a type of ester known as a lactone rather than an acetal or ketal. Ascorbic acid is prepared commercially from D-glucose.

Other Monosaccharide Derivatives

It is possible to replace one or more of the hydroxyl groups in a sugar molecule by an amino group. When replacement of the hydroxyl group at the anomeric carbon takes place, the resulting compound is called a glycosylamine. Examples of this type of replacement are found in the nucleotides which are used as monomeric units in the formation of nucleic acids; these compounds and their reactions will be taken up in Chapter 14. Replacement of hydroxyl groups at other positions generates a class of compounds known as amino

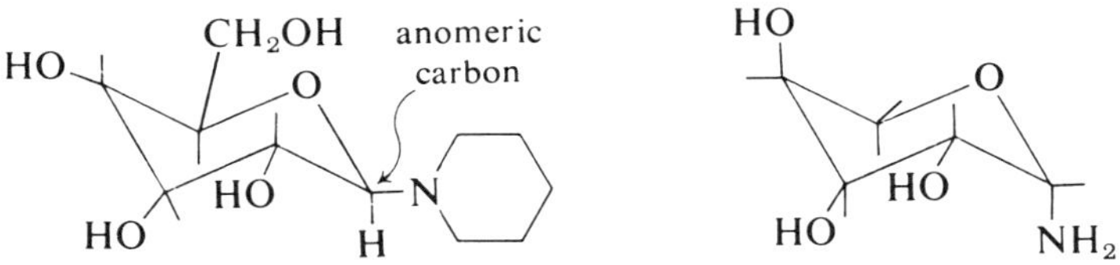

β-*N*-(D-Glucosyl)piperidine α-D-Arabinosylamine

Glycosylamines

sugars. Amino sugars occur widely in nature, especially in the shells of crustacea and in antibiotics such as streptomycin and neomycin.

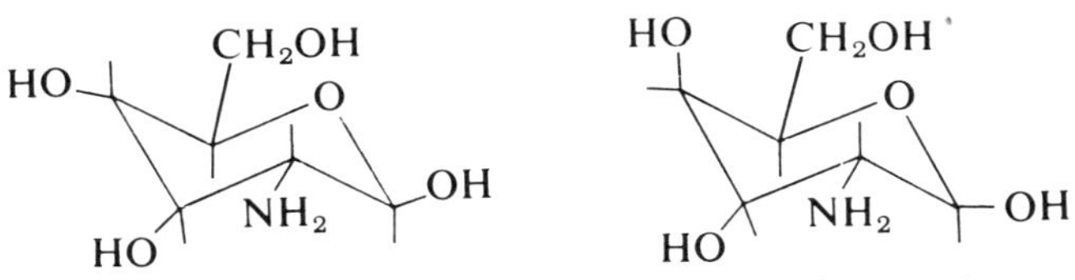

β-D-Glucosamine β-D-Galactosamine

Amino sugars

DISACCHARIDES

It has already been noted that glycosides can be formed when monosaccharides are treated with an alcohol in the presence of an acid catalyst. The alkoxyl group from the alcohol is placed in one of two orientations at the anomeric carbon so that each glycosidic bond can potentially give rise to two stereoisomers. In the disaccharides, two monosaccharides are joined by a glycosidic bond; a hydroxyl group of one sugar molecule becomes attached to the anomeric carbon of the second sugar molecule. Maltose, obtained by the enzymic degradation of starch, consists of two α-D-glucose molecules joined by a glycosidic bond between the hydroxyl at carbon atom 4 on one glucose and the anomeric carbon on the second glucose. Maltose can thus be regarded as a dimer of α-D-glucose.

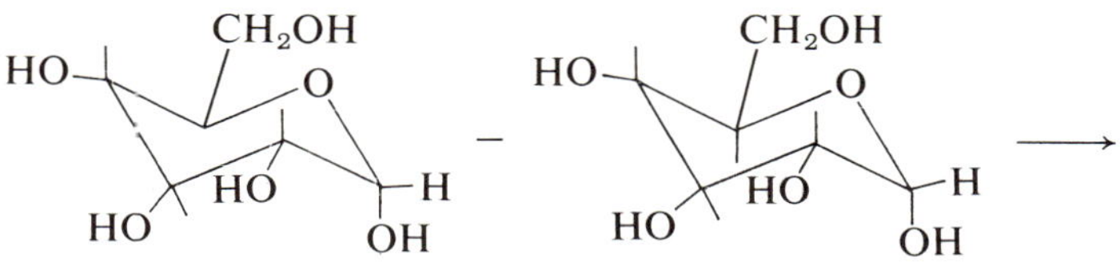

α-D-Glucose α-D-Glucose

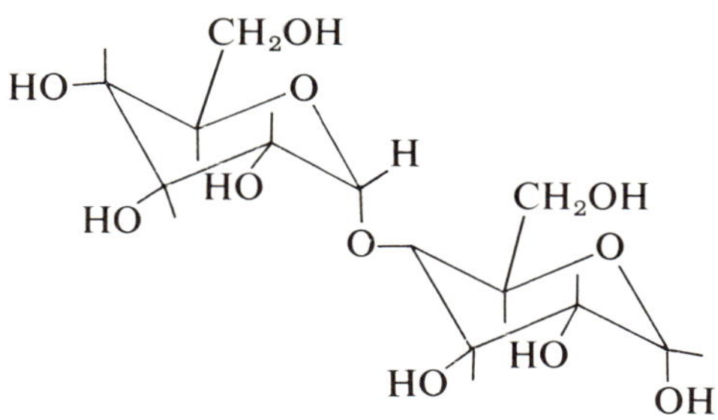

Maltose
(a disaccharide)

Two disaccharides occur abundantly in nature. These are sucrose, known most commonly as "sugar," and lactose, which is found in mammalian milk at levels of 4–8%. Like other acetals and glycosides, these disaccharides can be hydrolyzed to their component parts by dilute acids. Hydrolysis of sucrose thus shows that this molecule is made up of D-glucose and D-fructose, while lactose is broken down into D-glucose and D-galactose. The glycosidic bond in sucrose links the two anomeric carbons. In lactose the glycosidic linkage is between the hydroxyl on carbon–4 of glucose and the anomeric hydroxyl (in the equatorial configuration) of the galactose ring. Sucrose does not react with Tollen's or Fehling's reagent and is, therefore, a nonreducing sugar. There will be none

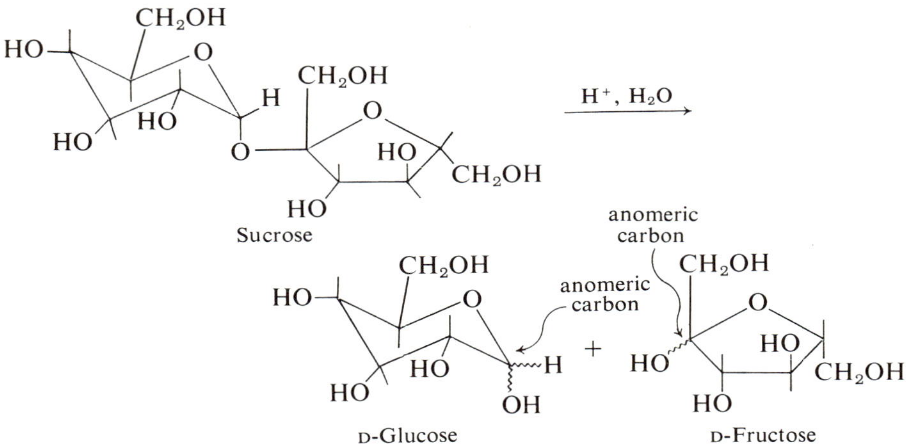

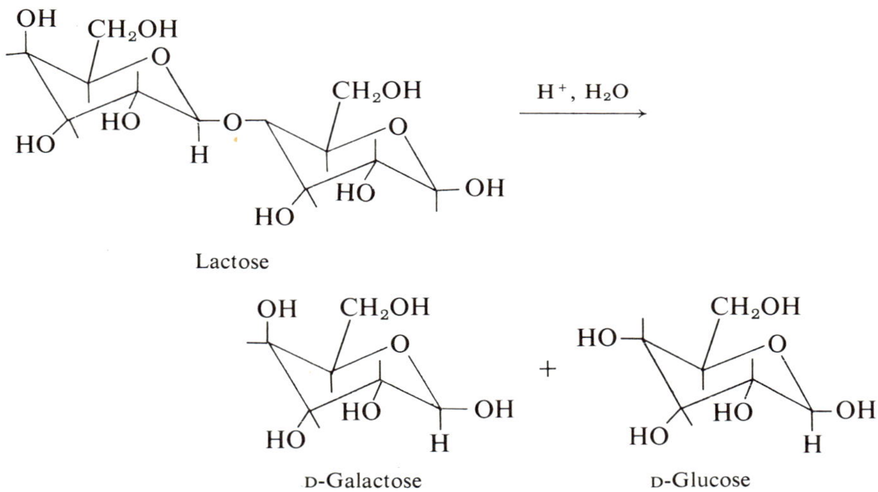

of the open-chain aldehyde or ketone forms of the two component sugars in sucrose available for reaction with these reagents since both potential carbonyl functions are tied together by the glycosidic bond. Equilibration between open-chain and cyclic forms is still possible in lactose and maltose where a D-glucose ring in each structure can open and reform. Both maltose and lactose are reducing sugars and will react with Tollen's or Fehling's reagent in the same way as the simple monosaccharides.

Most disaccharides are crystalline, water-soluble, sweet-tasting materials. Sucrose has been an important foodstuff for centuries and is now obtained commercially from sugar cane and sugar beets. The juice pressed from the crushed canes contains about 14% sucrose, while careful breeding has increased the sucrose content of the beet to about 20%. Large amounts of sucrose are used in the manufacture of candy, much to the joy of dentists. Special techniques are used to prevent the crystallization of sucrose in candies by inducing aqueous solutions of the disaccharide to become supersaturated or by adding ingredients such as gelatin or butter to inhibit the rate of crystallization.

There are several trisaccharides and tetrasaccharides found in nature. These molecules are formed by joining together monosaccharides via glycosidic linkages. Raffinose is the most abundant trisaccharide found in nature; it consists of D-glucose, D-galactose, and D-fructose linked in the manner indicated below. Raffinose is formed in sugar beets and is also a component of the exudates formed by some trees, especially eucalyptus.

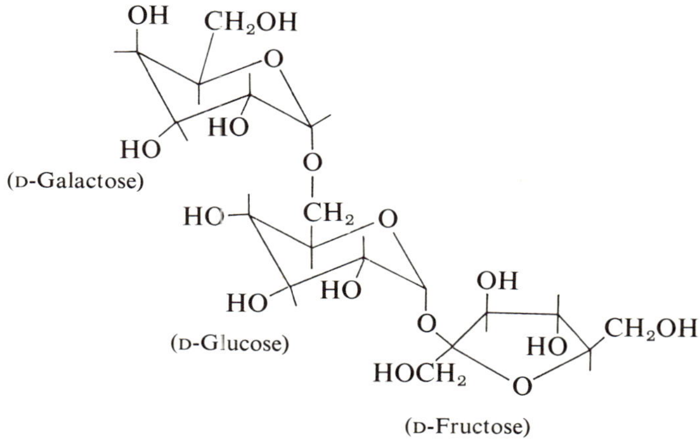

Raffinose

POLYSACCHARIDES

The idea that a large number of small molecules can be linked together to give high-molecular weight macromolecules was introduced in Chapter 3. Polysaccharides are macromolecules or polymers which are constructed from monosaccharides; glycosidic bonds hold the monomeric units together in the same way that the components of the di-, tri-, and tetrasaccharides described above are conjoined. Some polysaccharides are linear molecules with all of the monomer units bound in one continuous, unbranched chain. In others, a highly branched, cross-linked structure is evident. In both cases the monomeric units are linked together by the glycosidic functional group.

Celluose

Cellulose is the most widely distributed polysaccharide. Cotton is almost pure cellulose and the principal structural constituent of plants is cellulose; the paper this page is printed on was derived from the cellulose of trees. Cellulose is a homopolysaccharide in that all monomeric units are the same, in this case, D-glucose. Depending on the source, 300 to 10,000 glucose units may be joined with the bonds between units being formed at the anomeric carbon and carbon-4. The anomeric carbon has the same configuration (equatorial) as found in

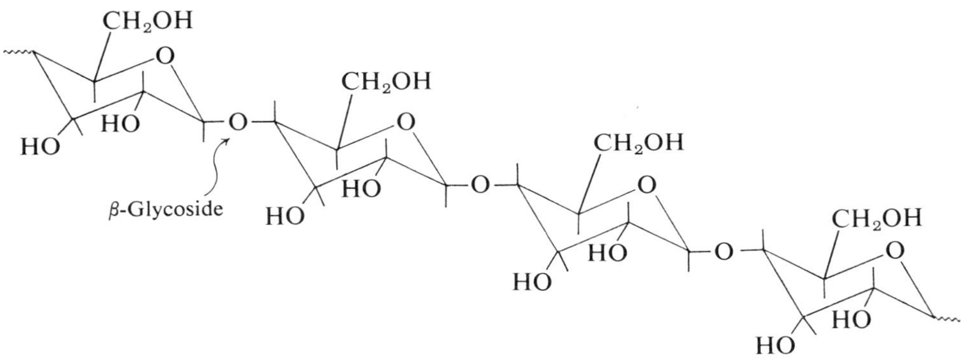

A portion of the cellulose structure

β-D-glucose. The arrangement of cellulose chains into fibers is facilitated by the presence of the hydroxyl groups (each in an equatorial environment) on the glucose units of the polymer so that interchain hydrogen-bond formation can take place. This effect, coupled with the highly regular nature of the cellulose chain, makes cellulose fiber a highly crystalline material and accounts for the great strength of these fibers.

Man and other carnivorous animals are not able to metabolize cellulose and it is useless as a foodstuff for these species. Some animals (ruminants) are able to digest cellulose because of the colonies of microorganisms that dwell in their intestines. Many bacteria and fungi are capable of degrading cellulose to glucose and hence are able to penetrate the membranes of plant cells (largely composed of cellulose) in order to infect them.

Cellulose can be modified by appropriate reactions to give new polymers with properties different from those of the natural cellulose. In a process discovered in 1850 and named after its inventor, cellulose from cotton is dissolved in sodium hydroxide solution for a few moments and is then washed free of the base and dried. The fibers swell, probably as a result of rearrangement of the hydrogen bonds between polymer chains, and are found to have greater tensile strength and capacity for the absorption of dyes; this is "mercerized" cotton.

Nitration of cellulose with a mixture of nitric and sulfuric acids leads to a polymer with many of the hydroxyl groups now converted to nitrate ester

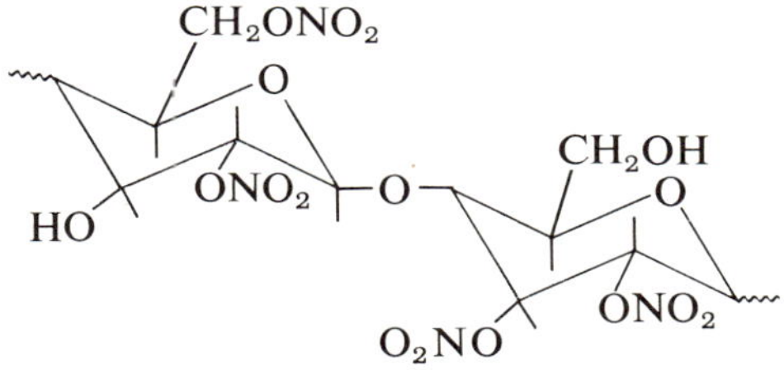

A section of nitrated cellulose

functions. If a large number of the hydroxyl groups are so esterified, an explosive material, nitrocellulose or "gun cotton" is the result. Less complete conversion of the hydroxyl groups on cellulose to nitrate esters (about 2 per D-glucose unit) gives a more stable material (pyroxlin or cellulose nitrate) which is soluble in organic solvents and is compounded into some plastics (celluloid).

A more common ester of cellulose is the one obtained by treating the polymer with acetic anhydride in the presence of sulfuric acid; the product is cellulose

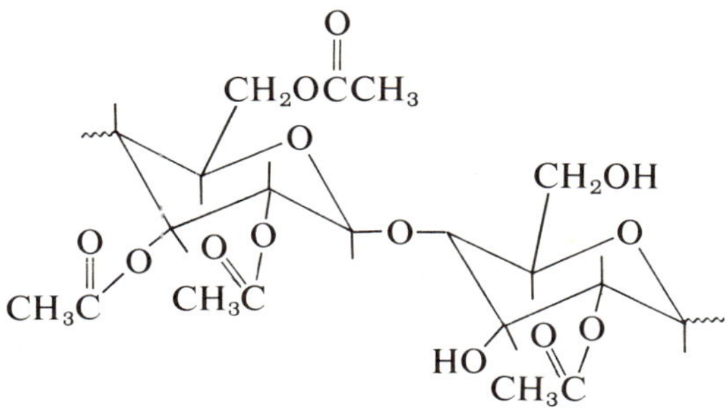

A section of cellulose acetate

acetate. Fibers formed from this and related cellulose esters are known as "acetate" or Arnel.

Starch

Starch, like cellulose, is a poly-D-glucoside but the glycosidic group at each anomeric carbon is in the axial position (α-glycosides). Starch is the carbohydrate reserve material of plants and is stored in roots, tubers, and seeds. It can be fractionated into two main components, amylose and amylopectin. Amylose is a homopolysaccharide formed from D-glucose and is essentially a linear molecule. The structure of amylose should be compared to that of cellulose. Molecular weight determinations of amylose give values between 10,000

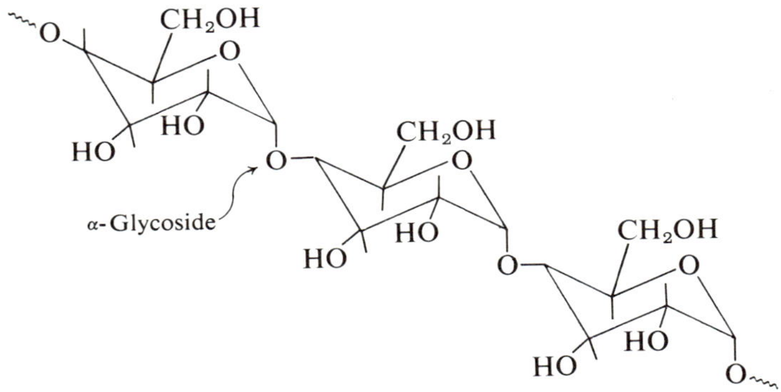

A section of amylose

and 50,000 depending upon the source of the material and the techniques used in its isolation. The α-glycosidic linkages do not allow the amylose polymers to assume a completely linear conformation so that strong, intermolecularly hydrogen-bonded fibers are not formed by this material. Instead amylose tends to wind into intramolecularly hydrogen-bonded helical structures. Amylose gives a characteristic blue color with iodine; the iodine molecules apparently fit inside the "tube" defined by the helical polymer and the polymer sufficiently perturbs the electronic structure of the iodine molecule so that its usual faint red color is replaced by the intense blue of the complex.

Physical methods indicate that the molecular weight of amylopectin is considerably higher than that of amylose and values up to 1 million have been reported. Amylopectin is much less soluble in water than is starch, but acid hydrolysis experiments indicate that this molecule is still composed only of glucose. The difference is the presence of cross-links in the polymer structure; in addition to the 1,4-α-glycoside bonds between D-glucose units, 1,6-α-glycoside bonds which serve to connect different polymer chains are found. The result is a rather tightly knit structure that is more stable than the simpler linear polymers of amylose. Amylopectin will not form fibers and is more resistant to enzymic or acid-catalyzed hydrolysis than is amylose.

Glycogen is the carbohydrate reserve material in animals found primarily in the muscles and the liver. Glycogen is structurally similar to amylopectin with D-glucose units joined by α-glycosidic bonds and a considerable amount of cross-linking. While the branched structure is similar to amylopectin, the polymer chains tend to be shorter.

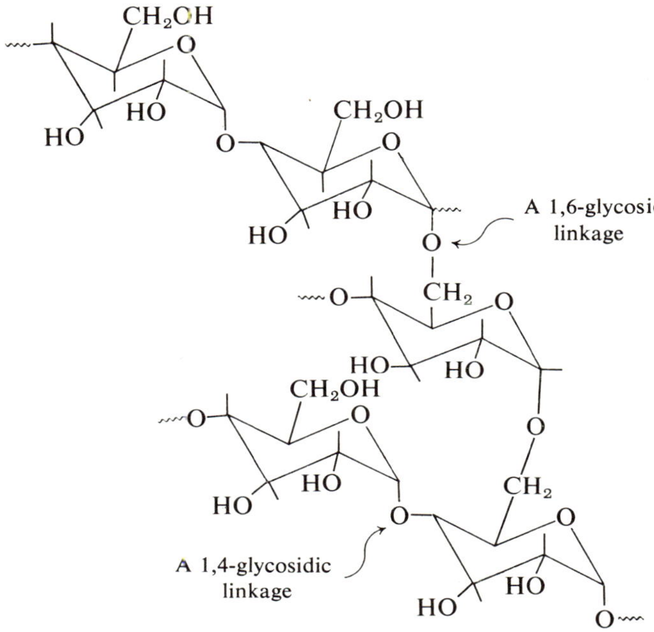

A section of amylopectin

Other Polysaccharides

A polysaccharide similar to cellulose but in which the monomeric unit is *N*-acetyl-D-glucosamine is the principal component of the shells of arthropods. This polymer is called chitin and can be obtained from the shells of these organisms by soaking them in dilute hydrochloric acid until the calcium carbonate matrix is dissolved. Chitin remains as a threadlike substance; x-ray structural work indicates that these fibers are very much like fibers of cellulose. Other polysaccharides that have amino sugars in their structure include heparin, a polymer of D-glucosamine and D-glucuronic acid that can prolong the clotting time of blood, and a group of complex saccharides known as the chondroitin sulfates. These latter materials contain 2-amino-2-deoxygalactose and D-glucuronic acid as well as the elements of sulfuric and acetic acid; they are found in skeletal tissues, especially cartilage, in nasal mucilage and in the umbilical cord.

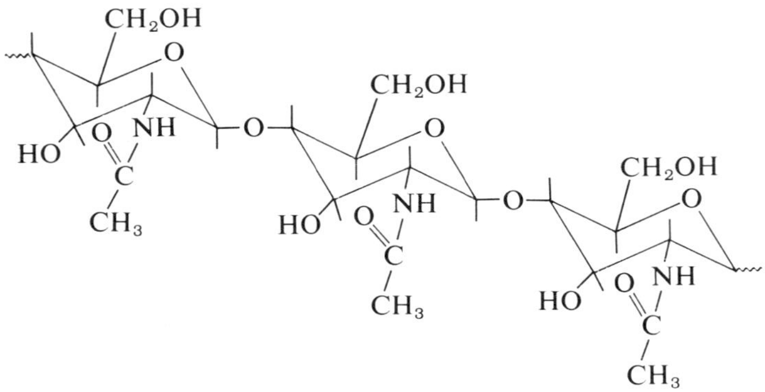

A section of chitin

NATURALLY OCCURRING GLYCOSIDES

In addition to the oligomeric and polymeric forms of sugar molecules that have been described and the nucleic acids, there are a large number of compounds found in nature which contain sugar molecules linked to other, nonsaccharide materials by glycosidic bonds. These substances are especially prevalent in plants and are generically referred to simply as "glycosides." The noncarbohydrate portion of the glycoside is called the aglycon, or aglucon if the sugar is glucose. The majority of the natural glycosides are β-glucopyranoses; some of these are depicted below. These materials are all acetals and can, therefore, be degraded to the sugar and aglycon components by acid hydrolysis.

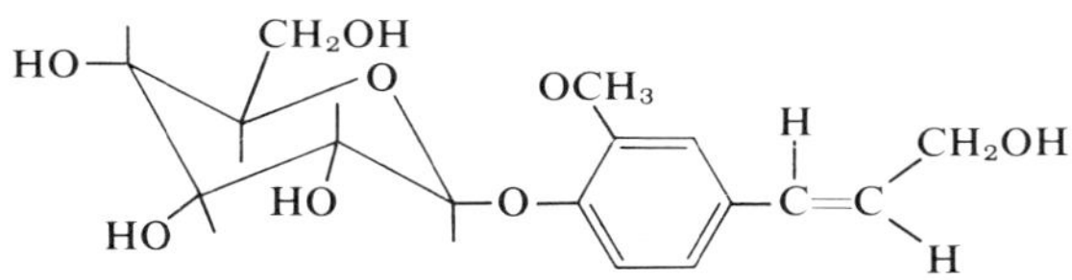

Coniferin
(pine trees)

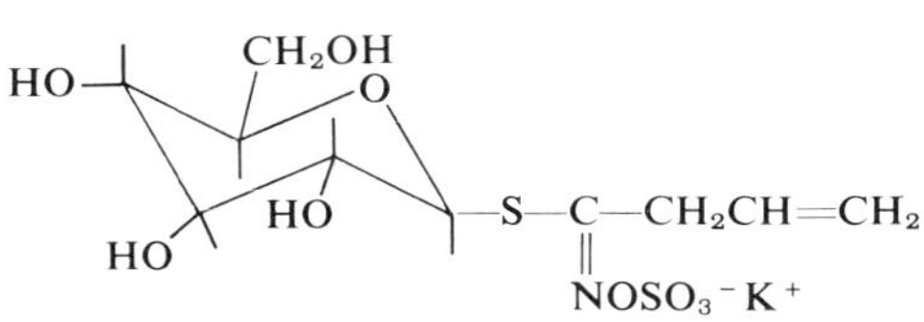

Sinigrin
(horseradish)

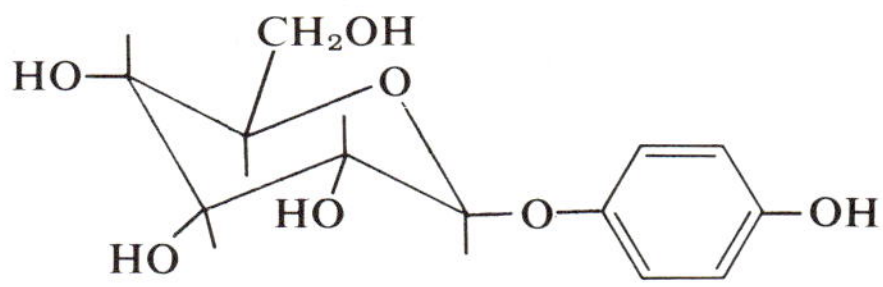

Arbutin
(blueberries)

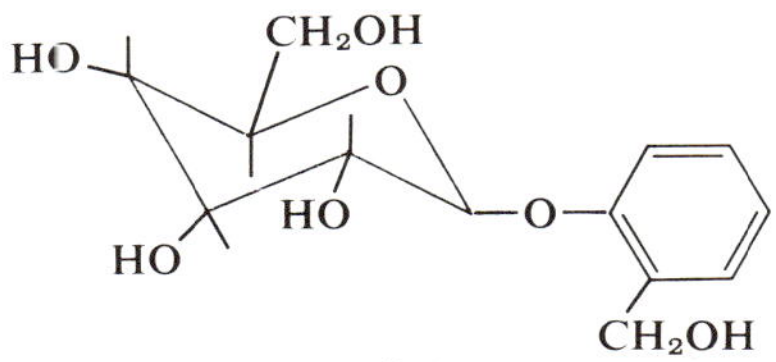

Salicin

PROBLEMS

1. Build molecular models of the two enantiomeric forms of the following molecules.
   (a) 2,2-Dimethylcyclohexanol
   (b) D-Glyceraldehyde
   (c) *meso*-Tartaric acid
   (d) 2-Chlorobutane
   (e) D-Ribose (D-ribofuranose)
   (f) α-D-Glucose (α-D-glucopyranose)
2. Ascorbic acid is a stronger acid than acetic acid (pK = 4.2 vs 4.7). Why?
3. Give the products of the following reactions.
   (a) α-D-Glucose + ethanol $\xrightarrow{H^+}$
   (b) D-Galactose + methyl iodide $\xrightarrow{Ag_2O}$
   (c) D-Ribose + phenylhydrazine $\xrightarrow{H^+}$
   (d) β-D-Glucosamine $\xrightarrow{H^+, H_2O}$
   (e) Maltose + $Ag(NH_3)_2^+$ $\longrightarrow$
   (f) Raffinose $\xrightarrow{H^+, H_2O}$
   (g) D-Ribose + acetone $\xrightarrow{H^+}$
4. Give a chemical test that could be used to distinguish between the molecules in the following sets:
   (a) D-Glucose and sorbitol
   (b) D-Glucose and D-fructose
   (c) D-Glucose and maltose
   (d) Starch and cellulose
   (e) D-Glucosamine and D-glucose

5. Which of the following molecules can exist in optically active forms?
   (a) Furan
   (b) 3-Bromooctane
   (c) Cyclohexanol
   (d) 1,2-Dimethylcyclobutane
   (e) 1,2,2-Trimethylcyclobutane
   (f) *trans*-Decalin
   (g) 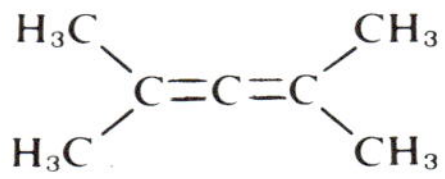
   (h) 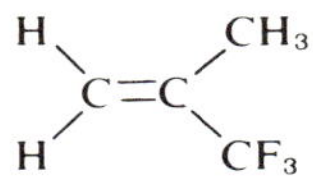

6. From the optical rotation data in the text, calculate how much of the two stereoisomers of glucose are present after the mutarotation process has reached equilibrium.

SUGGESTED READINGS

R. S. Cahn, C. K. Ingold, and V. Prelog, *Angew. Chem., Int. Ed.* **5**, 385 (1966).

E. A. Davidson, "Carbohydrate Chemistry." Holt, New York, 1967.

L. F. Fieser and Fieser, "Advanced Organic Chemistry." Van Nostrand-Reinhold, Princeton, New Jersey, 1961.

R. D. Guthrie and J. Honeyman, "An Introduction to the Chemistry of Carbohydrates." Clarendon, Oxford, 1968.

K. Mislow, "Introduction to Stereochemistry." Benjamin, New York, 1965.

See also the serial publication *Advances in Carbohydrate Chemistry*, Academic Press, New York.

11

Carboxylic Acids and Their Derivatives

CARBOXYLIC ACIDS

The layman's concept of an acid is usually a highly imaginative one. Largely as a result of old movies, he may associate the word with bubbling vats of foul-smelling liquids that are used by nefarious persons for ill-gotten gain. Organic acids are rarely this spectacular, although they can be as strong as sulfuric acid or so weak as to be hardly acidic at all. Here the words "strong" and "weak" refer to the relative ability of the acid to donate a proton to some common reference base. Carboxylic acids contain the carboxyl group (—COOH) in their structure. Upon loss of a proton, a *carboxylate anion* is formed and, since this ion can be stabilized by electron delocalization into both oxygen atoms as the resonance structures shown indicate, proton loss from a carboxyl group is more favorable than the corresponding ionization of a typical alcohol.

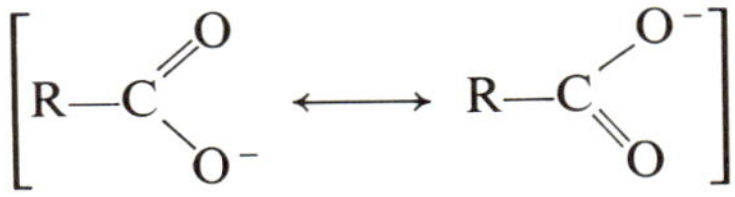

Resonance structures for the carboxylate anion

Table 11.1
Properties of Some Carboxylic Acids

Acid	*Structure*	*Mp* (°C)	*Bp* (°C)	pK_a (25°)
Formic	HCOOH	8.4	101	3.77
Acetic	CH_3COOH	16.6	118	4.76
Fluoroacetic	FCH_2COOH	33	165	2.66
Glycolic	$HOCH_2COOH$	79		3.83
Propionic	CH_3CH_2COOH	−22	141	4.88
n-Butyric	$CH_3CH_2CH_2COOH$	−5	162	4.82
n-Valeric	$CH_3(CH_2)_3COOH$	−35	187	4.81
Lactic	$CH_3CH(OH)COOH$	18	—	3.87
Malonic	$HOOCCH_2COOH$	135	—	2.80
Succinic	$HOOC(CH_2)_2COOH$	187	235	4.17

Table 11.1 lists a number of carboxylic acids in which the organic group (R) is aliphatic in nature. Although the low molecular weight acids are liquids, aliphatic carboxylic acids are waxy solids when the R group becomes sufficiently large. In spite of the presence of the polar carboxyl group, the carboxylic acids are essentially insoluble in water if the side chain contains more than six carbon atoms.

Nomenclature

In the systematic nomenclature of the carboxylic acids, the carbon atom of the carboxyl group is taken as the first carbon atom of the backbone chain. In naming a carboxylic acid, the name for the hydrocarbon corresponding to the backbone chain is first obtained and then the terminal *-e* of this name is replaced by *-oic* and the word *acid* is added. Examples of this procedure are found below.

Carboxylic acid	*Corresponding hydrocarbon*
$CH_3C(=O)OH$ Ethanoic acid (acetic acid)	CH_3CH_3 Ethane
$CH_3CH_2CH_2C(=O)OH$ Butanoic acid (butyric acid)	$CH_3CH_2CH_2CH_3$ Butane
$CH_3(CH_2)_{10}COOH$ Dodecanoic acid (lauric acid)	$CH_3(CH_2)_{10}CH_3$ Dodecane

A number of common or trivial names for carboxylic acids are accepted by the IUPAC; these are indicated in Table 11.1 and in parentheses with the structures below. The use of these latter names is preferred by the IUPAC for the simple acids. The positions of substituents on the back bone chain are indicated by the number of the carbon atom to which they are attached; the carboxyl carbon is taken as carbon-1 in the chain.

In trivial names there is another method used to indicate the position of the substituents. In this system, the carbon atom directly attached to the carboxyl group is designated the α-carbon and subsequent carbons are denoted by the next consecutive letter of the Greek alphabet. Applications of this nomenclature system are illustrated below. While the "α–β–γ" nomenclature has little to recommend it, it is in common use. Some examples of these nomenclatures for substituted carboxylic acids are presented below.

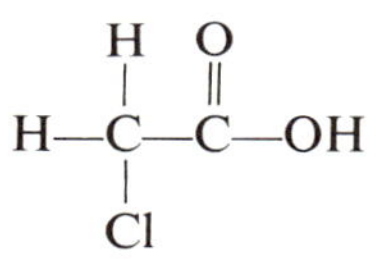

2-Chloroethanoic acid (IUPAC)
Chloroacetic acid (IUPAC)
α-chloroacetic acid (common)

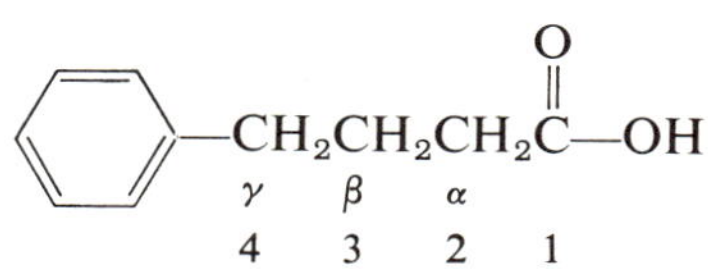

Carbon atom designations

4-Phenylbutanoic acid (IUPAC)
γ-phenylbutyric acid (common)

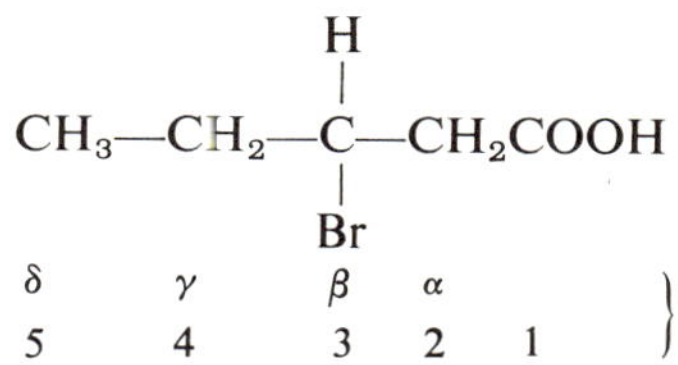

Carbon atom designations

3-Bromopentanoic acid (IUPAC)
3-Bromovaleric acid (IUPAC)
β-bromovaleric acid (common)

The carboxyl group may also be directly attached to an aromatic ring as it is, for example, in benzoic acid or 1-naphthoic acid. When substituents are present on the aromatic ring, the compound is named as a derivative of the corresponding unsubstituted carboxylic acid.

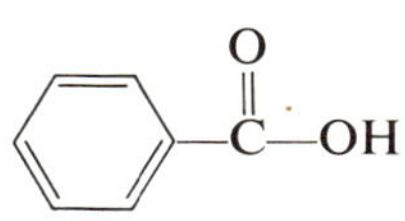

Benzoic acid

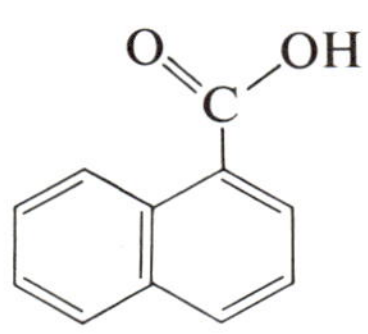

1-Naphthoic acid
(α-naphthoic acid)

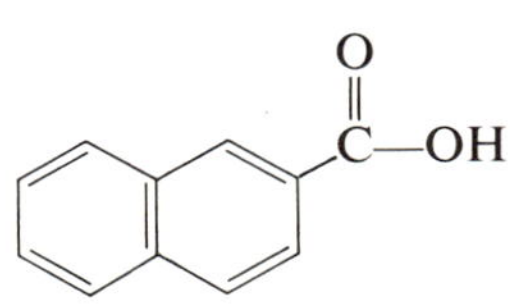

2-Naphthoic acid
(β-naphthoic acid)

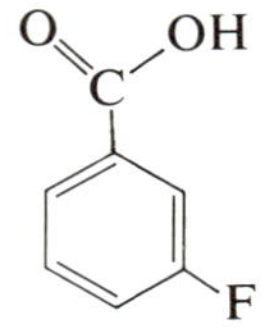

3-Fluorobenzoic acid
or
m-fluorobenzoic acid

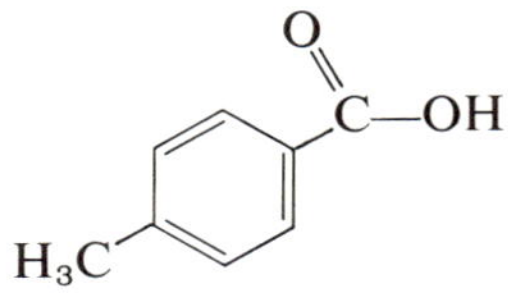

4-Methylbenzoic acid
p-toluic acid (common name)

Substituent Effects on Acidity

When seeking a quantitative definition of acidity, one usually refers to the ability of a molecule to donate a proton to a particular base, usually a water molecule. The relative concentrations of hydronium ion and carboxylic acid in the equilibrium shown below are thus a measure of the acidity of the acid.

$$\mathrm{R{-}\overset{\overset{\displaystyle O}{\|}}{C}{-}OH + H_2O \rightleftharpoons R{-}\overset{\overset{\displaystyle O}{\|}}{C}{-}O^- + H_3O^+}$$

As has been discussed in Chapter 1, the equilibrium constant for this system is given by

$$K_a = \frac{[H_3O^+][RCOO^-]}{[RCOOH]}$$

If the acid is very strong, then a large K_a is expected, while weak acids will have a corresponding small value for the equilibrium constant.

The K_a of a carboxylic acid can be very sensitive to the nature of the group attached to the carboxyl function. This group can affect the position of the acid–base equilibrium depicted above by acting to change the stability of the carboxylic acid or by influencing the stability of the anion that is obtained from the acid upon ionization. A given acid can be made stronger relative to some standard acid if a substituent on the acid acts to decrease the free energy difference between the carboxylic acid and its corresponding carboxylate anion.

Inductive Effects. This effect is illustrated in Fig. 11.1. Here we see that introduction of a chlorine atom at the α-carbon of acetic acid decreases the free-energy difference between the carboxylate (chloroacetate) and the un-ionized forms of the acid. The chlorine atom is powerfully electronegative so that the electronic charge of the carboxylate anion can be dispersed more completely in this ion than in the corresponding acetate ion, with the result that the total free

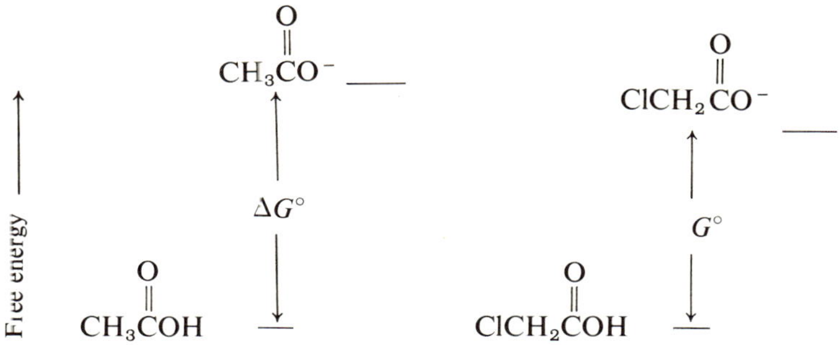

Fig. 11.1. A schematic representation of the effect of chlorine substitution on the acidity of acetic acid. $\Delta G°$ for α-chloroacetic acid is smaller than $\Delta G°$ for acetic acid.

energy of this species is lowered. The acid ionization equilibrium in chloroacetic acid therefore lies more to the right and chloroacetic acid appears as a much stronger acid. The electronic effect of the chlorine atom in the case above can be thought of as being transmitted through the σ-bond framework of the molecule. When a substituent modifies or stabilizes the electron distribution in a molecule by acting in this way, it is said to exert an *inductive effect.* The inductive effect of the halogen in the above example acted in a way that strengthened the acid, primarily by stabilizing the corresponding anion.

Inductive effects are cumulative. As the sequence of compounds below illustrates, when the α-hydrogen atoms of acetic acid are progressively replaced with chlorine atoms, the resulting acids become increasingly stronger. Inductive effects are rapidly attenuated by increasing distance between the substituent and the other groups involved. That is, carbon–carbon σ bonds behave as fairly

	CH_3COOH	$ClCH_2COOH$	$Cl_2CHCOOH$	Cl_3CCOOH
approximate K_a	0.00001	0.001	0.1	1

The effect of α-chlorine substitution on acidity

effective electrical insulators so that the farther an inductive effect must be transmitted, the smaller its final influence.

Resonance Effects. The carbon atom of a carboxyl group has an unhybridized $2p$ orbital perpendicular to the common plane through this atom and the two oxygen atoms of this functional group. This orbital can potentially interact with similar orbitals on the substituents attached to this carbon atom to give a resonance effect (delocalization effect) on the stability of the carboxyl group as well as the corresponding carboxylate anion. The differences in the K_a values of propionic, acrylic, and benzoic acids are consistent with this effect acting preferentially on the anion but the relatively minor changes that occur indicate that the resonance effect is not large, or is about the same for both the ionized and un-ionized forms of these acids.

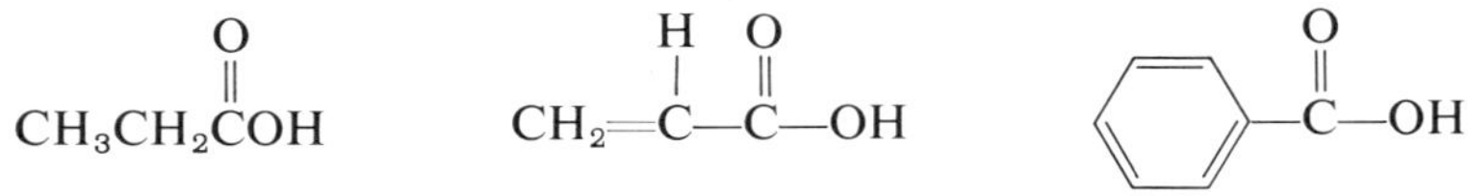

K_a 1.3 × 10^{-5} 5.5 × 10^{-5} 6.8 × 10^{-5}

Salts of Carboxylic Acids

The alkali metal salts of carboxylic acids can be obtained by treating the acid with a metal carbonate or hydroxide. The salts of the lower molecular weight acids behave as salts of inorganic acids; they are soluble in water and are dissociated extensively in aqueous solution. The alkali metal salts of aliphatic carboxylic acids with more than eight carbon atoms in the alkane portion of the

$$2\,CH_3CH_2COOH + Na_2CO_3 \longrightarrow 2\,CH_3CH_2C(=O){-}O^-Na^+ + CO_2$$

Propionic acid Sodium propionate

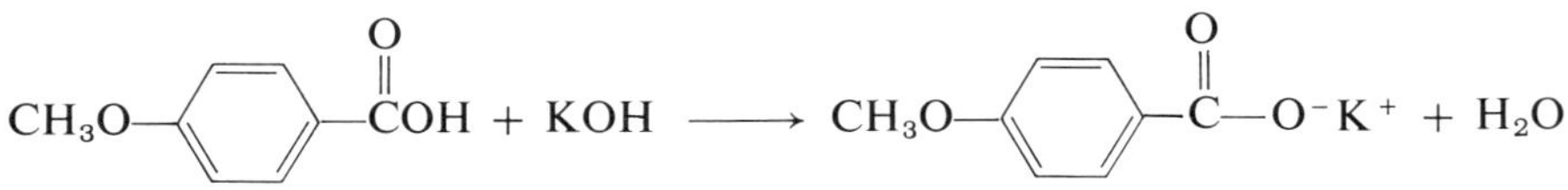

p-Methoxybenzoic acid Potassium *p*-methoxybenzoate

molecule form loose association complexes called micelles in an aqueous environment. Although these salts may be moderately soluble, their solutions have a cloudy appearance since the micellar structures scatter light.

In a gross way the carboxylate anions that take part in micelle formation can be regarded as long, nonpolar, hydrocarbon structures with a region of concentrated electrical charge at one end. The association takes place by mutual interaction of the hydrocarbon ends of the molecules. The resulting micelles can form into a spherical structure as shown in Fig. 11.2. (Cylindrical and rod-shaped micellar structures are also known.) The hydrocarbon chains are said to be hydrophobic ("water-fearing") and their free energy is less when they are self-associated than when the hydrocarbon parts of the carboxylates wander freely in solution. Micelle formation is thus another application of the adage, "Like dissolves like."

The spherical micelle structure consists of a nonpolar hydrocarbon-like core surrounded by a layer of high charge density, by virtue of the ionized carboxylate

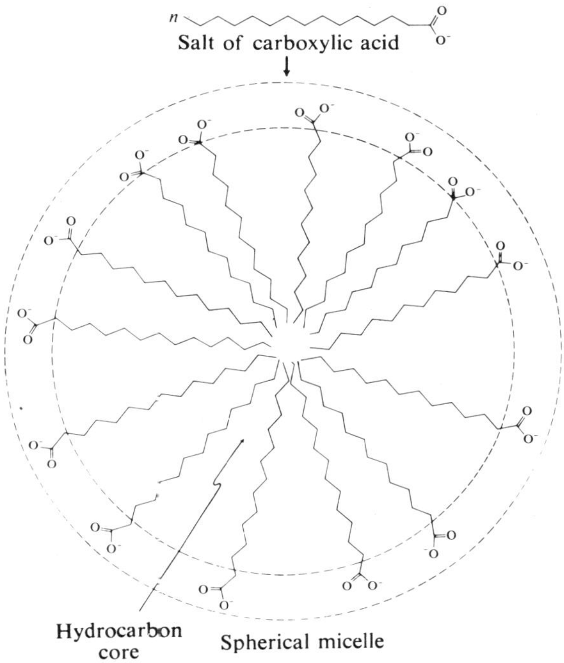

Fig. 11.2. Micelle formation in aqueous solutions of long-chain carboxylic acids.

groups on the surface of the structure. Each of these charges will have with it an associated counterion so that the ionic "coating" on the micelle is a highly polarized one.

As a result of these two very distinctive features—a nonpolar, hydrocarbon interior and a highly polarized shell—micelles often have interesting catalytic properties. Rate increases up to several thousandfold have been observed for organic reactions within a micelle relative to the corresponding rate in pure water.

The ability of long-chain carboxylate salts to solubilize organic molecules by micelle action makes them useful as cleaning agents. Soaps and detergents are basically long aliphatic chains of carbon atoms with a charged group at one end. These materials incorporate "dirt" (usually fats and oils from animal and vegetable sources) into micellar structures, thus making it soluble in water.

Naturally Occurring Carboxylic Acids

Carboxylic acids are widely distributed in nature and several examples are described here. Lactic acid is an acidic component of sour milk. Note that it is an asymmetric molecule; the naturally occurring isomer has the L configuration. Lactic acid was isolated from muscle tissue in 1807; it is one of the end products

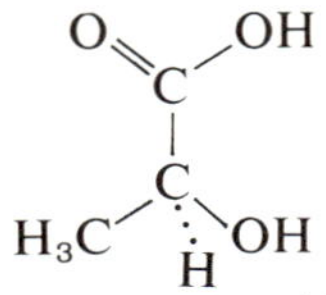

L-(+)-Lactic acid

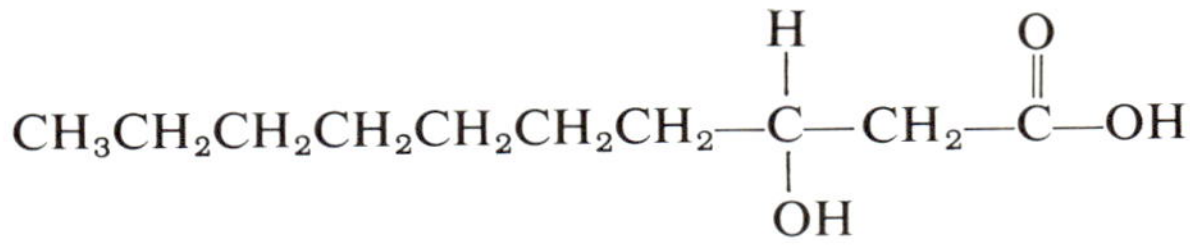

Myrmicine
(β-hydroxydecanoic acid)

in the series of biochemical reactions that converts the chemical energy of carbohydrates to the energy of muscular work. Lactic acid accumulates during strenuous muscular activity, especially if a person is "out of shape" and is responsible for the painful reminders of unaccustomed exertion the day after it occurs.

Myrmicine is a herbicide secreted by *Atta sexdens*, a leaf-cutter ant from South America. It was isolated and characterized in 1970 and was the first example of a herbicide produced by an insect. It selectively inhibits the growth of bacteria and some fungi, but does not disturb the fungi upon which the ant feeds. This compound appears to have some promise as a "natural" food preservative.

Oxalic acid is the simplest dicarboxylic acid. It is found in the stems of several plants and, as its monopotassium salt, is responsible for the tart taste of rhubarb. It also occurs in spinach. Glutamic acid is another dicarboxylic acid. It, too, is

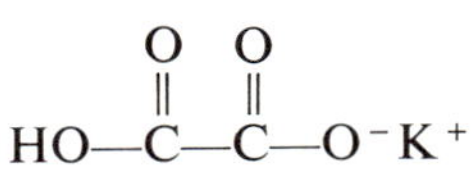

Monopotassium oxalate

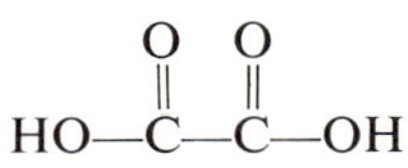

Oxalic acid

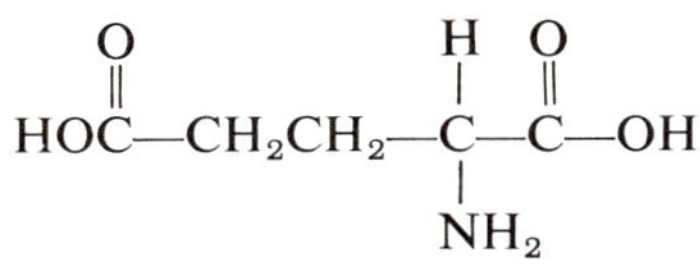

Glutamic acid

asymmetric and is one of the 20 or so amino acids used in the construction of protein molecules. Monosodium glutamate, the monosodium salt of this acid, sold commercially as Accent, is used as a flavor enhancer.

ESTERS

Esters are carbonyl compounds similar to carboxylic acids except that the hydroxyl group is replaced by an alkoxyl function. Esters are derived from an acid and an alcohol by the elimination of a water molecule. Esters are neutral though polar molecules and are more volatile than carboxylic acids of comparable molecular weights, since intermolecular hydrogen bonding is not possible in esters.

$$\underset{\text{A carboxylic acid}}{R_1{-}\overset{\displaystyle O}{\overset{\|}{C}}{-}OH} + \underset{\text{An alcohol}}{R_2OH} \xrightarrow{\text{catalyst}} \underset{\text{An ester}}{R_1{-}\overset{\displaystyle O}{\overset{\|}{C}}{-}OR_2} + H_2O$$

Formation of an ester

Molecules which have both a hydroxyl (or alcohol) group and a carboxyl group in close proximity can react to give internal esters. This type of ester is called a lactone. Lactones are formed most successfully when the resulting heterocyclic ring contains five or six atoms; attempts to form smaller rings usually lead to a dehydration reaction of the hydroxycarboxylic acid, while molecules which have the hydroxyl group more than four carbon atoms away from the carbonyl group tend to polymerize rather than cyclize to the lactone.

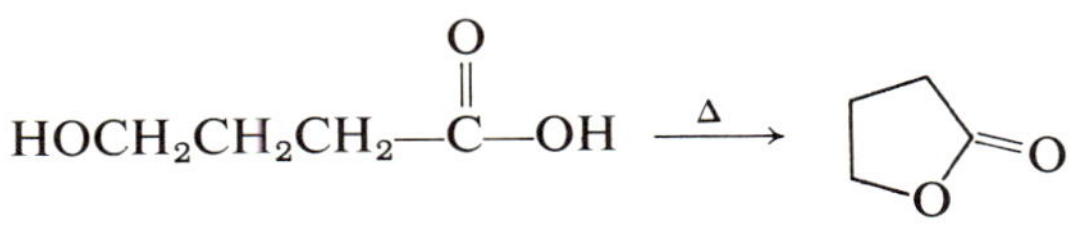

γ-Hydroxybutyric acid γ-Butyrolactone

Lactone formation

Nomenclature

The nomenclature system for esters takes into account both the name for the carboxylic acid and the alcohol when arriving at the name for the ester. First the alcohol *radical* (R_2) is named in the systematic way. Then the name of the carboxylic acid part is written but with the ending *-ic* changed to *-ate* and the word *acid* dropped. The two parts, the stem name for the alcohol and the modified name of the acid, are combined to form the name for the corresponding ester. Some examples of this nomenclature are found below.

$$\underset{\substack{\text{Ethyl alcohol}\\\text{(ethanol)}}}{CH_3CH_2OH} + \underset{\text{Acetic acid}}{CH_3\overset{\displaystyle O}{\overset{\|}{C}}{-}OH} \longrightarrow \underset{\text{Ethyl acetate}}{CH_3\overset{\displaystyle O}{\overset{\|}{C}}OCH_2CH_3}$$

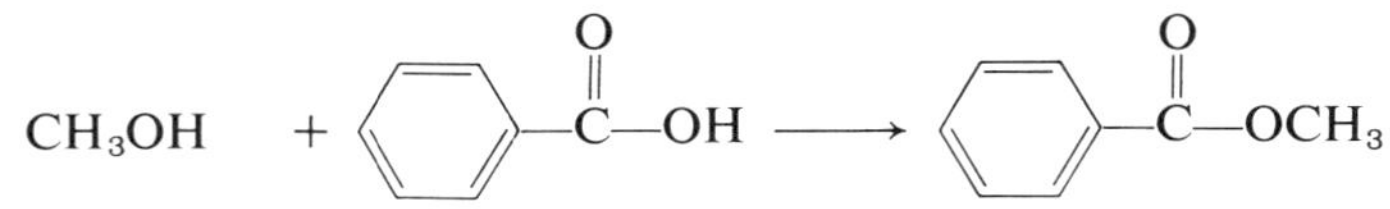

Methyl alcohol Benzoic acid Methyl benzoate

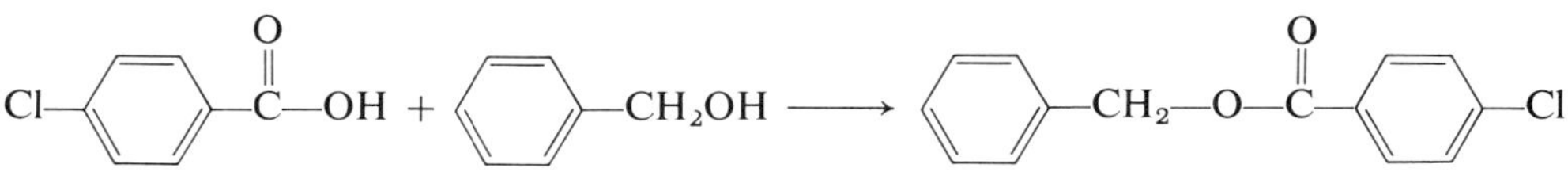

p-Chlorobenzoic Benzyl alcohol Benzyl *p*-chlorobenzoate

Occurrence of Esters

Many characteristic flavors and odors of natural products are due to esters. Among the esters which appear in this context are isoamyl acetate (banana oil), ethyl formate (rum), octyl acetate (orange), and butyl butyrate (pineapple).

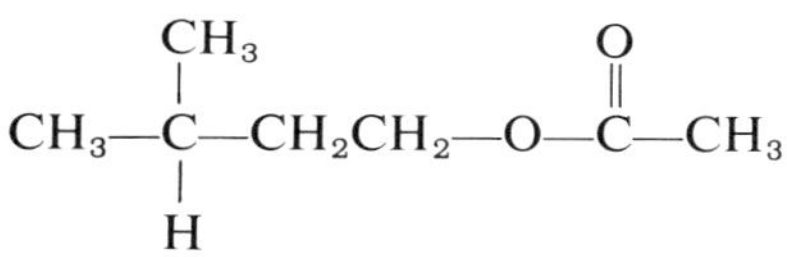

Isoamyl acetate

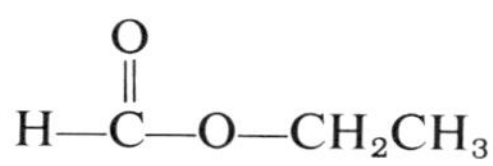

Ethyl formate

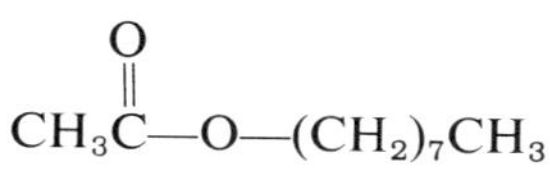

Octyl acetate

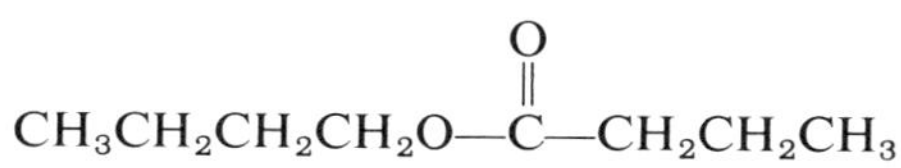

Butyl butyrate

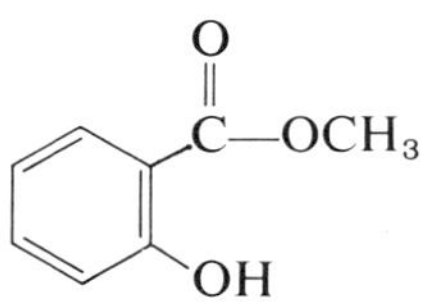

Methyl salicylate

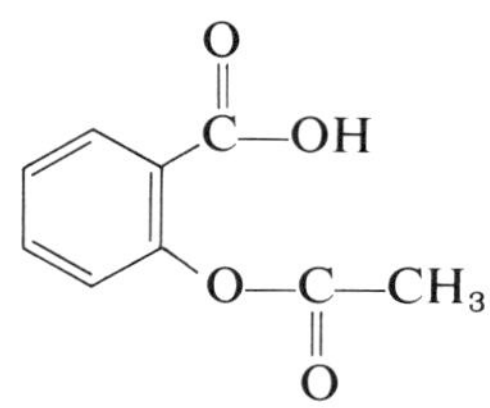

Aspirin

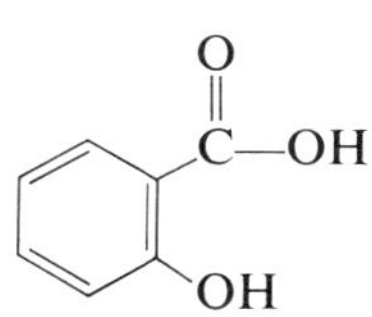

Salicylic acid

Oil of wintergreen is an ester (methyl salicylate) that is structurally related to aspirin which, in turn, is another ester. Natural flavors and fragrances are actually complex mixtures of many components; the characteristic odor of garlic is due to at least 40 different substances, many of them esters. Examples of biochemically interesting esters include Demerol, used as a morphine substitute, and propylure, a sex attractant for the female boll weevil used in traps for control of this insect. What are commonly regarded as fats, oils, and waxes are also esters; these will be considered in more detail later in the chapter.

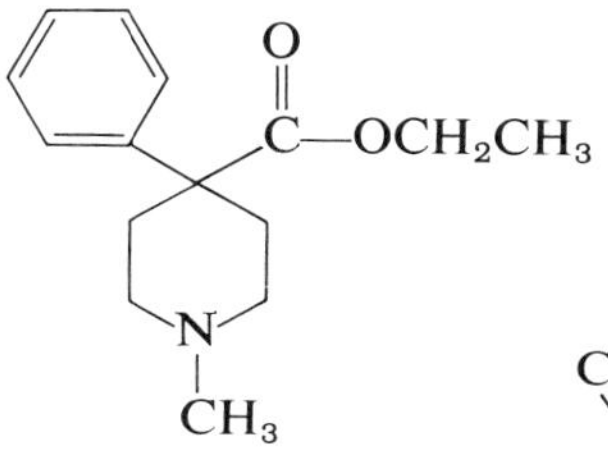

Demerol

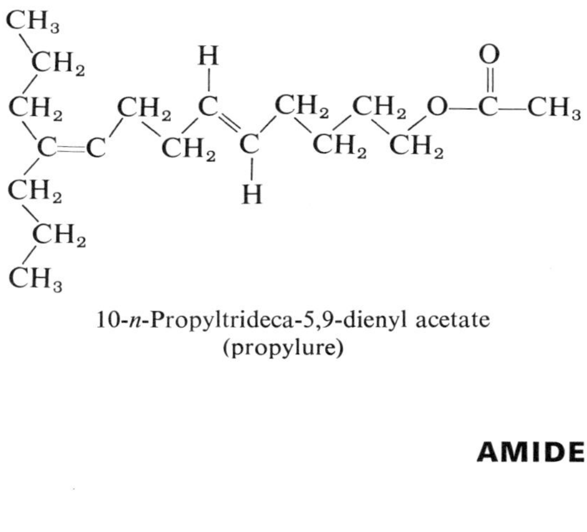

10-*n*-Propyltrideca-5,9-dienyl acetate
(propylure)

AMIDES

When the hydroxyl group of the carboxyl function is replaced by an amino group, another kind of carboxyl derivative, an amide, is formed. The replacing amino moiety may be derived from ammonia, a primary amine or a secondary amine.

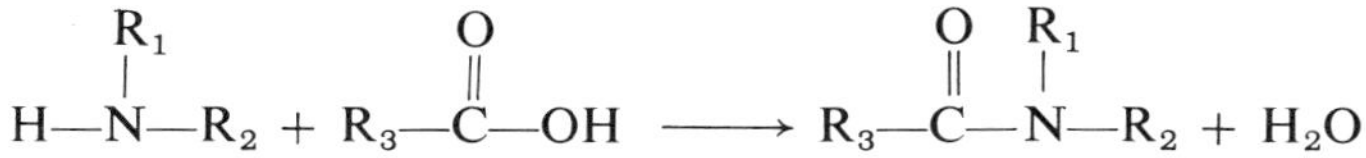

An amine — A carboxylic acid — An amide

Cyclic internal amides analogous to the lactones mentioned earlier can also be formed if a molecule has a carboxyl group and an amino group in a proximate arrangement. These cyclic amides are known collectively as lactams. Again ring formation is most favorable if a five or six-membered lactam ring is the result.

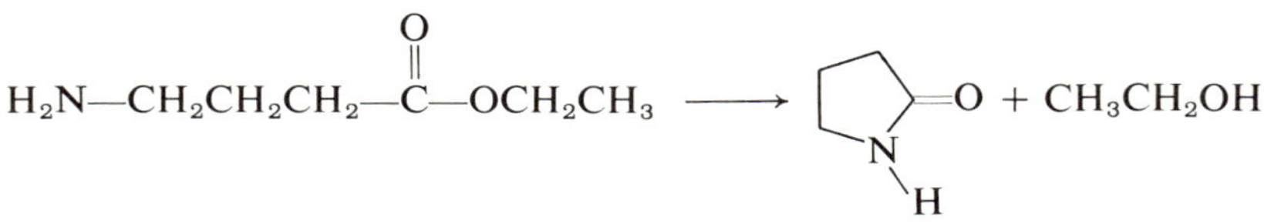

Ethyl γ-aminobutyrate

2-Pyrrolidone
(a lactam)

Lactam formation

The amide bond system is a particularly polar one and is able to accept hydrogen bonds at the amide carbonyl oxygen and, in the case of primary and secondary amides, to donate hydrogen bonds to appropriate acceptors as indicated in Fig. 11.3. Primary and secondary amides are strongly associated by these interactions.

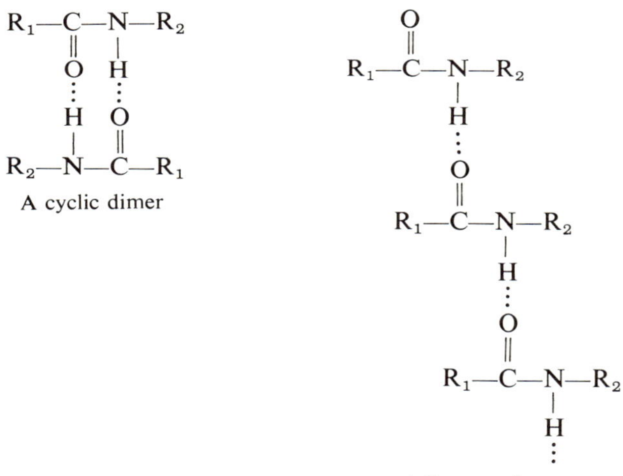

Fig. 11.3. Hydrogen-bonding in an amide.

Nomenclature

Amides are named by dropping the syllable *-ic* and *-oic* from the stem name of the acid and replacing it with the suffix *-amide*. (The word *acid* is also dropped.) The substituents on nitrogen, if any, are indicated by placing a capital *N* and the name of the substituent before the name of the corresponding unsubstituted amide. Examples of the use of this nomenclature are given with the representative compounds illustrated.

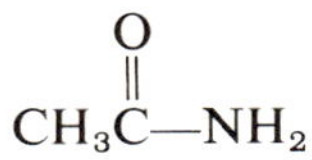

Acetamide

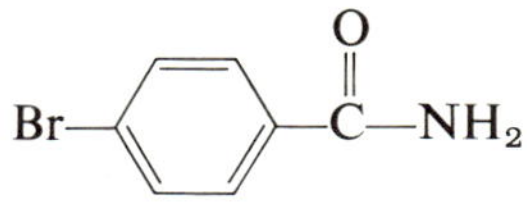

p-Bromobenzamide

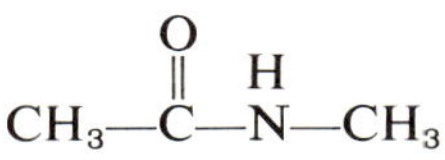

N-Methylacetamide

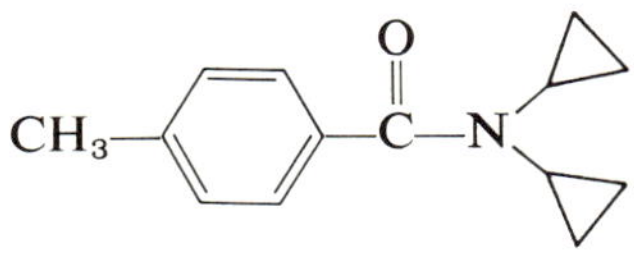

N,N-Dicyclopropyl-*p*-methylbenzamide
or
N,N-Dicyclopropyl-*p*-toluamide

Occurrence of Amides

Some possibly familiar examples of these various types of amides are given below. *N,N*-Diethyl-*m*-toluamide is sold commercially as an insect repellent ("6-12" or "Off") and is an example of a tertiary amide. ϵ-Caprolactam is a lactam (a cyclic secondary amide). Prepared from cyclohexanone in large quantities, ϵ-caprolactam is the monomer used in the preparation of some forms of nylon. Nylons are polyamides and can be converted to fibers or used to make molded articles. Probably no single synthetic organic polymer has had a greater impact on the American culture, and almost every activity from girl-watching to mountain-climbing has been influenced by the nylons. It should be noted that the name nylon is a generic one and refers to a family of polyamides, the difference between members of the family being the nature of the hydrocarbon segment between the amide groups.

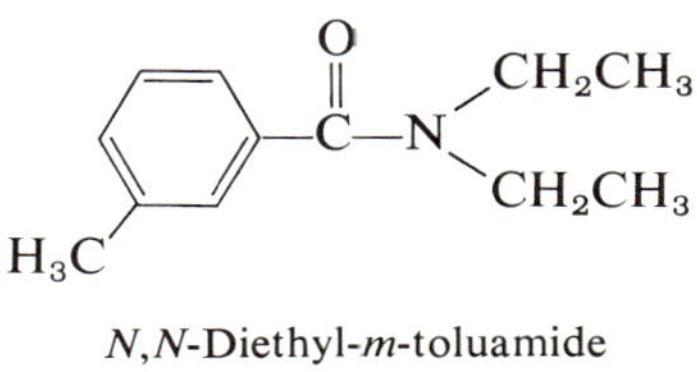

N,N-Diethyl-*m*-toluamide

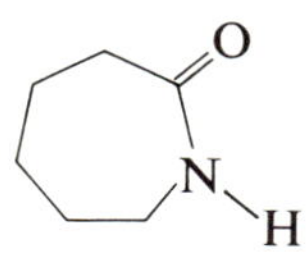

ϵ-Caprolactam

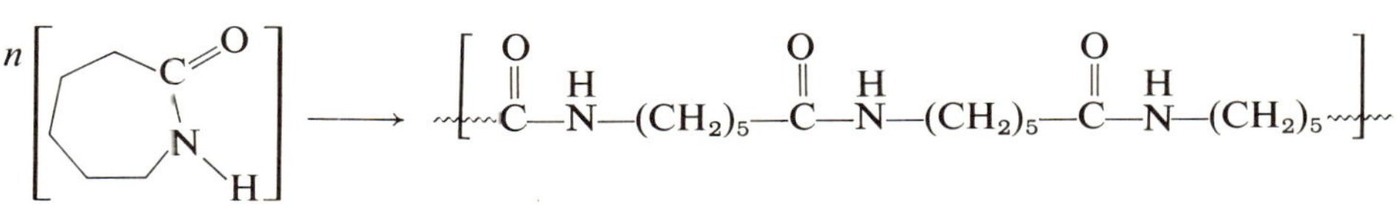

A nylon

Tolbutamide (orinase) has been used as an oral antidiabetic agent. It also contains a sulfonamide group ($R—SO_2\overset{H}{\overset{|}{N}}—R'$) in its structure.

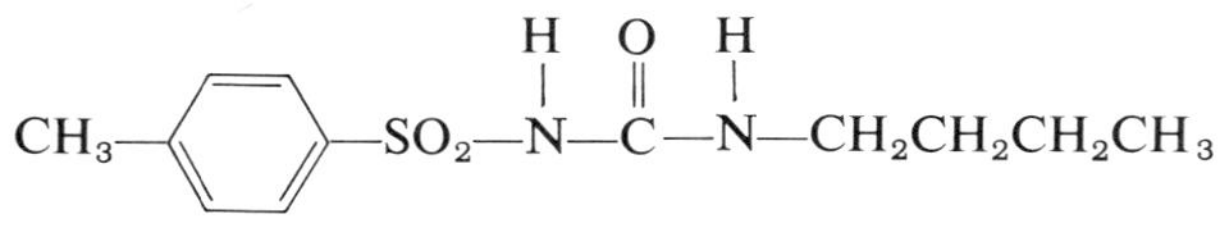

Orinase

Barbituric acid, made by combining urea and malonic acid, is the parent molecule for a class of drugs known as barbiturates. The amide bond systems are apparent in its structure; the heterocyclic ring in barbituric acid can also be regarded as a derivative of pyrimidine. Substitution on the barbituric acid ring leads to considerable variation in the properties of the drugs. Phenobarbital, with a phenyl substituent, is a long-acting sedative, while pentothal, familiar to most dental patients, is a short-acting anesthetic. Drug abuse utilizing barbiturates has increased recently, no doubt aided by the willingness of the drug

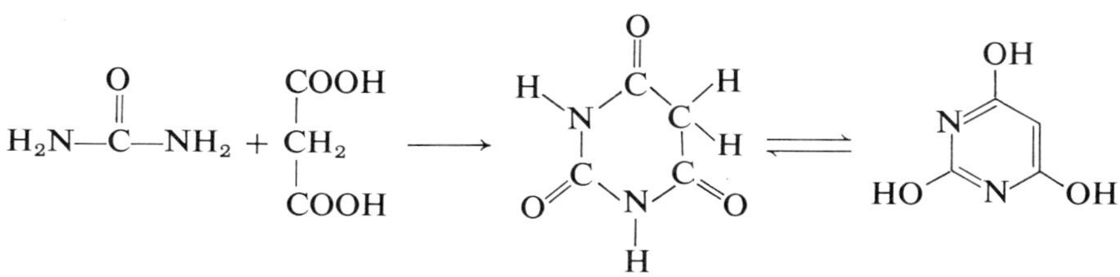

Urea Malonic acid Barbituric acid

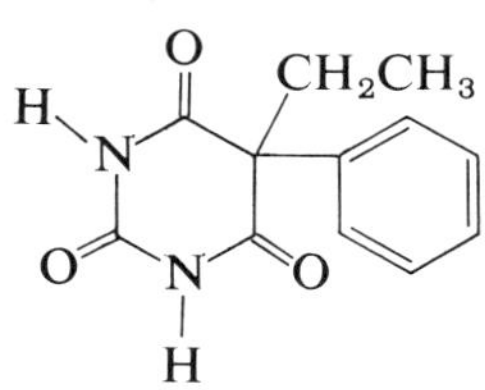

Phenobarbital

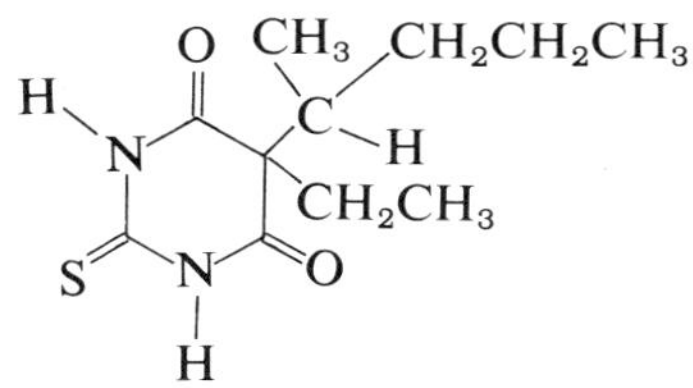

Pentothal

manufacturers to prepare many times the quantity of these agents than is needed for legitimate medical purposes.

The amide bond is an especially prevalent one in biological systems, being the fundamental linkage that holds proteins together. This facet of amide chemistry will be covered in a later chapter.

An interesting family of esters can be isolated from the flowers of *Chrysanthemum cinerariaefolum*; these are called pyrethrins and have considerable utility as insecticides. The pyrethrins are based on chrysanthemic acid or derivatives of this acid. They are relatively unstable in air and are easily bio-

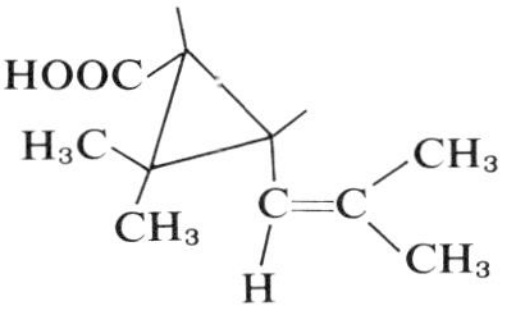

Chyrsanthemic acid

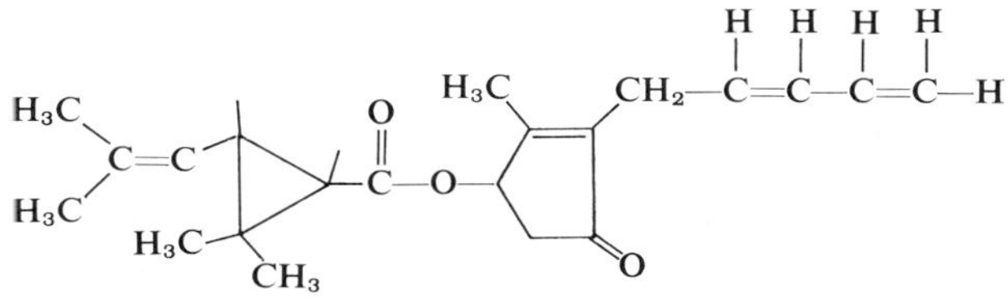

A pyrethrin

degradable. The toxicity of pyrethrins depends upon the age, sex, and local environment of the target insect and can be enhanced by the presence of synergistic molecules. Many of the synergists are amides such as the one shown below.

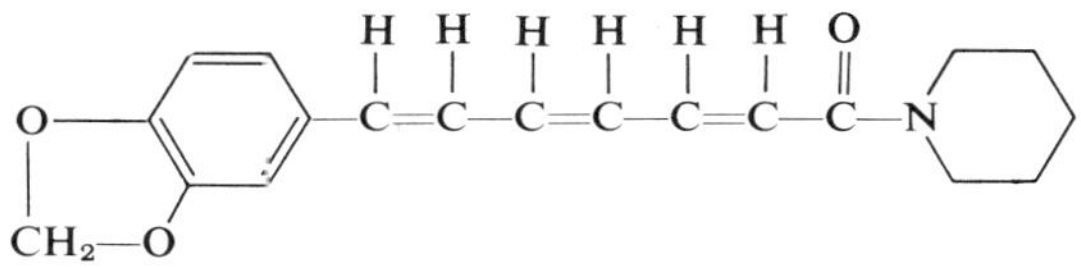

While potent insecticides, the pyrethrins are not toxic to humans or other mammals.

ACID HALIDES

In an acid halide the hydroxyl group of the carboxyl function has been replaced by a halogen atom. The resulting compounds are known as acid fluorides, chlorides, bromides, or iodides, depending upon the halogen that is used to replace the hydroxyl group. Volatile acid halides are usually disagreeable materials with powerful lachrymatory effects and pungent odors. The names for acid halides are derived from the name of the corresponding acid by removing the *-ic*, replacing it by *-yl* and substituting the name of the halogen for the word *acid*. Examples below illustrate the structure of some acid halides and their corresponding names.

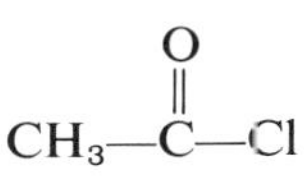

Acetyl chloride

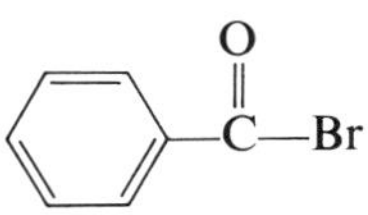

Benzoyl bromide

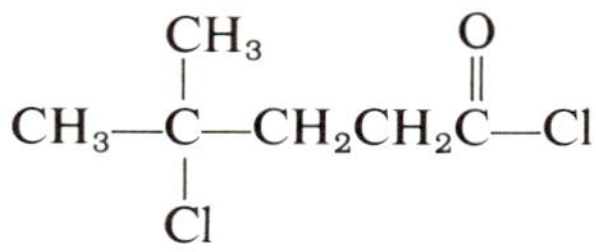

4-Chloro-4-methylpentanoyl chloride

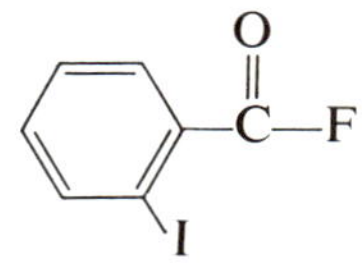

o-Iodobenzoyl fluoride

ACID ANHYDRIDES

Acid anhydrides can be thought of as being derived from two carboxylic acids by the expulsion of one molecule of water. Thus the name anhydride ("without water") is appropriately descriptive.

An acid anhydride

Acid anhydrides may be symmetric, that is both R and R′ groups are the same, or they may be mixed, in which case the two acid components have different structures. For symmetric anhydrides the nomenclature is simple: in the complete name of the acid the word *acid* is changed to *anhydride*. For mixed acid anhydrides the stem name for each acid component is listed and then the word *anhydride* added. Anhydrides with cyclic structures are formed from dicarboxylic

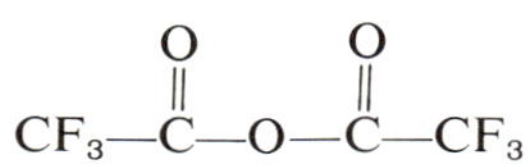

Trifluoroacetic anhydride

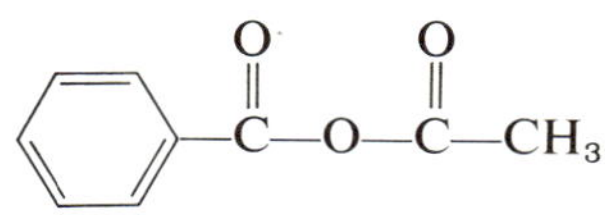

Acetic benzoic anhydride

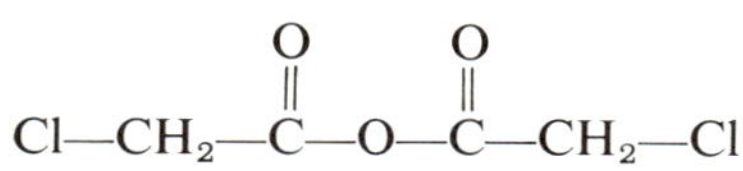

Chloroacetic anhydride

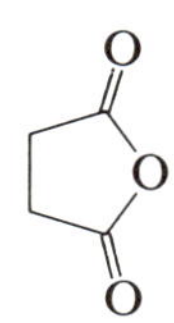

Succinic anhydride

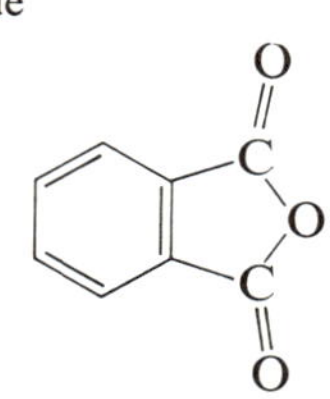

Phthalic anhydride

acids that can react to form five- or six-membered rings. Succinic acid, for example, preferentially forms a cyclic anhydride by intramolecular reaction rather than an open-chain anhydride by a bimolecular, intermolecular reaction. Both acid halides and acid anhydrides are principally used as intermediates in organic synthetic reactions.

REACTIONS OF CARBOXYLIC ACIDS AND THEIR DERIVATIVES

The carboxylic acid derivatives described so far are formally related to the corresponding carboxylic acid by replacement of the hydroxyl group of the carboxylic acid by another group. That is, esters, amides, acid halides, and acid anhydrides are obtained by substitution of an alkoxyl, amino, halogen, or carboxylate group, respectively, for this hydroxyl group. Both hydroxide and these substituting groups are reasonably good nucleophiles and the interconversion of these various acid derivatives occurs by nucleophilic substitution reactions. The mechanisms of these processes have common features and they may be considered together. However, before attacking this aspect of carboxylate chemistry, some useful reactions for the laboratory preparation of these compounds will be described.

Preparation of Carboxylic Acid Derivatives

Carboxylic Acids. It has been noted previously that an aldehyde is susceptible to oxidation to the corresponding carboxylic acid. Oxidation of alkenes with a strong oxidizing agent such as potassium permanganate is also a potential way to prepare carboxylic acids.

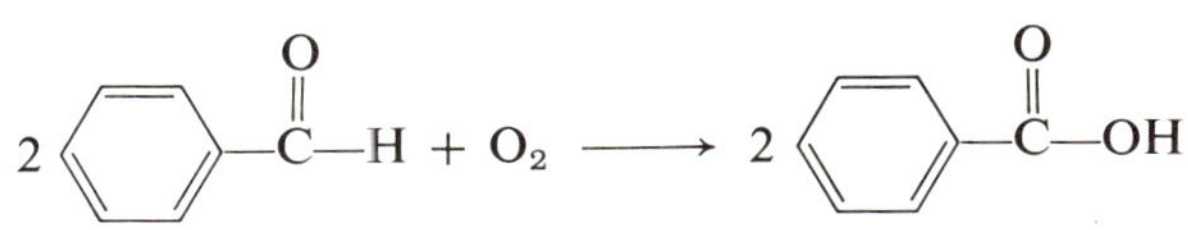

Formation of a carboxylic acid by oxidation of an aldehyde

The organometallic reagents discussed previously, especially the organolithium

$$CH_3-\overset{H}{C}=\overset{H}{C}-CH_3 \xrightarrow{KMnO_4} 2\ CH_3-\overset{O}{\overset{\|}{C}}-OH$$

Oxidation of an alkene to a carboxylic acid

and Grignard reagents, will react with carbon dioxide to afford the corresponding carboxylic acids. Hydrolysis of nitriles can produce amides which, upon

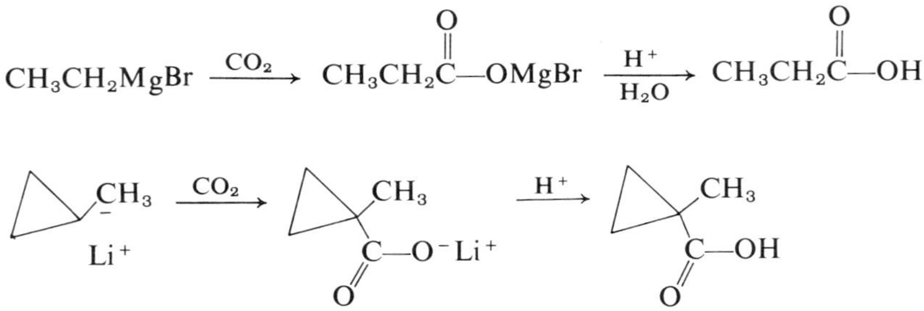

Formation of carboxylic acids from organometallic reagents

further hydrolysis, can be converted to carboxylic acids.

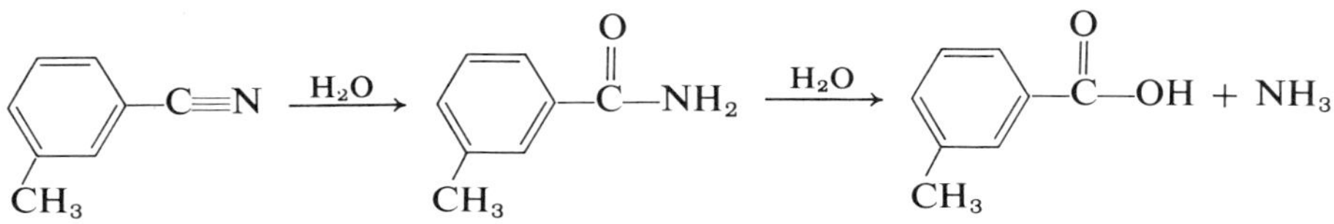

A nitrile An amide A carboxylic acid

Hydrolysis of nitriles

Esters. Esters can usually be prepared by heating a solution of the carboxylic acid in the appropriate alcohol. The mixture is saturated with hydrogen chloride gas since the esterification reaction is acid-catalyzed. The alcohol is used as a solvent because the reaction is an equilibrium that can be driven to favor product by a high concentration of this reactant.

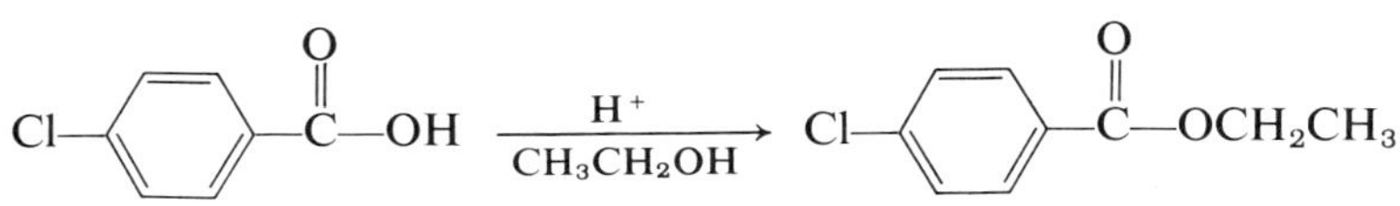

p-Chlorobenzoic acid Ethyl *p*-chlorobenzoate

Another general method for preparing esters involves the treatment of an acid halide (usually chloride or bromide) or anhydride with an alcohol.

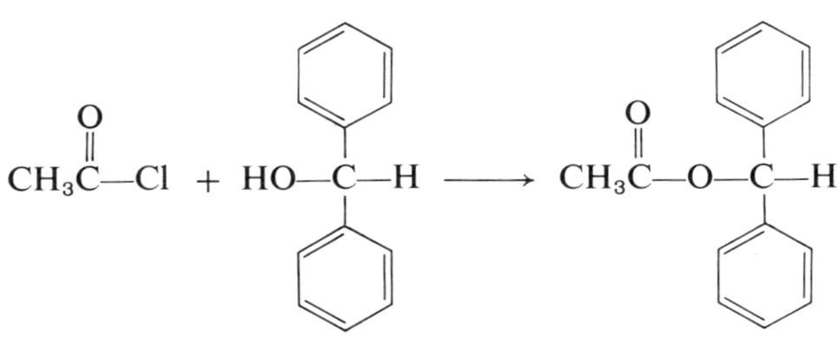

Preparation of an ester using an acid chloride

Acid Halides and Anhydrides. Carboxylic acids react with thionyl chloride ($SOCl_2$), phosphorus trichloride (PCl_3), and phosphorus pentachloride (PCl_5) to give the corresponding acyl chloride.

Propionic acid — Propionyl chloride

Preparation of acid chlorides

Acid anhydrides may be obtained by treating a carboxylic acid with an acid chloride.

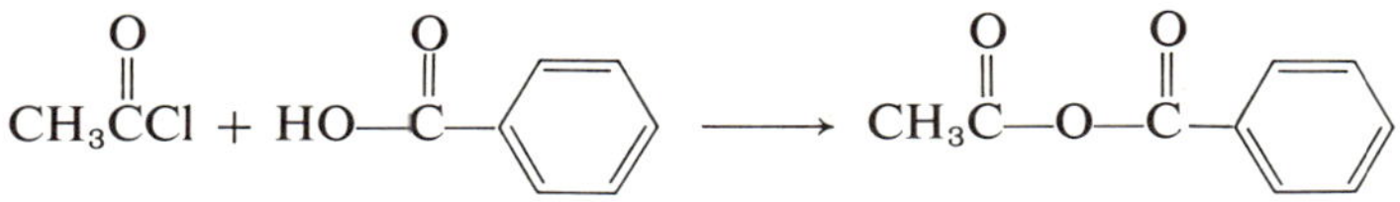

Formation of an acid anhydride

Alternatively, symmetrical anhydrides can be obtained by heating the carboxylic acid strongly enough to drive off water. This reaction can also lead to decarboxylation (loss of carbon dioxide) or to various types of cyclization reactions.

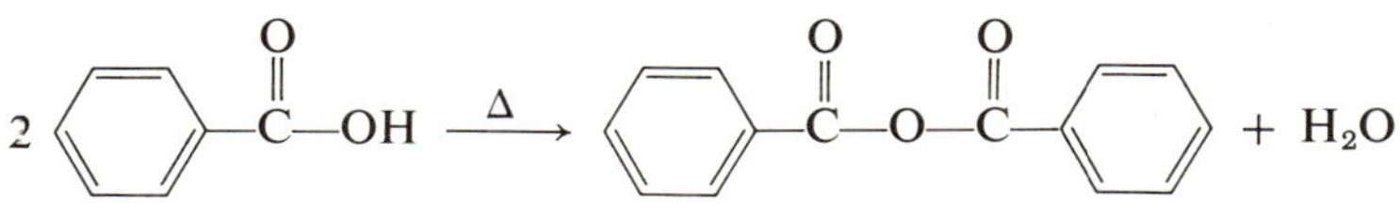

Preparation of acid anhydrides

Amides. Amides may be formed by treating an ester, an acid halide, or an acid anhydride with an amine.

Interconversion of Carboxylic Acid Derivatives

Reactions which interconvert carboxylic acids and carboxylic acid derivatives proceed by similar mechanisms and in each case a nucleophilic substitution reaction occurs. The substitution mechanism is different from the S_N1 and S_N2 reactions discussed earlier, however, and resembles more closely the substitution reactions of aromatic rings. We have previously seen how nucleophiles can add to carbonyl groups to give tetrahedral intermediates which can be stable (as in the case of hemiacetal formation) or undergo further reaction (as in hydrazone formation). A carboxylic acid or one of its derivatives has at least one good leaving group already attached to the carbonyl function in the form of a halogen, hydroxide, amine, or carboxylate. Addition of a nucleophile to one of these compounds affords a tetrahedral intermediate which may (1) expel the nucleophile which has just been added, or (2) expel the leaving group which is already

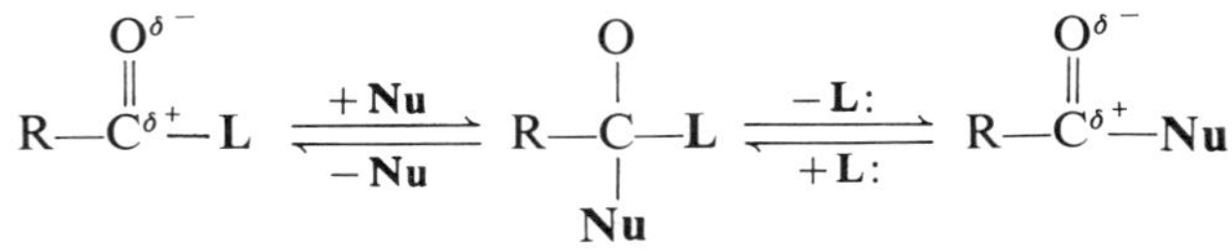

Fig. 11.4. A general mechanism for nucleophilic substitution at a carboxylic acid or carboxylic acid derivative. **L** represents a hydroxyl, alkoxyl, halide, amino, or carboxylate group (a leaving group) and **Nu** represents a nucleophile. Charges are not shown to preserve generality, although often **Nu**, the tetrahedral intermediate, or free **L** will be anionic.

present (Fig. 11.4.). This two-step reaction mechanism is illustrated by the attack of hydroxide on an acid chloride. The intermediate is characterized by an sp^3 carbon atom to which is attached both the halogen and the hydroxyl group. The intermediate is in equilibrium with the starting materials and the

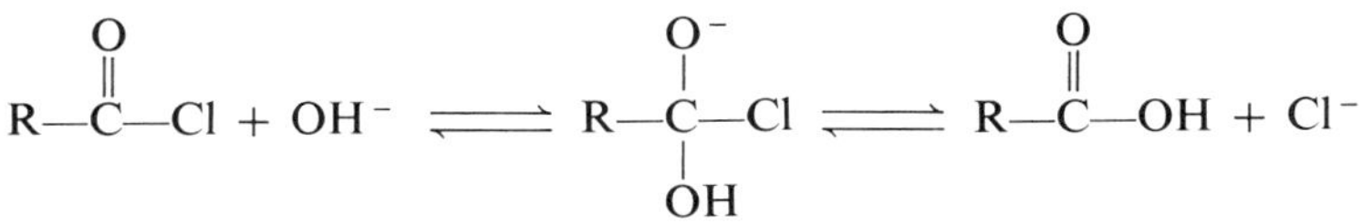

Formation of an acid from an acid halide

products of the reaction which can be derived by loss of halide ion from the intermediate structure. The position of this series of equilibria, i.e., the relative amount of products present at equilibrium, will depend upon the intrinsic leaving ability of the two nucleophilic species involved (chloride and hydroxide in the case discussed above). The ability of a group to leave the tetrahedral intermediate essentially parallels its leaving group ability in S_N2 reactions. Halide is lost more readily than hydroxide in the reaction above so that the carboxylic acid is the observed product.

A convenient, qualitative order of reactivity for carboxylic acid derivatives is given below. Acid halides and anhydrides are the most reactive or susceptible

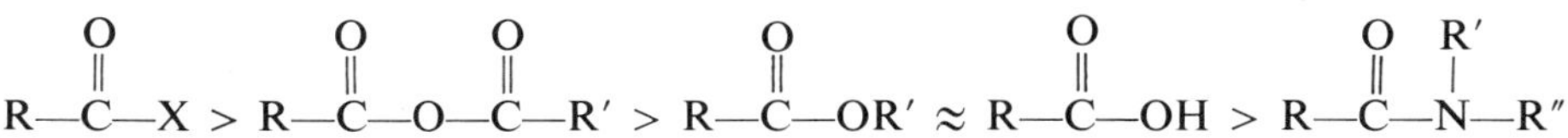

Qualitative order of decreasing carbonyl reactivity

to attack by nucleophilic reagents while esters and amides are less reactive. Usually a member of a less reactive class of compounds in this order can be prepared by treating a more reactive member of the series with the appropriate nucleophile. Acid halides and anhydrides are the best starting materials to choose for such syntheses because they are the most reactive.

Hydrolysis

All of the carboxylate derivatives can react with water under the appropriate conditions to generate the corresponding carboxylic acid. These hydrolysis reactions can take place in neutral water or may be catalyzed by the presence of acids or bases. Since they are very reactive, the acid halides undergo hydrolysis even when exposed transiently to water. In this case water is the attacking nucleophile. The hydrolysis reaction of acid halides has little practical value but rather

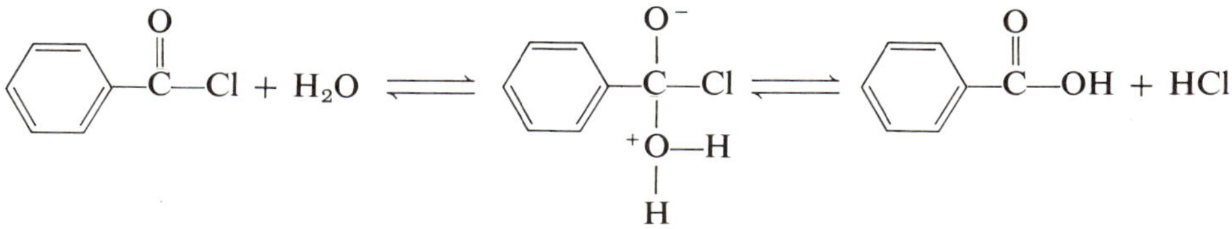

Hydrolysis of an acid chloride

is a nuisance, requiring that these materials be prepared and stored under anhydrous conditions.

Ester hydrolysis may be catalyzed by base or by acid but proceeds slowly under neutral conditions. In the case of base catalysis, the attacking nucleophile is hydroxide ion and the catalysis is a reflection of the greater nucleophilicity of this entity relative to water. In basic solution the acid product is present as a carboxylate anion, thus preventing the reverse reaction. An acid catalyst

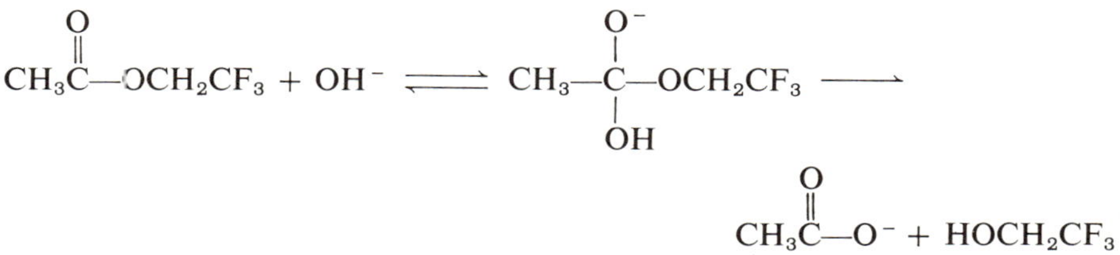

Basic hydrolysis of an ester (saponification)

functions by setting up an equilibrium protonation of the carbonyl group, thereby making it more susceptible to attack by the nucleophile water. (A similar catalysis was noted in the addition reactions of aldehydes and ketones.) It should be observed that this hydrolysis reaction occurs with the same catalyst that is used to prepare esters from alcohols and acids. The difference between the two reactions is the relative concentration of alcohol and water. The overall transformation of acid to ester is an equilibrium situation; to form the ester a large excess concentration of alcohol is used in order to shift the position of the equilibrium in favor of the ester, but when the hydrolysis of an ester is the desired course of reaction, an excess of water (as the solvent) is employed. In both cases, the acid functions only as a catalyst by accelerating the achievement of the equilibrium condition.

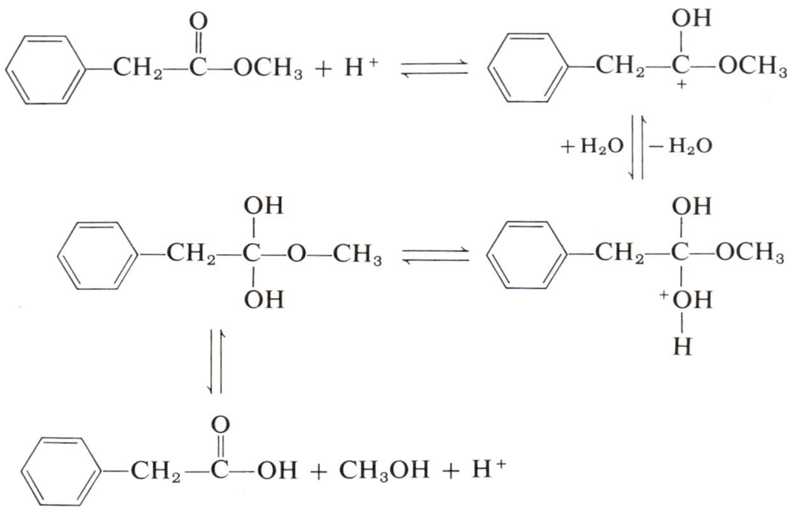

Acid-catalyzed hydrolysis of an ester

Amides are hydrolyzed by acid- or base-catalyzed reactions that are similar to the corresponding reactions of esters.

Reduction Reactions

Reductions of carboxylic acids and carboxylate derivatives are most conveniently carried out with lithium aluminum hydride (LAH) and its derivatives. Reduction of an acid, an ester, an anhydride or an acid halide with LAH leads to the same product, a primary alcohol. Amides, however, are reduced to the corresponding amines.

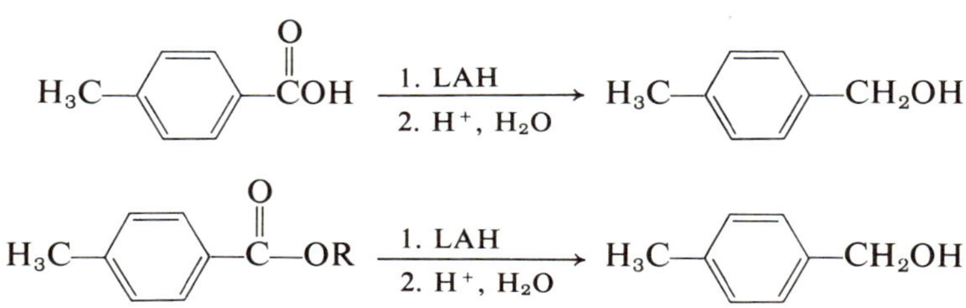

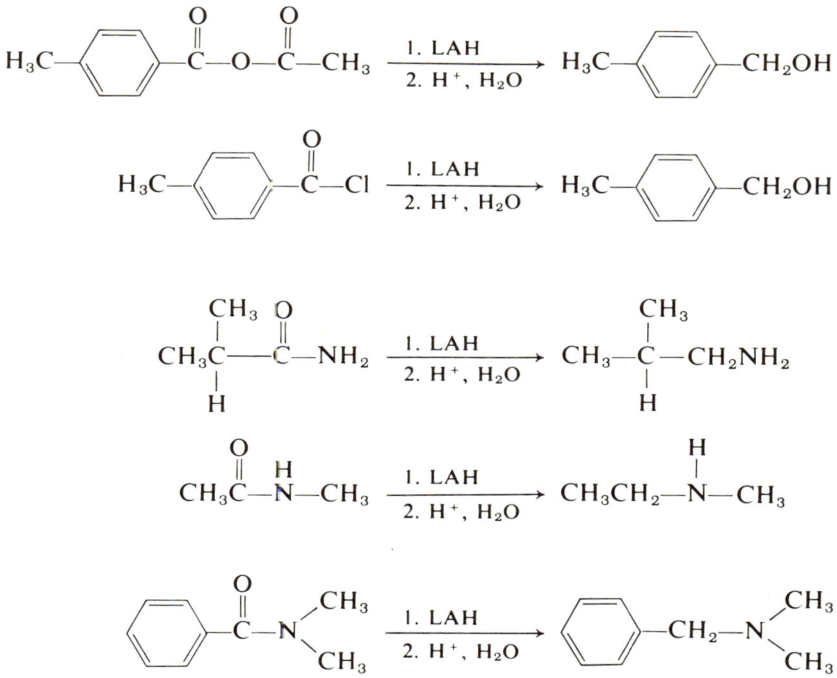

Lithium aluminum hydride reductions

The use of modified aluminum hydride reagents allows one to selectively reduce acid chlorides to aldehydes. The alkoxy groups attached to the aluminum atom in these reagents decrease the reactivity of the reducing agent sufficiently so that the reaction can be halted at this state.

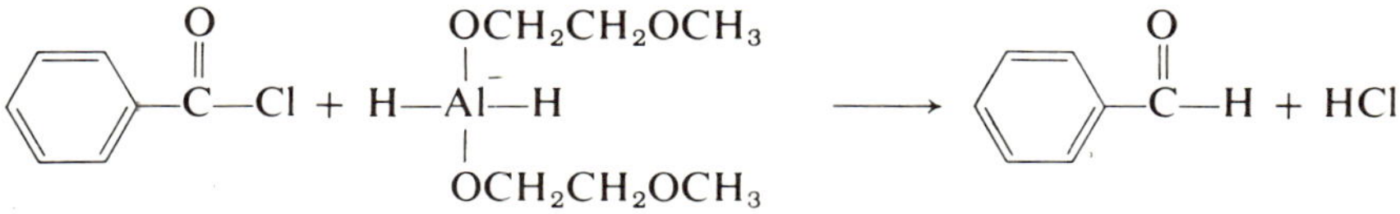

"Vitride"

Reduction of acid halides

Condensation Reactions

The hydrogen atoms adjacent (*alpha*) to the carbonyl groups of carboxylic acid derivatives resemble similarly situated hydrogen atoms in aldehydes and ketones. That is, these α-hydrogen atoms are sufficiently acidic that, in the presence of a strong enough base, carbanions may be generated at this position. Like the carbonyl group of an aldehyde or ketone, the carbonyl groups in these molecules can also be attacked by carbanions generated in this way. Condensation reactions involving the formation of new carbon–carbon bonds are thus possible. When an ester, say ethyl acetate, is treated with a strong base (usually the alkoxide corresponding to the alcoholic portion of the ester) and neutralized after an appropriate reaction time, a β-ketoester is formed by a condensation reaction according to the mechanism sketched below. Each stage in the mecha-

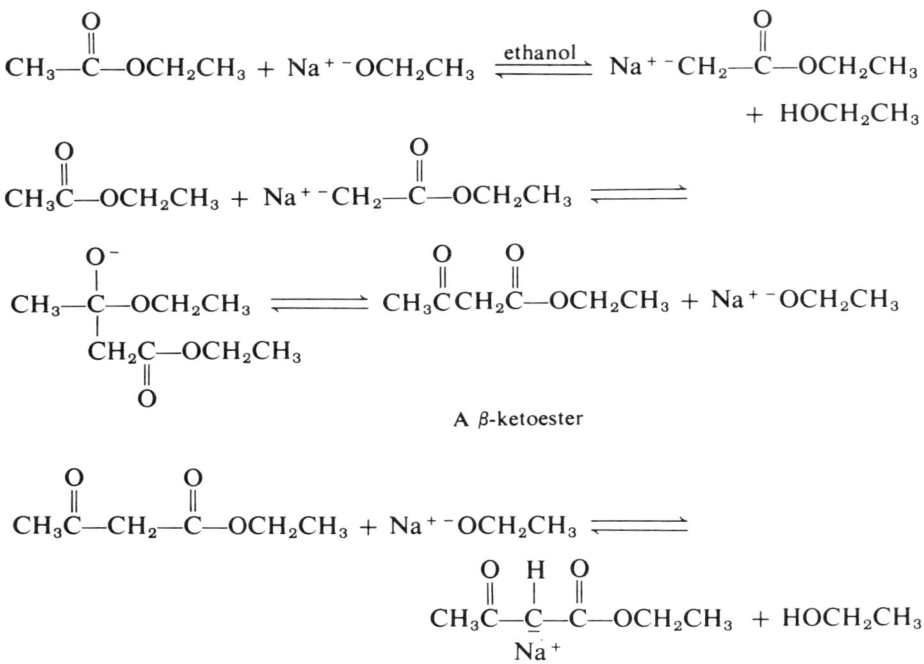

nism is an equilibrium. However, the last step, the formation of the carbanion from the β-ketoester, is highly favored so that this last equilibrium is shifted strongly to the right, thereby causing the series of equilibria to cascade in the direction of formation of this carbanion. Neutralization of the reaction mixture protonates this anion to give the observed product. For a condensation reaction of this type to proceed it is necessary that *two* α-hydrogens be present so that the last, salt-forming step can take place to drive the reaction toward products. This condensation reaction is known as the Claisen condensationn.

The carbanions that are formed by proton abstraction at α-carbon atoms can also function as nucleophilic reagents, and can take part in S_N2 displacement reactions. Thus, the carbanion from diethyl malonate can be alkylated by treatment with an alkyl halide.

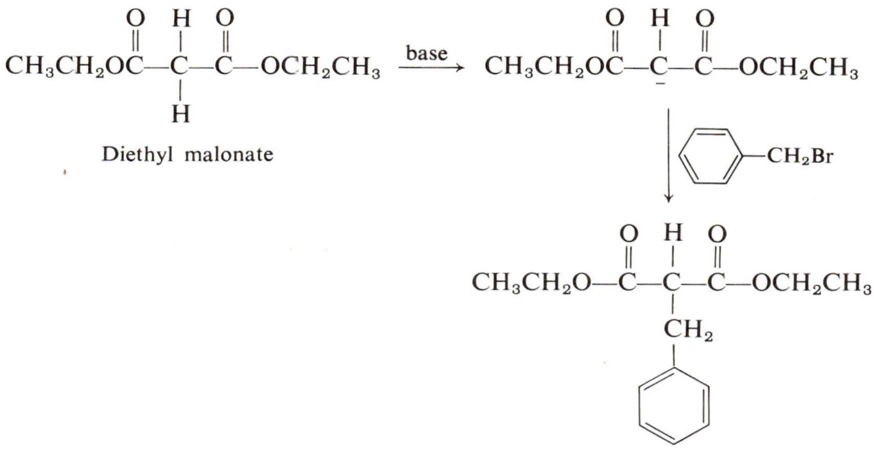

Alkylation of a carbanion derived from an ester

OTHER ACIDS

The acid derivatives—esters, anhydrides, acid halides, and amides—that have been described above can often also be obtained from other types of acids such as nitric acid, sulfuric acid, and the acids which contain phosphorus. The detailed chemistry of these acids and their derivatives is usually regarded as "inorganic chemistry" and in some cases has not been well explored.* Compounds analogous to the derivatives of carboxylic acids that were described above are known for inorganic acids as well. Alcohols can react to form esters

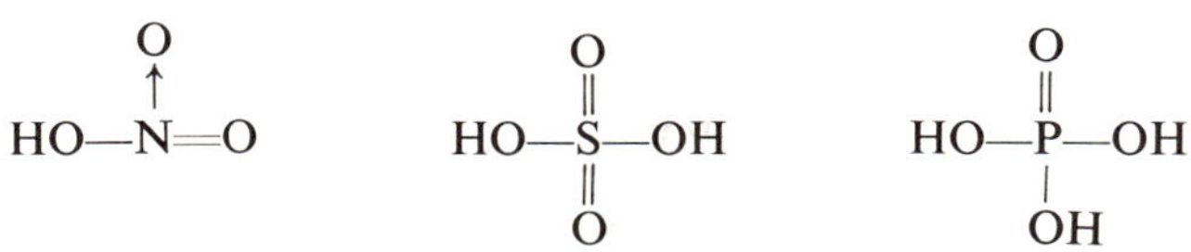

Nitric acid (HNO_3) Sulfuric acid (H_2SO_4) Phosphoric acid (H_3PO_4)

of these acids, as exemplified by the preparation of nitroglycerine. These reactions are carried out under roughly the same conditions used in the preparation of carboxylic esters.

---

* It should be mentioned that nitrogen, sulfur, and phosphorus have more than one valence state so that these acids are only some of those that are possible.

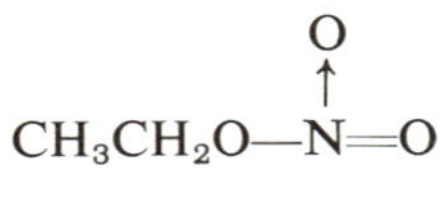

Ethyl nitrate

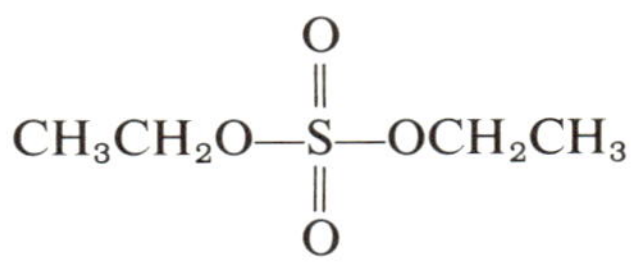

Diethyl sulfate

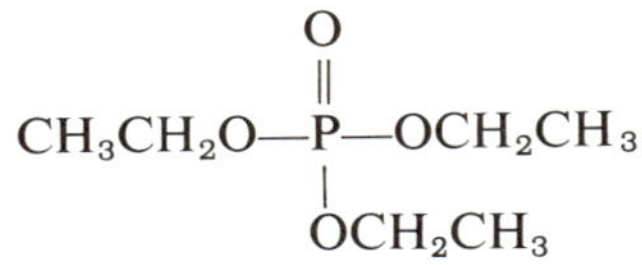

Triethyl phosphate

Ethyl esters of the acids on p. 285

Amides of these inorganic acids are also known. Nitramide, a weak acid in aqueous solution, decomposes slowly in water to nitrous oxide (N_2O) and water.

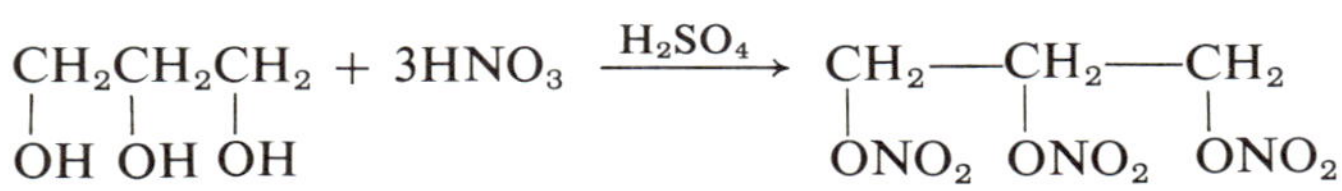

Glycerine

Glyceryl trinitrate (nitroglycerine)

The preparation of nitroglycerine requires carefully controlled reaction conditions. These were discovered and developed by the Swedish chemist, engineer, and philanthropist, Alfred B. Nobel. He found a way to incorporate the highly unstable nitroglycerine into solid diatomaceous earth, thereby increasing its stability and ease of handling. The commercial exploitation of this new explosive, called dynamite, made him a very wealthy man; he used his fortune to endow the prestigious international prizes in chemistry, medicine, physics, literature, and peace that now bear his name.

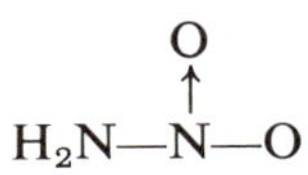

Nitramide

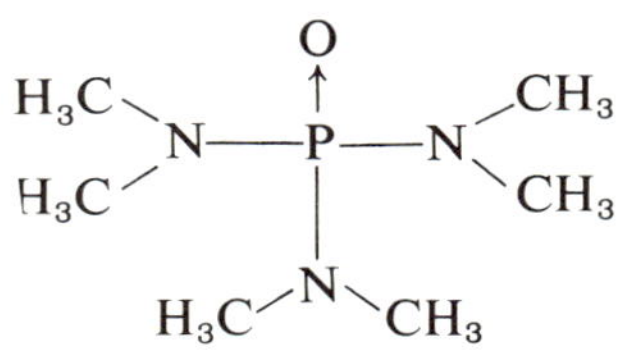

Hexamethylphosphoramide

Hexamethylphosphoramide is a unique solvent for many organic reactions. It is also able to dissolve many metals.

The phosphate group (PO_4) can combine with itself to give what are formally anhydrides of phosphoric acid. A large number of such compounds are known and they may be regarded as dimers, trimers, and higher polymers of phosphate. The condensed phosphates are not limited to linear arrangements, but, as the

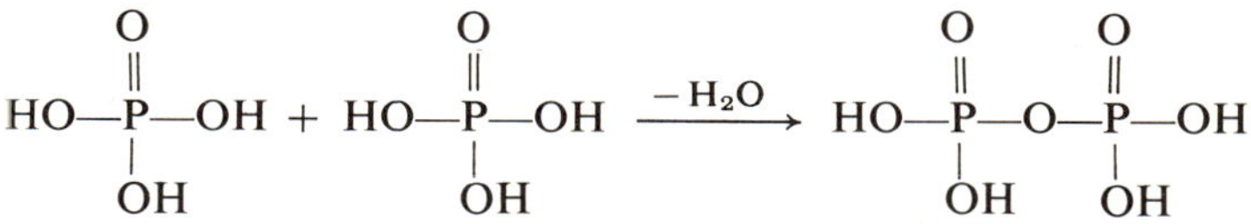

Pyrophosphoric acid (diphosphoric acid)

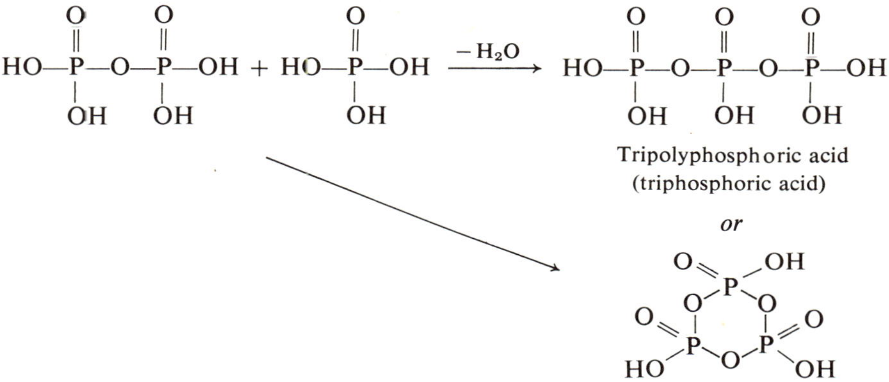

Tripolyphosphoric acid (triphosphoric acid)

or

Trimetaphosphoric acid

reaction above implies, can also take on cyclic structures. The $\left[—\overset{\overset{\displaystyle O}{\|}}{\underset{\underset{\displaystyle O}{|}}{P}}—O—\overset{\overset{\displaystyle O}{\|}}{\underset{\underset{\displaystyle O}{|}}{P}}—\right]$ bond system, like the $\left[—\overset{\overset{\displaystyle O}{\|}}{C}—O—\overset{\overset{\displaystyle O}{\|}}{C}—\right]$ bond in carboxylic anhydrides, is a "high-energy" bond, meaning that in the presence of nucleophilic reagents, the bond can be easily cleaved.

Pyrophosphoric and tripolyphosphoric acid can be esterified at the free hydroxyl groups to give phosphate esters, and esters of these acids, as well as the parent phosphoric acid, are essential components of all biological processes. The nucleic acids, adenosine triphosphate (ATP) and pyridoxal phosphate are examples of such molecules.

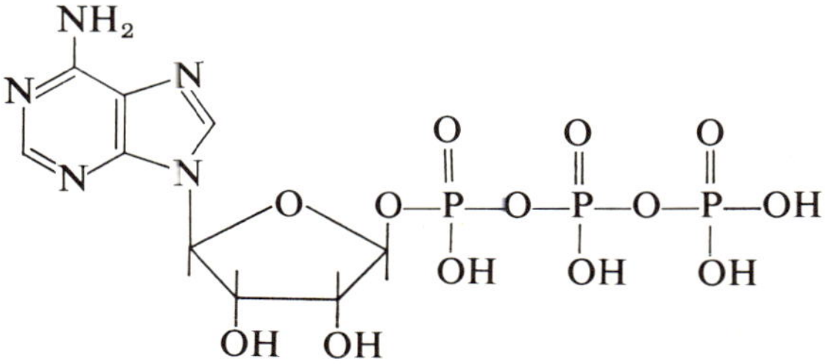

Adenosine triphosphate (ATP)

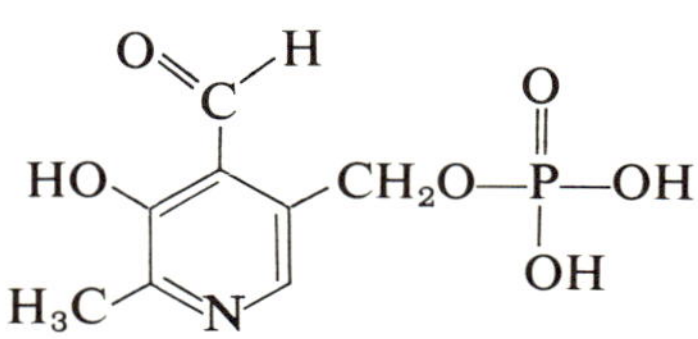

Pyridoxal phosphate

LIPIDS

Lipids are naturally occurring materials that are soluble in organic solvents such as chloroform, benzene, or hexane. The principal types of lipids are fats and oils from animal or vegetable sources, waxes, steroids, phospholipids, and sphingolipids. These substances are either esters of fatty acids or are closely associated in the natural state with esters of this type.

Names and condensed structures for the more common fatty acids are given in Table 11.2. Fatty acids isolated from animals or plants usually have several characteristics in common. They contain only one carboxyl group and consist of a single, unbranched carbon chain with an even number of carbon atoms present in the chain. It is not uncommon to find one or more double bonds in the backbone of these materials. As will be noted from the melting points listed in Table 11.2, there is a tendency for unsaturated fatty acids to be liquids at room temperatures.

Table 11.2
Some Fatty Acids

Name	*Formula*	*Melting point* (°C)
A. Saturated acids		
Caproic	$CH_3(CH_2)_4COOH$	−2
Caprylic	$CH_3(CH_2)_6COOH$	17
Capric	$CH_3(CH_2)_8COOH$	31
Lauric	$CH_3(CH_2)_{10}COOH$	44
Myristic	$CH_3(CH_2)_{12}COOH$	54
Palmitic	$CH_3(CH_2)_{14}COOH$	63
Stearic	$CH_3(CH_2)_{16}COOH$	70
B. unsaturated acids		
Palmitoleic	$CH_3(CH_2)_5CH{=}CH{-}(CH_2)_7COOH$ (*cis*)	32
Oleic	$CH_3(CH_2)_7CH{=}CH(CH_2)_7COOH$ (*cis*)	16
Linoleic	$CH_3(CH_2)_4CH{=}CH{-}CH_2CH{=}CH(CH_2)_7COOH$	−5
Linolenic	$CH_3CH_2CH{=}CHCH_2CH{=}CHCH_2CH{=}CH(CH_2)_7COOH$	−11

Waxes

Waxes are simple esters formed from high-molecular weight primary alcohols and the fatty acids. Naturally occurring waxes are mixtures of several such esters. Beeswax, for example, contains esters of the C_{26} to C_{28} fatty acids and C_{30} to C_{32} alcohols. In addition to these high-molecular weight esters, many natural waxes contain long-chain, solid alkanes.

Fats and Oils

What are commonly called fats and oils have similar chemical structures; they are triesters formed from glycerol and fatty acids.

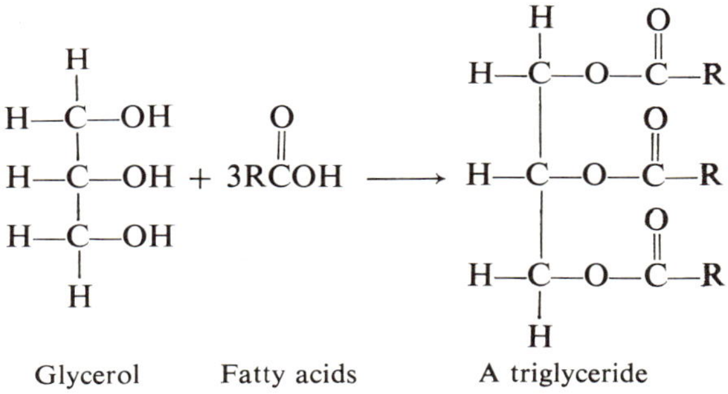

It is the degree of unsaturation of the constituent fatty acids which determines whether a triglyceride is classified as a fat (a solid) or an oil (a liquid). Figure 11.5 suggests why the triglycerides with mostly saturated fatty acids present in their structure tend to be solids, while highly unsaturated triglycerides are oils. A regular, pseudocrystalline arrangement of the hydrocarbon chains is possible with the saturated compounds. However, a double bond in the fatty acid skeleton results in a "kink" in the chain at that point, forcing the unsaturated fatty acids into a much more random conformation so that regular packing of these molecules is not as likely.

The most abundant and inexpensive triglycerides are the oils from corn, soybeans, peanuts, or cottonseeds. A large number of the fatty acid components of these materials are unsaturated. There exists a considerable prejudice to the use of oils in many forms of cooking in the United States so that these highly unsaturated oils are hydrogenated to the corresponding saturated forms in order to convert them to more acceptable solids, a process known as "hardening." This tendency to prefer solid fats is unfortunate since some evidence has been produced which links the amount of saturated fats in the diet to the presence of arteriosclerotic diseases. In recognition of this possible connection it

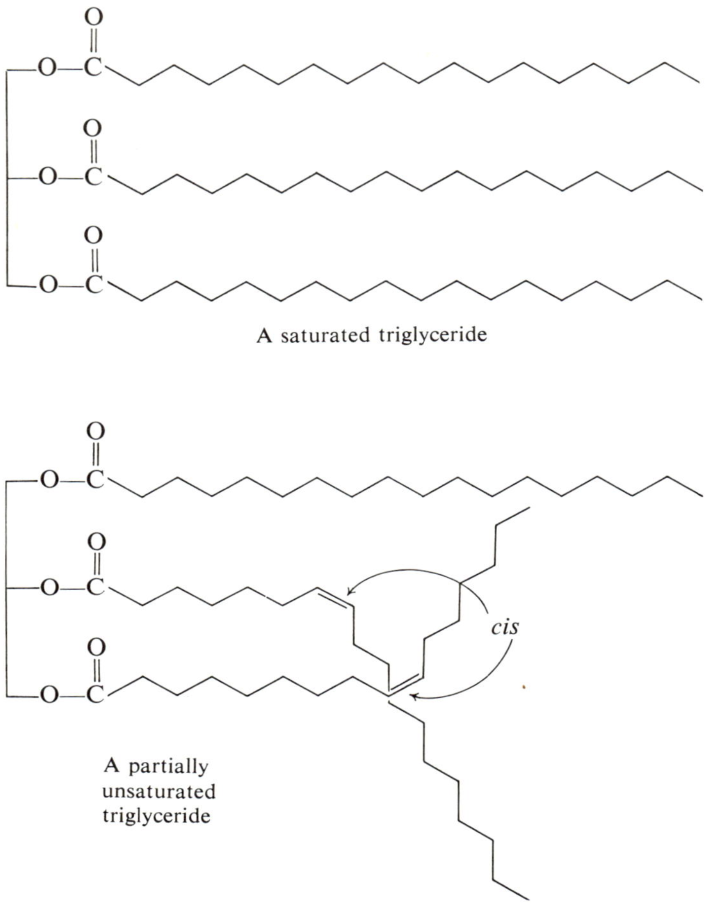

Fig. 11.5. Schematic representation of the conformations of the fatty acid side chain in a fat and in an oil.

has now become fashionable for some makers of margarine and cooking fats to emphasize the unsaturated properties of their products. New ways of formulating vegetable oils, especially corn oil and safflower oil, into materials which are solid at room temperature but which retain the unsaturation of the fatty acids in the oil have been developed and these products are now marketed.

The ester groups in triglycerides are chemically similar to ester groups found in simple compounds. These triesters can be hydrolyzed to the component fatty acids and glycerol with a strong base (potassium or sodium hydroxide). The product of the reaction with a natural fat or oil is a collection of salts of fatty acids. These products are called soaps. The utility of soaps as cleaning agents is considerably reduced in areas which have high concentrations of magnesium or calcium ions in their water supply. (This is called "hard" water.) The calcium or magnesium salts of the fatty acids are very much less soluble than the sodium or potassium salts, so that in "hard" water, an insoluble scum or curd forms.

Phospholipids

A number of important lipids are derivatives of phosphatidic acid. Phosphatidic acid contains a glycerol molecule as part of its structure but only one terminal hydroxyl group and the internal (secondary) hydroxyl group are

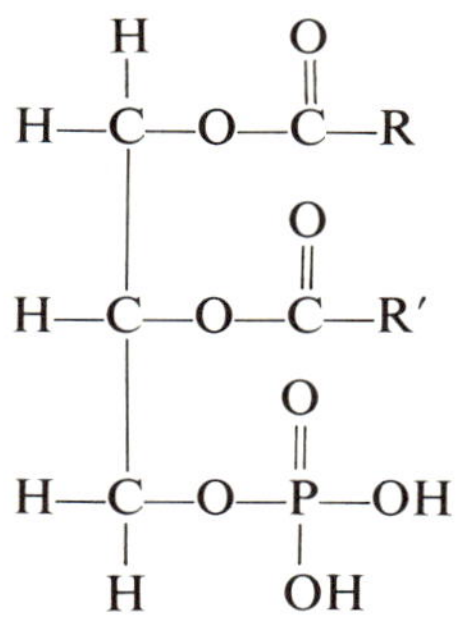

A phosphatidic acid

esterified with fatty acids. At the remaining terminal hydroxyl a phosphate ester is formed. Specific phospholipids are obtained by forming an additional ester bond at the phosphate group to other alcohols, most usually choline, ethanolamine, or serine. Materials containing choline are usually called *lecithins*, while derivatives based on the other two alcohols are sometimes called *cephalins*. Each of these names signifies a class of compounds, since the fatty acid groups,

Choline Ethanolamine Serine

R and R′, can be variable. The lecithins and cephalins are the major lipidic constituents of tissues and are an important part of mammalian cell walls and membranes.

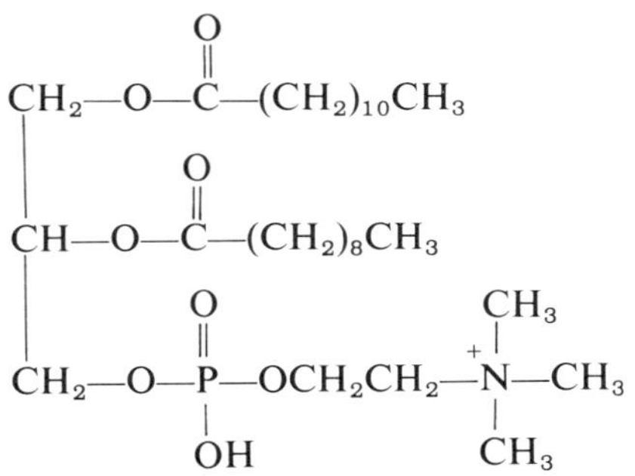

A lecithin

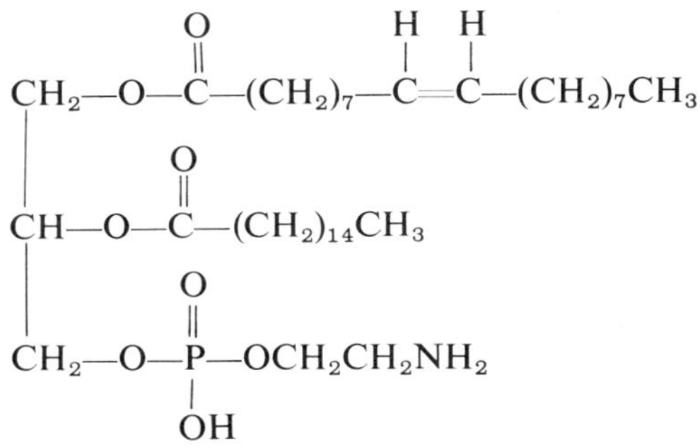

A cephalin

All of the lipids discussed in this chapter can be degraded to their component parts by acid- or base-catalyzed hydrolysis reactions since the ester and amide bonds involved in holding the compounds together are labile under these conditions. When brain tissue or other lipid-rich tissue is hydrolyzed, a fraction which is insoluble in water remains unhydrolyzed. This fraction consists of steroids and is predominantly cholesterol. The amount of cholesterol in the human body is quite large, being about 250 g in a 170-lb man. Cholesterol is incorporated into cell membranes and other lipidic structures in a regular way and apparently acts to strengthen these structures by synergistic interaction with the phospholipids present.

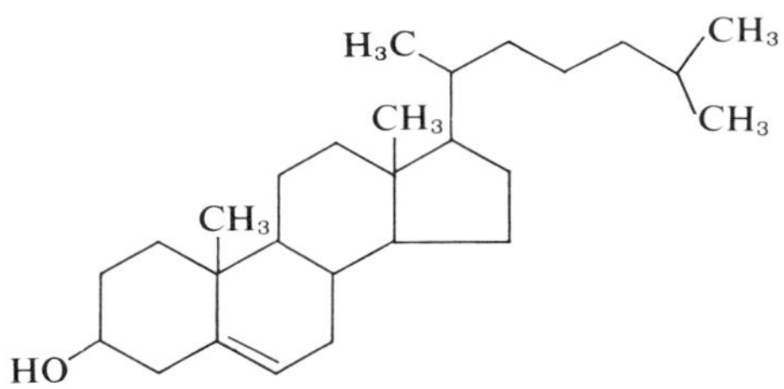

Cholesterol

PROBLEMS

1. How many grams of *m*-fluorobenzoic acid would be required to react with 2.5 g of sodium hydroxide?

2. How many grams of potassium hydroxide would be required to saponify (hydrolyze) 11 g of ethyl laurate?

3. What is the pH of a 0.1 *M* solution of chloroacetic acid?

4. What reaction(s) would you expect to take place when acetic acid is treated with methyl magnesium bromide?

5. Provide a common or systematic name for each of the following structures.

(a) 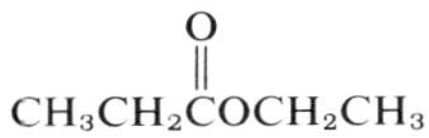

(b) 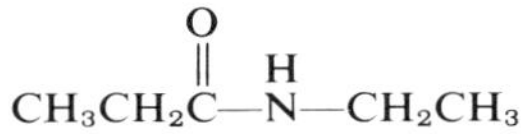

(c) 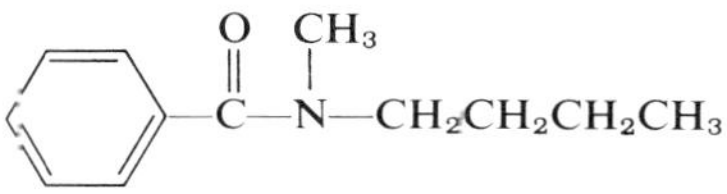

(d) $CH_3CH_2OCH_2CH_3$

(e) 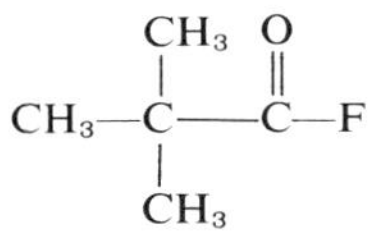

(f) 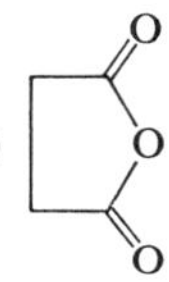

(g) 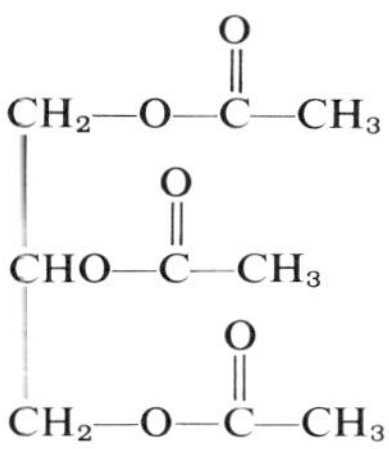

(h) 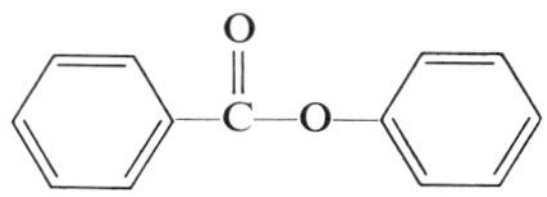

(i) 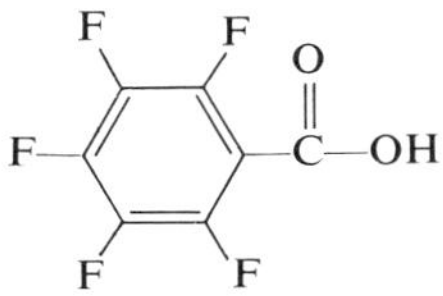

(j) 

6. Suggest a series of reactions that could be used to convert the starting material shown to the product indicated.

Starting material	Product
(a) Methyl bromide	Methyl acetate
(b) Benzoic acid	Benzoyl chloride
(c) 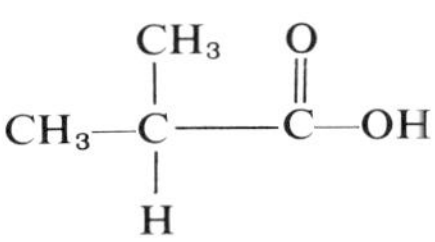	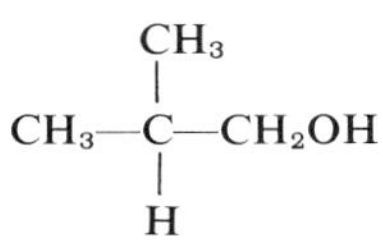
(d) 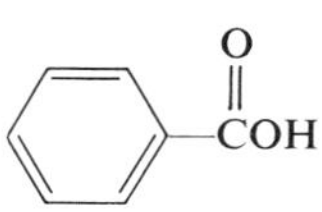	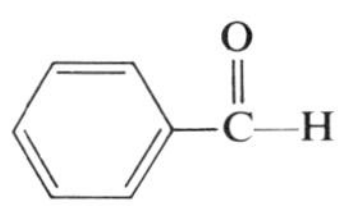

(e) $CH_3-\overset{\overset{O}{\|}}{C}-Cl$ $\qquad$ $CH_3CH_2-\overset{\overset{CH_3}{|}}{N}-CH_3$

(f) Butyryl bromide $\qquad$ Pentyl bromide

(g) Acetic acid $\qquad$ Acetic anhydride

(h) Propanoic acid $\qquad$ Butanoic acid

(i) 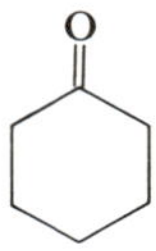 $\qquad$ 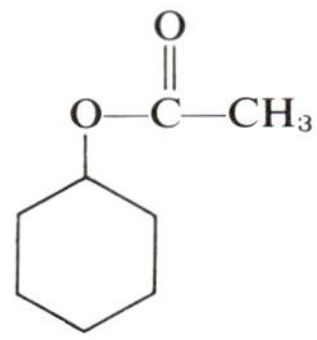

7. Give the products of the following reactions.
   (a) Acetic acid + calcium hydroxide $\longrightarrow$
   (b) Propyl alcohol + propionamide $\longrightarrow$
   (c) Butyric acid + phosphorus pentachloride $\longrightarrow$
   (d) Glycerol + stearoyl chloride $\longrightarrow$
   (e) Tribenzylphosphate + sodium hydroxide $\longrightarrow$
   (f) Cholesterol + acetic anhydride $\longrightarrow$
   (g) Butanoic acid + dimethylamine $\longrightarrow$

8. Compound A has a molecular formula C_8H_{10} and the pmr spectrum shown below. Upon oxidation with refluxing potassium permanganate, A was converted to B. The pmr spectrum of B is also given below. What are the structures of A and B?

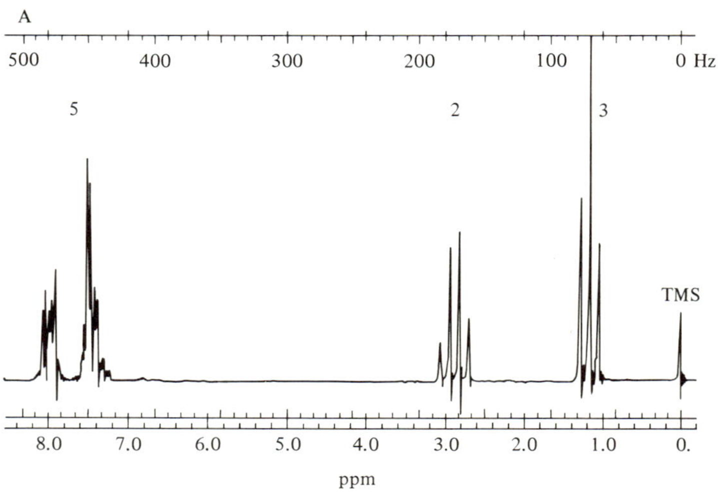

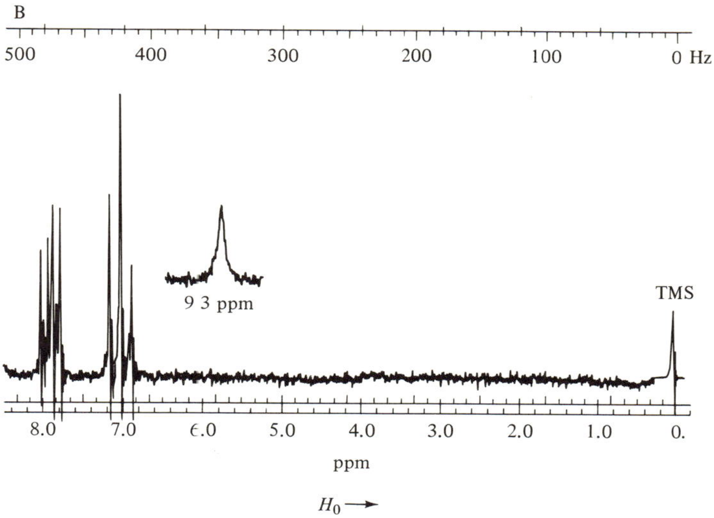

9. The pK_a of α-chlorobutyric acid is 2.85. What is the acid ionization equilibrium constant (K_a) for this acid?

10. *p*-Methoxybenzoic acid is a weaker acid than benzoic acid, while *p*-nitrobenzoic acid is a stronger acid than the unsubstituted material. Rationalize these observations.

11. Predict the relative order of acidities for 2-chlorobutanoic acid, 3-chlorobutanoic acid, and 4-chlorobutanoic acid.

12. While most esters hydrolyze by cleavage of the C—O bond between the alcohol part and the carbonyl group (acyl cleavage), if the alcohol (R′OH) is of certain structural types, the C—O bond between R′ and the carboxyl oxygen will

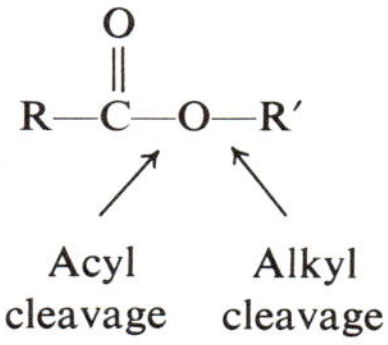

break (alkyl cleavage). Suggest an isotope labeling experiment that would enable one to distinguish between these two mechanisms for hydrolysis. (Hint: check Table 1.2.)

SUGGESTED READINGS

W. L. JOLLY, "The Inorganic Chemistry of Nitrogen." Benjamin, New York, 1964.

L. F. FIESER AND M. FIESER, "Advanced Organic Chemistry." Van Nostrand-Reinhold, Princeton, New Jersey, 1961.

S. PATAI, ED., "The Chemistry of Carboxylic Acids and Esters." Wiley (Interscience), New York, 1969.
H. RAKOFF AND N. C. ROSE, "Organic Chemistry." Macmillan, New York, 1966.
J. R. VAN WAZER, "Phosphorus and Its Compounds." Wiley (Interscience), New York, 1958.
J. ZABICKY, ED., "The Chemistry of Amides." Wiley (Interscience), New York, 1970.

12

Molecular Spectroscopy and Chemical Structure II

In earlier portions of this text we discussed the quantization of energy in molecules and their constituent atoms. In particular, it was noted that the electronic energy of an atom or molecule is quantized and, in Chapter 5, we discussed the way in which the quantization of nuclear energy can be put to use in the elucidation of the structure of organic molecules. The hydrogen nuclear magnetic resonance spectrum of a molecule was seen to be strongly characteristic of a given molecule and pmr spectroscopy provides a method by which the organic chemist can "visualize" the constitution of the hydrocarbon part of a molecule.

There are a number of other spectroscopic methods which utilize many of the same principles as nmr and which provide information of equivalent or higher quality about the functional groups of an organic structure. We shall briefly consider now the more important of these additional spectroscopic tools.

Review of the Principles of Spectroscopy

It is perhaps wise to survey again the fundamental ideas behind spectroscopic experiments. The precepts of quantum mechanics require that the energy of a molecule be quantized and, therefore, tell us that a molecule will be found only in certain discrete energy states. Upon stimulation by electromagnetic radiation of the correct frequency, it is possible to induce the molecule to assume a state that is higher in energy than was the one which characterized the molecule before the radiation impinged on the sample. Since the energy states are discrete, the energy differences between states must also be discrete, so that only energy (light) of certain frequencies can be absorbed. A plot of the frequency of the electromagnetic radiation incident upon a molecule against the quantity of light absorbed at this frequency defines a spectrum for the molecule. In our previous discussions of nmr spectroscopy the energy levels of interest were those which were characteristic of the protons (hydrogen nuclei) of an organic molecule. The electromagnetic radiation was in the radiofrequency range. Spectroscopic experiments are much wider in scope than the nmr phenomenon; the electromagnetic radiation spectrum spans a wide range of energies and can be used to examine nearly every aspect of a molecule's existence.

Infrared Spectroscopy

Viewed from a distance, the chemical bond between two atoms appears to be quite analogous to a spring which holds two balls together. A molecule behaves as an assemblage of balls and springs with all the balls (atoms) in constant motion relative to one another. These motions stretch bonds and deform the angles between bonds and are known collectively as molecular vibrations. In general, for a molecule made up of n atoms there will be $3n - 6$ unique molecular vibrations. The three possible vibrations of a simple molecule, water, are depicted in Fig. 12.1. The energy associated with each vibration of a molecule is quantized

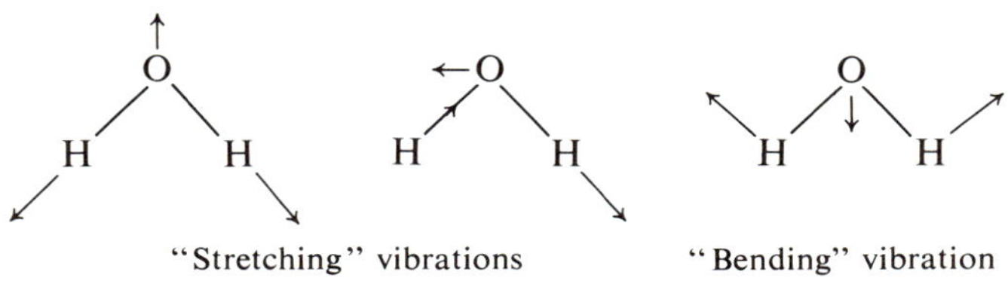

Fig. 12.1. The vibrational modes of water.

and each molecule has a characteristic set of vibrational energy levels. The electromagnetic radiation of proper energy to induce transitions between these vibrational energy levels is in the infrared region and has a frequency of ~300 to ~4000 cm^{-1}. A plot of the amount of light absorbed as a function of the wavelength of incident infrared radiation defines the *infrared spectrum* of a molecule. Figure 12.2 depicts the infrared spectrum of water; each band in the spectrum

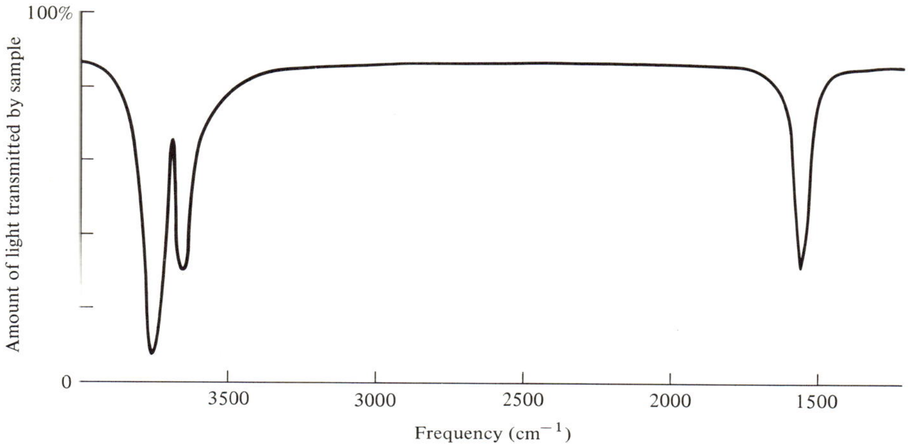

Fig. 12.2. The infrared spectrum of water.

corresponds to one of the vibrations described above. Since an infrared spectrum represents the vibrational behavior of a molecule and the vibrational behavior depends on the nature of the chemical bonds which hold the molecule together, an infrared (ir) spectrum can be highly characteristic of an organic structure.

Virtually all of the peaks in an infrared spectrum correspond to excitation of the molecule from its lowest, or ground, vibrational energy state and the quantities of energy involved in the excitation process are small, about 1–10 kcal/mole. Quantum theoretical considerations of the infrared absorption process indicate that vibrational excitation will be detectable in an ir spectrum only when the molecular motion corresponding to a vibration involves a change in the molecular dipole moment as the motion takes place. Thus, some vibrations in molecules which involve motions of high symmetry are not apparent in the ir spectrum of a molecule. A simple example of this effect is afforded by the nitrogen molecule. No matter how far the molecular bond along the nitrogen–nitrogen axis is stretched, the dipole moment of the molecule remains unchanged (Fig. 12.3). The nitrogen molecule has no infrared spectrum for this reason. Molecular motions which do not give rise to peaks in the ir spectrum are said to be *infrared inactive*.

Since the chemical properties of functional groups are normally not grossly

Fig. 12.3. The vibration of a nitrogen molecule. No dipole moment change is involved in this motion so that nitrogen has no infrared spectrum.

Table 12.1
Infrared Absorptions Associated with Common Functional Groups

Functional group	*Infrared absorption bands* (cm^{-1})[a]
CARBON–HYDROGEN BONDS	
—C—H (with two further single bonds on C)	2800–3000 1300–1500
=C(H)—	3000–3080 ~1400 ~900
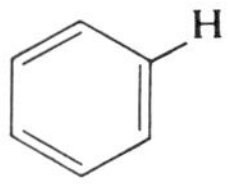 (aromatic)	3070 ~1200 700–900
C≡C—H	3300
—C(=O)H (aldehyde)	2700–2800
OXYGEN–HYDROGEN BONDS	
—O—H	3200–3600
NITROGEN–HYDROGEN BONDS	
—NH_2	3400–3500
—N⊕—H (with two further bonds on N)	2200–3000 1500–1600
SULFUR–HYDROGEN BONDS	
—S—H	2500–2600
CARBON–CARBON BONDS	
C=C	1640–1690
C=C (aromatic)	1500–1600 (weak)
C≡C	2080–2260
CARBON–NITROGEN BONDS	
C—N	1000–1200
C—N (aromatic amine)	1200–1400
C=N	1600–1700
C≡N (nitrile)	2210–2260

Table 12.1 *(Continued)*

Functional group	*Infrared absorption bands* (cm^{-1})[a]
CARBON–OXYGEN BONDS	
C—O	1000–1200
C=O (ketone)	1650–1720
C=O (aldehyde)	1720–1740
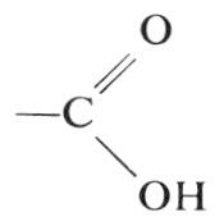 (carboxylic acid)	1700–1730
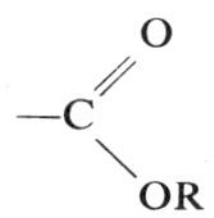 (ester)	1730–1750 1000–1300
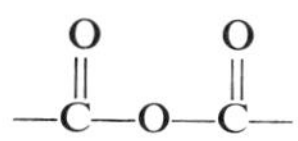 (anhydride)	~1760 ~1820
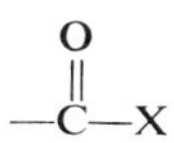 (X≡halogen)	~1800
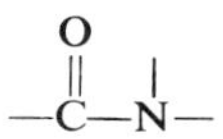 (amide)	~1300 1630–1690
MISCELLANEOUS BONDS	
—NO_2 (nitro)	1500–1600 1300–1400
—SO_2—N— (sulfonamide)	1330–1400 1150–1180

[a] Absorptions usually occur in the range given.

affected by the nature of the hydrocarbon skeleton to which they are bound, one finds that the infrared excitations associated with a particular functional group appear at nearly the same position in the ir spectrum, irrespective of the compound in which the functional group is found. Each functional group, thus, has a characteristic set of infrared frequencies which are diagnostic for the group and which can be used to identify the group.

An efficient way to utilize ir spectroscopic information in structural determinations is by the use of a correlation table. Over the past three decades, a great amount of information about the infrared properties of various functional groups has been accumulated, and correlations between the position of peaks in the ir spectrum and the properties of functional groups have been made. A correlation table presents in abbreviated form the results of studies of this kind. A short ir correlation table is given in Table 12.1. For long versions, the references at the end of this chapter can be consulted.

Infrared spectra may be obtained with gases, liquids, or solids. Spectra of solutions of organic molecules in various solvents (carbon tetrachloride is typical) can be obtained, although one must take care that the absorption bands of the solvent do not overlap the bands of the solute. The optical systems of infrared spectrometers cannot be constructed of glass or plastic since these materials absorb in the infrared and, instead, optics and sample containers (cells) made of crystalline sodium chloride, potassium bromide, silver chloride, or zinc sulfide are used in ir spectrometers.

The technology of infrared spectrometers has evolved remarkably, and relatively inexpensive instruments that can provide high quality spectra are available. A typical infrared spectrometer is shown in Fig. 12.4. It is now possible (and essential!) for virtually every organic chemist to obtain ir spectra of compounds that are important to him. These spectra, when reported in the chemical litera-

Fig. 12.4. A commercial infrared spectrometer. (Photograph courtesy of Perkin–Elmer Corporation.)

ture, are summarized in terms of the position of the important (and characteristic) absorptions given in frequency units of reciprocal centimeters (cm^{-1}) or the wavelength, expressed in microns (μ) or micrometers (μm). It is difficult to quantify the intensities of ir bands and the designations *s*, *m*, and *w* are usually used tc denote relatively strong, medium and, weakly absorbing transitions, respectively.

EXAMPLES OF THE USE OF INFRARED SPECTRA

A. A compound of molecular formula $C_3H_3NO_2$ gave the ir spectrum shown in Fig. 12.5. What is a likely structure for this compound? Referring to Table 12.1 it is found that the sharp bond at 2230 cm^{-1} is very characteristic of a nitrile (—C≡N) or possibly an acetylenic functional group. Table 12.1 indicates

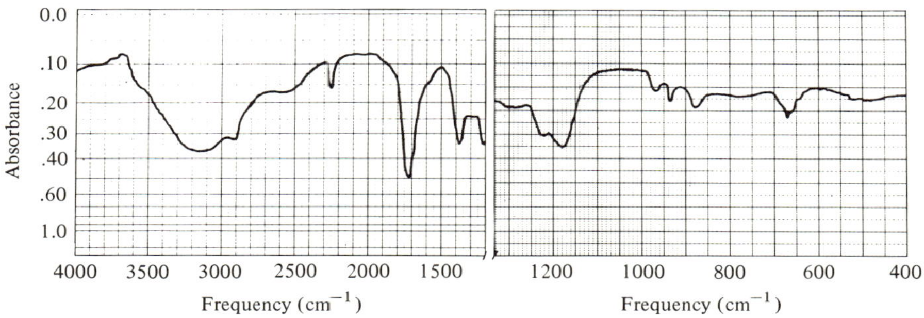

Fig. 12.5. The infrared spectrum of cyanoacetic acid.

that these are the only groups which absorb in this region. The strong bond at 1710 cm^{-1} likely signifies the presence of a carbonyl group in an aldehyde or ketone; a carboxylic acid might also absorb at this position and there is an indication of an —O—H group in the 3000 cm^{-1} region of the spectrum. As a trial structure for the unknown compound one can suggest cyanoacetic acid, N≡C—CH_2—COOH.

B. The infrared spectrum in Fig. 12.6 is that of propyl ether, $CH_3CH_2CH_2OCH_2CH_2CH_3$. The broad band absorptions between 2700 and 3000 and peaks in the 1300–1500 range are due to the C—H vibrations of the molecule. Most distinctive, however, is the absorption centered at 1120 cm^{-1}; this band is characteristic of the ether functional group.

C. A nitrogen-containing heterocycle which had an ir band at 1720 cm^{-1} gave a new compound when treated with acetic anhydride. The infrared spectrum of the new material is shown in Fig. 12.7. Mass spectral data indicated that the molecular weight of the product was 129 and that its molecular formula

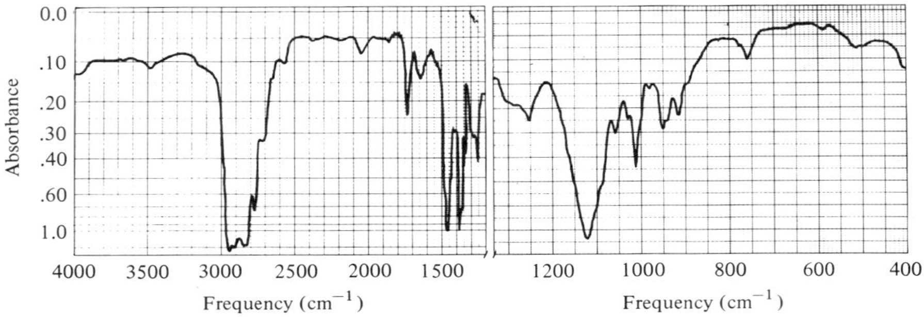

Fig. 12.6. The infrared spectrum of propyl ether.

was $C_6H_{11}NO_2$. The ir band at 1720 cm^{-1} indicates the presence of an aldehyde or ketone and is probably the same functional group that was present in the starting compound, while the adjacent carbonyl absorption at 1640 probably is due to an amide group formed in the acetylation reaction. The starting material was 4-piperidone and the product of the above-described reaction was *N*-acetyl-4-piperidone.

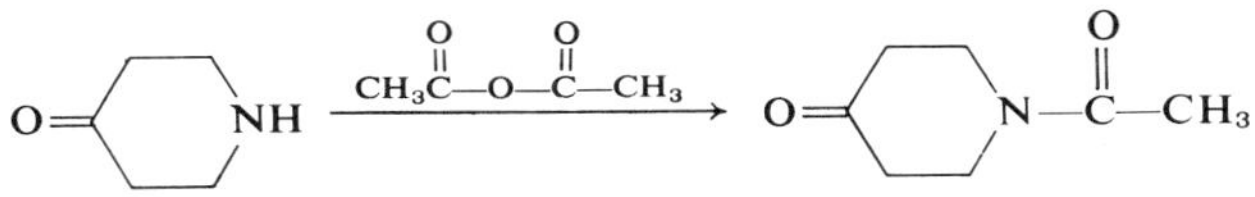

Unless an organic molecule is very symmetrical, the infrared spectrum of the molecule will have a large number of absorption bands. The positions and intensities of these bands are strongly characteristic of the molecule, so characteristic that, for small molecules, the ability to superimpose the two infrared spectra is considered sufficient evidence that two substances are identical. It has been suggested that identical ir spectra for two different compounds can be expected roughly once in ten thousand times.

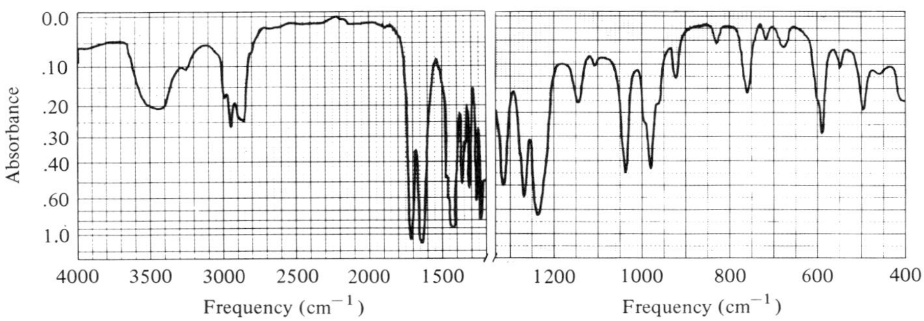

Fig. 12.7. The infrared spectrum of *N*-acetyl-4-piperidone.

Raman Spectroscopy

Raman spectroscopic experiments cover about the same frequency range as infrared experiments, but the two techniques are complementary rather than competitive. The Raman effect arises when a monochromatic beam of light strikes a molecule. Most of the light in the incident beam is scattered or reflected by the molecule. However, a small amount of the scattered radiation is at a different frequency than is the frequency of the incident beam (Fig. 12.8). It turns out

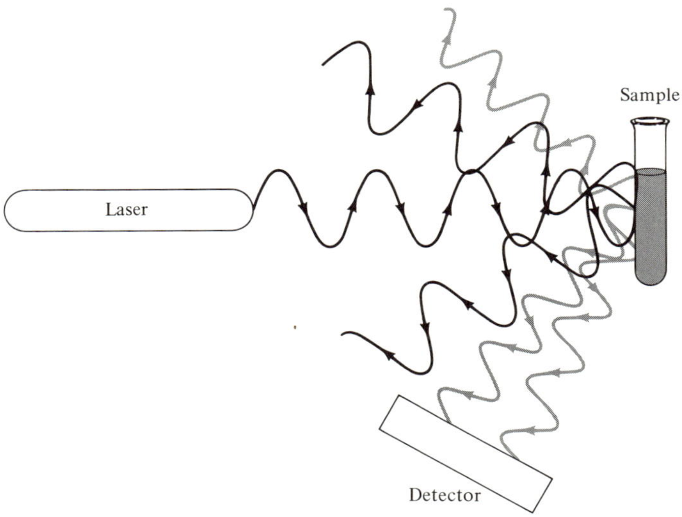

Fig. 12.8. A schematic representation of the Raman experiment. The light source is usually a laser. The black lines correspond to light scattered with the same frequency as the incident beam while the gray lines depict the light at other frequencies (Raman effect).

that the differences in frequency between this radiation and that of the incident beam corresponds to the vibrational absorption bands of the molecule. Thus, the bands in a Roman spectrum appear at the same frequencies as those in the corresponding ir spectrum. The quantum-mechanical rules which govern the intensity of these lines are different in the Raman effect than in ir spectroscopy and, in general, those lines which appear as only weak absorptions in the ir spectrum will be much stronger in the Raman effect and vice versa.

Until recently instrumentation for obtaining Raman spectra was crude. With the advent of inexpensive, high-powered lasers, several commercial instruments with very good performance have appeared on the market and one can expect

Raman spectroscopy to be a much more commonly applied spectroscopic tool in the future.

Ultraviolet and Visible Spectroscopy

A number of functional groups have rather loosely held electrons in their chemical bonding system. With the absorption of light energy in the ultraviolet (uv) or visible (vis) region of the electromagnetic radiation spectrum, the loosely held electrons can be promoted to higher energy states. A graph showing the amount of ultraviolet and visible light absorbed at each wavelength in this region defines the ultraviolet and visible spectra of the compound under examination. The positions of the absorption maxima are expressed in terms of wavelength, the units being the millimicron (mμ) or the nanometer.* The visible region of the spectrum extends from 350 to 800 nm, while the ultraviolet portion ranges from 100 to 350 nm. At wavelengths below about 200 nm it is necessary to perform the uv spectroscopic experiment with equipment that is evacuated because several of the gases in air absorb at these wavelengths.

The application of ultraviolet and visible spectroscopy by the organic chemist usually involves samples in solution, although spectroscopic experiments with gases or solids are quite feasible. The solution is contained in a glass or quartz cell (cuvette) which provides a path length of 0.5 to 10 cm. Commercial uv spectrometers permit highly accurate measurements of the degree to which light absorption takes place during the uv or visible spectroscopic experiment. The amount of light absorbed depends on the concentration of absorbing species, the pathlength of the sample cell and the inherent capability of the absorbing molecule to absorb light and utilize the incident radiation for electronic excitation. The equation below, called the Beer–Lambert law, summarizes the experimental dependence of the absorbance (A) of a sample on these parameters.†

$$A = \log I_0/I = \epsilon lc$$

Here I_0 is the intensity of the light incident on the sample, I is the intensity of light transmitted by the sample, c is the molar concentration of the absorber, l is the pathlength in centimeters, and ϵ is a proportionality constant, called the extinction coefficient or absorptivity, which reflects the inherent ease with which

---

* Recently the use of nanometers (nm) for these units has been encouraged. 1 mμ = 10^{-7} cm = 10^{-9} m = 1 nm.

† This same equation describes the absorption of infrared radiation but is utilized less in this context because quantitating the extent of absorption in their experiment is more difficult than in the case of uv-visible spectroscopy.

a molecule accepts a photon at a particular wavelength. The extinction coefficient is a function of wavelength and may range in magnitude from 0 to $\sim 10^5$.

Table 12.2 records generalized ultraviolet and visible spectral properties of the more common organic structures. It is noted that all of these groups have π-molecular orbital components in their electronic system. The absorption bands observed usually correspond to transitions to or from these π-orbitals, although the identification of the nature of an electronic transition is not always unambiguous. The exact position and intensity of a particular absorption is dependent on the structure of the molecule, on the solvent, and, occasionally, on the concentration of the chromophore itself. A number of correlations of these effects with structure are useful and are available in the references given at the end of this chapter.

In addition to the confirmation and elucidation of molecular structure, ultraviolet and visible spectroscopy provide the organic and biological chemist with

Table 12.2
Ultraviolet and Visible Absorption Bands

Type of functional group	λ_{max}	Approximate ϵ
Aromatic compounds	180–220	5×10^3
Olefin, >C=C<	210–230	3×10^4
Acetylenes, —C≡C—	170–175	5×10^3
Aldehydes, —C(=O)—H	~190	10^2
	~290	10^1–10^2
Ketones, —C(=O)—	~180	10^3–10^4
	~280	10^1–10^2
Carboxylic acids, —C(=O)—OH	~205	10^1–10^2
Esters, —C(=O)—OR	~205	10^1–10^2
Amides, —C(=O)—NH—	~210	10^1–10^2
Acid halides, —C(=O)—X	~235	10^1–10^2
Nitriles, —C≡N	~340	10^2
Nitro, —NO_2	~210	2×10^4
	~275	10^2
Azo, —N=N—	~350	2×10^4
	~440	1×10^3

an important tool for quantitative analysis. Because the extinction coefficients for many electronic transitions are large, it is possible to measure low concentrations of the chromophoric molecule. Typically, uv spectrometers can measure absorbances (A) of 0.005 to 3.0 with high accuracy. With reference to the Beer–Lambert law given above, one can calculate that for a species with $\epsilon = 10^4$, it is possible to measure concentrations as low as 10^{-6} *M* if a 1-cm pathlength cell is used. This sensitivity is ideal for numerous experiments involving biological macromolecules.

Optical Rotatory Dispersion

In an earlier chapter, the ability of certain types of organic compounds to rotate plane-polarized light was discussed. This phenomenon of optical rotation is found in materials which are asymmetric, that is, substances which are not identical with their mirror images. Most optical rotations are reported at a particular wavelength of observing light, usually the so-called D-line of the sodium emission spectrum (589 nm). However, the amount of optical rotation obtained with a given compound is a function of the wavelength of the light used to observe the rotation; study of this functional dependence makes up an investigation of the optical rotatory dispersion (ord) characteristics of the molecule.

Commercially available instruments permit the measurement of ord spectra between 230 and 700 nm. Thus, a chromophore that absorbs in the ultraviolet or visible portion of the electromagnetic spectrum can potentially give rise to effects in the ord spectrum if the molecule containing the chromophore is asymmetric. Optical rotatory dispersion spectra (a plot of optical rotation against wavelength) may exhibit maxima, minima, or inflections.

A common chromophore in the ultraviolet region which is important in much of organic chemistry is the carbonyl group. The ord spectrum of (+)-3-methylcyclohexanone is shown in Fig. 12.9. One notes a sharp maximum in the rotation at

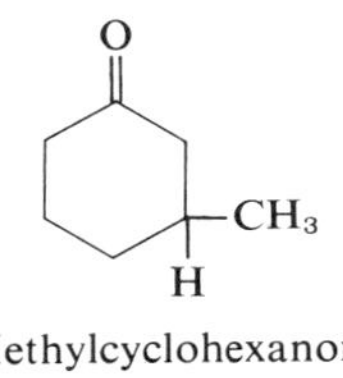

3-Methylcyclohexanone

~310 nm, a rapid drop through zero rotation, and, finally, a minimum at about 280 nm. The maximum is often called

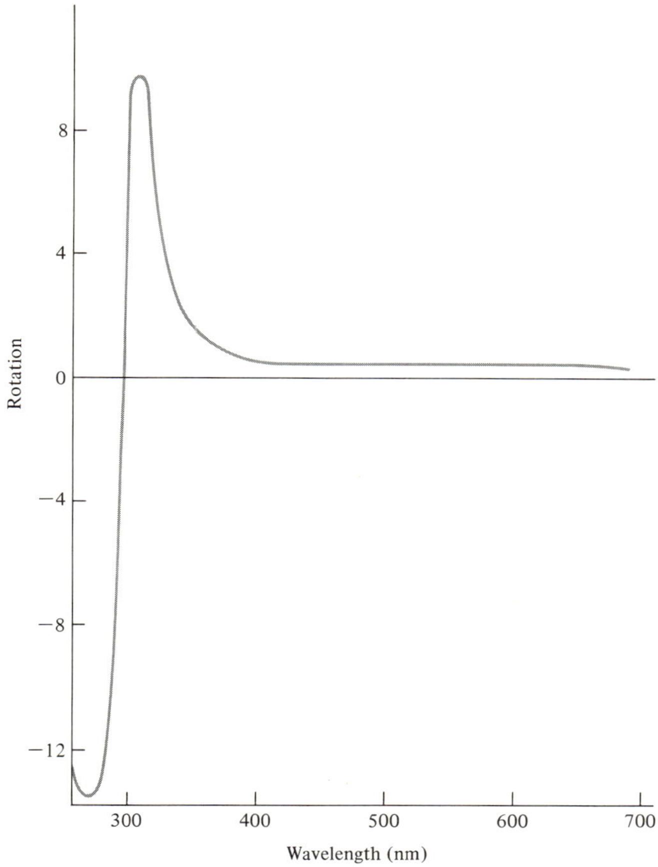

Fig. 12.9. The ord spectrum of (+)-3-methylcyclohexanone. (Spectrum is based on that appearing in C. Djerassi, *Optical Rotatory Dispersion*, McGraw-Hill, New York, 1960, p. 103.)

a peak, in order to avoid confusion with the concept of an ultraviolet spectral maximum (which, in this case, occurs at a different wavelength), while the minimum in the ord curve is referred to as a trough. The total differences in rotation between the peak and the trough is the *amplitude* of the effect and the distance, in wavelengths, between the peak and trough is the *breadth* of the effect. Figure 12.9 depicts a positive *Cotton effect* in that the peak occurs at a longer wavelength than the trough. If the opposite is true, then the curve corresponds to a negative Cotton effect. Finally, when a number of peaks and troughs and inflections are apparent in the spectrum, the curve is said to exhibit multiple Cotton effects.

An important use of ord spectra is the establishment or verification of the absolute stereochemical configuration in asymmetric molecules, and qualitative correlations between the shape of ord spectral curves and the configurations of asymmetric molecules have been obtained. A natural territory for the application of ord experiments is in the study of proteins and other biopolymers. An additional element of asymmetry may be introduced into these molecules by the formation of well-defined conformational features and optical rotatory dispersion is one of the most informative means available for detecting and quantitating changes in biopolymer conformation.

X-Ray Crystallography

Probably one of the most important discoveries of the twentieth century is the diffraction of x-rays by crystalline materials. X-rays are electromagnetic radiation of very short wavelengths; they are therefore highly energetic and can penetrate most matter with ease. When a beam of x-rays impinges upon a crystal of an organic substance it interacts with the electrons surrounding the atomic nuclei composing the molecule and is scattered in various directions relative to the direction of the original beam. The directions of the scattered x-rays are not random, however, but instead are rather well-defined and quite characteristic of the arrangement of the atoms in the molecule. A careful study of x-ray diffraction can thus provide detailed information about the precise positions of atoms in organic molecules—information that is impossible to obtain in any other way. Unfortunately, the interpretation of diffraction patterns in terms of crystal and molecular structure is not easy and, until recently, the deduction of an organic structure by crystallographic methods usually meant an investment of several man-years of tedious labor. Electronic computers have taken over a large part of this task and it has recently been possible in several cases to solve the structure of rather complex molecules in several weeks. Highly automated equipment for the collection of x-ray diffraction data has become available and the day when the organic chemist can use x-ray crystallography as routinely as nmr or ir spectroscopy is approaching.

Some of the most exciting investigations of molecular structure by x-ray methods have involved complex molecules of biological origin. Indeed, most of what is really known about biological macromolecules has been learned in this way. The three-dimensional structures of vitamin B_{12} (Fig. 12.10), myoglobin, several cytochromes and enzymes have been elucidated by x-ray diffraction techniques. A serious limitation to the use of x-ray crystallography is the requirement that the material to be investigated must be crystalline and very large biological molecules are usually quite resistant to attempts to crystallize them.

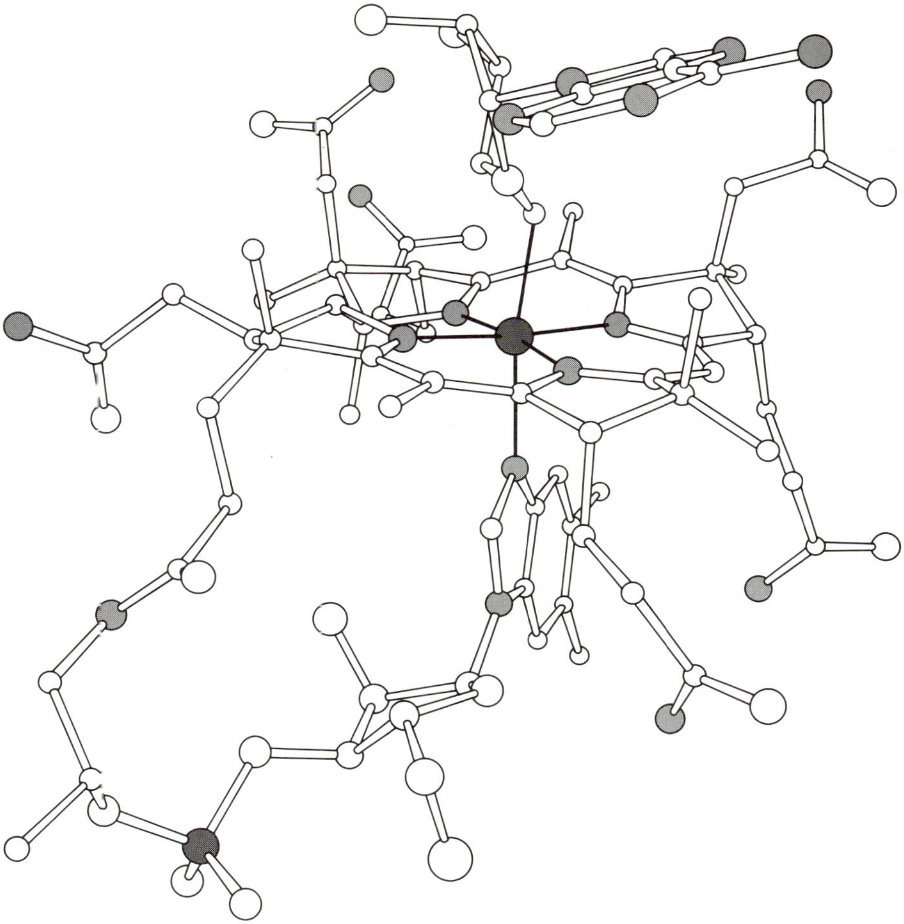

Fig. 12.10. The structure of vitamin B_{12} as deduced by x-ray crystallographic methods. The large central atom is cobalt; nitrogen and oxygen atoms are shaded. No hydrogen atoms are shown. (Drawing courtesy of Professor P. G. Lenhert, Vanderbilt University.)

We have tried to indicate briefly in this chapter and in Chapter 5 some of the instrumental techniques which provide structural information about organic molecules to the organic and biological chemist. The prognosis for the future is that much greater reliance will be placed on these and other perhaps yet undiscovered instrumental methods. Utilization of computers and forms of automation will extend the availability and applicability of these methods, and the use of instrumental techniques to probe biological macromolecules should provide new insights regarding the relations between structure and function of these materials.

PROBLEMS

1. In each of the problems below, sufficient spectroscopic data have been given to permit the identification of the unknown compound. Provide an organic structure that is consistent with the data and indicate the reasoning that led to your choice.

   (a) A compound of formula C_2H_3N gives only a single line in the pmr spectrum. The ir and Raman spectra of the compound show a strong band at 2250 cm^{-1}.

   (b) A colorless liquid has the formula C_2H_6O and shows strong infrared absorption at 3600 cm^{-1}.

   (c) A hydrocarbon with a molecular weight of 92 has an ultraviolet absorption band at 261 nm. The pmr spectrum exhibited singlets at $\delta = 7.75$ and $\delta = 2.65$ ppm from TMS.

2. Suggest prominent ir bands of the following molecules that would be characteristic and aid in their identification.

   (a) Nitromethane
   (b) Toluene
   (c) Acetophenone
   (d) Uracil
   (e) Ethanthiol
   (f) Benzaldehye

3. Suggest a spectral technique other than mass spectrometry that could be used to distinguish between the members of the following sets of molecules.

   (a) Acetone and acetic acid

   (b)

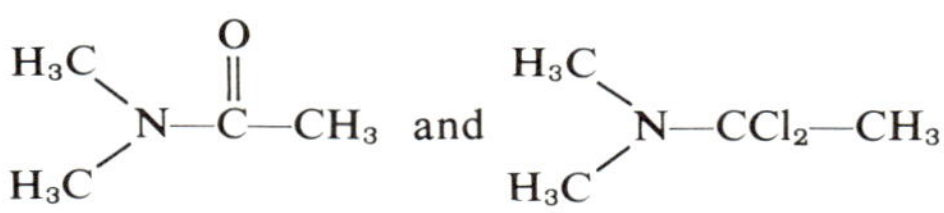

   (c)

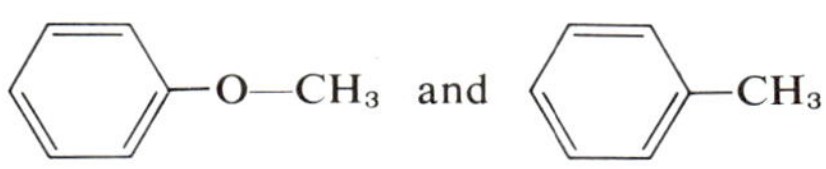

   (d)

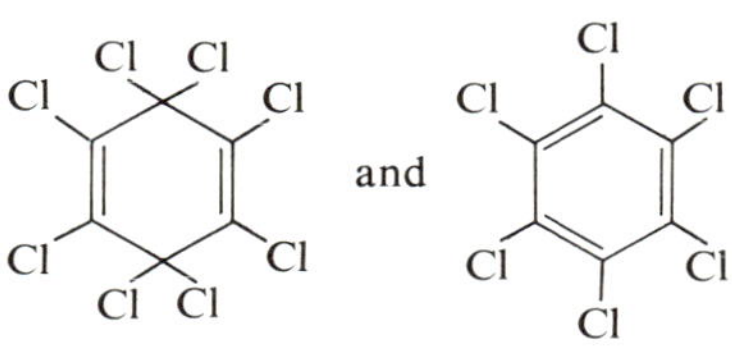

   (e)

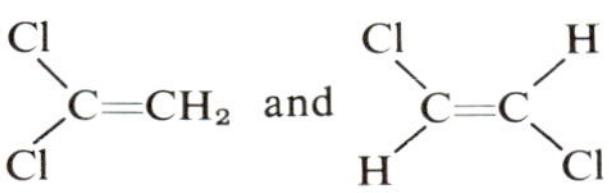

   (f) Furan and pyrrole

(g)

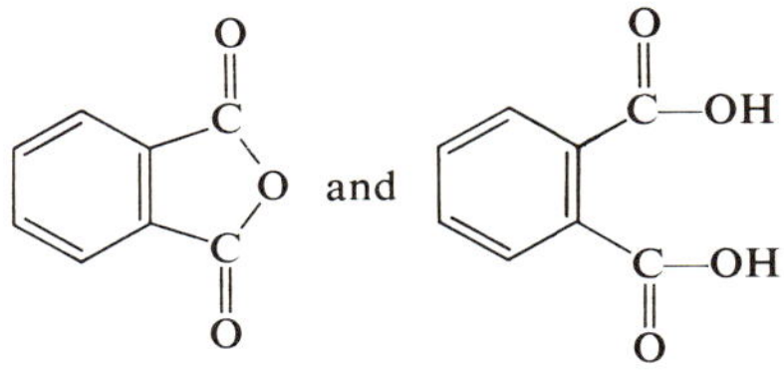

(h) D-Glucose and methyl-β-D-glucoside.

4. The pmr (a) and infrared (b) spectra of a compound (C_8H_5NO) are shown below. What is the compound?

(a)

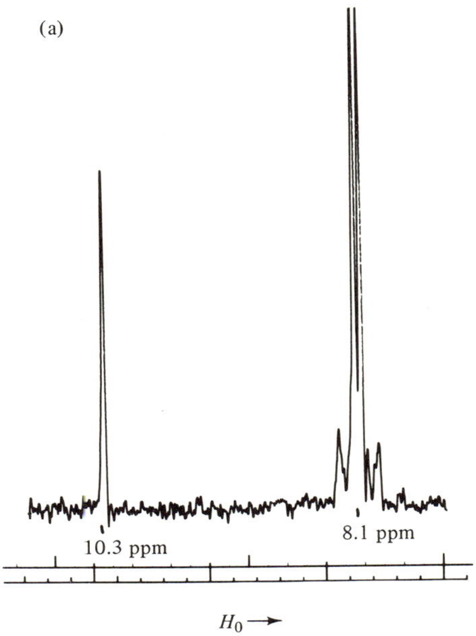

(b)

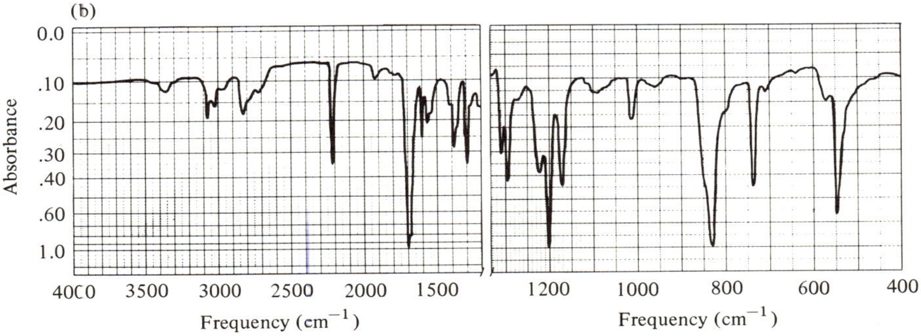

SUGGESTED READINGS

Introductory Treatments

E. F. H. Brittain, W. O. George, and C. H. J. Wells, "Introduction to Molecular Spectroscopy." Academic Press, New York, 1970.

J. R. Dyer, "Applications of Absorption Spectroscopy of Organic Compounds." Prentice-Hall, Englewood Cliffs, New Jersey, 1965.

D. J. Pasto and C. R. Johnson, "Organic Structure Determination." Prentice-Hall, Englewood Cliffs, New Jersey, 1969.

Infrared Spectroscopy

L. J. Bellamy, "The Infrared Spectra of Complex Molecules." Wiley, New York, 1958.

R. T. Conley, "Infrared Spectroscopy." Allyn and Bacon, Boston, Massachusetts, 1966.

C. N. R. Rao, "Chemical Applications of Infrared Spectroscopy." Academic Press, New York, 1963.

Ultraviolet and Visible Spectroscopy

A. E. Gillam and E. S. Stern, "An Introduction to Electronic Absorption Spectroscopy." Edwards Arnold, London, 1957.

H. H. Jaffe and M. Orchin, "Theory and Applications of Ultraviolet Spectroscopy." Wiley, New York, 1962.

F. A. Matsen, *in* "Technique of Organic Chemistry" (A. Weissberger, ed.). Wiley (Interscience), New York, 1956.

Optical Rotatory Dispersion

P. Crabbe, "Optical Rotatory Dispersion and Circular Dichroism in Organic Chemistry." Holden-Day, San Francisco, California, 1965.

C. Djerassi, "Optical Rotatory Dispersion." McGraw-Hill, New York, 1960.

X-Ray Crystallography

W. N. Lipscomb, *in* "Determination of Organic Structures by Physical Methods" (A. Weissberger, ed.), Vol. 1, p. 1641. Wiley (Interscience), New York, 1945.

J. M. Robertson, "Organic Crystals and Molecules." Cornell Univ. Press, Ithaca, New York, 1953.

D. E. Sands, "Introduction to Crystallography." Benjamin, New York, 1969.

M. M. Woolfson, "An Introduction to X-ray Crystallography." Cambridge Univ. Press, London and New York, 1970.

Raman Spectroscopy

T. L. Gibson and P. J. Hendra, "Laser Raman Spectroscopy." Wiley (Interscience), London, 1970.

H. J. Sloan, The technique of Raman spectroscopy: A state-of-the-art comparison to infrared. *Appl. Spectro.* **25**, 430 (1971).

H. Szymanski, ed., "Raman Spectroscopy." Plenum, New York, 1967.

13

Amino Acids, Peptides, and Proteins

The name *protein* was coined by Mulder in 1839 from a Greek word meaning "of prime importance," a choice that was singularly appropriate. Indeed, the chemistry of proteins is the chemistry of life itself. Virtually none of the thousands of organic reactions that take place in a living system would be possible without protein molecules present as catalysts; proteins also serve as structural materials and as messenger molecules (hormones), and they take part in the defensive systems that guard living organisms against foreign proteins.

Proteins are polymeric molecules that have molecular weights in the range 5000 to several million. The monomeric units for these polymers are α-amino acids, with the physical and chemical properties of the protein being determined by the properties of each amino acid and by the position of the amino acid in the polymer chain.

α-AMINO ACIDS

As the name implies, amino acids are organic molecules that feature both an amino group and an acid group in the same structure. The relative positions of the amino group and carboxyl group could be arbitrary in the molecule, but all amino acids important in the construction of proteins have the amino group attached to the carbon directly adjacent to the carboxyl carbon atom, i.e., to the α-carbon. This arrangement has one immediate and important consequence:

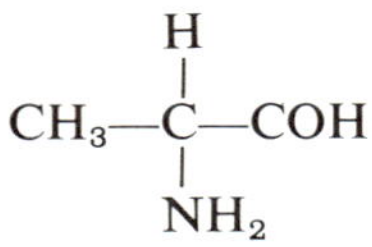

Alanine
An α-amino acid

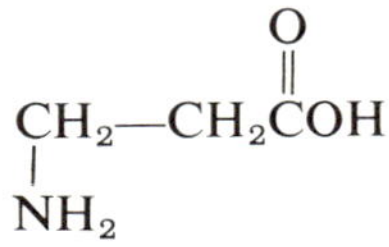

β-Alanine
A β-amino acid

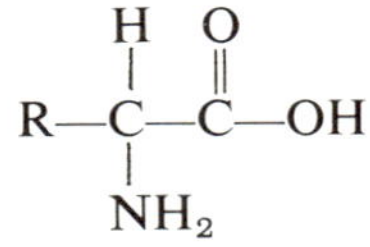

General form for
an α-amino acid

all α-amino acids except one (glycine, where R = H) can exist in optically isomeric forms, since the α-carbon atom has four different substituents. The configuration at the asymmetric carbon of naturally occurring serine (R = CH_2OH), for example, has been determined relative to glyceraldehyde, and has been

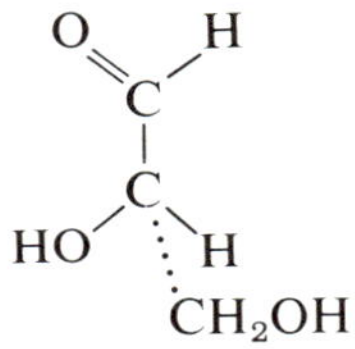

L-Glyceraldehyde

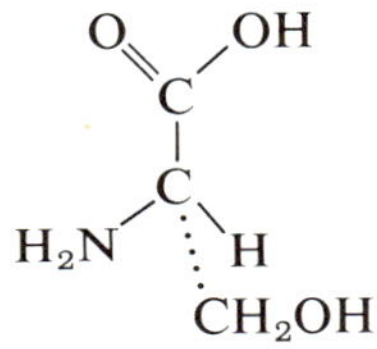

Natural serine
(L-serine)

found to be the L isomer. With a very few exceptions, all α-amino acids found in proteins have the L configuration. All amino acids written from this point on should be understood to be in the L configuration.

It is the structure of the amino acid side chains, designated R in the general formula above, that determines the properties of a given amino acid. About 20 different side chains are found in most proteins; the corresponding amino acids are grouped into a somewhat arbitrary classification scheme in Table 13.1. At this stage it is important to note that some side chains are rich in hydrocarbon character and are, therefore, hydrophobic while other side chains may contain acidic or basic functional groups. Some side chains have a hydroxyl group, making the corresponding amino acid also an alcohol or phenol. These side chains have greater dipolar character and water solubility than those of a more strictly hydrocarbon character. A thiol, a disulfide, and a sulfide are also among these 20 fundamental units of protein structure.

Amino acids are amphoteric organic molecules; that is, they can function as acids and as bases. This would be expected in light of the fact that the molecules have both a basic amino group and a proton-donating carboxyl group in their structure. The carboxyl groups of α-amino acids are characterized by acid dissociation constants of 10^{-3} to 10^{-2}, while the dissociation constants for the

Table 13.1
The Common α-Amino Acids[a]

Name	R—	pK_a *Amino group*	pK_a COOH *group*	*Iso-electric point*
HYDROCARBON-LIKE SIDE CHAINS				
Glycine	H—	9.60	2.34	5.97
Alanine	CH_3—	9.69	2.35	6.02
Valine	H | CH_3—C— | CH_3	9.62	2.32	5.97
Leucine	H | CH_3—C—CH_2— | CH_3	9.60	2.36	5.98
Isoleucine	H | CH_3—CH_2—C— | CH_3	9.68	2.36	6.02
Proline[b]	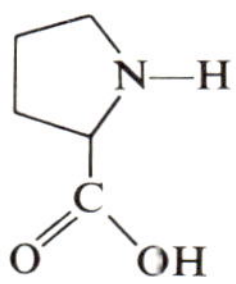	10.60	1.99	6.10
Phenylalanine	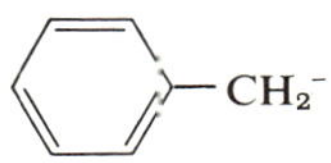	9.13	1.83	5.98
Tryptophan	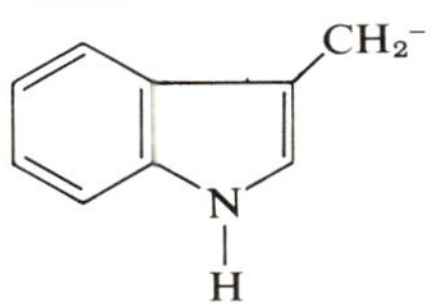	9.39	2.38	5.88
Methionine	$CH_3SCH_2CH_2$—	9.21	2.28	5.75
ACIDIC OR AMIDE SIDE CHAINS				
Aspartic acid	O ‖ HOC—CH_2—	9.82	2.09 3.86(β)	2.87
Asparagine	O ‖ H_2NC—CH_2—	8.8	2.02	5.41
Glutamic acid	O ‖ HOC—CH_2CH_2—	9.67	2.19 4.25(γ)	3.22

Table 13.1 (*cont.*)

Name	R—	pK_a *Amino group*	pK_a COOH *group*	*Iso-electric point*
Glutamine	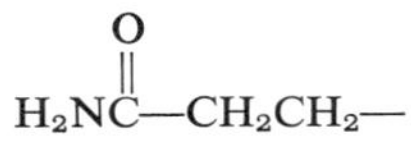	9.13	2.17	5.65
BASIC SIDE CHAINS				
Lysine	$H_2NCH_2CH_2CH_2CH_2$—	8.95 10.53(ε)	2.18	9.74
Arginine	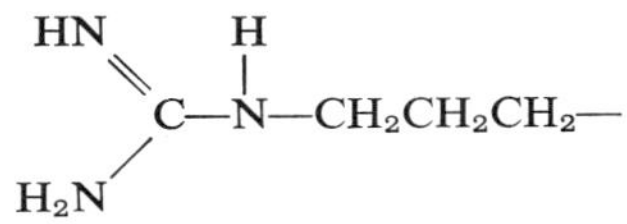	9.04 12.48	2.17	10.76
Histidine	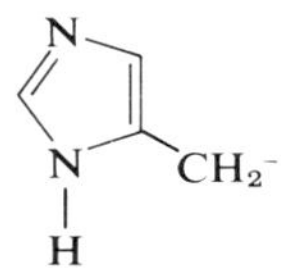	9.17	1.82	7.58
OTHER POLAR SIDE CHAINS				
Serine	$HOCH_2$—	9.15	2.21	5.68
Threonine	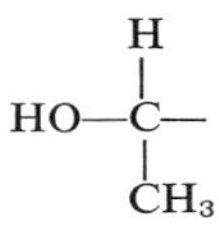	10.34	2.63	6.53
Tyrosine	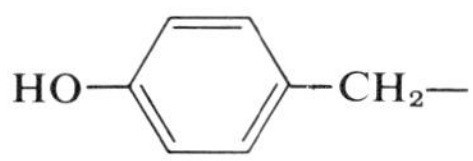	9.11 10.07(OH)	2.20	5.65
THIOL–DISULFIDE SIDE CHAINS				
Cysteine	$HSCH_2$—	10.78 8.33(SH)	1.71	5.02
Cystine	S—CH_2— | S—CH_2—	7.95 9.85	1.65 2.26	5.06

[a] Table is assembled from data given in H. R. Mahler and E. H. Cordes, "Biological Chemistry," p. 10. Harper & Row, New York, 1966.

[b] Proline is a cyclic amino acid: the complete molecular structure is drawn here.

conjugate acid of the amino group are 10^{-10} to 10^{-9}. In solutions where the hydronium ion concentration is between 10^{-9} *M* and 10^{-4} *M* (pH 9 to pH 4) essentially all of the carboxyl groups of an amino acid will be ionized to the carboxylate form and virtually all amino functions will be protonated. At these acidities, an amino acid thus can be considered to be a dipolar ion of the type shown below. This ion, often called a *zwitterion*, can also be regarded as a

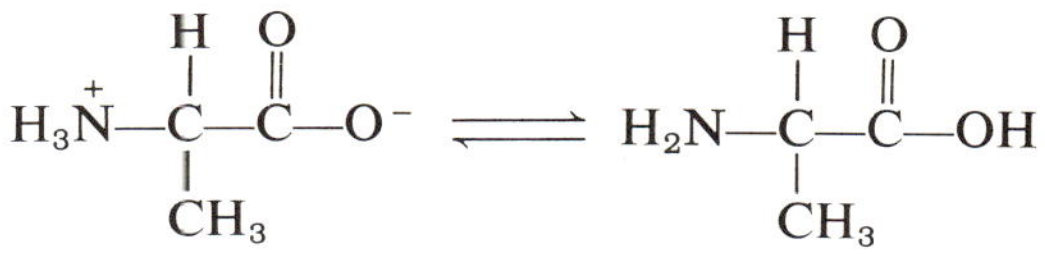

Dipolar form of alanine
at intermediate acidities

tautomer of the neutral amino acid. At some value of the pH, the amino acid molecules in a solution have no net charge; this pH value is called the isoelectric point. The isoelectric point for a particular amino acid will depend upon the acid-modifying effects of its side chain and whether or not the side chain itself has an acidic or basic group attached to it. Table 13.1 records the isoelectric points of the common amino acids.

Reactions of Amino Acids

α-Amino acids take part in the reactions expected for both the carboxyl and amino group. The acid function can be converted to an ester by heating the compound in an acidified alcoholic solution of the amino acid. In the process the amino group is converted to its ammonium form; this conversion is necessary

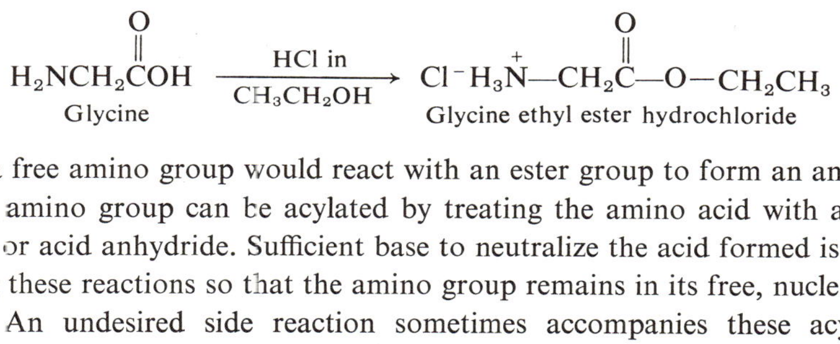

Glycine — Glycine ethyl ester hydrochloride

since a free amino group would react with an ester group to form an amide.

The amino group can be acylated by treating the amino acid with an acid halide or acid anhydride. Sufficient base to neutralize the acid formed is added during these reactions so that the amino group remains in its free, nucleophilic form. An undesired side reaction sometimes accompanies these acylation

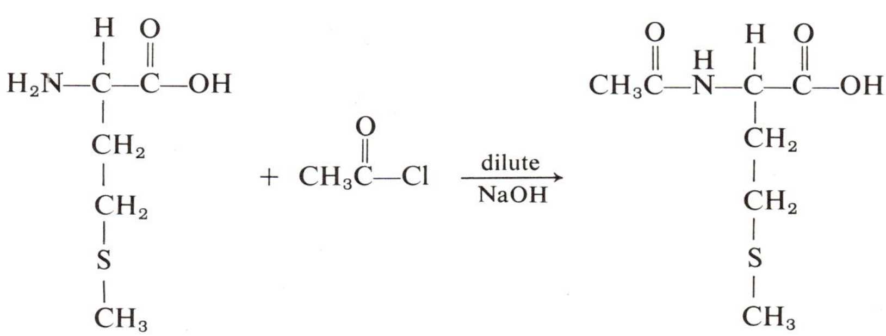

Methionine — Acetyl chloride — *N*-Acetylmethionine

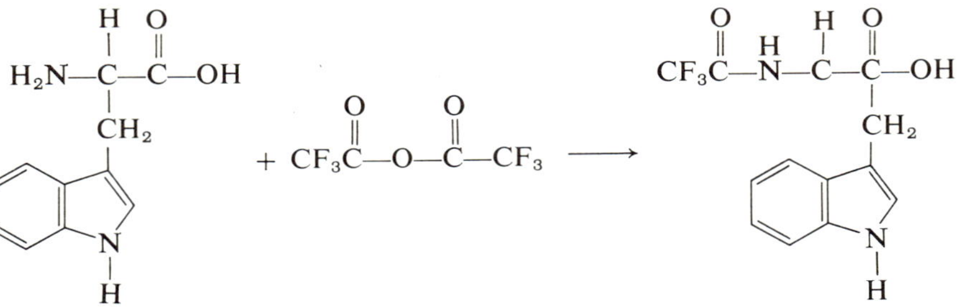

Tryptophan Trifluoroacetic anhydride *N*-Trifluoroacetyltryptophan

reactions, namely, racemization at the asymmetric carbon atom. Careful control of reaction conditions can minimize this complication, however.

The amino group of an α-amino acid will react with nitrous acid in a manner similar to that found with aliphatic amines. The products derived from an amino acid are complex and, although the α-hydroxy acid is usually the dominant one, the nitrous acid reaction has little preparative value. However, the evolution of nitrogen gas is quantitative, with 1 mole of N_2 being formed per mole of free amino group, so that the reaction has value as an analytical tool.

A simple color test for amino acids is afforded by the reactions of these compounds with ninhydrin, a carbonyl compound. Ninhydrin may be regarded as a hydrate of a triketone, indane-1,2,3-trione. Heating a solution of an amino acid with ninhydrin leads to a complex series of reactions in which all amino acids except proline give an intensely blue-violet product. The ninhydrin reaction

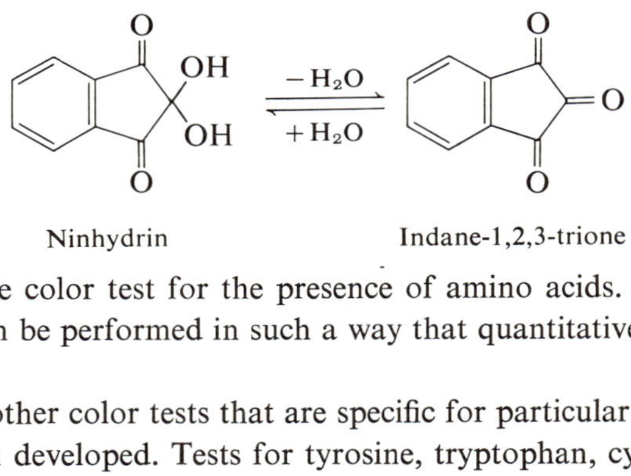

Ninhydrin Indane-1,2,3-trione

provides a simple color test for the presence of amino acids. It is simple, yet sensitive, and can be performed in such a way that quantitative determinations are possible.

A number of other color tests that are specific for particular amino acid side chains have been developed. Tests for tyrosine, tryptophan, cysteine, arginine, and histidine are available; the references should be consulted for the details of these procedures.

There are several important reactions wherein the amino group of an α-amino acid participates as a nucleophile. The fluorine atom in 2,4-dinitrofluorobenzene (Sanger's reagent) is particularly susceptible to displacement by amines and with amino acids gives *N*-2,4-dinitrophenyl derivatives. These derivatives are stable and have chromatographic behaviors that are unique for each amino acid. As we shall see, this reaction has value in the elucidation of protein structure.

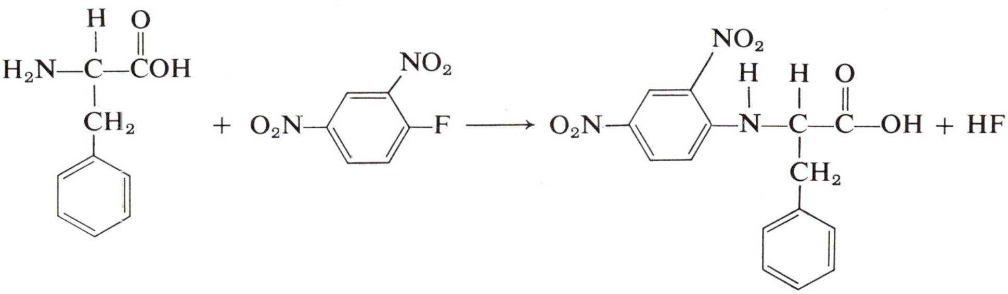

Phenylalanine 2,4-Dinitrofluorobenzene The DNP derivative of phenylalanine

The amino group of an α-amino acid will also react with the formal carbonyl group of isocyanates. The sulfur analogs of isocyanates, called isothiocyanates, react with amino acids in a similar manner. Again, these reactions have value in the elucidation of protein structure.

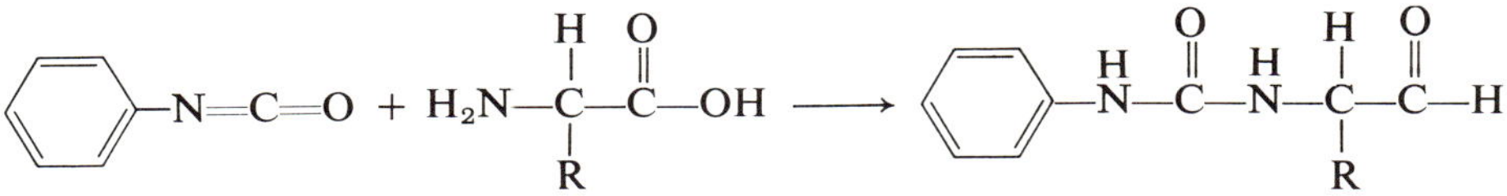

Phenyl isocyanate

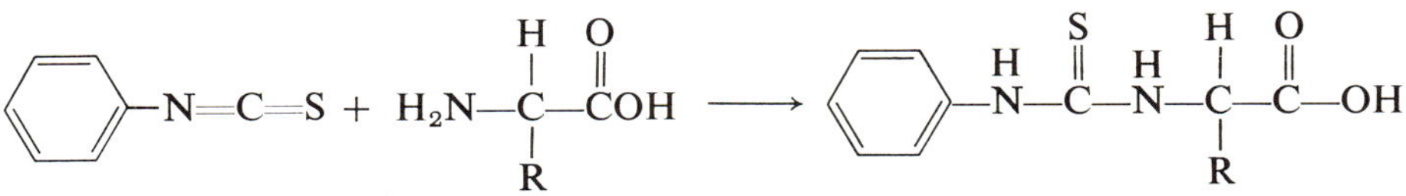

Phenyl isothiocyanate

Even at neutral pH, amino acids undergo racemization. However, the rate of racemization is slow and the relative amounts of the D and L isomers can be used to compute the "age" of the amino acid. Living things utilize only L-amino acids so that the amount of D-amino acid present in a sample is related to the length of time that has passed since the organism that used the amino acid died. The amino acids in fossil shells have been found to be partially racemized although racemization is essentially complete in shells from the Miocene age. As an example, the determination of the extent of racemization of isoleucine found in a sediment core taken from the Atlantic Ocean led to an estimate of 1.23 million years for the age of the core. [See J. L. Boda *et al., Science,* 170, 730 (1970).]

Meteorites which fall to the earth contain a number of amino acids of extraterrestrial origin. Over 14 different amino acids have been found in the Murchison and Murray meteorites, for

example. Nearly equal amounts of D and L isomers are present in these samples, but because of the intense heat generated when these materials enter the earth's atmosphere, reliable dating information probably cannot be obtained from the relative amounts of the two configurations.

PEPTIDES

The α-amino acids described above are the monomeric units from which proteins are assembled. In the polymer, these units are bonded together by an amide linkage formed between the amino group of one amino acid and the carboxyl group of another. In this context the amide bond system is often known as a peptide bond. When two amino acids are joined together in this manner

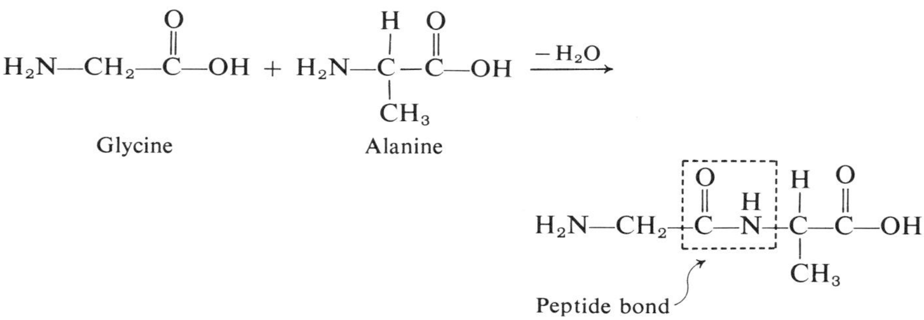

A dipeptide

the resulting compound is generically referred to as a dipeptide. The union of three amino acids gives a tripeptide, and so forth. When the molecular weight of polymers formed from many amino acids in this way is less than about 5000 (an arbitrary dividing line) the molecules are referred to as peptides or polypeptides; larger molecules are called proteins.

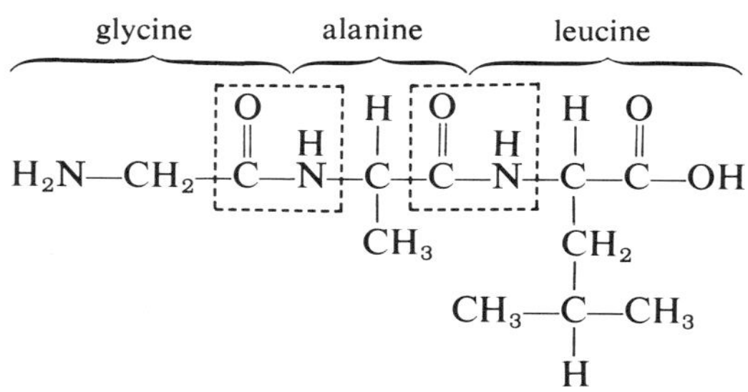

A tripeptide

The chemical reactivity of the peptide bond is essentially that which one would anticipate from the reactivity of other amides, as described in Chapter 11. The peptide linkage may be hydrolyzed to its component parts by the use of acidic or basic catalysts. Typically, the hydrolysis reaction is carried out in 6 *N* hydrochloric acid at about 110°C. Each amino acid in the polymer is thereby converted to its hydrochloride salt except tryptophan, which is destroyed during the reaction. Alkaline hydrolysis is used when it is essential that the tryptophanes in a substance be preserved.

Cis-Trans Isomers at the Peptide Bond

The amide or peptide bond has several important stereochemical features. In considering the generalized structure in Fig. 13.1 we see that the amide nitrogen

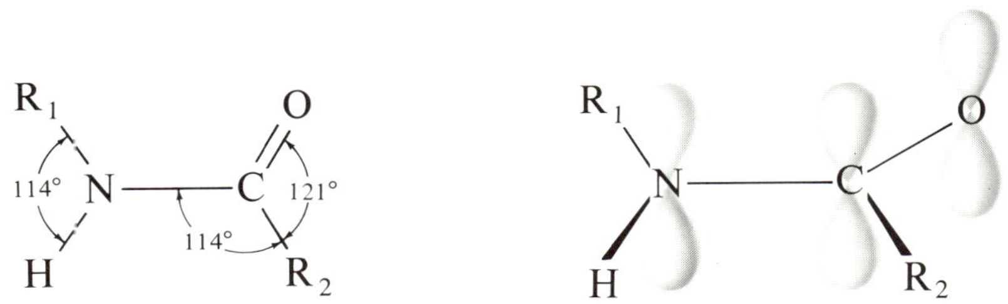

Fig. 13.1. The structure of the amide group.

can be regarded as essentially sp^2-hybridized. As a result, the carbonyl group, the N—H group and the first atoms of both substituents (R_1 and R_2) of an amide must be in a common plane. If we consider only the $2p$ orbitals in this group, we observe an electronic situation similar to that which obtains in the case of carbanions generated adjacent to carbonyl groups. Resonance delocalization of the unshared pair of electrons on the amide nitrogen into the carbonyl π system takes

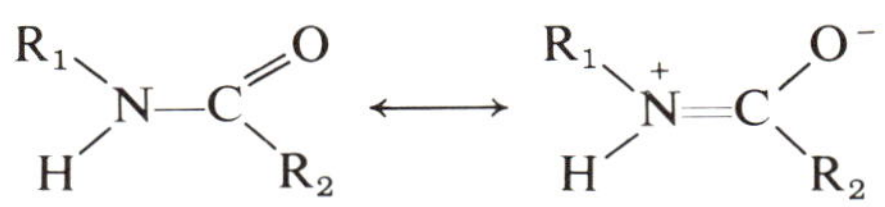

Resonance in the amide group

place and, as a result, there is appreciable double bond character in the amide carbon–nitrogen bond. The resonance energy of this system is appreciable. An important consequence follows from the planarity of the amide bond and the electron delocalization within this system: the amide may exist in geometrically isomeric forms, one with the N—H bond *cis* to the C=O bond and the other featuring a *trans* arrangement of these bonds. Interconversion (isomerization)

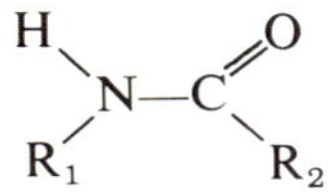

Cis isomer of an amide

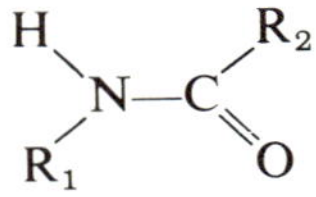

Trans isomer of an amide

of these two forms should be relatively slow since an amount of energy at least equal to the resonance energy must be introduced before rotation about the carbon–nitrogen bond can take place. To this extent an amide resembles an olefin. However, the double bond in an olefin is completely developed while it is only partial in the amide, so that the rate of isomerization in the olefin is much slower than in the amide.

In most peptide bonds, the amide linkage is predominantly in the *trans* form shown above. The groups R_1 and R_2 in a peptide are sufficiently bulky that steric interactions between them would be appreciable in the *cis* form, thereby raising the free energy of the *cis* isomer and favoring the *trans* form in an equilibrium.

Nomenclature

Peptides are named as derivatives of the amino acid which bears the terminal carboxyl group of the molecule. This amino acid is called the *C*-terminal acid. The amino acid found at the other end of the chain has a free amino group attached to it and is designated as the *N*-terminal acid of the peptide. The name of a peptide is generated by listing the amino acids present in the order they occur starting from the *N*-terminal acid. Usually the *-ine* found in the common name for the amino acid is replaced by *-yl*. The last name (that of the *C*-terminal amino

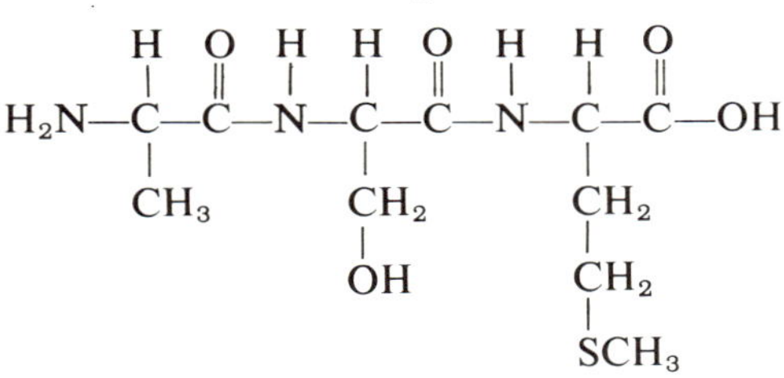

Alanylserylmethionine
(H–ala–ser–met–OH)

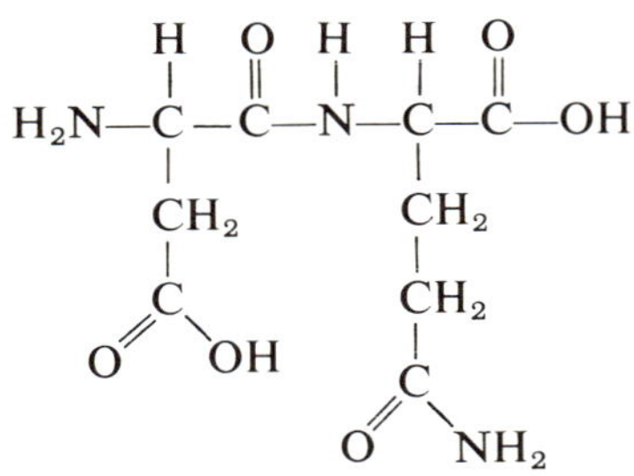

Aspartylglutamine
(H_2N–asp–glu–COOH)
with NH_2 attached to glu

acid) represents the parent molecule and is written unchanged. Some examples of this nomenclature are given below. Probably the longest words in the English language, peptide names formed in this way are very unwieldy. An alternative, shorthand notation uses only a three-letter abbreviation for each amino acid in the structure. The *C*-terminal is indicated by —*OH* or —*COOH* and the *N*-terminal by —*H* or —NH_2.

Structural Studies of Peptides

In seeking to define the structure of an unknown peptide, the amino acids contained in the polymer *and* their sequential arrangement from the *N*-terminal to *C*-terminal acids must be determined. Complete hydrolysis of a peptide gives a mixture of amino acids which can be quantitatively analyzed in order to find the number of each type of amino acid found in the polymer chain. Determination of the order of appearance of these amino acids in the chain is more challenging. One approach is to break the peptide into smaller fragments by partial hydrolysis. The amino acids present in each of these fragments can be determined by spectroscopic methods if enough pure material is available, or by one of the degradation reactions described below. Each peptide fragment corresponds to a piece of jigsaw puzzle and if enough fragments are identified, the sequence of amino acids in the polypeptide can be elucidated.

Analysis for the *N*-terminal amino acid of peptides is often accomplished by

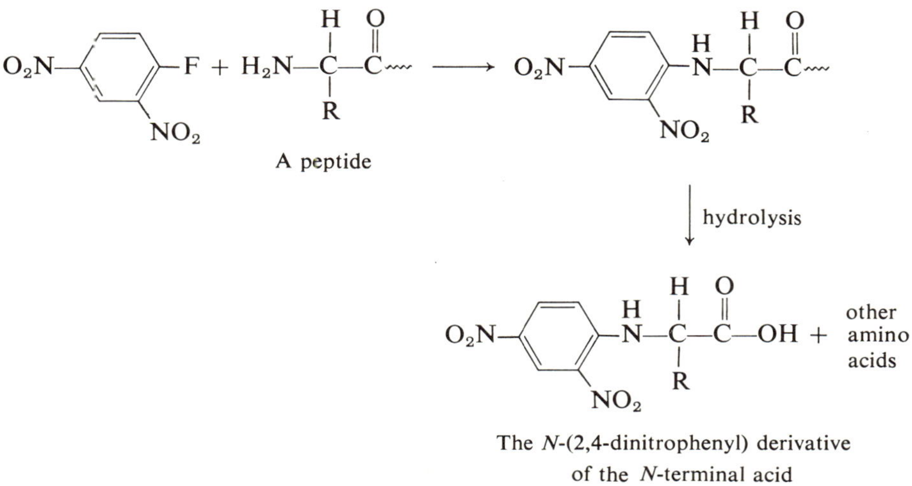

End group determination by Sanger's method

the Sanger method. This procedure takes advantage of the reaction of 2,4-dinitrofluorobenzene with the free, *N*-terminal amino group. After treatment with this reagent, the peptide is hydrolyzed to give as one of the hydrolysis products a *N*-dinitrophenyl amino acid. This particular secondary amide is stable to the hydrolysis condition and, easily recognized by its bright yellow color, can be identified by paper chromatographic procedures.

A second useful reaction is known as the Edman degradation and is based on the phenylisothiocyanate reaction of the *N*-terminal amino group. The peptide to be sequenced is treated with this reagent in dilute base to give the phenylisothiocyanate derivative of the *N*-terminal amino acid. Upon exposure of the resulting material to acid, the isocyanate splits from the peptide chain giving a phenylthiohydantoin derivative of the *N*-terminal amino acid (which can be identified) *and* a new peptide which is one amino acid shorter than the former one. This new peptide can be isolated and the sequence of reactions repeated. This step-by-step, amino acid-by-amino acid degradation procedure is currently the most powerful method for sequence determination of a peptide. Commercial automated equipment for carrying out this process repetitively has become available.

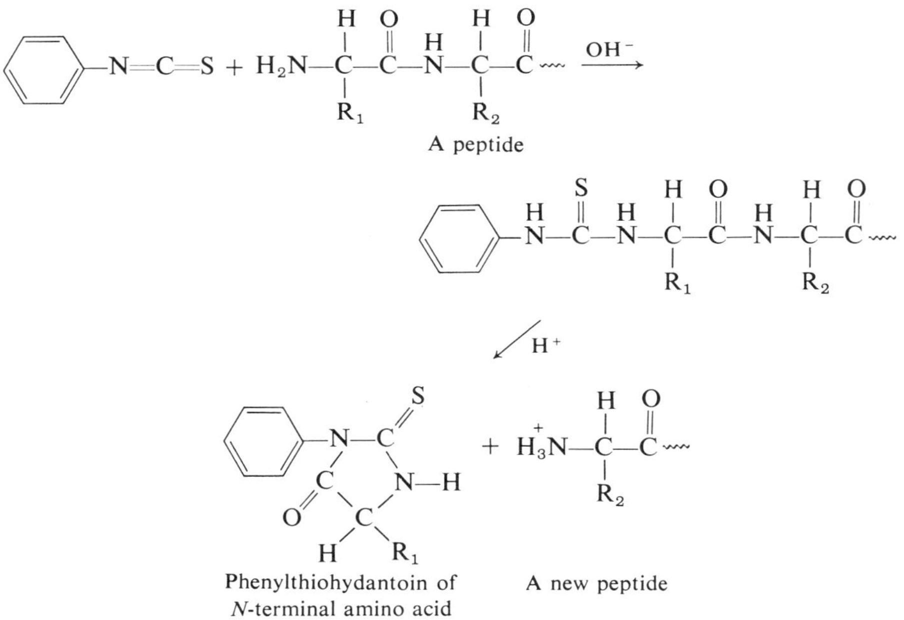

The Edman method for end group analysis and peptide sequence determination

Virtually all living tissues contain appreciable amounts of low molecular weight polypeptides. Many hormones in mammals are peptides, some of them surprisingly simple. The smallest hormone discovered to date is the thyrotropin-releasing hormone (TRH). TRH is one of at least seven hormones secreted by the hypothalamus that regulate the pituitary gland nearby. TRH is a tripeptide composed of histidine, proline, and pyroglutamic acid. (Pyroglutamic acid is a cyclized form of glutamic acid.) Isolation and identification of the hypothalamic hormones is very difficult, but research in this area is expected to lead to new birth control agents and a better understanding of the function of the pituitary system.

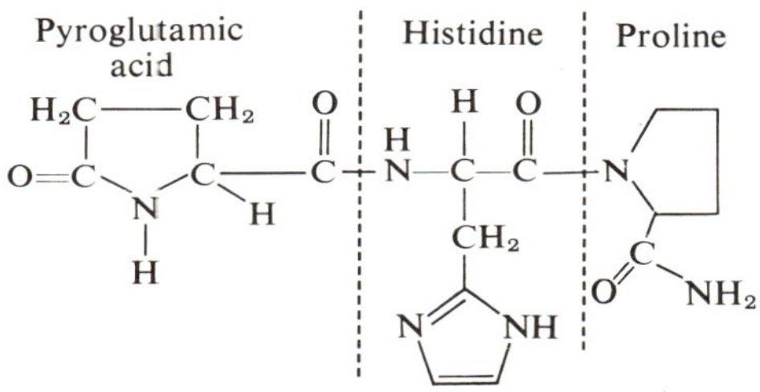

TRH

The pancreas is a large gland which secretes a peptide hormone, insulin, into the small intestine. In addition to stimulating enzymic activity related to the metabolism of glucose, insulin acts to promote the transfer of glucose from circulating blood into the cells of several kinds of tissue. When insulin is absent or insufficient in quantity the blood sugar level rises and a series of complications ensue, including excretion of glucose in the urine. This condition is known as diabetes mellitus. Insulin is a rather complex peptide formed from 48 amino acids arranged into two chains and connected by two disulfide bridges between cysteine residues. The elucidation of the structure of insulin was one of the major early triumphs in the sequencing of peptides.

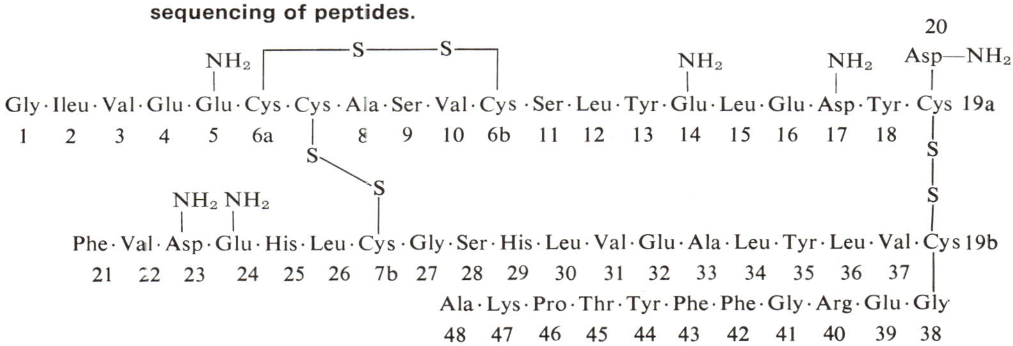

Insulin

Synthesis of Polypeptides

The laboratory synthesis of polypeptide molecules is equally as challenging as the determination of sequence in an unknown polypeptide. Part of the problem in synthesis derives from the fact that the amino acids each have at least two functional groups present (the amino group and the carboxyl group); a number of the amino acids have additional functional groups in their side chains. The chemical characteristics of each of these groups must be taken into account so that the net effect of reactions to generate peptide linkages does only that and does not racemize the asymmetric carbon atoms or modify amino acid side chains. A major consideration in constructing a chain of amino acids is the fact that very few organic reactions afford a 100% yield of the desired product. The overall fractional yield for a number of synthetic steps is the product of the fractional yields for each step. Suppose that a sequence of 200 reaction steps is to be carried out; this is not an unreasonable number for the synthesis of a peptide containing 100 amino acids. If each reaction in the sequence gives 95% yield, the total yield of product at the end of the sequence is $0.95^{200} \times 100\% = 0.004\%$. It is clear that to get even small amounts of synthetic peptides, very efficient, high-yield reactions are required.

Consider the synthesis of a dipeptide from its component amino acids, say glycine and alanine. In order to form the amide or peptide bond, the carboxyl group must be "activated."

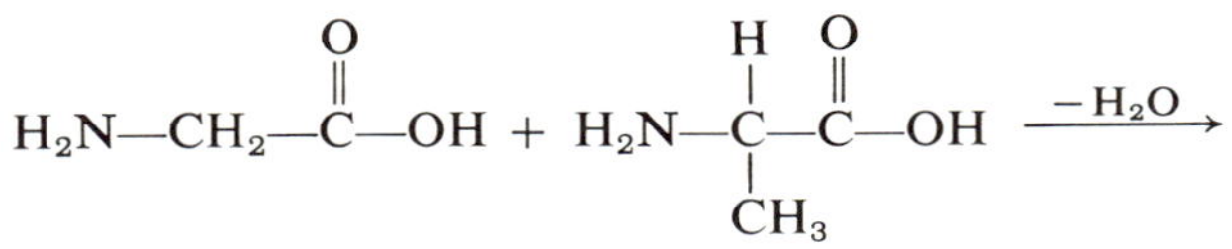

Glycine Alanine

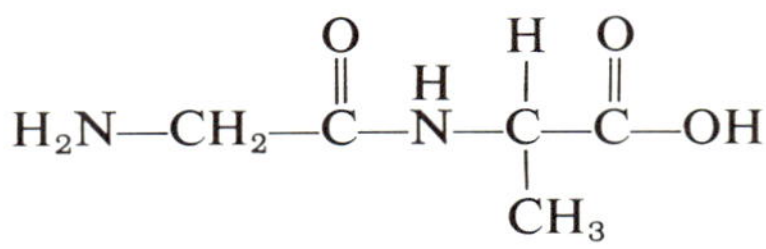

Glycylalanine

We have seen that converting a carboxyl group to an acid halide, an acid anhydride or an ester will serve this function so that what we wish to do is prepare one of these derivatives of glycine for the above reaction. However, the acid chloride of glycine will be able to react with the amino group of another molecule of glycyl chloride as well as with the alanine to give the intended product. To get only the desired product in good yield, it is necessary to diminish or eliminate the reactivity of the glycine amino group until the peptide bond has been formed.

$$H_2N{-}CH_2{-}\overset{\displaystyle O}{\overset{\|}{C}}{-}OH \xrightarrow{SOCl_2} H_2NCH_2\overset{\displaystyle O}{\overset{\|}{C}}{-}Cl$$

Will react with itself
as well as with
other amino acids

Possible means of activating carboxyl group for peptide synthesis

This is most conveniently accomplished by converting the amino group to a protected amide that will be stable during formation of the peptide bond, but which can be cleaved under conditions which will not break the newly formed peptide linkage when the reaction is over. One of several reagents that have been developed for this purpose is carbobenzyloxy chloride. The application of this protective reagent to the synthesis of glycylalanine is indicated below. The carbobenzyloxy amide formed at the *N*-terminal amino acid protects the terminal amino group from the reaction which forms the peptide bond; it is removed by catalytic hydrogenation, a reaction which does not perturb the more conventional amide linkage.

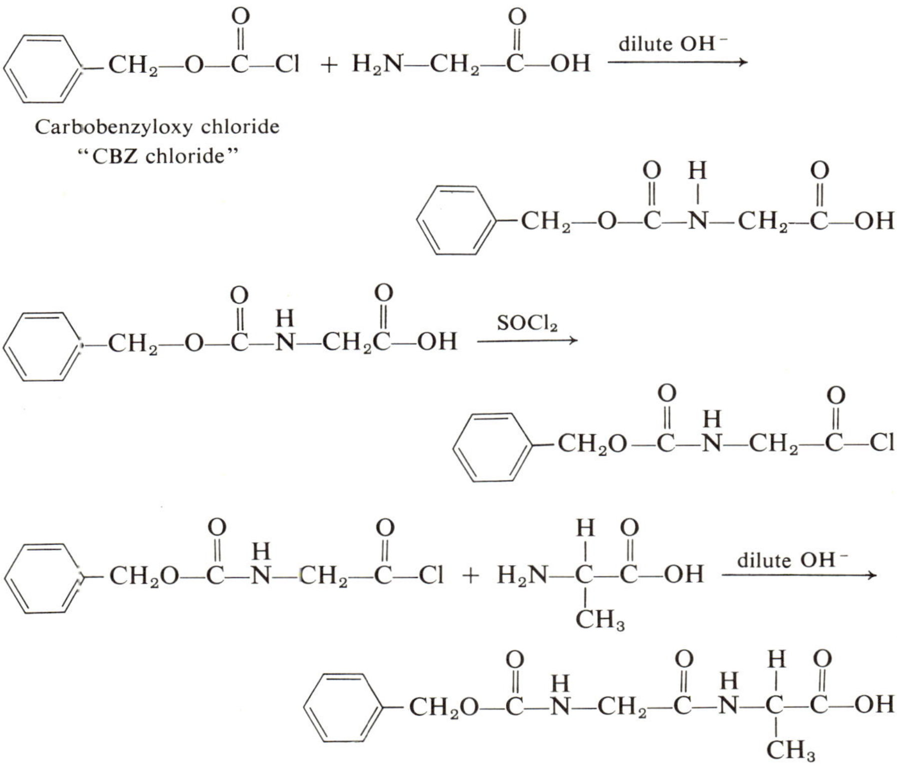

CBZ-protected peptide

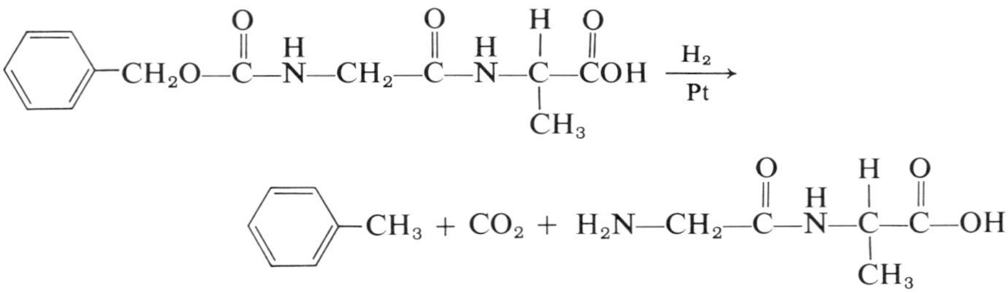

Glycyl alanine

Synthesis of a dipeptide utilizing the CBZ protective group

If the *N*-terminal amino group of one partner is protected appropriately and the *C*-terminal carboxyl group of the other is protected, two amino acids may be joined by exposing them to a dehydrating agent. The most common type of dehydrating agent for peptide bond formation is a carbodiimide, a molecule which features the —N=C=N— bond system. A commonly used carbodiimide is *N,N′*-dicyclohexylcarbodiimide. Upon adding a molecule of water, this com-

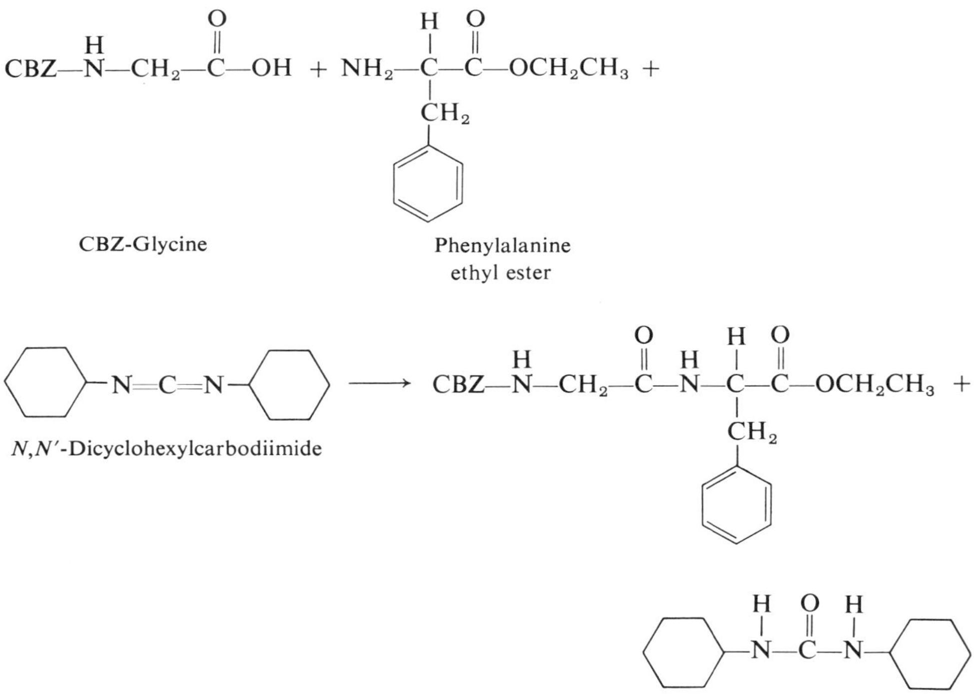

CBZ-Glycine

Phenylalanine ethyl ester

N,N′-Dicyclohexylcarbodiimide

N,N′-Dicyclohexylurea

Use of a carbodiimide as a dehydrating agent

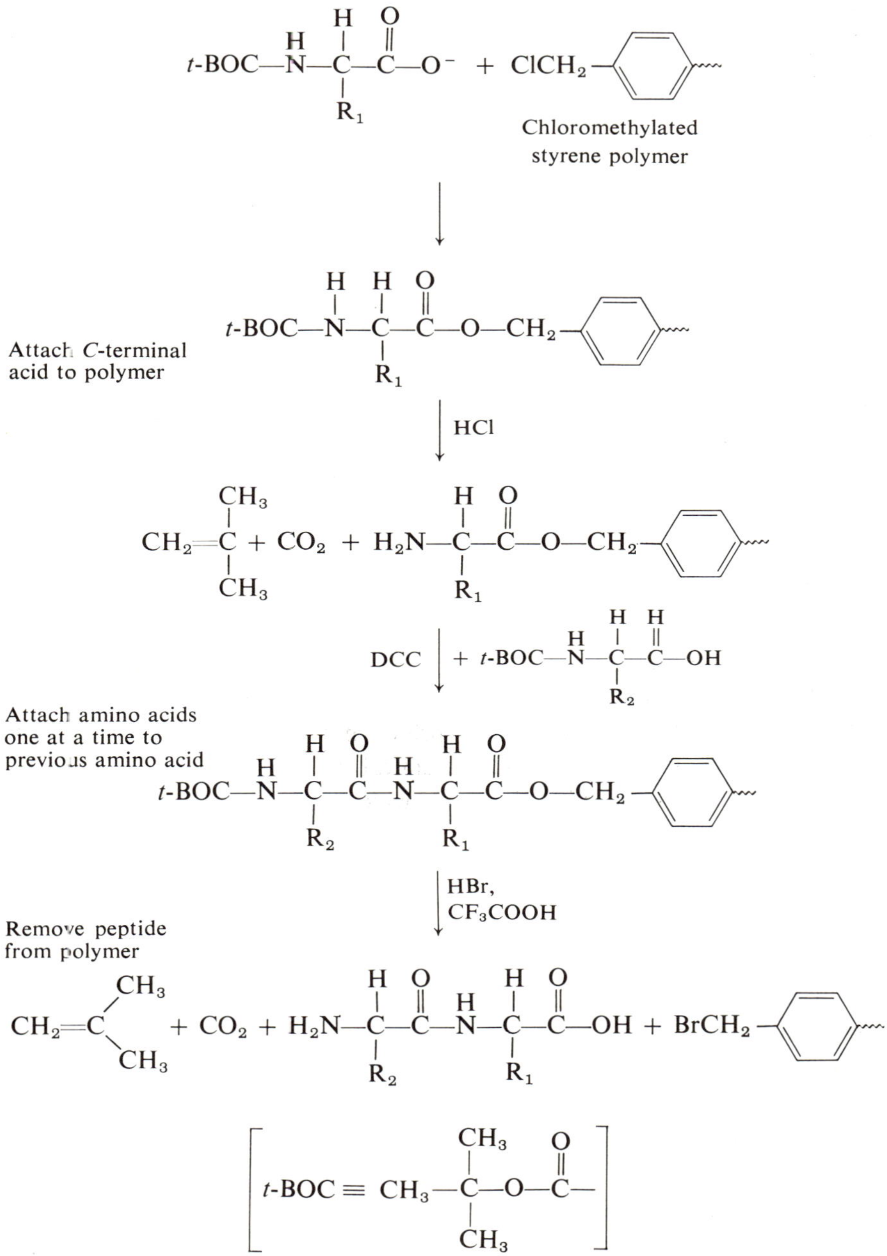

Fig. 13.2. Solid phase synthesis of peptides.

pound is converted to a urea; the urea is relatively insoluble in organic solvents and can be removed by filtration at the end of the reaction. The peptide is generated by deblocking the *N*-terminal amino acid in the usual way. Carboxyl groups are most commonly protected by esterification.

An important and often vexing problem in peptide synthesis is the isolation and purification of the products formed by each step of the synthesis. A significant advance has been proposed and exploited by R. B. Merrifield of Rockefeller University. The basic strategy is to prepare the desired peptide in such a way that it is attached to an insoluble polymer at each stage of the synthesis (Fig. 13.2). The only operation then needed to isolate and purify the growing peptide is to filter the solid polymer to which it is attached away from the reaction mixture and perhaps wash it with pure solvent. After the polypeptide has been constructed stepwise from its component amino acids, it is cleaved from the polymeric support. The entire sequence of operations from the Merrifield method of peptide synthesis has been automated and a wide range of peptides has been prepared by this procedure, including a synthesis of an enzyme, ribonuclease (124 amino acids).

PROTEINS

Proteins are polypeptides that, by an arbitrary decision, have molecular weights greater than about 5000. Each protein has a definite molecular shape and a characteristic set of physical and chemical properties which are determined by the number and kind of amino acids which form the polypeptide chain.

Classification of Proteins

Proteins isolated from living organisms may consist of simply one or more polypeptide chains, or may be more complex structures formed from polypeptides and a nonprotein, or prosthetic, group. The first type of protein is referred to as a simple protein, while the second is called a conjugated protein. The conjugated proteins are classified according to the nature of the prosthetic group; the major categories of conjugated proteins are listed below.

Nucleoproteins. The nucleoproteins are found in close if not covalent association with nucleic acids; the prosthetic groups in these cases are the nucleic acids. Two types of nucleoproteins are the histones and the protamines. Both classes of nucleoproteins are extremely basic molecules. The histones are rich in lysine and arginine and have molecular weights between 10,000 and 20,000. They exist in highly ordered conformations when in combination with a nucleic acid. The protamines are considerably smaller molecules (MW $\sim$5000) but also are exceedingly basic; up to 80% of the amino acids in the protamines is the basic amino acid arginine. The extreme basicity of the nucleoproteins facilitates their interaction with the very acidic nucleic acids.

Phosphoproteins. Phosphoproteins are proteins which contain amino acids linked to phosphoric acid. Casein from milk in which serine residues are found as phosphate esters is an example.

Glycoproteins. In these proteins the prosthetic group is carbohydrate in nature. Usually a hexosamine is present, but the structure and amount of the carbohydrate component is highly variable. Mucin in saliva and the mucoids in tendons and cartilage are examples of glycoproteins.

Lipoproteins. The lipoproteins are polypeptides combined with lipidic material such as fatty acids, lecithins, fats, or steroidal molecules. The amount and type of the lipidic prosthetic group(s) are highly variable. Lipoproteins are prevalent in brain and nerve tissue and are an integral part of cell membranes.

Chromoproteins. Chromoproteins are simple proteins bound to a pigment. The prosthetic group has a characteristic color but can be of various structural types. Hemoglobin and the cytochromes are examples of chromoproteins.

Proteins are also classified according to their gross structural features and their solubility. In this case two major classes are defined: fibrous and globular proteins. Fibrous proteins, as the name implies, exist largely in elongated conformations. They are formed into tough, interconnected sheets as well as fibers and employed for architectural purposes by organisms and for interconnecting or protecting other tissues. The collagens, a class of fibrous proteins, are major components of cartilage and other connective tissue and make up about one-third of the protein in the human body. Collagens are rich in glycine, proline, and hydroxyproline. Another type of fibrous protein is the keratins. These proteins have an unusually large amount of sulfur-containing amino acids in their structure; they are a major constituent of the hair, nails, and skin. Fibrous proteins are not soluble in water.

Globular proteins are much more soluble in water than are fibrous proteins. Each protein molecule in this class is coiled in a rather specific way into a gross structure that is spheroidal or egg-shaped. The polar or ionizable amino acid side chains of the polymer tend to congregate on the surface of these structures, while the hydrocarbon-like, nonpolar side chains tend to become buried within the interior of the protein. Globular proteins are sensitive to external conditions and the carefully structured native conformation of the polypeptide chain can rather easily be disrupted. When this happens the protein becomes essentially a randomly coiled polymer with very different physical and chemical properties. The unwinding process is called *denaturation*.

Structural Studies of Proteins

Each protein molecule has a definite three-dimensional structure that is precisely appropriate for its biological function. There are a number of factors which determine the exact shape of a protein. These have traditionally been considered in terms of four levels of structural organization called the primary, secondary, tertiary, and quaternary structures of the protein. Each succeeding level of organization is more complex than the previous one and is a direct result of the chemical features of the previous levels.

Primary Structure. The primary structure of a protein is a specification of all of the *covalent* bonds which hold the polypeptide chain together. It includes definition of the sequence of amino acids along the peptide chain and indicates the location of any disulfide (S—S) bridges formed between cysteine side chains. We have previously discussed approaches to the elucidation of the amino acid sequence of a peptide. These methods have been applied to a number of rather complex proteins and several generalizations regarding the primary structure of proteins have emerged. First of all, only L-amino acids have been found in proteins, and connection of these units by anything other than a peptide bond is very rare. Second, the sequence of amino acids along the polypeptide chain is essentially random; there is no evidence for a regular, repeating sequence of amino acids. However, proteins that have similar functions tend to have somewhat similar sequences. Finally, it has been observed that small variations in amino acid sequence can have pronounced effects on the chemical and physical properties of the protein.

Secondary Structure. The secondary structure of a protein refers to the local conformation of the polypeptide chain and essentially is a result of the way the polymer chain is coiled. The conformation of the chain cannot be completely random since the peptide or amide group must remain planar in this structure. Current understanding of the secondary structure of proteins is in large measure due to a number of investigations by L. Pauling and R. B. Corey. These workers showed that there are only a limited number of secondary structures available to a polypeptide which retain the stereochemical features of the *trans*-peptide bond and place the amino acid side chains in sterically acceptable environments. It was found that these structures, two of which are the β-sheet and the α-helix,

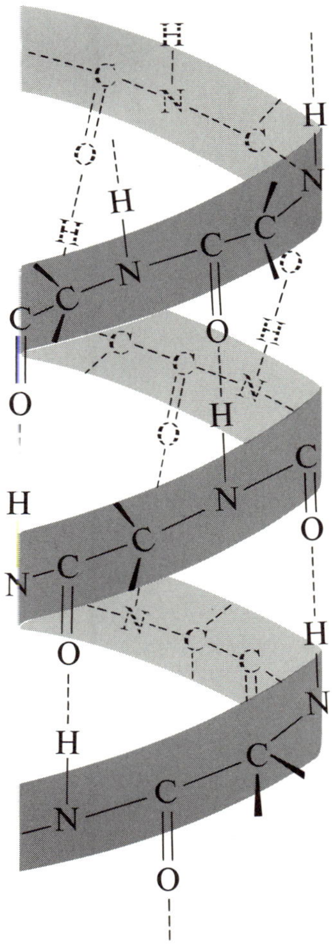

Fig. 13.3. A drawing of the α-helix, a type of protein secondary structure. Note the N—H····O=C hydrogen bonds between turns in the helix.

also permit formation of extensive networks of hydrogen bonds between the amide groups. The N—H of one amide group is hydrogen-bonded to the amide carbonyl oxygen of another peptide linkage. The α-helix is sketched in Fig. 13.3.

The β-sheet structure is illustrated in Fig. 13.4. This secondary structure is found in the fibrous protein of silk and, like the α-helix, is stabilized by hydrogen-bond formation between the polypeptide chains. Regions of β-sheet structures are also found in globular proteins.

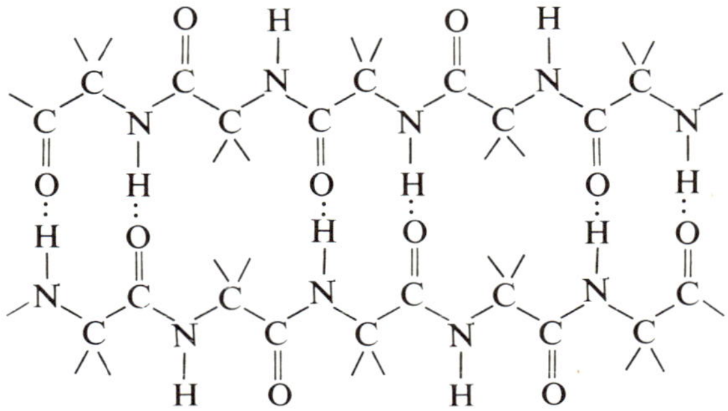

Fig. 13.4. A sketch of the β-sheet arrangement for polypeptide chains. Side chains on α-carbons not drawn.

Tertiary Structure. The conformation or secondary structure defined by the local arrangement of a polypeptide chain is not the final form of a protein. Secondary structures, no matter what their exact form, can be folded or twisted into rather compact structures; it is the form of this final arranging and packing that is called the tertiary structure of the protein. There is only one physical technique that can give reliable information about the tertiary structure of a protein. That method is x-ray crystallography. X-ray determinations of tertiary structure have been done for about 30 proteins; progress in this area is not always rapid because it is not easy to force a molecule as complex as a protein to crystallize into solids which are amenable to crystallographic analysis and because a tremendous amount of computational labor is required to interpret the results. As more powerful digital computers and greater chemical sophistication become available, one can expect to see the tertiary structures of increasingly more complex proteins unraveled by x-ray methods. Some examples of protein tertiary structure are illustrated in Fig. 13.5.

Quaternary Structure. Very large proteins are often formed from a set of subunits rather than a single very long polypeptide chain. Each subunit is in effect a complete protein with a characteristic primary, secondary, and tertiary

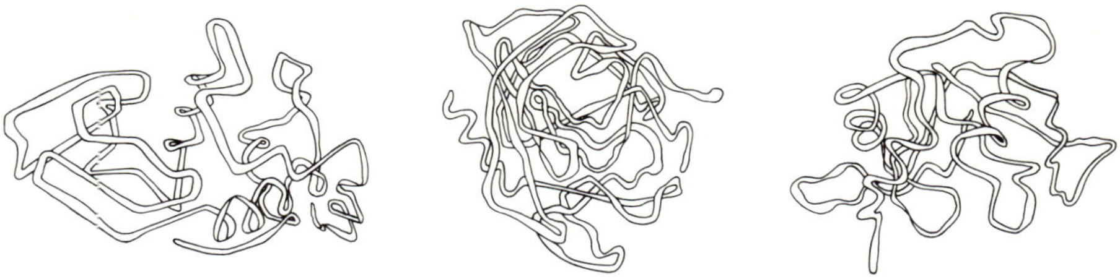

Fig. 13.5. Crude representations of the tertiary structures of several globular proteins. The "folded cylinders" are intended only to give relative orientations of the peptide chain in the molecule and imply nothing about the nature of the interaction between chains or secondary structural features.

structure. The way in which the subunits of a complex protein are associated defines the quaternary structure of the protein.

Amino Acid Side Chain Interactions

In addition to intra- and interchain hydrogen bonding via the peptide group, several other nonbonded interactions between amino acid side chains are considered to be responsible for the generation and maintenance of protein secondary, tertiary, and quaternary structures. These are (a) the hydrophobic (or nonpolar) bond, (b) the salt bridge or ionic bond, and (c) hydrogen bonding between functional groups on side chains.

We have already described the tendency of hydrocarbons to agglomerate in water solution when discussing micelle formation. For similar reasons nonpolar, hydrocarbon side chains on amino acids tend to group together. X-ray studies on the tertiary structure of proteins show that in many proteins there are clusters of nonpolar side chains packed tightly enough together that water is excluded from between them. Nonpolar amino acid side chains are usually found directed toward the interior of a protein's tertiary structure, away from the solvent, water.

Salt bridges are electrostatic interactions between oppositely charged amino acid side chains. For example, a lysine residue ($R = —CH_2CH_2CH_2NH_2$) and a glutamic acid $\left(R = —CH_2CH_2CH_2\overset{\overset{\displaystyle O}{\|}}{C}—OH\right)$ may be located close together in a protein structure. At some value of the solution pH, the amino and carboxyl groups of these side chains will be ionized into their ammonium and carboxylate forms and the attraction between the positively and negatively charged groups will tend to hold these side chains together. The salt bridge is useful in

Nature since it provides a mechanism by which protein conformations can be made pH-dependent.

Even when complete ionization of polar groups is absent, these groups can hydrogen-bond with one another. This type of hydrogen-bond formation must compete with hydrogen-bond formation to solvent molecules (water) and in the proteins that have been studied so far, few of the total possible side chain hydrogen bonds that are possible appear to be fully realized.

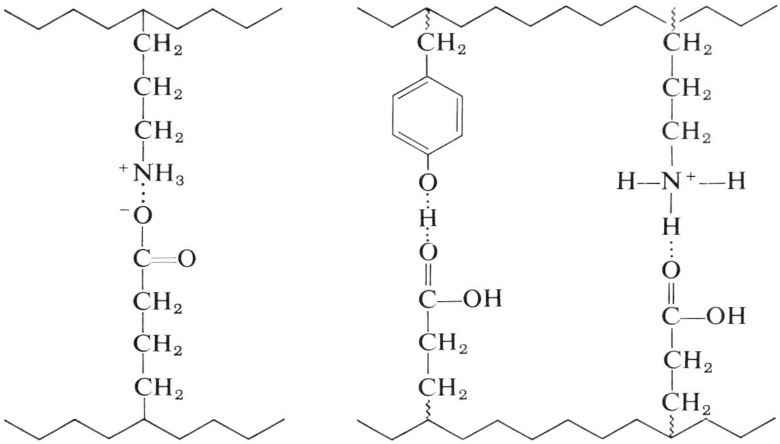

A salt bridge in a protein structure

Side chain hydrogen bonds in a protein structure

Denaturation

When a protein denatures much, if not all, of its secondary, tertiary, and quaternary structure is destroyed. The biological activity of the protein is also lost in the process. Common reagents or conditions that can denature proteins are heat, solutions of urea or guanidine, organic solvents such as ethanol or acetone, thiols, detergents, and extreme variation in pH. The chemical basis for denaturing ability is not well-understood although much experimental work in this field has been performed; many of the denaturing conditions listed above could potentially disrupt hydrogen bonds within the protein structure as well as hydrogen bonds between water and the amino acids on the surface of the protein. We should remember that many proteins as they exist in organisms are enmeshed in a complex matrix of other proteins and possibly lipidic material. These environments were designed by Nature to provide a stabilizing milieu for proteins and it should not be alarming that a protein is susceptible to degradation or denaturation when removed from its natural environment. Understanding the biological action of a protein *in its natural state* is an important experimental undertaking, one that has barely begun.

Possibly no subject has engendered more discussion, consternation, and condemnation recently than hair. Hair is largely protein and some forms of hair, especially wool, have been carefully studied. The amino acid composition of wool, given in Table 13.2, is probably representative of the composition of other types of hair. The striking thing about this data is the large amount of cystine present, indicating that the hair fiber is very rich in disulfide bonds. A considerable amount of work has been invested in attempts to elucidate the secondary, tertiary, and quaternary structures of hair and α-keratins, in general. Most of the present ideas regarding hair structure are based upon electron microscopic and x-ray experiments.

Examined with an electron microscope, a cross section of a hair fiber appears to consist of rodlike microfibrils about 75Å in diameter (Fig. 13.6). The density of microfibrils and their arrangement vary from source to source in a characteristic way.

Fig. 13.6. Electron micrograph of merino wool. From *Advan. Protein Chem.* **20,** 289 (1965). Photo courtesy of T. P. MacRae.

X-ray diffraction studies indicate that there is a substantial amount of α-helical structure within the microfibrils. The following proposed structure for them appears to be consistent with most available data. The protein chains within the microfibril are in an α-helix conformation and it appears fairly certain that three such chains are wound together into a ropelike structure that has been called a protofibril (Fig. 13.7). Eleven

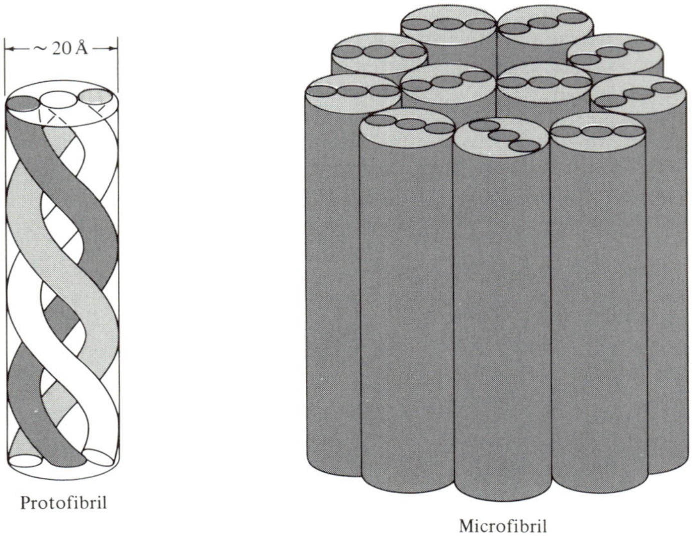

Fig. 13.7. The probable structure of the hair protofibril and microfibril.

protofibrils are bundled together into the "9–2" arrangement depicted in the figure. This 11-stranded structure called a microfibril is apparent in the electron micrograph. The microfibrils are packed in an amorphous matrix of additional proteins and several hundred of these structures are combined in an irregular way into macrofibrils which, in turn, are packed together to form the hair cell.

The properties of the hair fiber are a direct result of the way its component protein molecules are assembled into the completed structure. A wool fiber is extensible and can be stretched to about twice its resting length. The local unwinding of coiled polypeptides probably accounts for this. The flexibility of the fiber is also a consequence of the deformable helical structures which can bend but not break when stress is applied.

The disulfide bonds which must be very prevalent in the protein structure of the hair provide dimensional stability and resistance to flow. Most "permanents" for human hair consist of chemicals to destroy some of the disulfide bonds by reduction reactions and then to encourage them to reform after the hair fibers have been arranged into the desired shape.

Table 13.2
Amino Acid Composition of Wool[a]

Amino acid	%
Alanine	3.5
Arginine	18.9
Aspartic acid	4.3
Cystine	7.1
Glutamic acid	8.0
Glycine	6.3
Histidine	1.6
Isoleucine	2.6
Leucine	5.2
Lysine	3.8
Methionine	0.4
Phenylalanine	1.8
Proline	5.0
Serine	6.2
Theonine	4.9
Tryptophan	Trace
Tyrosine	2.0
Valine	4.5
Unidentified	16.8

[a] Data are taken from W. G. Crewther *et al.*, *Advan. Protein Chem.* **20**, 228 (1965) and are the mean of eight determinations using various samples.

Enzyme Action

Protein molecules play many roles in a living organism but possibly the most crucial is the ability of some proteins to catalyze chemical reactions. These molecules are called enzymes. The magnitude of the catalytic effect exerted by an enzyme is not small and, under optimal conditions, an enzyme-catalyzed reaction can be 10^8 to 10^{11} times more rapid than the uncatalyzed process. Enzymes catalyze a wide variety of reactions including hydrolysis reactions, oxidation–reduction reactions, condensation reactions, dehydrations, and polymerizations. Their special catalytic powers are a direct consequence of their secondary, tertiary, and quaternary structures and some small progress has been made toward understanding the relation between enzyme structure and enzyme function. We shall describe here in a very abbreviated way some facets of enzyme catalysis as they are currently perceived.

A molecule which is subjected to an enzymic reaction and undergoes transformation is called the substrate of that enzyme. Some enzymes require that, in addition to the substrate, other organic molecules or inorganic ions be bound to them before a reaction takes place. Occasionally these nonprotein species are covalently bonded to the enzyme as a prosthetic group. In either case, the required molecule is called a coenzyme or cofactor. Without the cofactor the enzyme is catalytically inactive and is often unstable; a cofactor-requiring enzyme that does not have its cofactor(s) is called an apoenzyme.

A typical enzyme is a globular protein in which the polypeptide chain or chains are arranged in a highly specific, and usually complicated, way. This complex structure cannot be radically perturbed without denaturing the enzyme. Enzymic catalysis must, therefore, take place near the surface of the protein. The locale at which the catalytic effect is exerted is not arbitrary but, rather, it is found that the substrate (and cofactors, if any) bind to the enzyme at particular locations in a very specific way. The locations on the protein surface where the substrate binds and reacts is called the active site. It is defined by the particular amino acid side chains in this vicinity and by their three-dimensional arrangement in space; the high catalytic efficiency and pronounced specificity exhibited by enzymic reactions must be directly related to the structure of the enzymic active site involved. One of the major goals of current research in enzyme chemistry is the identification of the amino acids at the active sites of enzymes and the determination of their arrangement in space. With this information in hand, it may be possible to explain the specificity of the enzyme and rationalize the roles of these side chains in the catalytic mechanism.

Reaction rate studies of enzyme-catalyzed reactions show that, qualitatively, the rate of the reaction varies with the concentration of substrate in the manner

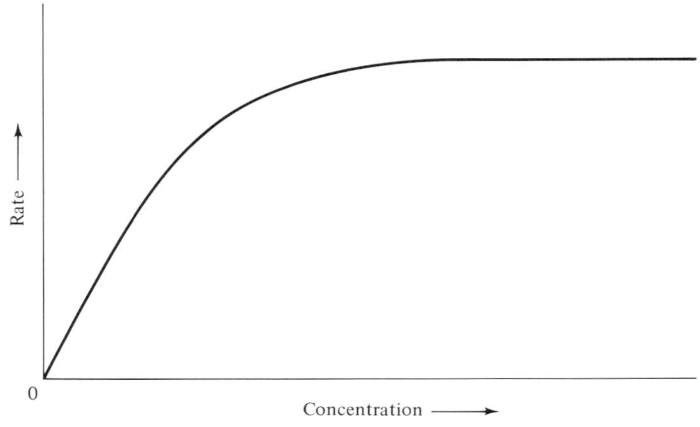

Fig. 13.8. The dependence of reaction rate on substrate concentration in an enzyme-catalyzed reaction (schematic).

depicted in Fig. 13.8. It is seen that the reaction rate does not increase indefinitely with increasing substrate but rather increases initially and then reaches a plateau. The simplest mechanism that can account for this behavior is the one shown below. In this mechanism, the substate and enzyme combine to form an enzyme–substrate complex which, in a second step, goes on to give the products of the reaction. This mechanism is consistent with the rate behavior depicted

$$\underset{\text{Enzyme–substrate complex}}{E + S \rightleftharpoons ES} \quad , \quad ES \longrightarrow \text{products} + E$$

A simple mechanism for enzyme catalysis (E represents the enzyme and S stands for the substrate)

in Fig. 13.8; initially an increase in substrate concentration will result in an increase in the concentration of the enzyme–substrate complex and a corresponding increase in reaction rate. However, at a high enough concentration of substrate, virtually all of the available enzyme will be in the complexed form and further increases in substrate availability will have little effect on the observed reaction rate. This phenomenon is called saturation. The mechanisms of most enzyme reactions are more complex than the simple scheme written above but will, nevertheless, normally show this saturation effect. The important point is that rate studies of enzymic reactions indicate that an association complex between protein (enzyme) and substrate must be formed before catalysis of the reaction takes place.

There are several characteristics that are common to all enzyme-catalyzed reactions. The first is specificity. Most enzymes will react with only one substrate or, at least, one class of molecules. They can distinguish between optical and geometrical isomers of a substrate. For example, the enzyme fumarase catalyzes the addition of water to the double bond of fumaric acid but has no effect on maleic acid, the *cis* isomer of fumaric acid. Some enzymes will transform one and only one substrate; these enzymes are said to have absolute specificity. Second, the rates of enzyme-catalyzed reactions are much more rapid than the

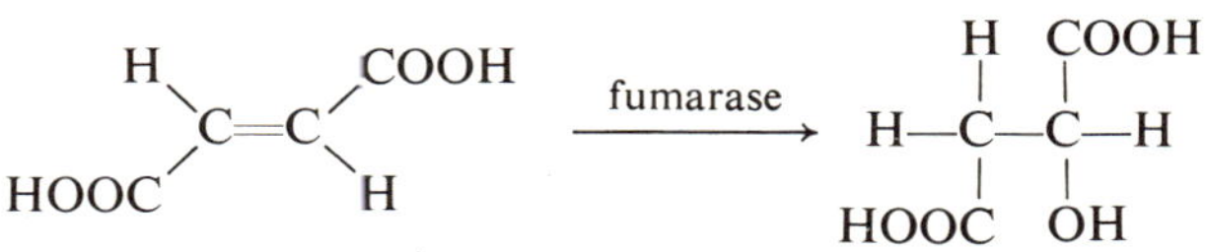

Fumaric acid   Malic acid

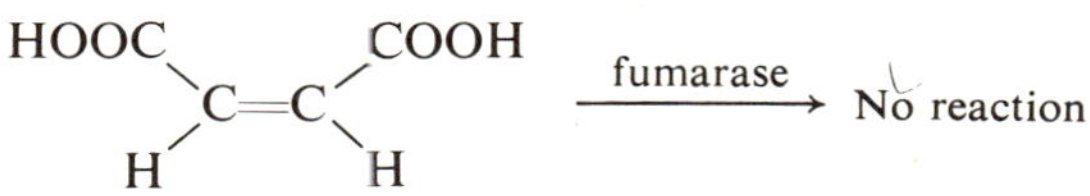

Maleic acid

rates of similar processes in the absence of protein. The number of substrate molecules that react with a single enzyme molecule in 1 minute ranges from 10^3 to 10^6 and often the rates of some steps in an enzymic reaction are limited only by the ability of the substrate molecule to diffuse to the enzyme. This remarkable catalysis obtains even though enzymes act under conditions of nearly neutral pH and moderate temperature. Finally, the activity of enzymes is subject to external or cellular control. The concentrations of cofactors, substrates and other small molecules, the rate of accumulation of products, and the regulation of pH are aspects of the cellular environment of enzymes which can be combined into elaborate "feedback" networks for the control of enzymic activity.

PROBLEMS

1. A hexapeptide was hydrolyzed under acidic conditions and, upon amino acid analysis, was shown to contain 1 glycine, 2 serine, 2 threonine, and 1 tyrosine residues. Subsequent work showed that the following fragments were present upon partial hydrolysis: Ser–Ser, Ser–Gly, Ser–Gly–Thr, Thr–Tyr, Tyr–Thr. What is the sequence in this peptide?

2. Devise reaction sequences which could be used to prepare the following materials.
   (a) *N*-Acetyltryptophan ethyl ester from trytophan
   (b) Prolylglycine
   (c) Alanol from alanine

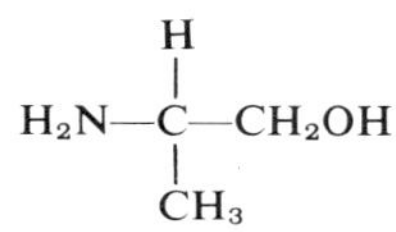

Alanol

   (d) Glycylalanylmethione
   (e) Aspartic acid from asparagine

3. With the aid of molecular models examine the effect of having one amino acid of the D configuration in a polypeptide chain of L-amino acids when the material is in an α-helix.

4. The presence of mercaptoethanol ($HOCH_2CH_2SH$) can lead to denaturation of a protein. Why?

5. Write out the complete structures for the following peptides.
   (a) H_2N–Ala–Gly–Aly–Pro–Leu–COOH
   (b) H_2N–Try–Ile–Met–Gly–COOH
   (c) H_2N–Phe–Pro–Leu–Val–Ala–Cys–Asp–COOH

6. An octapeptide was hydrolyzed into the smaller peptides: Ala–Gly, Ala–Met–

Gly, Gly–Gly–Ser, Ala–Gly–Gly, Ser–Leu, and Leu–Ala–Met. Alanine was identified as the *N*-terminal amino acid. What is the structure of the octapeptide?

7. Suggest a reason for the observation that essentially equal amounts of the *cis* and *trans* forms of glycylproline are present at equilibrium.

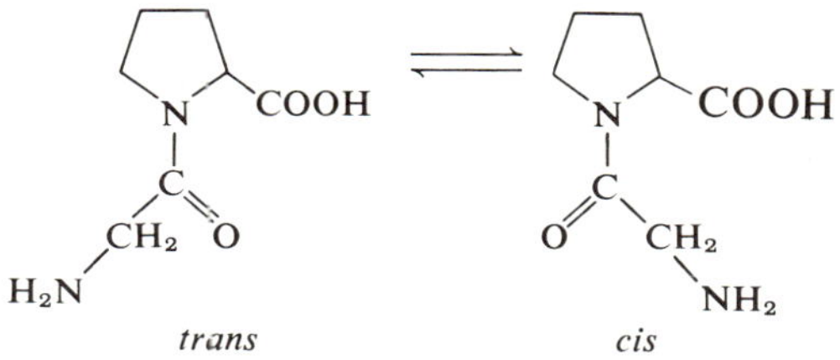

Glycylproline

SUGGESTED READINGS

W. G. CREWTHER *et al.*, *Advan. Protein Chem.* **20**, 191 (1965). (The chemistry of keratins.)

R. E. DICKERSON AND R. GEIS, "The Structure and Action of Proteins." Harper, New York, 1969.

S. W. FOX AND J. F. FOSTER, "Introduction to Protein Chemistry." Wiley, New York, 1957.

H. D. LAW, "The Organic Chemistry of Peptides." Wiley (Interscience), New York, 1970.

W. A. SCHROEDER, "The Primary Structure of Proteins." Harper, New York, 1968.

J. M. STEWART AND J. D. YOUNG, "Solid Phase Peptide Synthesis." Freeman, San Francisco, California, 1969.

A. WILLIAMS, "Introduction to the Chemistry of Enzyme Action." McGraw-Hill, London, 1969.

14

Nucleic Acids

Patiently working in a monastery during the mid-nineteenth century, an obscure Austrian monk, Gregor Mendel, carried out experiments that led to the foundations for classical genetics. Many of his conclusions regarding the transmission of genetic information from one generation to the next are familiar. Taking the broad view that all of life is concerned with chemistry and chemical reactions, one should immediately be curious regarding the physical and chemical bases for the transmission of genetic information. The fascinating details of this process are under vigorous investigation now; our knowledge in this field is expanding rapidly and in the limited amount of space available here, we can only suggest some of the more important chemical aspects.

The discussions of the previous chapter should have provided some indication of the pervasiveness and importance of proteins to the life process. So important and so pervasive, in fact, that if one is able to specify all of the proteins of an organism, he has, in fact, defined the organism. Particularly important are the enzymes since these molecules control virtually every reaction of importance to the life process. The problem of transferral of genetic information largely is a problem in telling an offspring how to make the enzyme and other protein systems that are necessary for maintaining life.

By the time Mendel's ideas were gaining acceptance it was becoming clear to biologists that genetic information was contained in threadlike strands of matter found in cell nuclei. These were called chromosomes; during cell division these structures duplicate themselves so that each daughter cell gets an equal share of the information that is contained within them. Chromosomes are found to be rich in a polymeric substance called deoxyribonucleic acid as well as various protein structures. It has been shown that it is the deoxynucleic acids (DNA's) that carry the genetic message from one generation to the next. Hydrolysis of these polymers under various conditions shows that they are constructed from several heterocyclic molecules, a sugar (D-ribose), and phosphoric acid. It is our intention here to discuss briefly some of the chemistry of these

$$\text{Nucleic acid} \xrightarrow[H_2O]{H^+} \text{sugars} + \text{purines} + \text{pyrimidines} + \text{phosphate}$$

heterocyclic molecules, to describe how these components are assembled into the nucleic acid polymers, and to indicate the probable mechanism by which protein structure is coded by these molecules. The references at the end of this chapter provide additional details.

PURINES AND PYRIMIDINES

The heterocyclic molecules found in the nucleic acid polymers are derivatives of the nitrogen-containing heterocycles, pyrimidine and purine. Pyrimidine may be regarded as a derivative of pyridine or, alternatively, as a derivative of

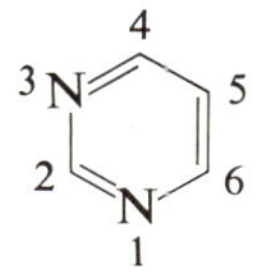

Pyrimidine

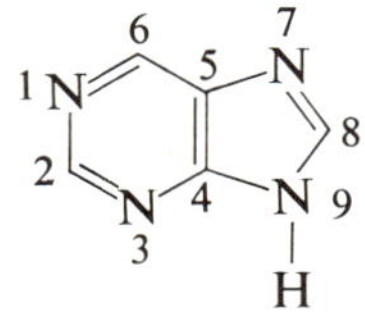

Purine

benzene. It is a planar array of sp^2 hybridized atoms with a well-developed π-electronic system; the resonance energy of pyrimidine is about 26 kcal/mole. The second nitrogen in the ring acts as an electron-withdrawing substituent and decreases the basicity of the molecule to less than that of pyridine. Purine is also a potentially aromatic molecule and can be looked upon as being formed by fusing an imidazole ring to a pyridmidine ring. Purine is a weaker base than imidazole but is a stronger acid.

Neither of these parent heterocyles are found in nature, although a number of molecules, in addition to those found in nucleic acids, are based upon these ring systems. Some of these are sketched below.

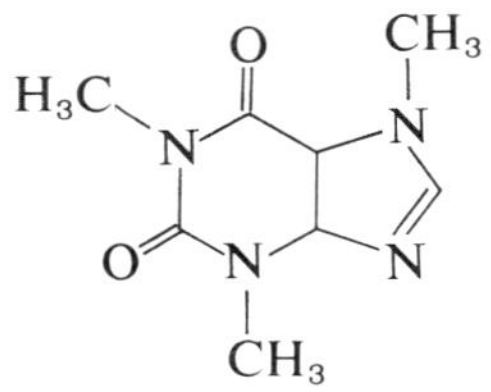

Caffeine

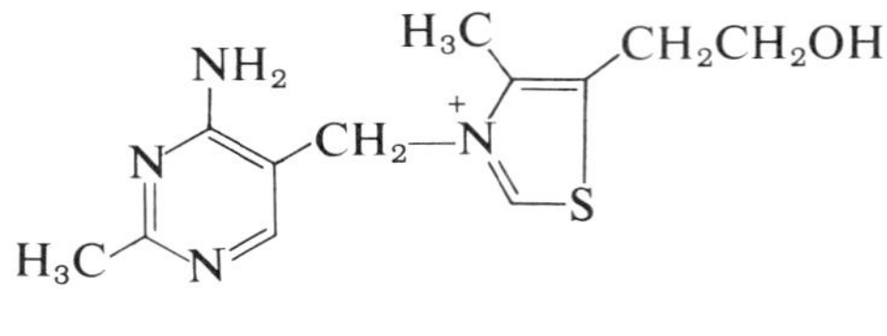

Thiamine
(vitamin B_1)

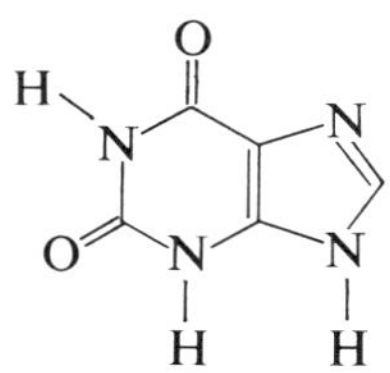

Xanthine
(found in tea and
animal tissues)

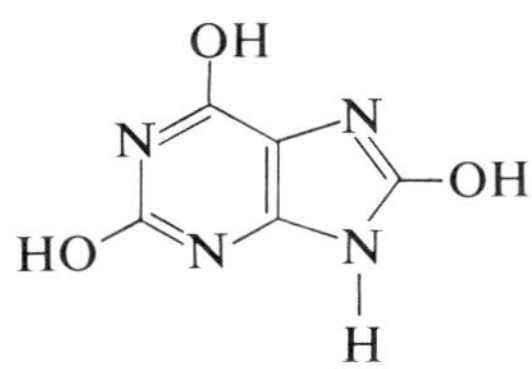

Uric acid
(found in guano)

Some Reactions of Purines and Pyrimidines

Electrophilic Substitution. The presence of nitrogen atoms in purine and pyrimidine has the effect of decreasing drastically the rate of electrophilic substitution on the heterocyclic rings. In this sense these nitrogen atoms function

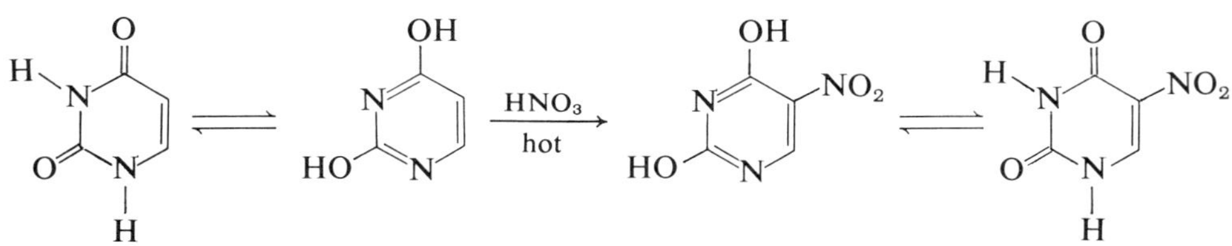

(Tautomerism) (5-Nitropyrimidine)

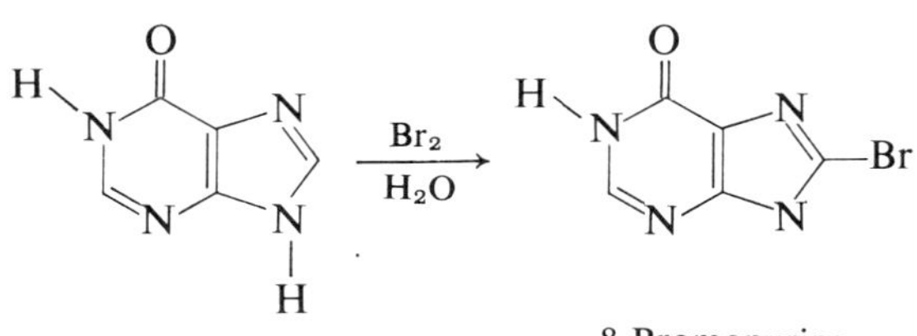

8-Bromopurine

similarly to nitro- groups. In order to bring about electrophilic substitution one or more electron-donating or activating groups such as —OH, —NH_2, or alkyl must be present. Substitution in a suitably activated pyrimidine occurs at carbon-5, while carbon-8 of the purine ring is most susceptible to electrophilic attack.

Nucleophilic Substitution. Like pyridine, purines and pyrimidines can undergo nucleophilic substitutions with a variety of reagents in an appropriately constituted molecule. Direct displacement of a ring hydrogen atom is not common but substitution reactions involving halogens or other good leaving groups take place readily. The halogen commonly involved in these reactions is chlorine; chlorine can be attached to the ring by treating the proper hydroxy compound with a phosphorus oxychloride ($POCl_3$) or with phosphorus pentachloride.

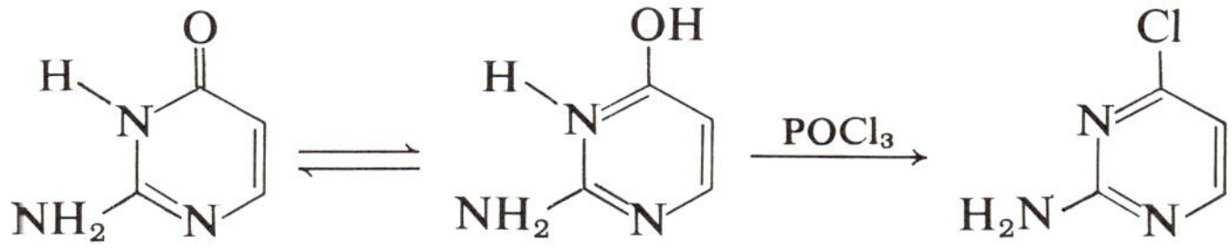

2-Amino-4-chloropyrimidine

Nucleophiles which can displace chlorine include hydroxide, alkoxides, mercaptides, and amines.

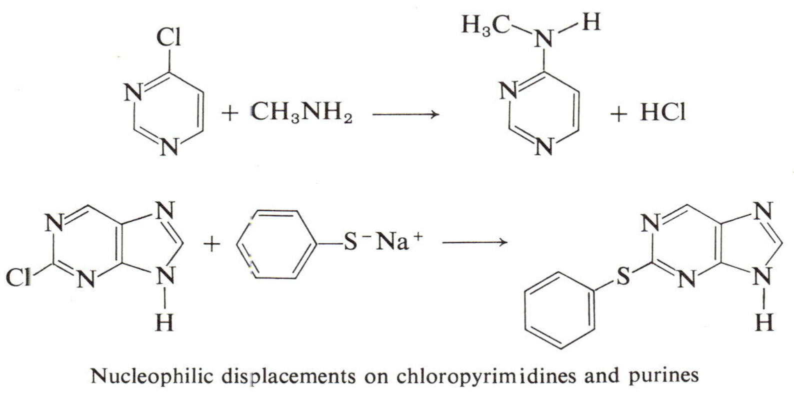

Nucleophilic displacements on chloropyrimidines and purines

Heterocycles Found in the Nucleic Acids

There are two specific purine derivatives, adenine and guanine, which are involved in nucleic acid architecture as well as three particular pyrimidines: uracil, cytosine, and thymine. Various spectroscopic experiments have demonstrated that the molecular structure of uracil is best represented by the tautomer shown here.

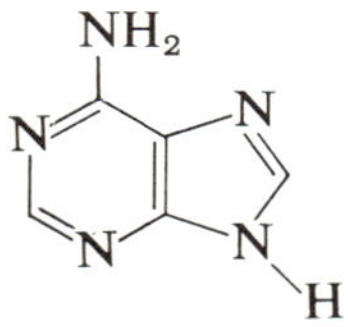

Adenine

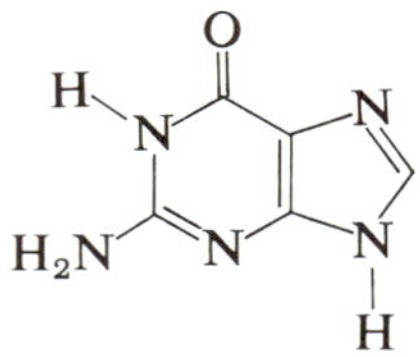

Guanine

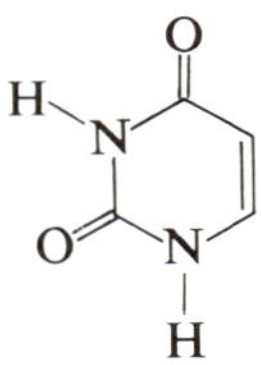

Uracil

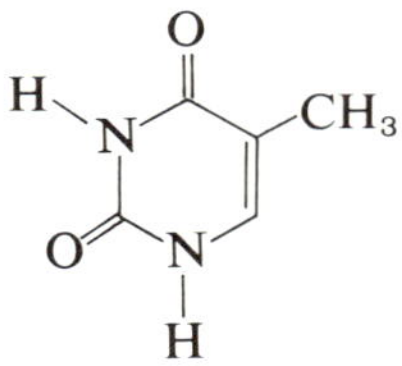

Thymine

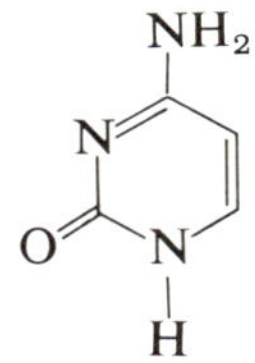

Cytosine

Thymine is also found in the corresponding tautomeric form. However, tautomerism at the amino group of cytosine does not take place and this molecule is best represented by the structure shown.

Uracil and cytosine undergo electrophilic substitution; the substitution takes place at carbon-5 of the ring. This carbon may be regarded as being activated by the potential hydroxyl groups of the ring. Electrophilic reagents such as nitric acid and diazonium ions react readily with uracil and cystosine at this position to

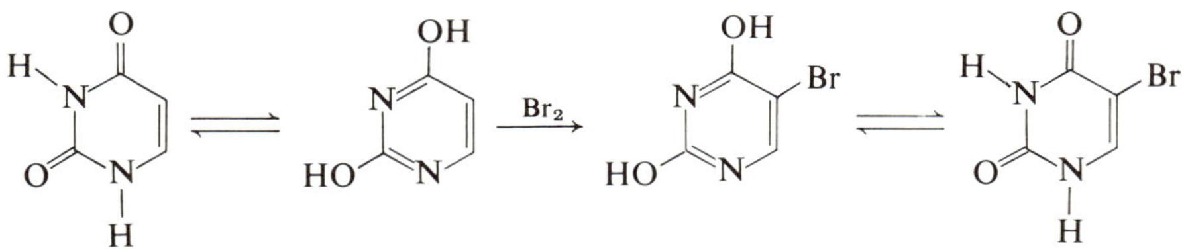

5-Bromouracil

(Both hydroxyl groups are *o–p* directors with activation)

give the corresponding substitution products. These reactions do not take place, however, with thymine or the purines since the activated carbon at position-5 in these molecules is already occupied.

An important aspect of the chemistry of these three pyrimidines is the acidity of the hydrogen at the 1-nitrogen. This hydrogen atom is in a position analogous to the α-hydrogens of other carbonyl compounds and the anion formed upon ionization of this group can be stabilized by a similar type of electron delocalization. The pyrimidine anions are good nucleophiles and will execute displacement reactions on alkyl halides; the 1-nitrogen atom is alkylated in the process. Anions can also be formed at the 9-position in the purines by

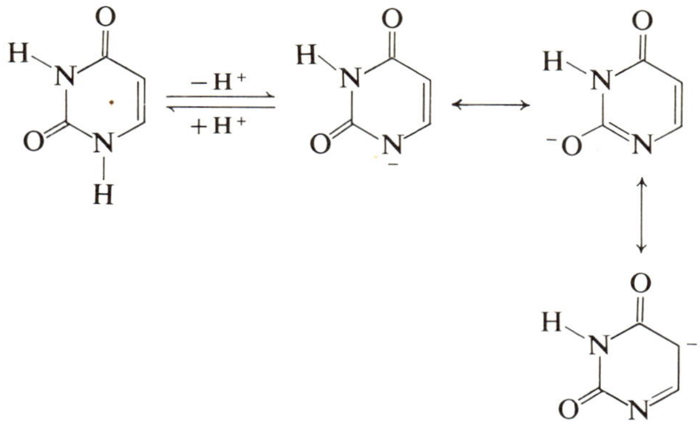

Uracil as an acid

treating adenine or guanine with a base. Again the anions formed are good

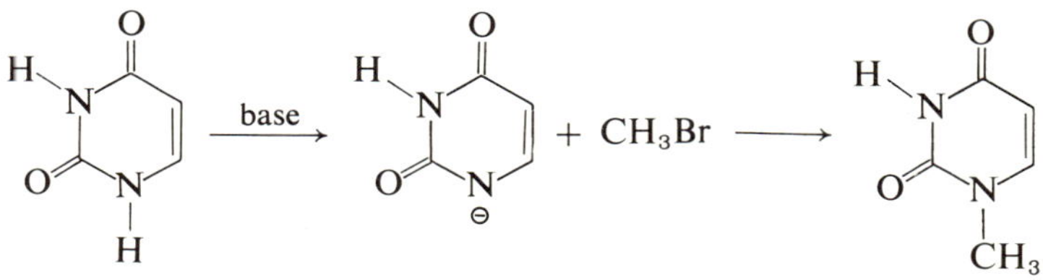

Uracil 1-Methyl uracil

nucleophiles and can take part in displacement reactions in the same way as the pyrimidines.

NUCLEOSIDES

In the nucleic acids the five heterocyclic molecules (or bases) mentioned above are found attached to sugar molecules through a glycosidic bond. The sugar involved is either D-ribose or D-2-deoxyribose. Both sugars are in the furanose form; the deoxy sugar is lacking an oxygen atom found in ribose. In the

D-Ribose β-D-Ribofuranose

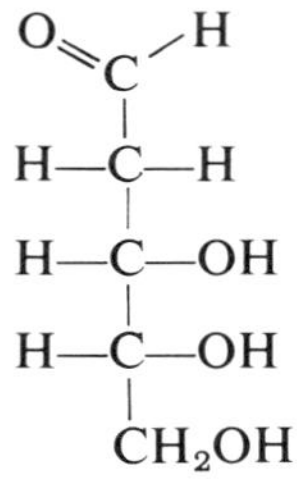

D-2-Deoxyribose

or

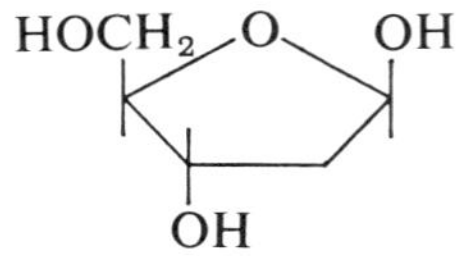

β-D-2-Deoxyribofuranose

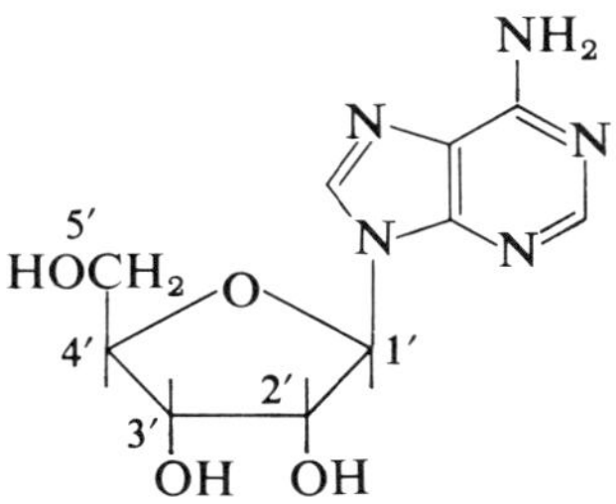

Adenosine
(9-β-D-ribofuranosyladenine)

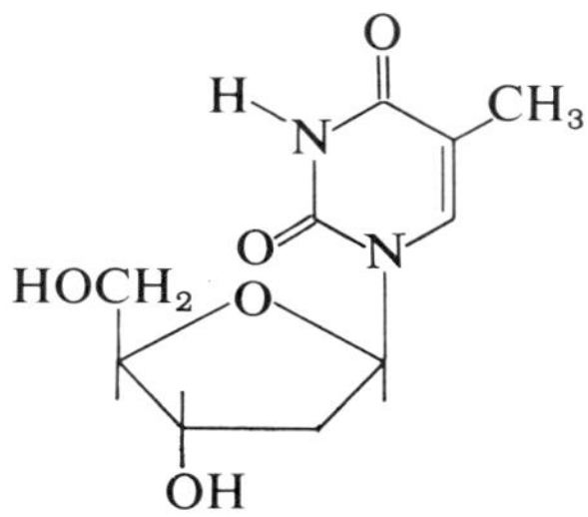

Thymidine
(1-β-D-2-deoxyribofuranosylthymine)

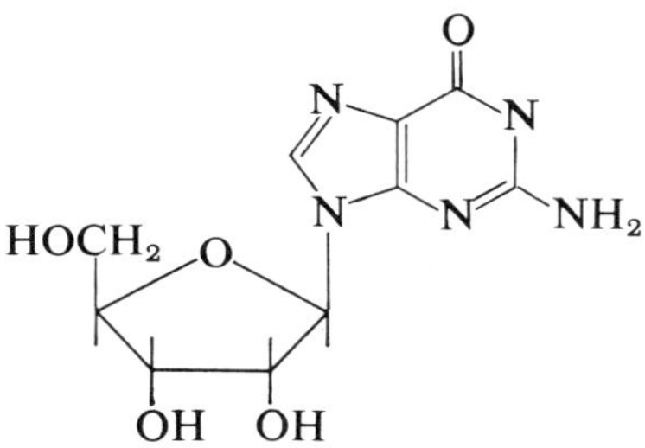

Guanosine
(9-β-D-ribofuranosylguanine)

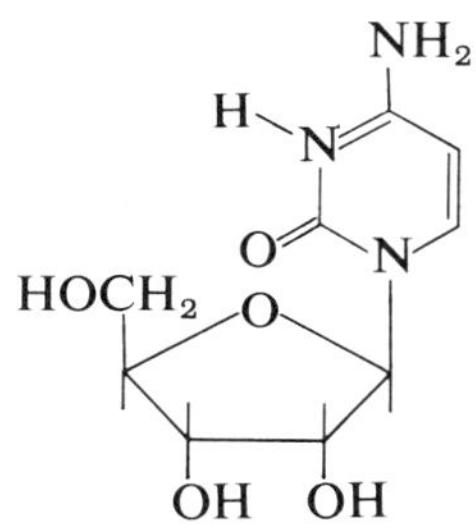

Cytidine
(1-β-D-ribofuranosylcytosine)

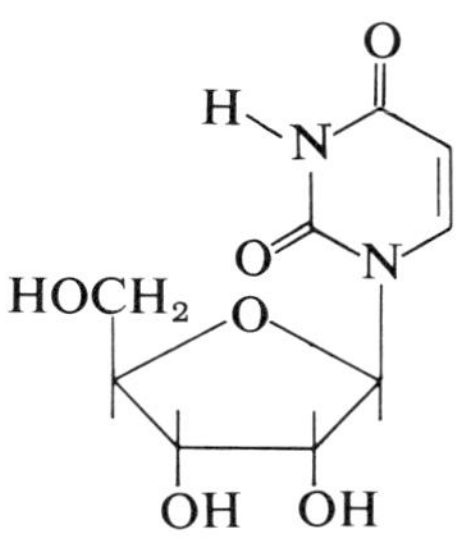

Uridine
(1-β-D-ribofuranosyluracil)

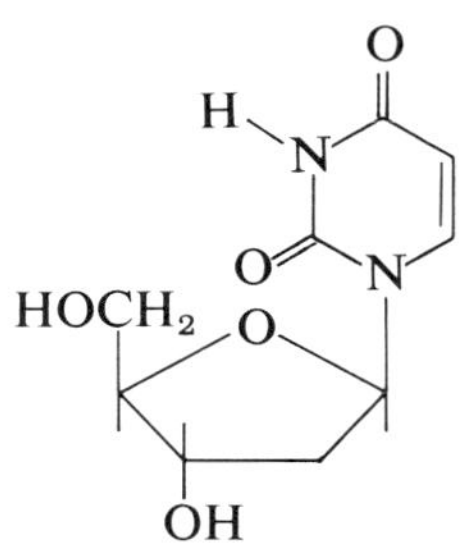

Deoxyuridine

broadest sense, a nucleoside is an *N*-glycoside formed between one of these sugars (or a close relative to them) and a heterocyclic base. Some examples of these structures are presented on p. 352.

Those *N*-glycosides formed by combining adenine, guanine, uracil and cytosine with a D-ribose ring are called adenosine, guanosine, uridine, and cytidine, respectively. This particular group of four nucleosides occurs in ribonucleic acids (RNA's) and are known collectively as ribonucleosides. *N*-Glycosides which are found as part of the deoxyribonucleic acids (DNA's) are called deoxyribonucleosides. This latter group of compounds includes deoxyadenosine, deoxyguanosine, deoxycytidine, and thymidine. Each of these is an *N*-glycoside formed with 2-deoxyribose and the appropriate heterocycle.

The anomeric carbon in these nucleosides is designated as the 1′ position; the other carbons in the sugar ring are numbered consecutively around the ring as indicated above. It is important to note that the 5′-hydroxyl group is not directly attached to the furanose ring.

The chemical and physical properties of the nucleosides reflect the presence of both the sugars and the heterocyclic bases involved. Acid hydrolysis of a nucleoside liberates the base and the sugar, although the rate of hydrolysis depends considerably upon the particular structure of the nucleoside.

Nucleosides have an important role in medicinal chemistry. Many antibiotics are nucleosides and some are effective in halting the growth of cancers. As the name indicates, antibiotics are compounds which can inhibit growth or destroy microorganisms. Some examples of these compounds are shown here. Psicofuranine was first isolated from cultures of

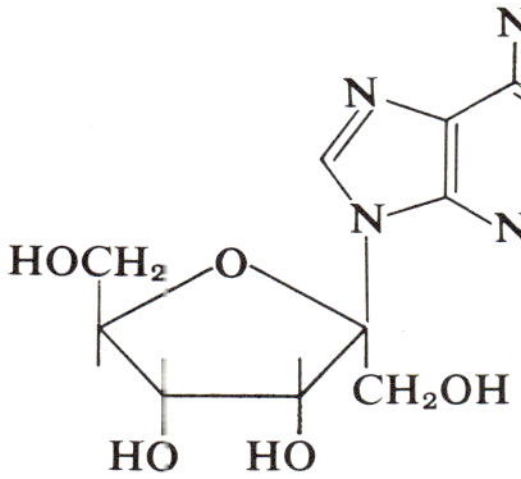

Psicofuranine

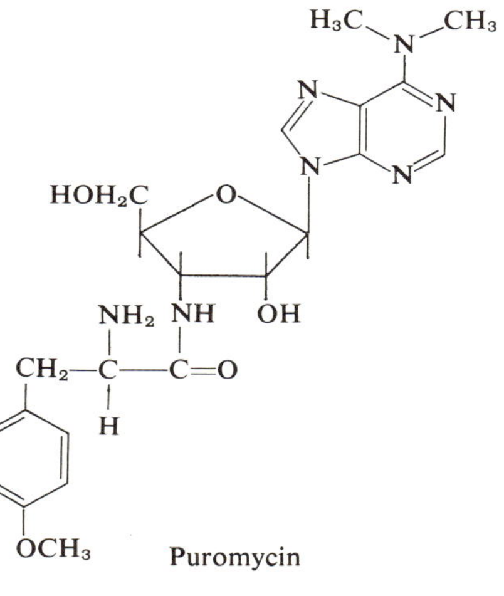

Puromycin

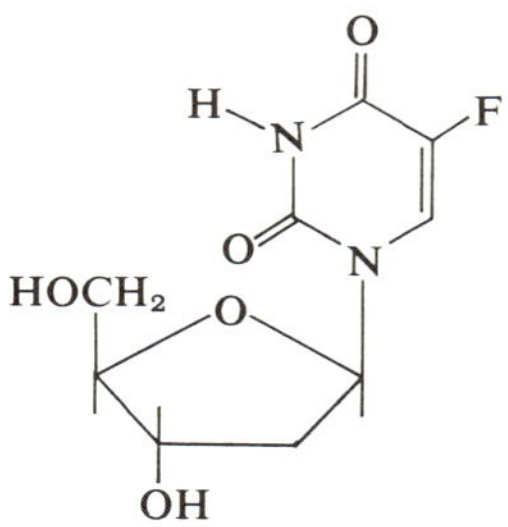

5-Fluorouracil deoxyriboside
(FUDR)

the fungus *Streptomyces hydroscopins*; it shows considerable antibacterial and antitumor activity. Puromycin, formed by *Streptomyces alboniger*, is active against protozoa and has inhibited the growth of experimental tumors in mice. Fluorine-containing FUDR is another useful anticancer agent. This nucleoside is a powerful inhibitor of the enzyme thymidylate synthetase, which is essential for continued replication of the cell.

The preparation and testing of nucleosides have turned up a number of new medically useful compounds and research continues in this area. It is usually true that it is the corresponding nucleotide which is truly the active material, the nucleoside having been phosphorylated by an enzymic system of the organism. However, nucleotides cannot penetrate cell membranes because of the ionic or polar character of the phosphate group, so that the corresponding nucleosides are the compounds which are administered.

NUCLEOTIDES

The term nucleotide is used to denote a phosphate ester of a nucleoside. Three possible esters could be formed from a ribonucleoside and two possible esters from a deoxyribonucleoside. The point of esterification is indicated by the position number of the sugar hydroxyl group involved in ester formation. Some examples of nucleoside monophosphate esters include uridine-2′-phosphate, adenosine-5′phosphate, and thymidine-3′-phosphate. A three-dimensional drawing of the structure of adenosine-5′-phosphate (5′-AMP) is shown in Figure 14.1. In neutral or basic solution, these phosphate esters exist as dianions. Nucleotides are stable in basic solution but in acidic solution the phosphate ester linkage is hydrolyzed to give a phosphate ion and the nucleoside.

More complex phosphate esters are formed from the nucleosides and the condensed phosphoric acids. Two important examples of this type of nucleotide are

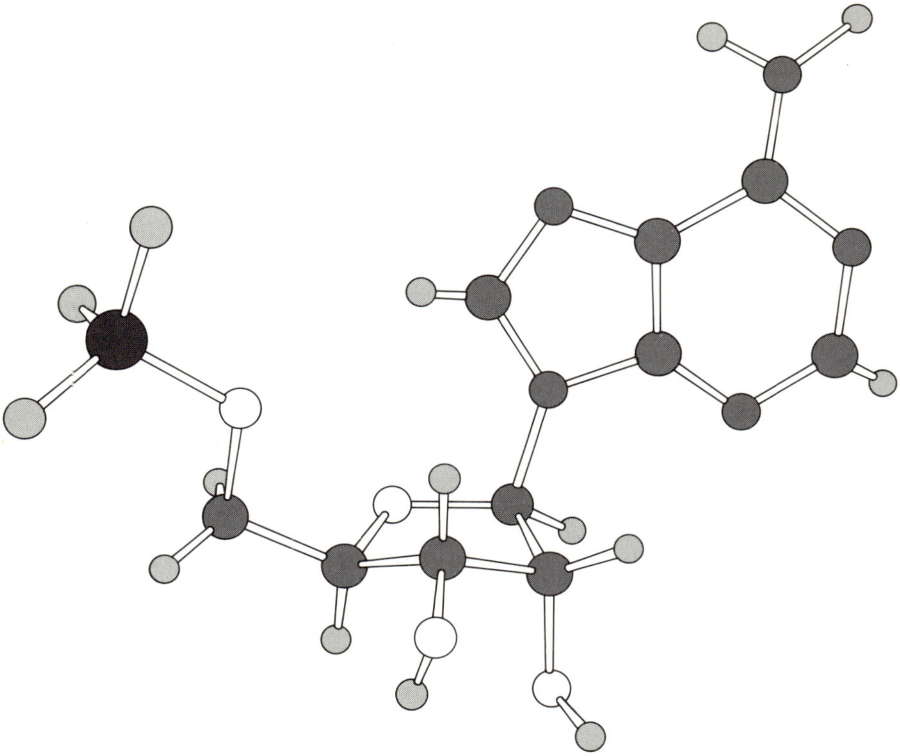

Fig. 14.1. A three-dimensional view of AMP. This is the conformation of the molecule in the solid state as determined by x-ray crystallography; the solution state conformation may be somewhat different. Some hydrogen atoms of the structure are not shown.

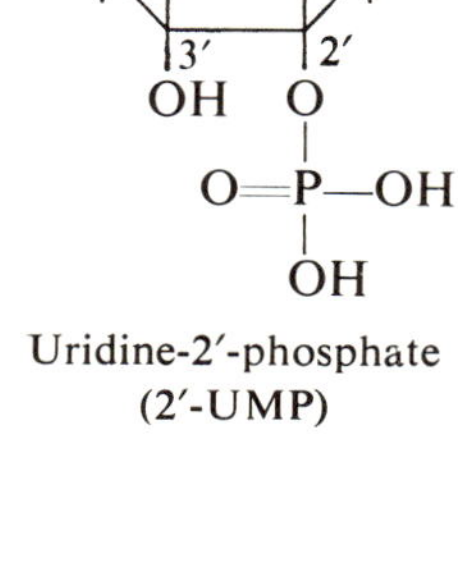

Uridine-2′-phosphate
(2′-UMP)

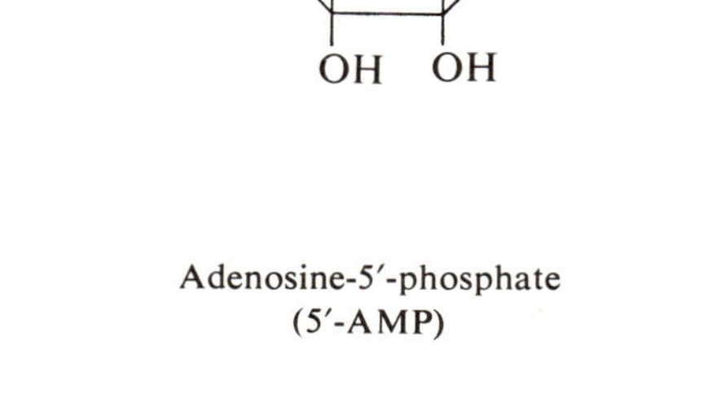

Adenosine-5′-phosphate
(5′-AMP)

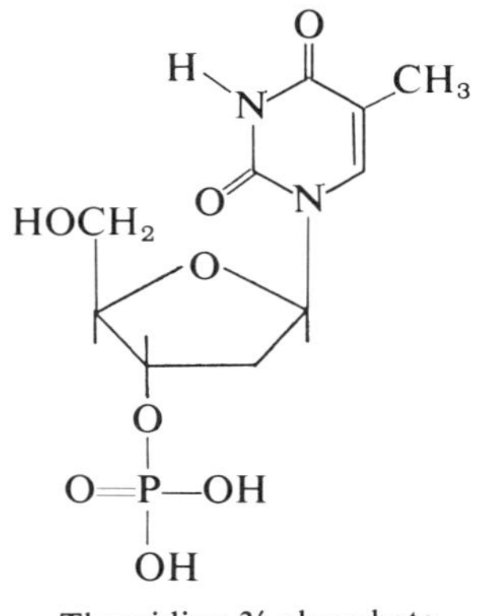

Thymidine-3′-phosphate
(3′-TMP)

adenosine-5′-diphosphate (5′-ADP or simply ADP) and adenosine-5′-triphosphate (ATP). These compounds are key intermediates in the biochemical transfer of energy within living organisms. ATP is most stable under alkaline conditions where the ionization of the phosphate groups is maximal. It is hydrolyzed

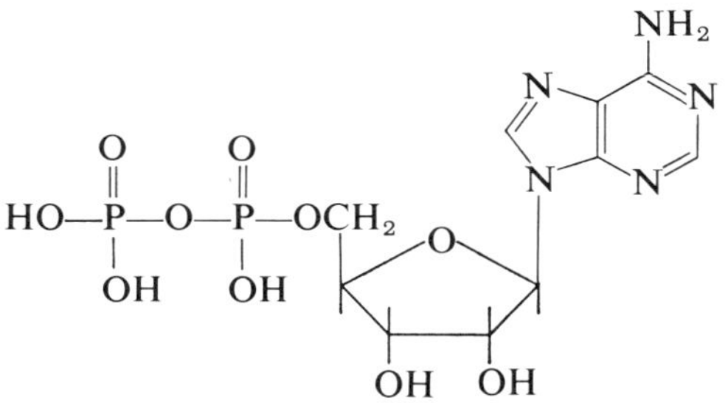

Adenosine-5′-diphosphate
(ADP)

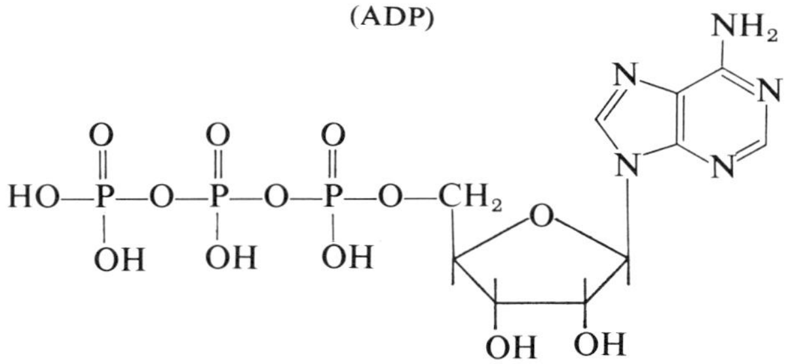

Adenosine-5′-triphosphate
(ATP)

rapidly at neutral or acid pH. ADP is somewhat more stable than the triphosphate. The 5′-diphosphates and 5′-triphosphates of cytidine, guanosine, uridine, and thymidine are also of widespread prevalence in nature; these nucleotides are intermediates in the biosynthesis of the nucleic acids.

Several cofactors or coenzymes required by certain enzymes are derivatives of nucleotides. Nicotinamide adenine dinucleotide (NAD) is a cofactor utilized by many enzymes which catalyze oxidation–reduction or hydrogen transfer reactions. The easy reducibility of the pyridine ring in this nucleotide is an an essential aspect of these reactions.

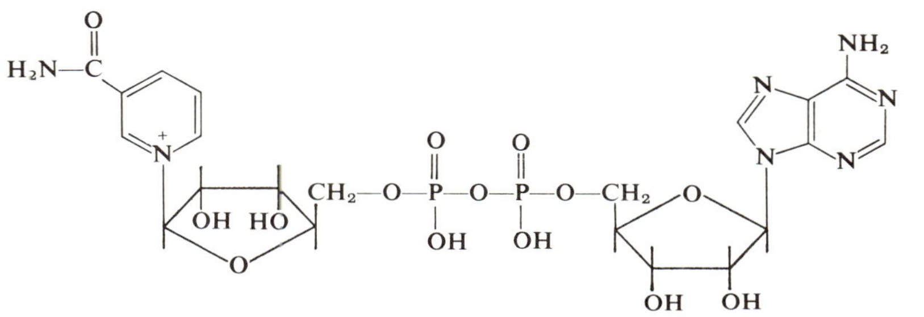

Nicotine adenine dinucleotide (NAD)

Also involved in oxidation–reduction reactions is the enzyme cofactor, flavin adenine dinucleotide (FAD). This coenzyme is synthesized enzymically by a reaction involving phosphorylation of riboflavin with ATP.

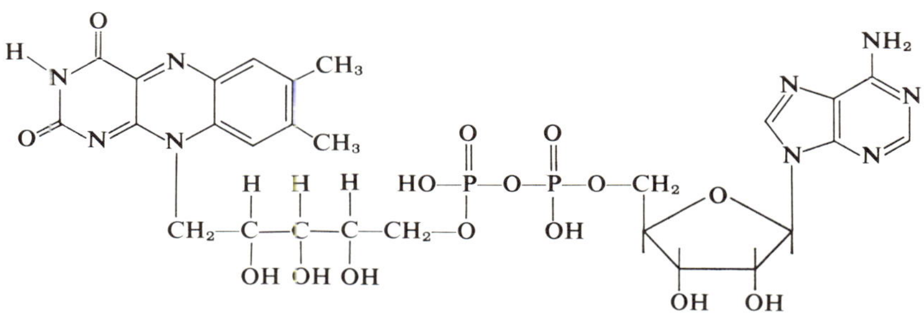

Flavin adenine dinucleotide (FAD)

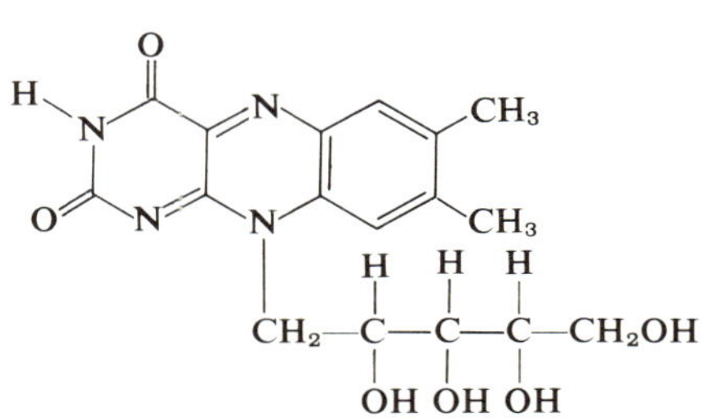

Riboflavin (vitamin B_2)

NUCLEIC ACIDS

Nucleic acids are polymeric molecules which can be degraded to nucleotides by proper choice of hydrolysis conditions; the nucleotides are the monomeric units from which the nucleic acid is constructed. Virtually all animal and viral nucleic acids are found in close association with proteins, the nucleic acids taking the role of prosthetic groups. The protein parts of these nucleoproteins are the histones and protamines that were briefly mentioned in the last chapter.

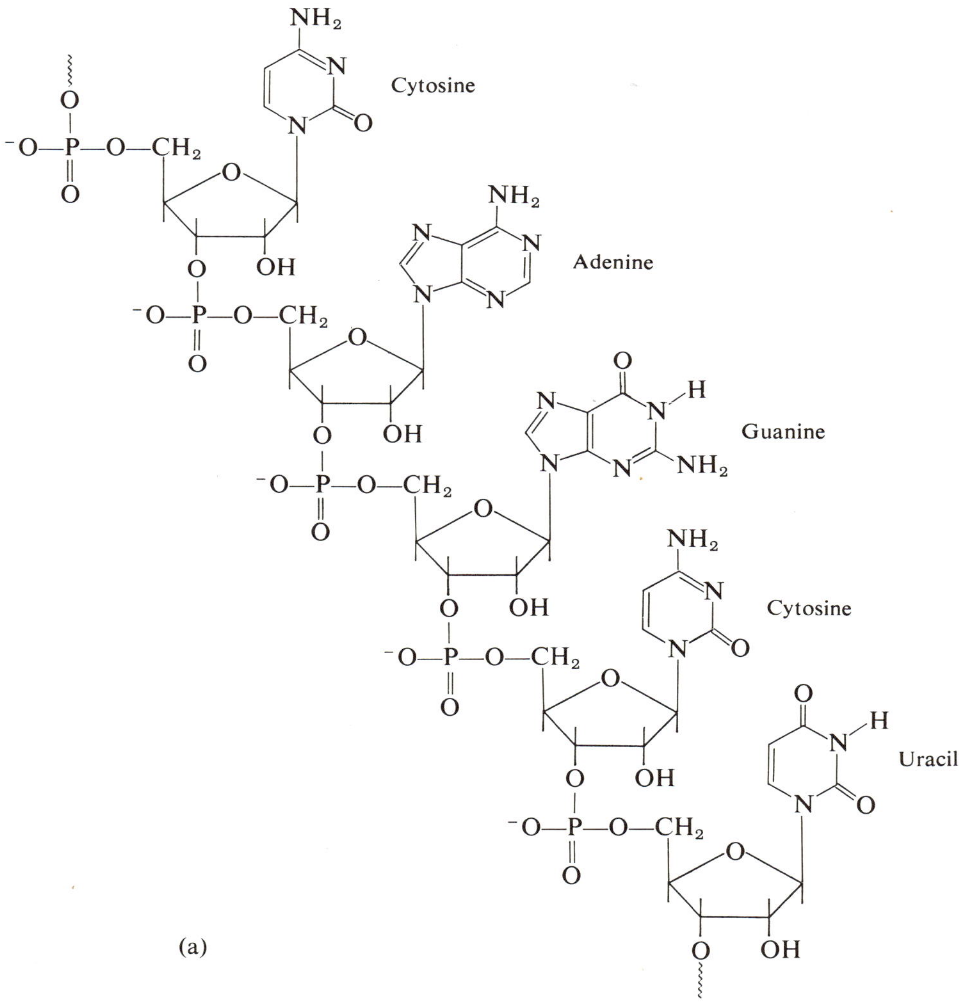

(a)

Measurements of the physical properties of the nucleic acids have demonstrated that these polymers have molecular weights ranging from 10^5 to 10^9. A very regular structure or conformation is found for them under conditions of moderate pH and temperature but when the temperature is raised or the medium is made very basic or very acidic, or if simple organic molecules like alcohols, amides, or ureas are present, a nucleic acid will undergo conformational changes to a more random or denatured state. This process is sometimes called "melting."

One class of nucleic acids appears predominantly in the nuclei of cells; exhaustive hydrolysis of these materials gives phosphate, the heterocyclic bases

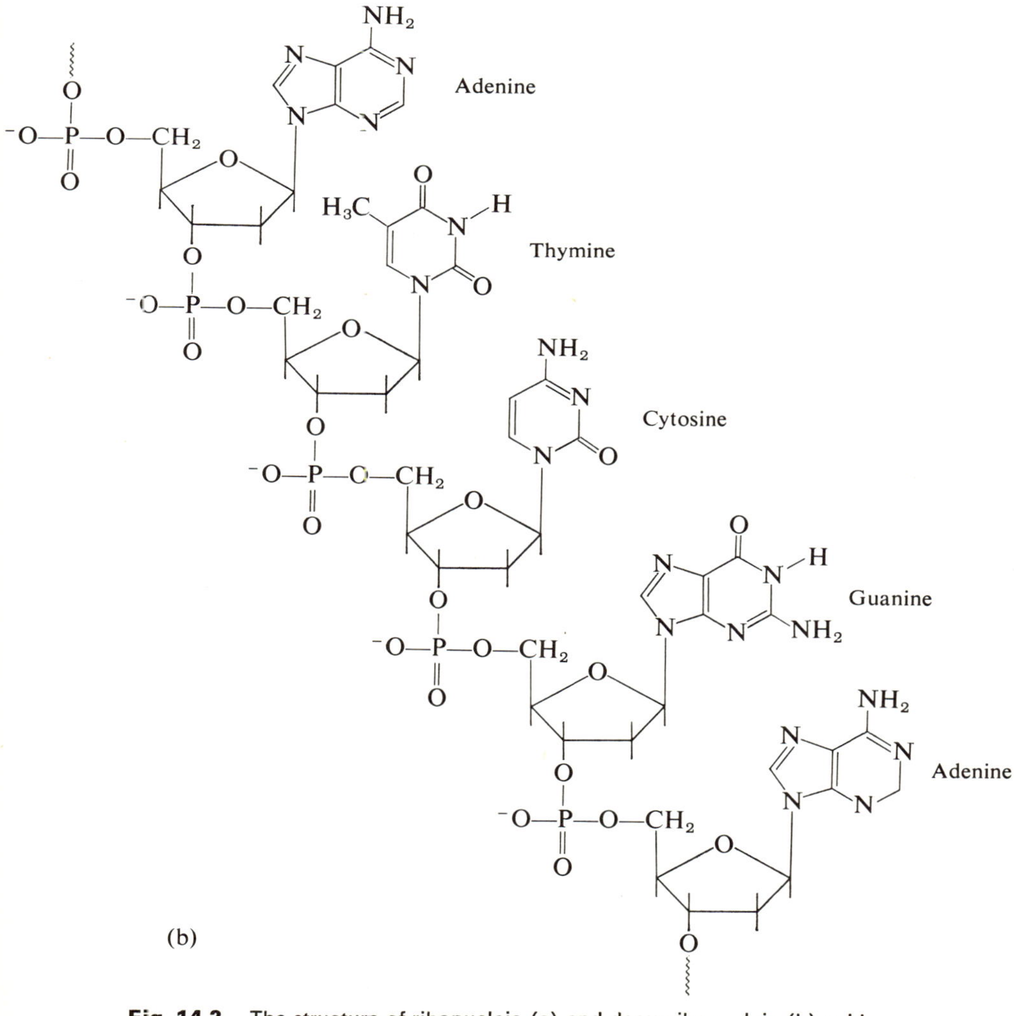

(b)

Fig. 14.2. The structure of ribonucleic (a) and deoxyribonucleic (b) acids.

guanine, cytosine, adenine, and thymine, and the sugar 2-deoxyribose. A member of this class of nucleic acids is referred to as a DNA, for deoxyribonucleic acid. A second major type of nucleic acid is found mainly in the cytoplasm of cells. Upon hydrolysis of polymers of this kind, phosphate, guanine, cytosine, adenine, and uracil and the sugar D-ribose are found; they are called ribonucleic acids (RNA). It is thus seen that the major distinctions between these two classes of nucleic acids are the nature of the pentose found upon hydrolysis and the presence of the pyrimidine, uracil, in one substance and the pyrimidine, thymine, in the other.

Chemical studies have shown that both DNA and RNA are linear polymers; that is, there are no cross-links between polymer chains such as those found in vulcanized rubber or in many proteins. Furthermore, the fundamental covalent bond system holding the monomeric units together is the phosphate ester group. The formation of ester bonds between the 5′-hydroxyl group on one sugar ring to the 3′-hydroxyl group on the next via a phosphoric acid molecule is the basic step in polymerization. Sections of a DNA and RNA polymer are shown in Fig. 14.2. The molecular weights for RNA's range from about 10^5 to 10^6, while DNA molecules can be significantly larger with molecular weights in the 10^5 to 10^9 range. The primary structure of the nucleic acids is clear, but, as we should now suspect from previous experience with other biopolymers, the primary sequence of the nucleic acids is only the beginning of a description of the structure and function of the nucleic acids.

Structural Features of DNA's

An important clue regarding the conformational organization of the nucleic acids comes from studies on the number of the various heterocyclic bases which appear in a given DNA or RNA. DNA's from a large selection of organisms have been analyzed for the proportion of the various heterocyclic bases present and it has been generally found that (a) the base composition of a DNA is characteristic of the organism, (b) different cells of the same organism have DNA's with identical or very similar base compositions, (c) cells from closely related organisms have similar base compositions, (d) the total number of purines (adenine + guanine) in a DNA is equal to the total number of the pyrimidines (cytosine and thymine), and (e) the amount of guanine present is equal to the amount of cytosine. The constancy of the A/T and G/C ratios strongly suggests that these pairs of heterocyclic bases appear in sets within the DNA structure; that is, each time an adenine (A) is found in the polymer, a

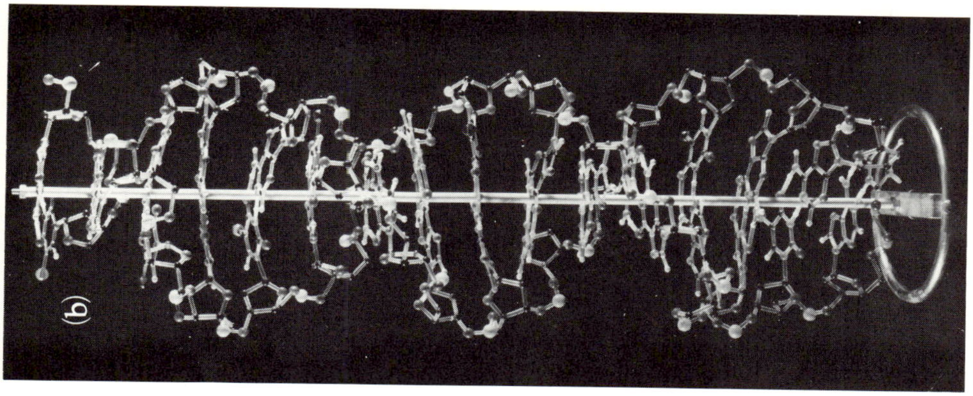

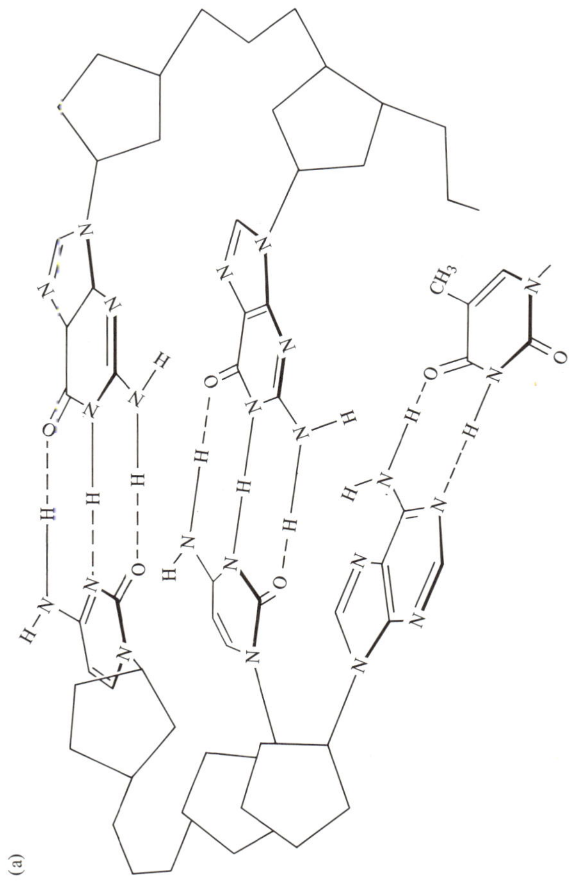

Fig. 14.3. (a) A small section of the DNA molecule showing the pairwise interaction of the heterocyclic bases. (b) Overall expanded view of the DNA polymer indicating the overall molecular organization.

thymine (T) must be closely associated with it. Similarly, guanine (G) and cytosine (C) must be intimately associated with one another.

Considerable difficulty was experienced trying to account for these observations within the framework defined by the sugar–phosphate polymer structure for the nucleic acids until the general features of the DNA structure were elucidated by Watson, Crick, and Wilkins in 1953 using x-ray crystallographic techniques. The molecule is constructed of not one linear, polymeric chain but two. The bases attached to one chain interact with the bases attached to the second sugar–phosphate chain (Fig. 14.3). Moreover, the double-stranded molecule is not in a random linear conformation but rather in a regular structure; the two polymer chains are coiled around the same axis into a double helix (Fig. 14.3b). The sugar and phosphate parts of the polymer are the most polar and water-soluble; they appear on the outside "surface" of the double helix. The pyrimidine and purine bases are relatively nonpolar and hydrophobic and they stack within the core of the helix, forming what appear to be the steps in a circular staircase. The base pairs, which carry all of the genetic information contained within the DNA, are thus protected from external influences by the structure of the polymer and by the basic proteins which are attracted to the surface of the helix by its polar nature.

Careful examination of the bases that are paired in DNA shows how well-matched these heterocycles are. As sketched in Fig. 14.4, the distance across an

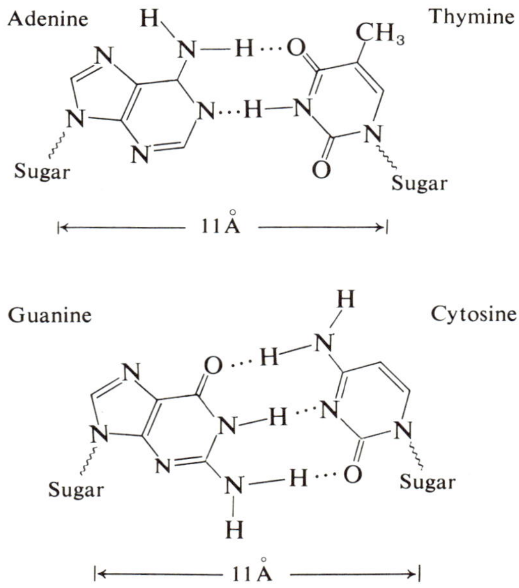

Fig. 14.4. The dimensional aspects of A–T and G–C base pairing.

A–T pair is essentially the same as that across a G–C pair. When the bases are in their keto tautomeric forms, well-developed hydrogen bonds can form between the bases in a pair; two such hydrogen bonds are found in the A–T base pair and three hydrogen bonds in the G–C arrangement. It is unlikely that hydrogen bonds contribute much to the stability of the double helix but rather primarily determine its form. Hydrophobic bonding between the potentially aromatic, planar heterocyclic rings by a mechanism similar to that found in protein structures probably accounts for the overall stability of the helix.

Denaturation of a DNA molecule corresponds to disruption of the highly ordered double helix. The disruptions can be localized along certain sections of the polymers or may involve the separation of both strands. Because of the complementary nature of the base pairs along each strand, DNA molecules may often be renatured; that is, reassembled into their native form, by removal of the conditions which initially led to denaturation.

Structural Features of RNA's

The ribonucleic acids (RNA's) differ from the DNA's significantly in their chemical and physical properties. The molecular weight difference has already been mentioned. RNA's are also more widely distributed within a cell and can be found in almost all cell fractions. Three types of cellular RNA have been distinguished: ribosomal RNA (rRNA), messenger RNA (mRNA), and transfer RNA (tRNA). A fourth type of RNA is found in viruses. Ribosomal RNA's are the largest molecules of this type, having a molecular weight of about 1 million; they are found associated with protein molecules in the cellular granules called ribosomes. The ribosomes are the site of protein synthesis in the cell, a process we shall take up shortly. Messenger RNA's are highly variable in length, while tRNA's are small molecules with molecular weights between 25,000 and 30,000. Because of their relatively small size, the tRNA's are soluble under conditions when the other forms of RNA are not and, as a result, they are sometimes called soluble RNA's (sRNA's).

The very striking complementary nature found in the heterocyclic bases of DNA is absent in the RNA's. There is a tendency for the sum of the number of guanidine and uracil bases to be equal to the sum of the adenine and cytosine bases, but this trend is not universal. (It should be recalled that uracil replaces thymine in the RNA's.) The sequence of bases in one tRNA has been determined and crystalline tRNA is under investigation by x-ray techniques, so that the three-dimensional arrangement of this molecule in space should be available within the near future. However, it is clear from other experiments that the

RNA's are conformationally different from DNA. RNA molecules consist of only single strands of the polymer; they are highly flexible and behave as random coils, although short stretches of helical conformations may be present. Their conformational properties depend strongly on temperature and the presence of electrolytes. This greater freedom of conformational change is very likely an advantage to the molecules as they play their roles in protein synthesis.

THE GENETIC CODE AND PROTEIN SYNTHESIS

It is now universally agreed that the long, highly structured DNA molecules carry the master copy of all genetic information needed to direct the manufacture of proteins by an organism. The genetic message to be passed from one generation to another (whether cells or organisms) is contained within the *order* of the purine–pyrimidine base pairs as they appear along the double helix. During mutation the order of a few base pairs is disturbed by a chemical reaction or some other event and the protein ultimately defined by the DNA is deficient in some respect, usually one that is fatal to the organism. The DNA is, therefore, structured in such a way as to protect as completely as possible the integrity of the base pair arrangement. But the double helix structure is also ideally suited for reproduction of itself when cell division takes place. At the completion of mitosis, the daughter cells must contain a copy of the genetic heritage of the parent cell so that their own progeny may be defined. The replication of DNA during cell division involves a temporary unwinding of the double helix and a separation of the double-stranded structure into single strands. A new deoxy sugar–phosphate polymer containing base pairs complementary to each separated strand is synthesized from appropriate nucleotide monomers by the involvement of enzymes (Fig. 14.5). It appears that the synthesis of the complementary strand has a particular starting point and may proceed simultaneously in both directions from the point. The result of this mode of replication is two copies of the original, double-strand DNA, one for each of the daughter cells. Evidence suggests that several genes are located in a single DNA helix with very few base pairs separating them.

There are 20 amino acids found regularly in protein molecules but only four nucleotide bases in DNA.* If it is the order of appearance of the heterocyclic bases along the DNA molecule which determines the order of appearance of amino acids in a protein, it is clear that more than one type of base must be used

* Only 20 amino acids are coded by DNA; other amino acids are synthesized by enzymic modification of one of these twenty.

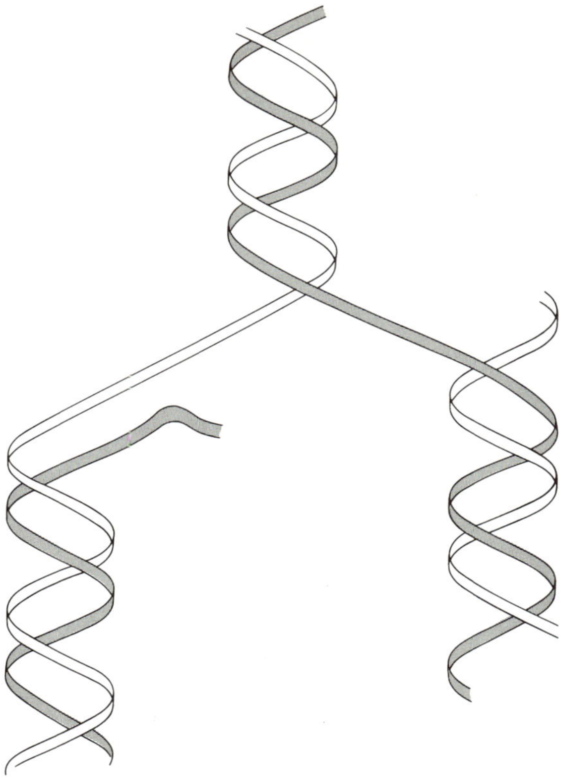

Fig. 14.5. The replication of DNA. It now seems likely that synthesis of the two new double-stranded structures proceeds as the original helix unwinds.

to specify an amino acid. Physicist George Gamow proposed that at least three bases were needed to specify a single amino acid, a suggestion that was confirmed in 1961 by Crick. Taking three base pairs at a time, there are $4^3 = 64$ code "words" (codons) that can be used to specify an amino acid. This is more than enough (only 20 "words" are really needed, at least at first glance) so that there is some redundancy possible in the system. Most amino acids are specified by more than one three-base code word. A necessary feature for delineating genes along a DNA strand and for bringing to a halt the protein-synthesizing machinery is the message "stop;" three codons produce this message.

Although it is the triplets of base pairs along a DNA helix which specify the order in which amino acids appear in a protein molecule, DNA itself is not the template which directs protein synthesis. Instead, the genetic information contained in a DNA is transferred to an RNA molecule, which then serves as an intermediate template. Many experiments have shown that protein manufacture can go forward in the absence of DNA but not in the absence of RNA. In a process called transcription, a RNA complementary to the DNA which specifies a protein is made. An enzyme selects the gene to be used as the master template from a DNA molecule and causes the DNA helix to partially unwind and separate at that point. As the DNA strands separate, nucleotides in solution in the cell become attracted to the exposed bases. An enzyme, RNA polymerase, then links together the ribonucleotides to form a RNA molecule which has a sequence of heterocyclic bases along its length that is a direct reflection of the sequence of bases found in the DNA molecule. As the RNA polymerase moves along the DNA template, growing RNA moves away from the DNA and the DNA helix closes up and reforms behind it. The RNA synthesized in this way is messenger RNA and it thus is imprinted with the genetic code that is carried in the DNA. It acts in much the same way as a punched paper tape that is used to provide instructions to an automated machine in that it provides explicit directions to the protein-synthesizing machine, telling it the exact structure of a protein to be assembled.

The sites of protein synthesis are the nucleoprotein granules called the ribosomes. These particles have molecular weights in the range of 3 million to 4 million and are approximately 50% protein and 50% ribonucleic acids. The exact function of the ribosomal-RNA is not clear; it may be present to provide a set of hospitable binding sites for mRNA and tRNA. In any event, a ribosome has binding loci for a mRNA molecule, a site for tRNA molecules and the enzymes necessary to guide the synthesis of protein.

Several steps in the protein synthetic mechanism can be delineated. First, a complex between a mRNA molecule and a ribosome must form. Second, a set of 20 activated amino acid monomers must be made available in the correct order to the proper enzymic system. Third, peptide bond formation between one carboxyl-activated amino acid and the amino group of the amino acid prior to it in the protein chain must be completed and, fourth, the completed protein must be released from the ribosome. It requires 10–20 seconds to assemble a polypeptide of 300–400 amino acids under optimum conditions in bacterial systems. This incredible efficiency must be due to the exquisite specificity of the intermolecular interactions involved in the process. We can note that this sequence of steps only defines the primary sequence of the newly prepared protein and one might be concerned with regard to how the secondary and tertiary structures of the peptide are established. Nature seems to have taken care of this latter problem by utilizing the inherent chemical features of the amino acid sidechains. When the enzyme ribonuclease was prepared synthetically in the laboratory, only the primary sequence was specified. However, upon dissolving the material in water under essentially physiological conditions, the polypeptide spontaneously assumed a catalytically active form. It seems safe to conclude that once a protein has been prepared by and released from the ribosomal machinery, it will find the correct secondary and tertiary structure by itself.

The above description of the biological role of nucleic acids and of protein synthesis has necessarily been brief and oversimplified. This area of scientific endeavor continues to be one of great vigor and excitement, with new and sometimes contradictory results becoming available at a rapid rate. The laboratory synthesis of a gene has recently been reported, for example. It is fairly clear that, like it or not, it will soon be possible to manipulate genetic molecules to alter the characteristics of what we regard as human beings. Evolutionary developments of this type might well herald the birth of new, efficient, and possibly superior species. Like everything connected with new scientific knowledge, the ability to reengineer human genetics will also bring with it in equal measure the possibility of terrifying abuses. It is a situation that must be dealt with morally and legally, soon.

PROBLEMS

1. Caffeine, a purine derivative found in coffee and tea, can induce mutations. Suggest a mechanism by which this heterocycle could be mutagenic.

2. Some forms of intracellular communication involve cyclic adenosine-3′,5′-monophosphate (cyclic AMP). What is the structure of this molecule?

3. The enzyme L-lactic dehydrogenase consists of four identical subunits of about 35,000 molecular weight. About how many amino acids would be found in each 35,000 MW subunit? Suggest a reason why nature does not create a single polypeptide of 140,000 MW.

4. Why might certain nucleotide analogs be effective antibiotics?

5. Iodine can be used for the conversion of thymidine to the cyclic form shown below. Postulate a mechanism for this reaction.

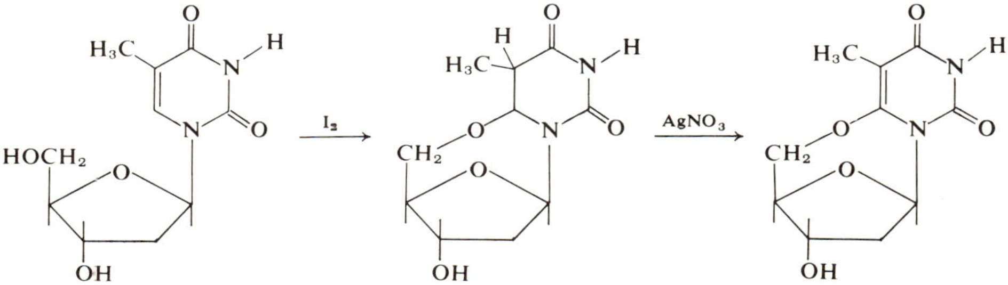

6. Write the structure of the product expected when cytidine is treated with nitrous acid.

7. Adenosine can form a ketal with acetone. Suggest a structure for this product.

SUGGESTED READINGS

C. B. Anfinsen, ed., "Aspects of Protein Biosynthesis." Academic Press, New York, 1970.

D. J. Brown, "The Pyridmidines." Wiley (Interscience), New York, 1962.

H. Chantrene, "The Biosynthesis of Proteins." Pergamon, Oxford, 1961.

H. R. Mahler and E. H. Cordes, "Biological Chemistry." Harper, New York, 1966.

A. M. Michelson, "The Chemistry of Nucleosides and Nucleotides." Academic Press, New York, 1963.

T. L. V. Ulbricht, "Purines, Pyrimidines and Nucleotides." Pergamon, Oxford, 1964.

J. D. Watson, "Molecular Biology of the Gene." Benjamin, New York, 1970.

15

Chemical Evolution

Although everyone knows what the terms *life* or *living* mean, it is very difficult to define completely the characteristics of a living thing. A kitten is alive but a stone is not. Although we might understand in detail various physiological processes in the kitten—respiration, growth, reproduction and so forth—and define life in these terms, if we consider both objects at an atomic level it is difficult to state clearly why one is alive and the other inanimate. Both have common origins in that both are composed of collections of electrons and atomic nuclei and these atoms must have been formed about the same time and must have appeared on earth nearly simultaneously. In the billions of years that have elapsed since that event, some kind of reorganization of atoms into systems of such exquisite complexity that they are "alive" began. The term "chemical evolution" has come to mean the chemical events that took place on the primitive Earth which led to the appearance of the first living cells. Darwin and others in the late nineteenth century recognized that the logical backward extension in time of biological evolution would ultimately lead to a study of evolution on the molecular level. It was not until the early twentieth century that sufficient experimental data were available to permit speculation about the nature of chemical evolution. A. I. Oparin, a Russian biochemist regarded now as the father of chemical evolution, published "The Origin of Life," which trenchantly described reasonable chemical reactions by which organic compounds could have been formed on the earth before the advent of life. The recent interest in interplanetary explorations has given new impetus and financial backing to investigations related to questions regarding the origin of life both on the planet Earth and other celestial bodies.

The part of the universe most accessible to us is that part near the surface of the earth. Most known chemical elements are found in the earth's crust, but oxygen and silicon make up more than 70% of its total mass. Aluminum, iron, magnesium, and titanium are the most abundant metals, while calcium, sodium, and potassium are the most abundant alkaline earth metals. Carbon, hydrogen, sulfur, and phosphorus, all elements essential for biological processes, are found in only trace amounts on a percentage basis.

Exactly how the earth began is a matter of controversy and conjecture. A widely accepted theory holds that the entire local solar system was formed from a common cloud of dust and gas. This process probably began over five billion years ago; current evidence suggests that the earth is 4.5–4.8 billion years old and recent data from the Apollo missions indicate that the earth's moon is 4.7 billion years old. Sometime after the contraction of this cloud began, it is likely that the temperature within the center may have reached several million degrees centigrade. At these temperatures, sufficient energy becomes available for nuclear reactions which, in turn, can produce more energy. The cloud then became our sun, a relatively minor star, and additional aggregations of the cloud-matter formed the planets around the sun, including our own. The original elemental composition of the Earth was probably close to that of our Sun, consisting of mostly hydrogen and helium, but as the young planet plowed its way around the newborn sun, it probably accumulated more and more material, in the process increasing its gravitational field and its surface temperature. As the planet cooled, the lighter elements were the first to leave the planet's gravitational field and eventually the gaseous atmosphere with which we are more familiar was formed.

The synthesis of the organic compounds which in time led to the emergence of living creatures probably began during this period and those compounds with low molecular weights could have also escaped the earth. In fact, diatomic molecules containing carbon were the first interstellar molecules observed by radioastronomers. These included C—H, CH^+, and C≡N. In 1963, hydroxyl (OH) was discovered in space and in 1968 interstellar ammonia (NH_3) was observed. Since then clouds of H_2O, C═O, CH_3OH, $H{-}\overset{\overset{\displaystyle O}{\|}}{C}{-}H$, H—C≡N, CH—C≡N, $H{-}\overset{\overset{\displaystyle O}{\|}}{C}{-}OH$, CH C—CH, HNCO, and HNC have been discovered.

These clouds are numerous and very large; formaldehyde has been identified in more than 50 radiosources and one cloud of formaldehyde has been reported to be several light-years along its major axis. The density of molecules in these clouds of organic molecules is not large, perhaps one molecule per cubic centimeter, but the total number of molecules within them must be very large indeed, given the overall dimensions. It is clear, then, that the primitive earth easily could have been characterized by conditions conducive to the synthesis of organic molecules.

For organic syntheses to go forward, the prebiotic Earth must have provided both the materials and the conditions necessary. It is generally accepted that the primitive atmosphere of the earth was a highly reducing atmosphere because it was so rich in hydrogen molecules and atoms. Our present atmosphere is, of course, a highly oxidizing one. It is likely that the change occurred over a considerable period of time about 3.5 billion years ago and was possibly the result of decomposition of water in the upper atmosphere by ultraviolet radiation from the sun. The hydrogen, because of its lighter mass, then preferentially escaped the planet's gravitational field. The upper atmosphere of ozone (O_3) that was developed during this period is now essential to the survival of life since it absorbs much of the ultraviolet radiation reaching the earth. There have been serious suggestions that this "shield" is in danger from the supersonic aircraft now under development.

The early reducing atmosphere assured that most elements capable of covalent bond formation existed as their hydrides during this period. Thus, carbon was primarily in the form of methane and other small hydrocarbons, nitrogen was in the form of ammonia and hydrazine, oxygen, and sulfur in the form of water and hydrogen sulfide, respectively. Carbon and nitrogen probably coexisted in hydrogen cyanide (HCN) and its polymers. Various metals in low oxidation states and mineral particles in the muck and clays formed during the geological evolution of the earth's surface could have provided catalytic surfaces for the subsequent reactions of these materials.

Current opinion favors the proposition that life on earth began in the oceans of the planet. It has been estimated that over a period of 300 million years after the Earth cooled sufficiently to permit the existence of liquid water on its surface, oceans with up to 1% of dissolved organic compounds could have been formed. (Some workers believe that this estimated concentration is much too high.) With plenty of energy available from the sun, from electrical discharges of the atmosphere, and from radioactive and volcanic action on the surface of the earth, and with an abundance of catalytic surfaces available, the synthesis of those organic molecules necessary for life could have begun.

AMINO ACIDS

A number of experiments supporting the speculations described above have been reported. Especially important is an experiment carried out by S. L. Miller and reported in 1953. He passed a silent electrical discharge through a mixture of methane, ammonia, hydrogen, and water vapor and obtained among the products of the reaction several amino acids, including glycine, alanine, aspartic acid, sarcosine, and α-aminobutyric acid, several carboxylic acids, aldehydes, and hydrogen cyanide (HCN). This mixture of gases in equilibrium with liquid water resembles what is believed to have been the composition of the primitive Earth's atmosphere, while the electrical discharge through the mixture can be assumed to simulate lightning. When experiments of this type are carried out with ionizing radiation or intense ultraviolet irradiation as the sources of energy, amino acids are also formed. There have been at least four mechanisms proposed to account for the formation of amino acids in experiments like Miller's. The appearance of aldehydes in the reaction mixture suggests that a sequence of reactions known commonly as the Strecker synthesis of amino acids could explain the appearance of amino acids; it is particularly consistent with

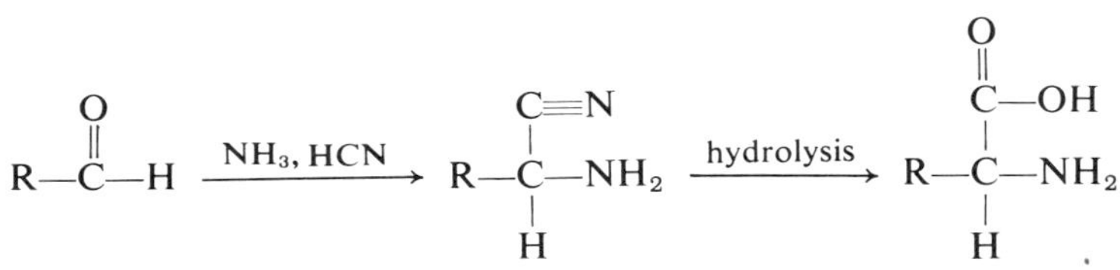

An aminonitrile

Strecker synthesis of amino acid

the observed predominant formation of α-amino acids. Discharge experiments in mixtures of methane and ammonia have since shown that aminonitriles can be formed in the absence of water. Subsequent hydrolysis of these materials would give amino acids. Alternatively, hydrogen cyanide and polymers of hydrogen cyanide might have been intermediates in the formation of amino acids. It has been shown that glycine can be formed from hydrogen cyanide and that ultraviolet irradiation or strong heating of a water solution of HCN polymers produces a number of amino acids including glycine, alanine, glutamic acid, and serine. There seems to be no doubt that the simpler amino acids could have been formed under the conditions that probably existed in the prebiotic Earth. Table 15.1 summarizes the amino acids that have been isolated from primitive atmosphere simulation experiments. It should be noted that high molecular weight amino acids such as tyrosine and phenylalanine have been observed

Table 15.1
Amino Acids Formed in Primitive Earth Simulation Experiments[a]

Amino acid	*Input energy[b]*
Glycine	E, I, H, U
Alanine	E, I, H, U
β-Alanine	E
Sarcosine (*N*-methylglycine)	E, H
Serine	E, H
Threonine	E, H
α-Aminobutyric acid	E, H, U
Asparaginine	E
Aspartic acid	E, I, H
Glutamic acid	E, H
Valine	H
Leucine	E, H
Isoleucine	E, H
Lysine	E
Phenylalanine	H
Tyrosine	H

[a] Taken from a similar table compiled by R. M. Lemmon, *Chem. Rev.* 70, 95 (1970).

[b] Conditions used: E, electrical discharge; H, strong heating to 1000°C; I, ionizing radiation; U, ultraviolet radiation.

only when methane–water–ammonia–hydrogen mixtures are heated to high temperatures, above 1000°C.

It has been mentioned earlier in this text that the molecules involved in biochemical processes usually have a very stereospecific nature. For example, virtually all α-amino acids in living things have the L-configuration. Even though it is possible to conceive of ways in which amino acids could initially have been formed stereoselectively, it is very likely that the vigorous chemical and physical conditions existing on the young earth were more than sufficient to ensure the racemization of these amino acids to D,L mixtures. (Racemates of amino acids are thermodynamically more stable than are either the pure D or L isomers.) It has been shown that one enantiomer will crystallize preferentially from a solution of racemic amino acids when the mixture is seeded with a trace amount of that enantiomer. Thus, the appearance of a trace excess amount of one isomer of an amino acid might have triggered a separation of enantiomers based upon solubility with only one optical isomer remaining in the reaction mixture. It has also been suggested and subsequently demonstrated that polarized radiation can selectively destroy one isomer in a racemic mixture.

The spontaneous generation of optically active material in a closed system with no disymmetric reagents present has recently been observed. This system involved the crystallization of 1,1′-binaphthyl. The interaction of the hydrogen atoms *ortho* to the carbon–carbon bond between naphthalene rings in this compound is strong enough to drastically slow the rate of rotation about this

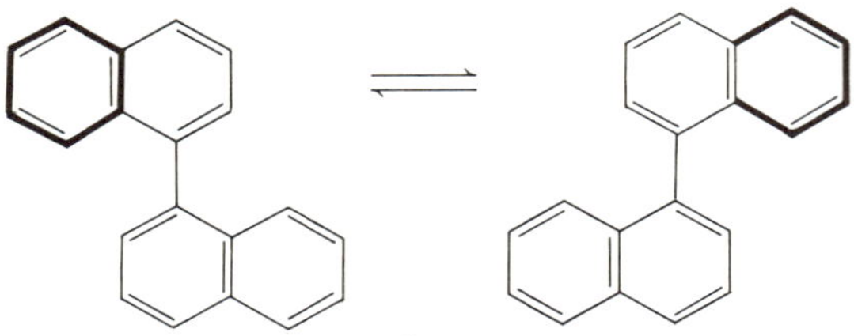

1,1′-Binaphthyl

bond and, as a result, the hydrocarbon is asymmetric. It was shown that crystallization of this compound from a racemic melt preferentially produced one *or the other* enantiomer in solid form. About 200 determinations were made with essentially half of them resulting in the levorotatory isomer with the other half giving the dextrorotary form. Extraneous sources of asymmetry such as dust or particles from the bodies of the experimenters seem to be ruled out as possible starting centers for preferential crystallization of one isomer since these sources of contamination should have the same symmetry properties from experiment to experiment. Although simple, this binaphthyl system convincingly demonstrates that in a configurationally mobile system one enantiomer can be formed from a racemic mixture by crystallization.

It was also demonstrated that the introduction of trace quantities of optically active molecules into the racemic binaphthyl mixtures brought about the formation of one form of solid binaphthyl. Thus, when a small amount of α-mandelic acid (C_6H_5—CHOH—COOH) was dissolved in the racemic melt only dextrorotatory binaphthyl was formed. These observations lend support to the suggestion that small quantities of optically active material, once formed, would induce optical activity in other asymmetric molecules by preventing them from crystallizing at random.

Once an excess of one configuration of an amino acid was established in a prebiotic reaction system, that configuration of the amino acid would be incorporated more frequently into a growing polymer. Polyamino acids which contain only all L or all D optical isomers of the individual amino acid are conformationally more stable than polymers in which the amino acids randomly have either of these configurations. Thus, given an excess of L-amino acids, the eventual development of polymers (proteins) based exclusively on L-amino acids would be thermodynamically favored even though initially both D and L isomers of the amino acid were available.

SUGARS

Formaldehyde, identified as a product formed when an electrical discharge is passed through methane–water–ammonia mixtures, may have been the precursor to the saccharides formed on the prebiotic earth. Over a century ago it was shown that formaldehyde dissolved in dilute aqueous base polymerizes into a number of sugars including fructose, cellobiose, galactose, mannose, and ribose. Several tetroses were also found in these experiments. This condensation reaction has been studied further and the sequence of reactions shown below accounts for the formation of tetroses; further condensations of the four-carbon aldehyde would lead to pentoses and hexoses.

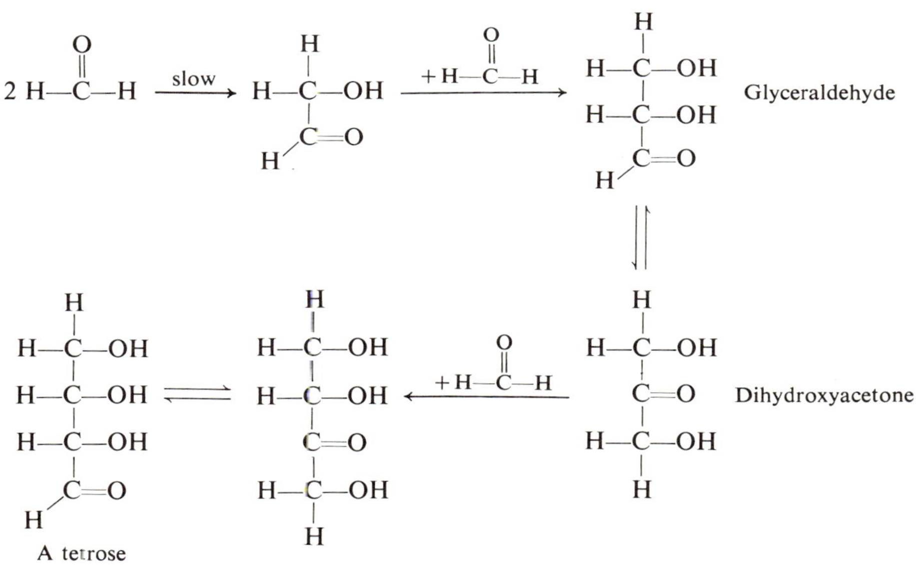

The conditions for these reactions do not resemble those that were found on the primitive earth. However, experiments have demonstrated that formaldehyde or simple aldehydes derivable from formaldehyde will condense to sugars under these conditions. For example, heating a dilute aqueous solution of formaldehyde with the mineral kaolenite (aluminum silicate) gives about a 4% yield of ribose. Formaldehyde heated with calcium oxide or ammonium hydroxide (ammonia gas dissolved in water) leads to the formation of deoxyribose. Moreover, irradiation of formaldehyde with ultraviolet light or γ-rays produces a number of pentoses and hexoses including ribose and deoxyribose. Finally, these two sugars have been identified as products formed during "primitive earth atmosphere" experiments in which gaseous mixtures of methane, water, ammonia, and hydrogen are exposed to an electrical discharge. It thus seems quite clear that the sugars needed for the synthesis of nucleic acids, for use as structural materials and for important roles in metabolic pathways arose on the surface of the earth long before the appearance of life.

PURINES, PYRIMIDINES, PORPHYRINS

Heterocyclic molecules based upon the purine and pyrimidine ring systems are essential for the formation of nucleic acids. No great successes in synthesizing these molecules under primitive Earth conditions have been achieved. Adenine is the only nucleic acid base formed during the experiments with the mixtures

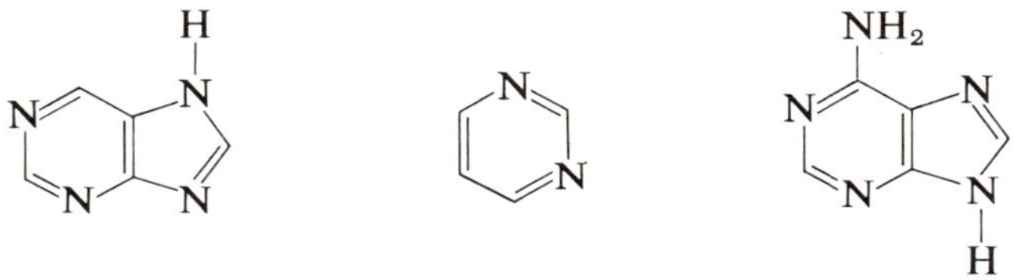

Purine Pyrimidine Adenine

of gases described above and the yield was very low. It is likely that HCN, a product formed in "primitive Earth atmosphere" experiments, is the precursor to adenine since it is known that oligomers of hydrogen cyanide form adenine in basic aqueous solution as indicated in the reactions below. Trimerization

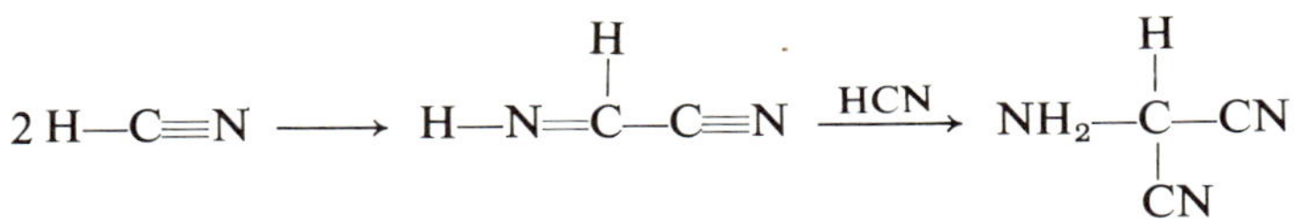

A dimer of HCN Aminomalononitrile

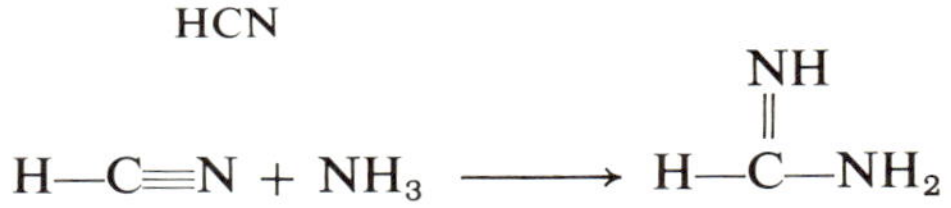

Formamidine

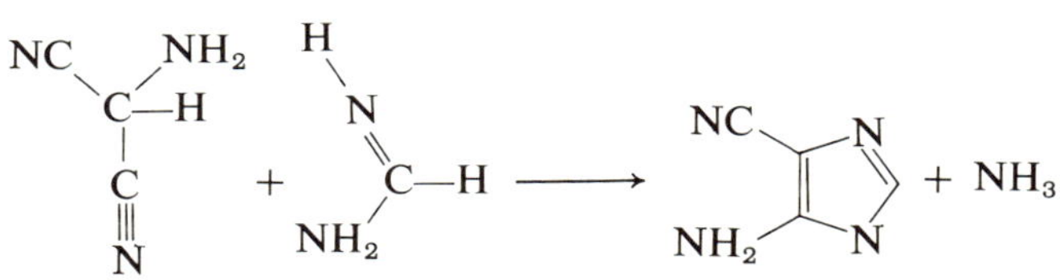

An imidazole

of HCN gives aminomalononitrile which can condense, stepwise, with two

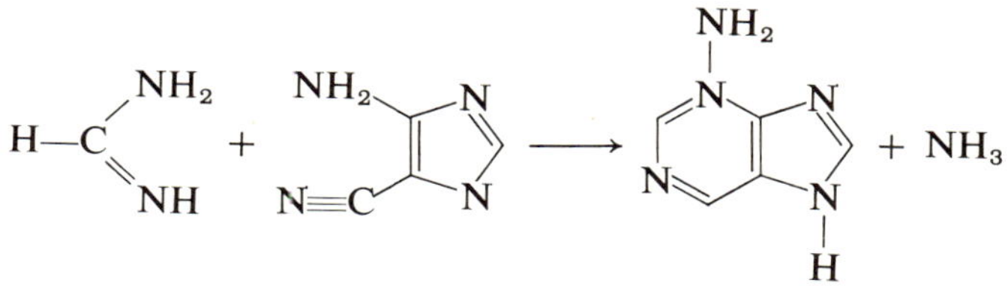

Adenine

A possible method for the formation of adenine from HCN

molecules of formamidine to give, first, an imidazole and then, adenine. Alternative mechanisms have been proposed for this process but all have in common the conversion of five molecules of hydrogen cyanide into a molecule of adenine.

Adenine and the other purine base found in nucleic acids, guanine, have been synthesized in the laboratory by irradiating dilute solutions of HCN with ultraviolet light. Photochemical reactions of this type may provide an alternative pathway for the formation of these purines, although the stability of these compounds to subsequent photochemical reactions after their formation is not clear.

Of the pyrimidines that appear in nucleic acids only cytosine has been observed

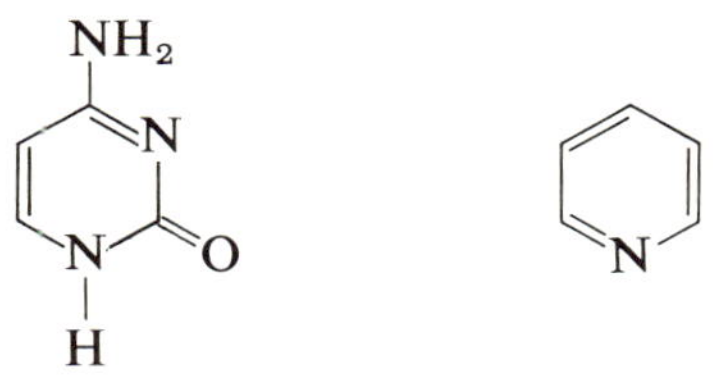

Cytosine Pyridine

in what can be regarded as a primitive atmosphere experiment. Cyanoacetylene (H—C≡C—C≡N) is formed when a mixture of methane and nitrogen is exposed to an electrical discharge; when an aqueous solution of cyanoacetylene and KCNO is heated, a small yield of cytosine is observed. Uracil has been prepared by heating urea, malic acid, and strong inorganic acids but these conditions probably have little relation to conditions on the early earth. There has been no reported "primitive Earth" synthesis of thymine. The formation of pyrimidine bases is a perplexing aspect of the genesis of living things, in that no completely convincing reactions for the synthesis of all of these necessary starting materials for nucleic acid synthesis have been reported.

The apparent ease of adenine synthesis may be related to the fact that adenine plays a large number of biologically important roles. In addition to being a constituent of both ribonucleic and deoxyribonucleic acids, adenine is found in a number of enzyme cofactors such as ATP, ADP, and coenzyme A. Adenine is the most stable of the nucleic acid bases in the thermodynamic sense and is the most resistent to degradation by radiation.

Other Heterocycles

The formation of the biologically important molecules derived from the pyridine ring skeleton probably involves reactions with cyanoacetylene and

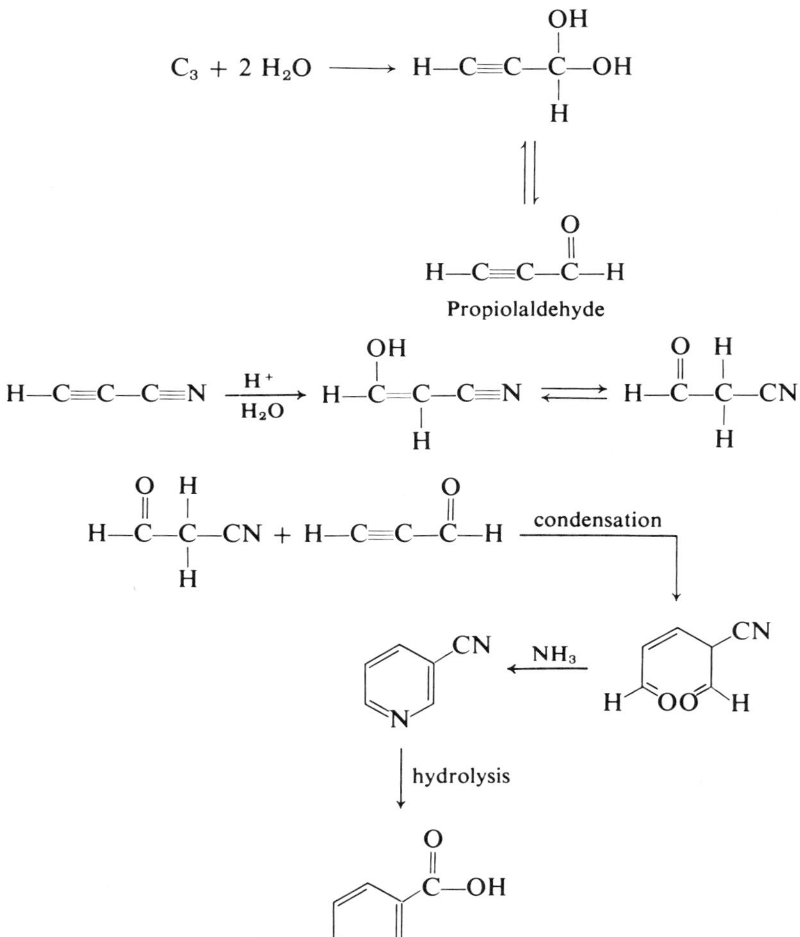

Nicotinic acid

A possible synthesis of pyridines under prebiotic conditions

propiolaldehyde. It has been shown that triatomic carbon (C_3) is formed by the action of an electric discharge on methane. When methane is sparked above a layer of water, a small quantity of propiolaldehyde is formed in solution, presumably by the reaction of C_3 with water. It appears that nicotinic acid could have been formed under prebiotic conditions by (1) hydrolysis of cyanoacetylene to cyanoacetaldehyde, (2) condensation of this aldehyde with propiolaldehyde, (3) ring formation, and (4) hydrolysis of the nitrile group.

The porphin ring is a fundamental structural unit in the cytochromes, hemoglobin and chlorophyll. Aldehydes were possibly involved in the primordial synthesis of porphyrins, although it has been claimed that detectable amounts of these important pigments can be formed during the passage of an electrical discharge through water–ammonia–methane mixtures. The porphin skeleton can be formed by heating an aldehyde with pyrrole; pyrroles can be synthesized by passing an electrical discharge through mixtures of acetylene and ammonia.

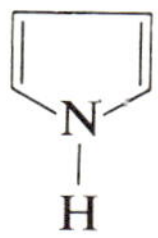

Pyrrole

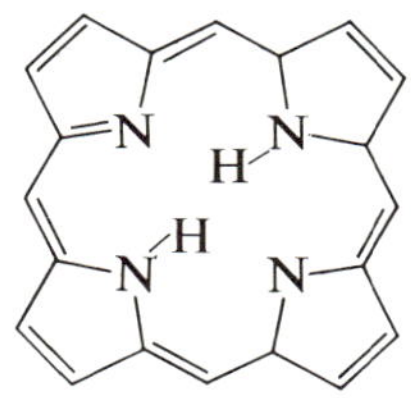

Porphin ring

FATTY ACIDS

The appearance of fatty acids on the earth is easily imagined in terms of the ideas described above. A variety of experiments have shown that the passage of high-energy, ionizing radiation through solid or gaseous methane leads to the formation of a large variety of hydrocarbons, some of very high molecular weight. It has also been demonstrated that radiation can cause the attachment of carbon dioxide to alkanes to form fatty acids, and that exposure of methane and water vapor to an electrical discharge produces monocarboxylic acids containing up to 12 carbon atoms.

The generation of glycerol, the polyhydroxylic alcohol esterified by fatty acids in the formation of fats and oils, on the primitive earth is something of a mystery and no experiments have yet been reported which indicate how this event might have taken place.

DEHYDRATION AND POLYMERIZATION

The discussion so far has shown us that there are reasonable ways in which most of the amino acids, sugars, heterocyclic molecules and fatty acids necessary for life processes could have been formed on Earth. Previous chapters in this book have tried to explain how the macromolecules of living systems can be thought of as polymers of these smaller molecules and, therefore, the next aspect of chemical evolution which we must consider is the way in which polymer formation from these available monomeric units might have taken place.

The common feature in the formation of (1) proteins from amino acids, (2) polysaccharides from simple sugars, (3) nucleic acids from nucleosides and phosphoric acid, and (4) fats and oils from fatty acids and alcohols is the elimination of water during the generation of the polymer. The elimination of water

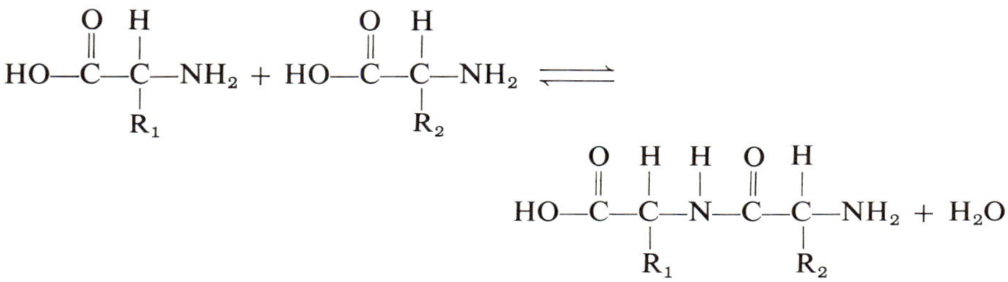

Peptide, protein formation

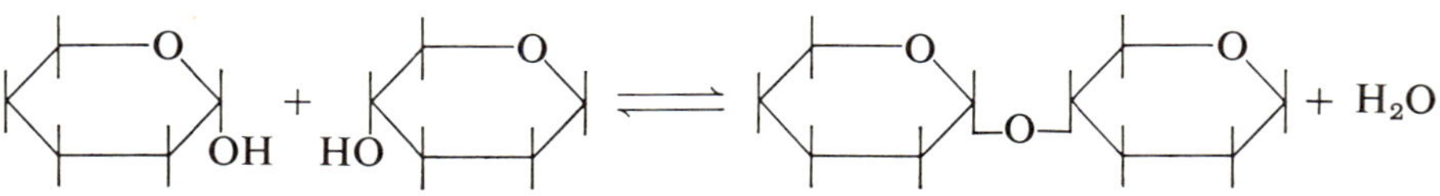

Glycoside, polysaccharide formation

in each of these reactions is not a process favored by thermodynamic considerations since the product molecules are 3–5 kcal/mole higher in free energy than are the reactants. One way of getting around this difficulty is to carry out the dehydration reaction at high temperatures so that the water formed will boil away, thereby favoring a shift of these equilibria to the right by Le Chatelier's principle. High temperatures and a relatively anyhydrous environment could have been found in the vicinity of volcanic activity on the primitive Earth. Alternatively, the elimination of water could be carried out in the presence of a molecule which reacts readily with water so that the overall free-energy change for the dehydration–hydration set of reactions is favorable. Molecules of this type would be especially useful if the dehydration reactions they catalyze could be carried out in aqueous solution. Chemical evolution experiments based upon variants of both of these approaches to macromolecule formation on the prebiotic Earth have been carried out. Most of these have been directed toward understanding the generation of nucleic acids and proteins.

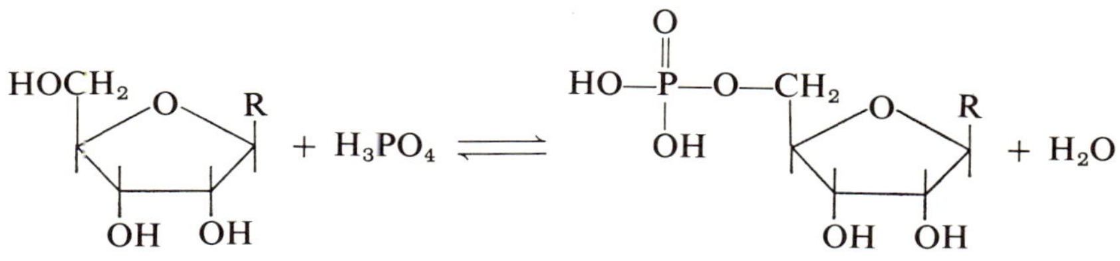

Nucleotide formation

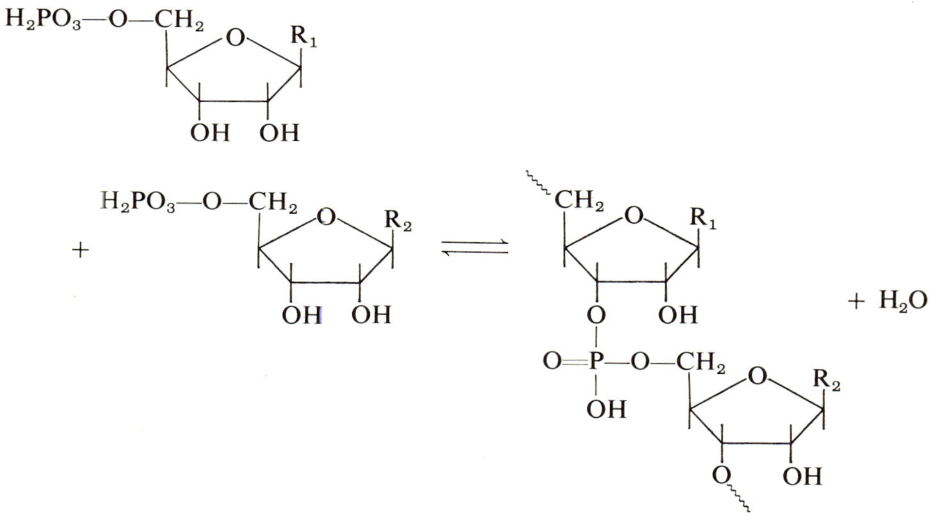

Nucleic acid formation

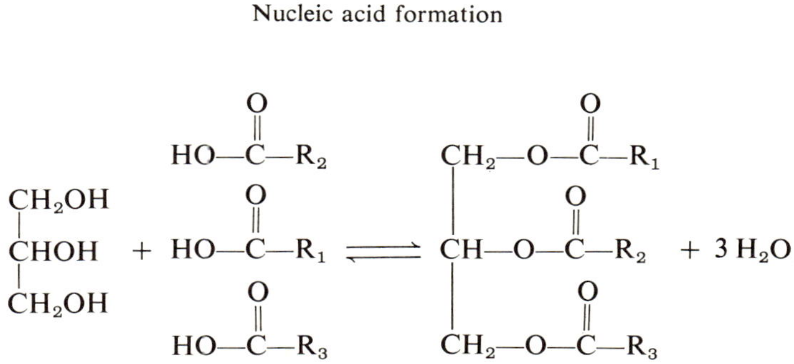

Formation of fats and other lipids

Formation of Proteins

Simple amino acids are high-melting crystalline solids. Attempts to heat amino acids past their melting points (~200°C) in order to effect elimination of water and the formation of peptide bonds have usually resulted in the production of discouraging black tars. Surprisingly, when one or two amino acids are present in large excess so as to form a "solvent" for the reaction, protein-like polymers in the molecular weight range 3,000–11,000 can be formed. Aspartic and glutamic acids have been used as the solvent amino acids; when the other amino acids are present, the mixture melts at about 150°C and water distills from it. If polyphosphoric acid is included in this system, the condensation reaction will proceed at temperatures below 100°C. The amino acid polymers formed in these reactions have been called proteinoids. They have infrared spectra and other physical properties similar to those of proteins.

A significant aspect of proteinoid formation in melts of amino acids is that it is not a random or statistical reaction. The crude products from these reactions would be expected to have a smooth, symmetrical distribution of molecular weights if the coupling of pairs of amino acids occurred indiscriminately (Fig. 15.1). As indicated in Fig. 15.1, the experimental distribution of molecular

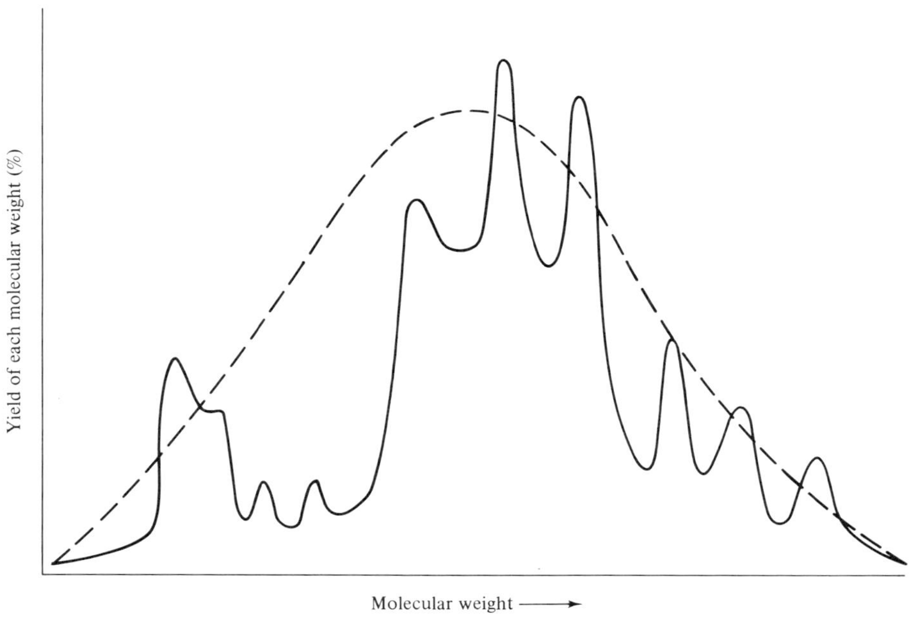

Fig. 15.1. Schematic representation of molecular weight distribution in proteinoid-formation experiments. If the coupling reactions were completely random, the distribution indicated by the dotted line would be expected.

weights is not symmetrical; it has been shown that in some proteinoid preparations the distribution of molecular weight resembles that found with blood serum proteins to a remarkable degree.

Another indication of lack of homogeneity in the proteinoids synthesized by thermal dehydration of amino acids is the distribution of amino acids in the products. In a particular reaction the starting material consisted of 33% aspartic acid and 17% glutamic acid with the remaining 50% consisting of about 3% each of 17 other amino acids. Table 15.2 shows the amino acid composition of the proteinoid product. It is clear that the combination of amino acids even

Table 15.2
Distribution of Amino Acids in Proteinoid Formed by Dehydration of Amino Acid Mixture[a]

Amino acid	*Percent in proteinoid*	*Percent starting material*
Aspartic acid	52	33
Glutamic acid	13	17
Threonine	0.6	3
Serine	0.6	3
Proline	1	3
Glycine	3	3
Alanine	1	3
Leucine	3	3
Phenylalanine	6	3

[a] Data taken from S. W. Fox *et al.* (see Suggested Readings).

under these rather brutal conditions is not a process involving the random formation of peptide bonds. Although the various amino acids surely have different reactivities under conditions such as these, it is also possible that the growing polymer chain can selectively influence the type of monomers attached to growing polymer chains.

Protein formation may also have occurred in the primitive oceans, although this would involve elimination of water from amino acids within an aqueous environment. We have previously noted that coreactants which selectively absorb water from the reaction partners could make the dehydration reaction energetically feasible. These dehydrating agents must be stable to or at least react much more slowly with liquid water to be of any value. In fact we have already examined such a reagent: the carbodiimides used in the preparation of peptides. An example of the use of this dehydrating agent is the synthesis of the peptide glycylalanine described in Chapter 13. The reaction of carbodiimides with water itself takes place slowly relative to the rate of this coupling reaction and several successes in the formation of peptides in aqueous solution at room temperature using carbodiimide reagents have been described.

Cyanamide, $NH_2—C≡N$, can be formed from ammonia and hydrogen cyanide. This molecule is a tautomer of carbodiimide, the parent of the carbodiimide mentioned above. The dimer of cyanamide, a material produced during

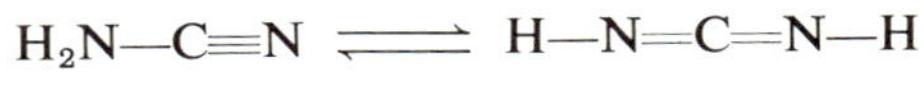

Cyanamide Carbodiimide

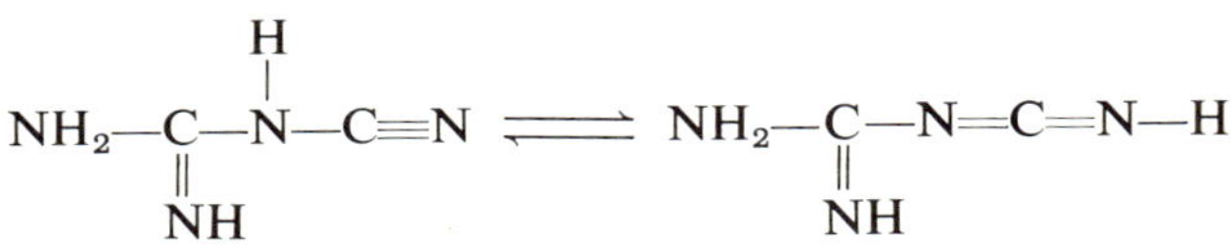

Cyanamide dimer

primitive Earth atmosphere experiments, can also be written in a tautomeric carbodiimide form. It has been demonstrated that peptide bond formation can be promoted by cyanamide and cyanamide dimer in aqueous solution. It is quite reasonable, thus, to suspect that peptide bond formation could have taken place in the aqueous solution of amino acids provided by the oceans of the primitive Earth.

Experiments have also been reported which indicate that ionizing irradiation of amino acids can lead to the formation of peptides. Some workers believe that polypeptides may have been formed on the primitive Earth directly by the hydrolysis of polymers of hydrogen cyanide. There is nothing known that precludes the synthesis of protein molecules by several independent or even interrelated processes of the types described above. Since the information required to generate the secondary and tertiary structure of polypeptides is contained in the linear sequence of amino acids in the polymer, it is reasonable to expect that once formed, by whatever method, primitive polypeptides would have rearranged themselves into the energetically most favored conformations. A wide selection of proteins and protein conformations similar in many respects to present-day biologically synthesized proteins were likely present on prebiotic Earth.

Nucleic Acids

The formation of nucleic acids involves the polymerization of nucleotides by the formation of phosphate esters with the concurrent elimination of water. The monomeric nucleotide units themselves must be formed from phosphoric acid and a nucleoside in another dehydration reaction. In order for nucleotide synthesis to proceed a suitable form of phosphoric acid must be present. In an aqueous environment the most abundant form of phosphorus is phosphoric acid, but the solubility of most phosphate salts in water is very low. Hydroxyl apatite (calcium phosphate), the most common phosphorus-containing mineral on the earth's crust, is very insoluble in water and, at present, the concentration of phosphate in the ocean and most surface waters is about 10^{-6} *M*. A major obstacle to effective phosphorylation reactions in aqueous solution, then, is the low concentration of a necessary reactant that possibly prevailed at the time prebiotic synthesis of nucleotides took place. However, organic materials in contact with solid hydroxyl apatite or other phosphate minerals could have been absorbed by them with the phosphorylation reaction taking place at or near the sites of absorption. Recent experiments have shown that mixtures of urea, ammonium chloride, and sodium phosphate will smoothly phosphorylate nucleosides in the solid state when warmed to about 100°C. When hydroxyl apatite was substituted for the sodium salt of phosphoric acid the reaction proceeded more slowly but in reasonable yield. Irradiation of ammonium cyanide solutions (NH_3 + HCN) gives good yields of urea $\left(NH_2{-}\overset{\overset{\displaystyle O}{\|}}{C}{-}NH_2\right)$ so that all of the materials needed for production of nucleosides by this method were available on primitive Earth. The phosphorylation reaction described above could easily have taken place on beach sand or in desert areas where surface temperatures in excess of 70°C are not uncommon.

The formation of polynucleotides as precursors to nucleic acids may have taken place by much the same chemistry as that used in assembling the proteinoids. Strong heating of mononucleotides has been reported to give polynucleotides, although the nature of the phosphate linkages between the various sugar rings may not be the natural 3′,5′-arrangement. There is evidence that the primitive "carbodiimide" reagents mentioned above may have been able to assist in the formation of sugar–phosphate ester bonds to produce polynucleotides.

THE DEVELOPMENT OF COMPLEXITY: TOWARD THE CELL

The evidence briefly described so far indicates that all classes of compounds required for the formation of contemporary biological molecules could have been present on primitive Earth; reasonable reactions for the formation of monomeric molecules and the dehydration–polymerization of these molecules can be postulated, although our ignorance of many of the details of the processes is still very great. Paleontological evidence suggests strongly that the first unicellular organisms appeared on the earth about 3.1 billion years ago. That means that over a period of 1 to 2 billion years from the formation of the earth, chemical evolution, which began with methane, ammonia, water, and hydrogen gases, and some powerful sources of energy, had reached a state of sophistication or complexity such that chemical processes within well-defined, bounded regions called cells, were appreciably different from the processes that took place exterior to these regions. Two plausible collecting mechanisms by which an enhanced degree of organization, leading eventually to a living cell, could have been obtained are the coacervation of colloidal droplets, proposed by Oparin, and the formation of proteinoid microspheres, suggested by Fox.

Colloids are dispersions of particles of about 10^{-7} to 10^{-4} cm in diameter in a second phase. Fog is a dispersion of small droplets of a liquid (water) in air; smoke is a colloidal dispersion of solid material in air. Particles of this size are easily seen by an electron microscope. The dimensions of protein or nucleic acid molecules are such that their solutions can be regarded as colloidal.

A coacervate is a collection of colloidal particles that is held together by some mutually attractive interaction between the molecules. A micelle formed by the aggregation of detergent or fatty acid molecules can be regarded as a coacervate. In this case the hydrophobic interaction of the hydrocarbon "tails" of the molecules leads to their self-association. Coacervates of protein molecules can also be formed although in this case it is the attractive interactions between positively and negatively charged amino acid side chains that lead to aggregation. In either case, the effect of coacervation is to build up a larger organic structure than existed before, with the exclusion of water molecules. By selection of the large organic compounds that went into the coacervate, structures of increasing stability and complexity could have developed until a cell-like organized unit evolved.

There are several flaws in the coacervation model for the organization of component macromolecules into a cell. No coacervate has yet been found which shows the regular, well-defined boundaries found in living cells and coacervates are not stable to centrifugation. Moreover, the stability and structure of a coacervate is very dependent upon the conditions of temperature, pH and salt concentration under which it was formed; it is difficult to imagine a variety of specialized coacervate structures existing simultaneously.

Formation of proteinoids by high temperature condensation reactions has already been mentioned. When proteinoids are boiled with water and allowed to cool, small globules called microspheres appear. Typically the microspheres are about 2×10^{-4} cm in diameter and have the same shape, range of size, uniformity, and tendency to associate as are found with some bacteria. Electron microscopic examination of these globules reveals that these materials are essentially hollow spheres and that the wall of the sphere consists of a double layer of protein molecules (Fig. 15.2). It has been shown that these microparticles

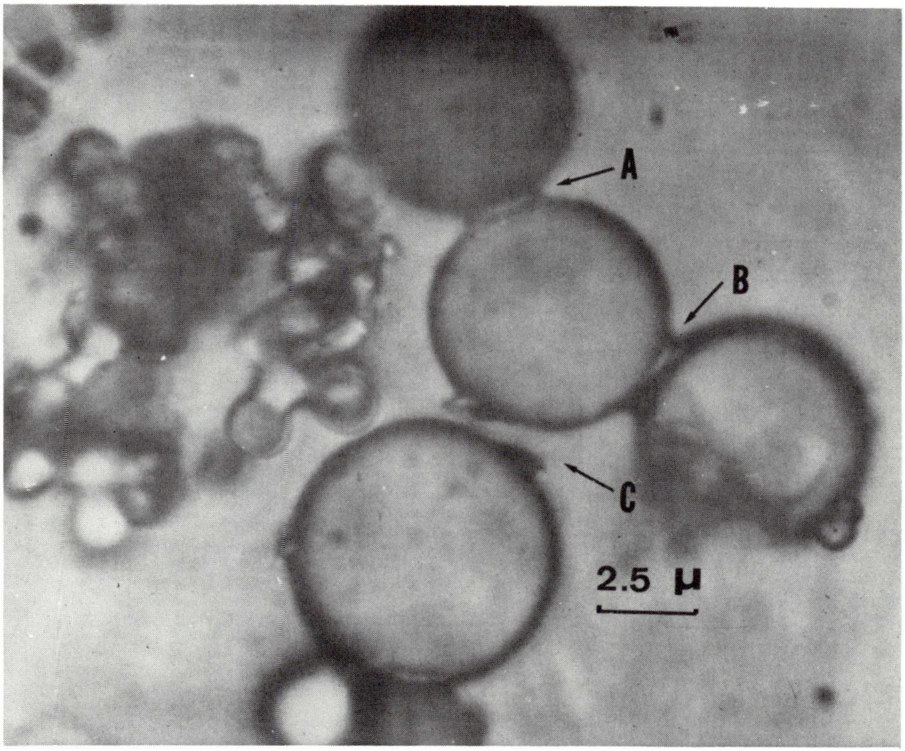

Fig. 15.2. Electron micrograph of proteinoid microspheres. Courtesy of Dr. S. W. Fox, University of Miami.

(1) are stable to centrifugation, (2) can selectively entrain certain sugars, (3) can enlarge themselves by accretion with other microspheres, and (4) can generate other microspheres by a kind of budding. Thus, proteinoid microspheres strikingly possess many of the properties one finds with contemporary cells. The same thermal environment that led to the formation of proteinoid molecules on the primitive Earth, if occasionally washed by a heavy rain, could have been the origin point for the self-assembly of amino acid polymers into primitive protocells similar to the microspheres.

Experiments with proteinoid molecules support the idea that, given the conditions that existed on the young earth, cell-like structures could have appeared.

Fox has postulated the "evolutionary flowsheet" shown in Fig. 15.3 to summarize possible stages in the development of cellular systems. The first steps have been successfully modeled in the laboratory, but the subsequent ones are more speculative. There is, as yet, no clear indication of how the exterior wall of the protocells which consisted solely of amino acids evolved into the lipid–protein matrices that make up contemporary cell membranes. The highly complex interrelated functions of a living cell are not found in these model structures and it is not clear how this much complexity would have developed from these protocells to what it is today. The microsphere experiments do suggest that on the evolutionary time scale the formation of nucleic acids for use as information

Formation of amino acids

↓

Formation of proteinoids ("proto proteins")

↓

Formation of protocells

↓

Protocellular proliferation and selection

↓

"Living" cells

↓

Proliferation and selection of cells

Fig. 15.3. An evolutionary flow sheet, as suggested by Fox.

carriers occurred at a later time than the formation of protein molecules; it has been suggested that DNA- and RNA-like polymers appeared as byproducts of a metabolic process in an early protocell and that their interaction with selected proteins led to biologically useful insoluble complexes. The use of these complexes as information storage devices may have presaged the eventual formation of the nucleic acids.

It is a giant step from the rather crude ideas that have been generated regarding the origin of life to an understanding of what life is. Man goes on, aware that he is the most complex organism on this small planet, but blithely ignoring or ignorant of the physical, chemical, and biological factors which brought him to his present state and which determine the environment in which he continues to function. We may grandly dream that one day we shall understand all of the chemical reactions which led to the evolution of man and other living things from a mixture of gases and all of the reactions which sustain them in their present state. But even then, shall we know what is man?

SUGGESTED READINGS

M. CALVIN, "Chemical Evolution." Oxford Univ. Press, London and New York, 1969.

M. J. DOWLER *et al.*, Prebiotic synthesis of propiolaldehyde and nicotinamide. *Science* **169**, 1320 (1970).

S. W. FOX, Origin of the Cell: Experiments and Premises. *Naturwissenschaften* **60**, 359 (1973).

S. W. FOX *et al.*, Chemical origins of cells. *Chem. Eng. News.*, June 22, 1970, p. 80, and Dec. 6, 1971, p. 48.

D. H. KENYON AND G. STEINMAN, "Biochemical Predestination." McGraw-Hill, New York, 1969.

J. KEOSIAN, "The Origin of Life." Van Nostrand-Reinhold, Princeton, New Jersey, 1964.

R. M. LEMMON, Chemical evolution. *Chem. Rev.* **70**, 95 (1970).

A. I. OPARIN, "Origin of Life," 2nd ed. Dover, New York, 1953.

D. M. RANK, C. H. TOWNES, AND W. J. WELCH, Interstellar molecules and dense clouds. *Science* **174**, 1083 (1971).

INDEX

C

F

G

H

I

O

P